GIS/LIS '91
PROCEEDINGS

Volume 1

28 October - 1 November 1991
The Inforum
Atlanta, Georgia

Sponsored by:
American Congress on Surveying and Mapping
American Society for Photogrammetry and Remote Sensing
Association of American Geographers
Urban and Regional Information Systems Association
AM/FM International

ISBN 0-944426-75-1
ISBN 0-944426-76-X

Published by

American Society for Photogrammetry and Remote Sensing
American Congress on Surveying and Mapping
5410 Grosvenor Lane
Bethesda, Maryland 20814

Association of American Geographers
1710 16th Street, N.W.
Washington, D.C. 20009

Urban and Regional Information Systems Association
900 2nd Street, N.E.
Suite 304
Washington, D.C. 20002

AM/FM International
14456 East Evans Avenue
Aurora, Colorado 80014

Printed in the United States of America

C O N T E N T S

VOLUME 1

AUTHOR INDEX

GIS/LIS AND THE NEW GLOBAL GEOGRAPHY

Nigel Calder

Investigative journalism has at least as much to do in research and education as in the more obvious areas of politics and current affairs. For the news correspondent the question is, "Who's concealing what?" For a science writer like me, it's more a matter of "Who's ignoring what?" Nor is this just a matter of informing the general public and high-school teachers about progress in science. Even among professionals, a grasp of revolutionary innovations in their field can lag years behind events. My theme today is a revolution in geography, and how this will affect the development of GIS and LIS. I have spent the past few years investigating matters geographic, in connection with a TV series called Spaceship Earth. The need to tour the world and to speak with academics, researchers, and users, in deciding what story to tell, gave me an overview of the revolution.

Geographers may own the academic geography of the classroom, but they do not own the world, or have any monopoly of descriptions of the world. One feature of GIS is that it puts geographic tools in the hands of many people who a not geographers. And GIS is not the only area where gatecrashers intrude. The main tools of the revolution of which I speak are remote-sensing satellites and supercomputers. That makes the chief gatecrashers in geography the designers of spacecraft instruments, the interpreters of remote-sensing data, and the conceivers of global computer models that describe the physical systems of our planet.

Other fields of knowledge have experienced the shock of intruders. For example, the radio experts who built the first radio telescopes transformed astronomy. X-ray experts who figured out the structure of DNA revolutionized biology. Closer to geography, geophysicists measuring the magnetism of the sea floor proved that the continents really move about the face of the Earth. That made all the professional geologists who had denied any such a possibility look pretty foolish. With continental drift and plate tectonics, whole libraries of textbooks went out of date.

The historian of science, Thomas Kuhn, describes the structure of scientific revolutions. He tells of science lurching through a succession of "paradigm shifts", when the assumptions and habits of normal science are brutally superseded by the new ideas of small groups of individuals. To me as a science writer, Kuhnian shifts are a fact of life, which is why, quite often, I find myself in this embarrassing position of a non-expert explaining a subject to its experts.

In this talk, I shall clarify what I mean by the new global geography. I shall sketch what the fundamental researchers are up to in their efforts to understand the Earth system. Then I shall discuss the present and future roles of human geography, and of GIS and LIS, in developing the human side of the story. To bring our thoughts into focus, I shall speak of a global geography machine of the 21st Century. I shall comment on analytical problems created by quirks of human nature, and end with remarks on some public-policy issues that seem likely to arise from this scientific revolution.

If I want to provoke my geographer friends, and give them a sense of the professional challenge they now face, I tell them that the new global geography is geography done by physicists. In a revolution in Earth studies, the physicists are setting the pace. For example, the NASA Laboratory of Terrestrial Physics in Maryland is a prime center. It is from the ranks of physicists, or from people in other disciplines who have sat down to master the physics, that the satellite technologies and the computer models come. A few academic geographers have played a prominent part in these innovations, but typically one talks to physicists, biologists, oceanographers, geologists, and space scientists - and to the climatologists who are now physicists in disguise and make intricate computer models of the fluid dynamics of the atmosphere. Such people have hijacked geography and are busy turning it from a descriptive science into an analytical and predictive science.

The global character is there from the start. Geography till now has usually recapitulated exploration. It starts with the home region and spreads its horizons gradually to other regions and continents. Minimal geography of the traditional kind would be a plan of the building where it is taught. The new global geography inverts the process and asserts that you can understand no locality properly without a global overview. Its minimal geography would reduce the planet to a single dot, and list some of its global properties: mean temperature, gravity, chemical composition, age, human population, combined GNPs, and so on. But even at the minimal scale there is attention to the fourth dimension, time, and to variations in the global properties--to global change.

The new global geography then proceeds to explore the regional details on an increasing scale, set by the resolving power of satellites or by the grid intervals of a computer model. At present, for many real-time, day-by-day purposes, the pixel sized or grid interval is on the order of a few kilometers, up to a hundred kilometers or so. In terms of legible detail, that corresponds to a mapmaker's globe about a meter in diameter. For comparison, even a modest scale for local geography, a mile to an inch or 1:50,000, corresponds with a globe 200 meters in diameter.

The exemplary topic is the atmosphere. Nobody would kid himself these days that he could understand the weather here in Atlanta without knowing the circulation of the entire atmosphere, and the influences of such features as the Rocky Mountains, the Himalayas, the polar ice, and the great warm pool of ocean water off new Guinea. Meteorology is already a fully-fledged global science. Daily data from world-wide ground stations and a swarm of satellites feed the supercomputers that generate our daily weather forecasts. Forecasts for a few days ahead are now as good as those for 24 hours, a quarter of a century ago.

The plate-tectonics revolution made geology a truly global science too. The precise measurements of relative positions using the satellites of the Global Positioning System makes possible the direct measurement of movements of plates of the Earth's outer shell. Radar altimeters in space can in effect drain the ocean and lay bare the ridges and trenches that drive the machinery of the plates. Other satellites can observe volcanic activity even in uninhabited places, and monitor effects of eruptions on the atmosphere. The drive is on to explain plate movements by convection inside the Earth using subterranean remote-sensing with seismic waves and gravity.

The world's ice and snow cover are now well observed by satellites. The timings of the coming and going of the ice sheets, in ice ages, are now understood from astronomical theory, but the interactions of the ice with the composition of the atmosphere are somewhat enigmatic, and so is the local behavior of sea ice and glaciers.

Two other elements in the Earth system are in a rapidly evolving phase of science: the ocean, and terrestrial vegetation. The ocean is the largest geographic feature of our planet and observations from satellites can, at least in principle, provide far more information about the ocean surface than all the world's research ships. Weather satellites measure the surface temperature of the sea. The European ERS-1 satellite, launched three months ago, picks up where the short-lived American Seasat left off, in monitoring sea waves and ocean eddies and currents by radar. The U.S. Nimbus 7 satellite showed what can be done to observe the pastures of the sea, by measuring the chlorophyll in the microscopic green plans of the phytoplankton. But that instrument is now defunct, and marine scientists are waiting impatiently for a successor. Meanwhile, they are busy with seagoing expeditions, to measure temperatures, salinities and currents far below the surface, and to study details of the interactions between the ocean surface and the atmosphere.

On the modeling side, the ocean has a prodigious appetite for supercomputing power. The eddies of the ocean are on a much smaller scale than the atmospheric depressions of which they are the marine equivalents, so the mesh of grid points has to be much finer to capture realistic detail. At present, even with the most powerful Cray machines, you can either have a full global model of oceanic water movements which lacks the all–important eddies, or you can have a model of a part of the ocean which resolves the eddies.

The work on global vegetation deserves the name of revolution. It is a good example of getting the scale right because the breakthrough came about 20 years ago when Jim Tucker at NASA's Laboratory of Terrestrial Physics abandoned the comparatively high–resolution but low–frequency views from Landsat in favor of the low–resolution, high–frequency data from the NOAA satellites. In either case, the characteristic signature of vegetation – dark by visible light, and very bright in the infra–red––is registered by spaceborne instruments. The selection of the NOAA satellites cured at a stroke the problem of clouds, because with daily observations you were much more likely to hit a cloud–free day at any locality. It also made possible the routine observation of seasonal changes of growth rates on all the world's land masses.

At a resolution of a few kilometers, or at best 1100 meters, this system of vegetation monitoring is of no use for gardens or individual fields, but is ideal for regional, continent–wide and global studies. In operational monitoring of vegetation in Africa, on a ten–day cycle by the UN Food and Agriculture Organization, vegetation maps provide early warning of a famine. They also pick out adventitious blooms in the desert, of just the kind that nurture locust plagues. Correlations are emerging between the vigor of vegetation and the risk of insect–borne disease.

Another side of the vegetation story concerns modeling. Vegetation interacts strongly with the weather and climate. While it is obviously affected by rainfall, frost and so on, vegetation reacts upon the atmosphere – for example, by sucking carbon dioxide out of the air, by pumping water from the soil into the air, by trapping the Sun's heat, and by acting as a brake on the wind. In the mid–1980s, scientists at the University of Maryland and elsewhere came up with a Simple Biosphere Model that, in 105 equations, describes how the vegetation and soil of any living landscape interact with the weather.

I mentioned earlier "people who have sat down to master the physics". An Englishman called Piers Sellers, who was prominent in the development of the Simple Biosphere Model, is a case in point. As a student at the University of Edinburgh, Sellers wanted to be a scientific ecologist, but he became profoundly discontented with the ecologists' strategy of describing life in small patches and immense detail. He switched to graduate work in geography at Leeds, but he taught himself physics and then went to the University of Maryland to work on climate modeling by computer. Casual talk of multidisciplinary research may overlook the contortions in the careers of individuals that are often involved.

After the Simple Biosphere Model, Piers Sellers became a pioneer of ecology from space. He joined forces with Tucker at Goddard, to think through the fundamentals of what the space observations of vegetation could really measure. On the Kansas prairie, Sellers masterminded a NASA experiment called FIFE to verify emergent principles of ecology from space, by comparing observations from space, from the air and at ground level. Similar experiments are planned for other biomes, but it is already clear that all ecosystems are in basic respects simpler and more predictable than you would imagine from conventional ecological science.

An excessive fascination with the glorious diversity of species in a living landscape may be misleading, in much the same way as the glorious complexity of minerals and rock formations on a regional scale blinded geologists to the overriding facts of continental drift and continental collisions. Just to hint at what is emerging: Measurements of vegetation from space turn out to be good indicators of the biomass of animals living on the vegetation--though you'd need the resolution of a military photoreconnaissance satellite to tell whether they were elephants or kangaroos.

The atmosphere, the solid Earth, the ice and snow, the oceans, and vegetation--scientific developments in all these areas share a common grounding in satellite observations, computer modelling, the global view, and geography treated as a branch of physics. As fundamental research, judged by the norms of scientific history, progress in all these areas during the past 30 years has been breathtakingly fast. Judged by the political demands put on the emergent science of the Earth system, it is all painfully slow. Scientists find themselves under intense pressure to deliver premature verdicts on environmental issues, on global change, and especially on global warming. On the other hand, without the political motivation of environmental concern, scientists would find it harder to get their satellites and supercomputers funded.

In the International Space Year, 1992, some 28 national space agencies are joining together with UN agencies and other institutions to emphasize the role of satellites in the global-change enterprise. Among the lasting products will be an animated Encyclopedia of Global Change showing similar materials in book form. Both will comprise data sets in a geographic framework on all feasible aspects of environmental monitoring from space--but aimed, I think it's fair to say, at students and the general public rather than professionals.

Global Change

Global change research will attract billions of dollars of funding in the years ahead. Governments and international agencies want to bring brainpower to bear on environmental issues where the clamor for prescriptions runs far ahead of reliable diagnoses. The World Climate Research Program and the International Geosphere-Biosphere Program are coordinating research activities world-wide on global change. These are backed up by data bases maintained at a network of World Data Centers.

At the center of concern is the prediction of a global warming due to manmade carbon dioxide and other "greenhouse" gases. A rapid increase in carbon dioxide in the atmosphere is a fact. A general warming of the world during the 1980s is a fact. Whether these facts are connected is not yet clear. The climate modelers run in their supercomputers models very similar to those of the weather forecasters, but they are so far only moderately good at simulating even the present climate. In forecasts of future climate with doubled carbon-dioxide, there is consensus among the models about a general trend to warming, but regional forecasts are often contradictory.

Commonplace clouds remain jokers in the climatic pack. Locally, they can either warm or cool the surface, but no one knew whether their overall effect was plus or minus until space observations showed that clouds strongly cool the planet. The assumptions about cloud behavior in the climate models have big effects on the predicted temperature changes and on the regional patterns. The ocean acts as a thermostat for the planet. The more realistically the effect of the ocean is introduced into the climate models, the slower the warming predicted for the 21st Century becomes. Part of the folklore of the global warming is a rise in sea-level that will flood low-lying coastal areas like Bangladesh and the southern U.S. with salt walter, and extinguish some oceanic island states. But while some experts wrangle about just how great and how fast the sea–level rise will be, others offer a scenario in which the sea–level actually falls in a warmer world. How? By increased snowfall in Antarctica adding to the ice–sheets which could keep water in deepfreeze and withhold it from the ocean.

Ecology from space may give us a handle on how vegetation may respond to changes in the Earth's atmosphere and climate, to ameliorate global change, or make it worse. Nature melds physics and biology, and so must the human mind: hence the Geosphere–Biosphere Program. But physics and biology are only part of the story, if the issue is global change, supposedly due to human activities.

Satellites like Landsat and Spot monitor human activity globally. They show the patchwork fields of the farmers, city smog, the estuarine plumes of effluents rich in nutrients that support unnatural blooms of marine organisms. And even before glasnost allowed journalists to visit Soviet Central Asia's mismanaged irrigation schemes at ground level, western satellites observed the shrinkage of the Aral Sea, as the most conspicuous manmade change to the face of the planet. Robbed of its river water, the once–bountiful Aral is a new Dead Sea, and salt from the dried–up seabed pollutes the winds of Central Asia.

Damage to forests in Europe and North America, attributed to air pollution from industry and car exhausts, is also visible from space. So are fires caused by human action in the tropical rain forests. On the other hand, the evidence from space contradicts some of the shriller assertions of environmentalists. For example, Brazilians have been angered by wildly exaggerated statements about the loss of the Amazon rain forest, which is still part of the environmental folklore.

The satellites see the destruction of forests plainly, and can measure the growth of cleared areas from year to year. So far, recent clearances in Amazonia amount to 5 or 7 per cent of the forest, as compared with 44 per cent predicted for the late 1980s by an outside expert. At the edge of the Amazon basin, the Bolivians are using satellites purposefully to help distinguish forest zones that may be suitable for long–term farming from those that are too fragile.

The myth of the marching Sahara is an another example. The fearful droughts of the southern edge of the desert, in the Sahel, gave rise to assertions, which you'll still hear, that the Sahara is advancing southward at 5 kilometers a year. The satellites that now measure the vigor of African vegetation routinely see, in years when the rain is good, normal vegetation reappearing in drought–afflicted areas. There is no sign of a general advance of the desert. The real facts are more interesting. Degradation of land ("desertification") occurs not at the desert's edge but in moister districts where intensive grazing or attempts at cultivation overtax poor soil. The affected pockets are well–observed by satellites.

A project on assessing "desertification" world–wide by combined satellite and ground studies is one of the tasks assigned to the United Nations Environmental Program. This UN program, by the way, has a notable GIS system called GRID based at Geneva. Besides maintaining a growing data base on the global environment, with some emphasis on Africa, the GRID outfit offers training on GIS and remote sensing to students from the Third World.

The most dramatic example of a supposedly manmade change is the ozone hole over Antarctica. The depletion of ozone in the upper atmosphere was detected by British scientists from the ground at Halley Bay, but the U.S. Nimbus satellite mapped the extent of the hole and monitored its growth during the 1980s. The world was shocked into curbing ozone–unfriendly chemicals. But atmospheric chemistry is another complex subject, not yet as well understood as it should be. The Earth's busy geology, the photochemical action of the Sun and the planet–wide biochemistry of living things all contribute to the chemical complexity of the atmosphere. They did so, long before the first puff of smoke from a manmade fire, or the first squirt from an aerosol spray.

How do industrial activities compare with agriculture, and with natural processes, in altering the composition of the air? Attempts to blame tropical deforestation for a major part of the current increases in carbon dioxide, in the air, now seem to fail. Ground-level measurements of carbon dioxide world-wide show little net input from the tropics. But forest fires, whether caused by nature or by human action, put carbon monoxide, methane and ozone into the lower atmosphere. Ozone, by the way, is considered "good" in the upper atmosphere, but "bad" in the lower atmosphere, where it attacks living things.

Volcanoes pump huge quantities of dust, hydrochloric acid and sulfuric acid into the atmosphere. What do these do to the ozone layer, to the climate, and to the pollution of forests and seas? Most cryptic is the role of the ocean, as a source and sink for chemical agents in the atmosphere. For example, living organisms in the ocean release compounds of sulphur and the halogens into the air. There is a long catalogue of chemical questions to be added to the other uncertainties about the Earth system and the human role in global change.

Another example of observable change is the alteration in the course of a river during floods. An impressive use of the French Spot satellite was the rapid remapping of the nation of Bangladesh in the aftermath of the severe river floods of 1988. This revealed large changes in the river beds, which wiped out many farms and villages and created new land other places. In such a case there is interaction between natural forces and prior human efforts to build levees and other river defenses.

In speaking of global change, scientists slither between the natural science of the Earth system and attention to the human factor. This is reckless as a technique of analysis, and a weakness of the emergent science of the Earth system. Here is the opportunity to assert a role for the geographic tradition. Indeed, in our age of over-specialization, the very idea of putting together physics plus biology plus the human factor seems mind-boggling, until you remember the great integrating science that has been uniting these elements for 2400 years. Herodotus, the founding father of geography, pondered the creation both of the Nile Delta and of the Pyramids.

GIS and the Global Geography Machine

The most striking satellite product so far, for portraying the human factor, is the image of the Earth at night. It shows the gas flares of the oil fields, a Japanese fishing fleet using lamps to lure the squid, and especially the city lights of bright, rich urban areas in Japan, Europe, and North America. The northern cities radiate extravagant amounts of light into the universe, which are pretty good surrogate data for the rates at which regional energy systems are putting carbon dioxide into the air. The nocturnal image of the Earth also reveals differences in living standards from continent to continent. Africa, for instance, is literally a dark continent. This is a vivid picture of a world divided between rich and poor.

To remind you again of who is setting the pace, this is the work not of a geographer but an astronomer. For certain global maps in human geography, I have to apply to the space scientists. NASA's Goddard Institute for Space Studies in New York City produced the world's first map of carbon dioxide emissions form fossil-fuel combustion. It was not available from the geographers or economists, so the space scientists created it for themselves. They also made the first fully consistent world map showing the intensity of cultivation, world-wide. To solve the problem of differences in the quality and criteria of data from different nations, they used Landsat for spot checks on the proportion of land under cultivation.

Five years ago, at a planning meeting in Stockholm for the International Geosphere–Biosphere Program, I noticed an almost total absence of social scientists. The natural scientists seemed content to reduce all human life to a puff of carbon dioxide, or to a fall–off in transpiration from a block of denuded soil. I asked some Earth–system pundits how they could investigate global change due to human activity without help from experts on human activity. Their answer was that the social scientists lacked the intellectual rigor of physics and biology.

The social scientists may have been slow off the mark, but they did not take the snub lying down. They began shadowing the International Geosphere–Biosphere Program of the physicists and biologists, with their own program called Human Dimensions of Global Environmental Change. A token of geography's special place on the intellectual map is that the International Geographical Union belongs both to the International Council of Scientific Unions, which masterminds the Geosphere–Biosphere program, and to the International Social Science Council, which leads the Human Dimensions program. When there are deals to strike across the academic boundary, geographers are often the most effective brokers.

Geographic information systems have a special role in the Human Dimensions program. A working group on data and information systems wants to collaborate with the Geosphere–Biosphere program's group on data and information systems, not just to exchange data but to solve technical problems of common concern and create parallel global networks. A more elementary part of the attention to GIS in the Human Dimensions program is simply to coax those social scientists who have been slower than the geographers to adopt GIS, to learn to use it in human–dimensions research. Physicists and biologists seem readier to grab GIS whenever they want it, and to learn the packages quickly, than do political scientists, sociologists, and anthropologists. There are some new fields for you to conquer, in making the world GIS–minded. Perhaps you can make the social scientists satellite–minded too.

I am glad to say that the academic Berlin Wall is crumbling. The natural scientists of Geosphere–Biosphere and the social scientists of Human Dimensions have identified a first core area for practical research where they can collaborate in earnest. You will not be surprised that the subject is changes in land cover and land use.

Thinking of the Earth system as a whole, and human activity as a whole, the most extensive interface is the farmland that replaces natural vegetation with crops or managed rangeland. Urban development has more intense effects but over similar areas. Land cover affects the climate regionally and globally by determining what proportion of the Sun's energy is absorbed, and how much is reflected into space. Land use, of which the land cover is the outward sign, also affects the exchanges of moisture, carbon dioxide and other greenhouse gases between the atmosphere and the biosphere. With environmentalists' attention focused hitherto on the trouble spots of desertification and deforestation, there has been a danger of neglecting the possibly greater influence of changes in normal farming. These include the increasing use of irrigation and fertilizers in the Third World, and some farmland going out of production in the richest countries, which find themselves producing more food than they can eat.

Research on land–use change is therefore very timely. Existing programs in North America and the European Community for satellite–aided monitoring of agricultural production include crop–identification and assessment from space. These help to get such research off to a flying start. More fundamentally, satellites can make direct observations of changes in the albedo and in vegetation growth–rates as the land–cover changes. Computer models can sketch the consequences for climate, water supply and so on.

But to know why the land cover changes requires a thorough, global understanding of human factors that alter land use. The joint working group of natural and social scientists for the land-cover/land-use core project want field research for case studies that look for causes of, and responses to, land-use change. These will follow standard protocols, so that results from different regions and for different kinds of change can be aggregated and merged with global computer models. Work is due to start next year.

GIS itself is developing extremely rapidly, both in technique and in content. At such times it can be hard to know which way everything is going - or should be going. A conference of communications engineers a century or so back might have found life similarly confusing. The electric telegraph had put the Pony Express out of business, but here was this teacher of deaf children claiming to send his voice by electricity. The global telecommunications networks of today, with their glass-fiber and satellite links carrying images and data as well as messages and speech, would have been hard to imagine.

In his keynote address to your conference at Anaheim last year, John Borchert spoke of the piecemeal development of GIS, and a need for coherence, both intellectual and organizational. As an urban geographer, he sought intellectual coherence in regional analysis. He added the fourth dimension of time to the three dimensions of space, to see how the structure of a region unfolds in time. He looked for organizational coherence in the creation of coordinated and standardized data sets.

Nothing I wish to say contradicts any of that, but I should like to nest it in a larger framework. Where Professor Borchert spoke of the regional, I stress the global. And beyond the data sets I stress the numerical models. More quickly than the growth of the telecommunications industry, information systems in the geographic domain will evolve, I believe, into a global geography machine. It will monitor, model and predict every readily measurable aspect of the natural world and of human social and economic life.

To exaggerate the promise of computer modeling in the natural-science sector would be difficult. Take the growth of plants, on which all life depends. I have already mentioned the NASA work that combines satellite observations of vegetation with computer models of interactions between the plants and the weather. But there are also computer models of plant growth, notably in The Netherlands, that describe the life cycle of an individual plant: how it puts out roots and shoots, grows in mass and leaf area and comes to maturity--all the while responding to changes in day-length as the season advances, to the exigencies of the weather, and to limiting factors such as lack of nutrients in the soil. By the way, it was the Dutch modelers who showed that, contrary to expert opinion and popular belief, food production in the Sahel of Africa is limited less by rainfall than by soil fertility.

Meanwhile the molecular biologists are beginning to understand the apparatus of genes, enzymes and hormones that govern the life of plants. This system too will be amenable to computer modeling, linking like a cog in a train of gears to the plant-growth models, the plant-weather models and on up the space scale of regional and global ecosystems, in which even the animal life supported by the vegetation will be to a large degree predictable. That will open the door to the realistic models of the biosphere in different climatic settings--in the ice ages, for example, or in the far warmer world that existed in the days of the dinosaurs, when the continents were grouped in a supercontinent. A deeper understanding of the evolution of the biosphere will give a more vivid explanation of the present biosphere. We badly need it, if we are to say with more confidence how tropical deforestation will compare in its effects with an ice age, or with the deforestation of Europe, India and China in the Bronze Age.

I could give similar examples of gear–trains from the chemical to the global scale in models of the world's ice cover, say, or of the solid Earth.

A model is of course a theory of a kind, and the capacity to model realistically is a test of understanding a system, as opposed to merely describing it. When physicists complain about a lack of intellectual rigor in the social sciences, what they really mean is a lack of computable and testable models to compare with, say, the meteorologists' models––in which elementary laws of physics act on the parcels of air to generate observable depressions and jet streams. Such are the norms in modeling being set by the physicists and biologists.

Can geographers and other social scientists match those norms, not only in the quality of the models but in compatibility too, so that human dimensions models can usefully mesh with the global change models? I must return later to an analytical problem about human choice, which I tag as the Cussedness Factor. But the general answer to the question must surely be, we don't know, but can with optimism try.

Especially since the eclipse of Marxism there are no overarching theories of human affairs to compare, say, with quantum theory and thermodynamics in physics. On the other hand, there are powerful commonsensical ideas in geography, in particular the notion that a person will travel further to buy a jacket or a car than to buy a newspaper. This underlies the classic German theories of von Thünen's rings, Weber's cost–minimizing theory of industrial location, and Christaller's theory of the central place and his honeycomb patterns of towns. These theories are not so much falsified as swallowed up in modern computer models in human geography. They have much more subtlety and geographic realism, and are dynamic rather than static, being concerned from the outset with geographic change.

The state of the art is well represented in the spatial–interaction models developed by Alan Wilson at the University of Leeds in England, as an aid to marketing strategies. Combined with GIS and tailored to run on PCs, these models use census data, local economic data, and equations that describe the fall–off in attraction with distance, to predict sales at a new store. The Leeds models have been adopted by several large retailing chains in the U.K. and also by car dealers and water utilities. They are now coming into use in other parts of Europe.

Similar models can offer to predict demands for medical services in a pattern of hospitals, or traffic flows after the construction of a new road. So far, geographic models of such kinds typically operate at the level of postal districts, cities and counties. There is no difficulty in principle in extending them to national, continental, and global scales, although the conceptual problems, the data assembly and assimilation tasks, and the sheer number–crunching, will all increase formidably.

Starting from the other extreme, some socio–economic models are inherently global in scope. These go back 20 years to the Limits to Growth study at MIT, which predicted a population crash in the 21st Century as the world converted its natural resources into industrial pollution. Psychologically, it was part of the environmentalist surge of the early 1970s. Technically, there was no regional detail: the plant was treated as a geometric point. The extreme simplicity of its assumptions made Limits to Growth hopelessly flawed, according to its critics. Since then, global models have evolved with more regional detail, to describe and predict changes in population, agriculture, energy use, and so on. On an intermediate scale there are national economic models, marine pollution models on the scale of Europe's North Sea and Adriatic Sea, and continent–wide models of air pollution.

Already, then, there is a small repertoire of models concerning the human factor that might usefully be plugged into the natural–science models. I hope that this will happen in the new international core project on land cover and land use. It could be a first small step towards the global geography machine I have in mind.

Apart form the full-range of natural-science elements I have mentioned, the global geography machine will model human interactions with vegetation, air quality, river runoff and pollution, and so on, and all the internal elements of the human factor--demography, the urban systems, transport, and the general living standards and other socio-economic factors which together determine the intensity of the human interactions with the environment. For example, the world's population has doubled in the past 40 years, but the fact that the number of automobiles has multiplied seven-fold may be more significant.

As with present-day weather models, routine observations from space and data-collecting on the ground must verify or modify the contents of the machine, and prevent it running away into fantasy. The global geography machine will offer great practical benefits. For example, I visualize it telling instantly how a change in the price of rice in Bangkok, Asia, will affect the demand for fertilizers by the rice-growers of West Africa. Or what a sea-temperature change in the Atlantic implies for umbrella-makers' profits in Hungary. Or simply how many tourists Alaska can expect next week. No one has the imagination to guess at all the linkages and unexpected feedbacks that such a global geography machine will comprehend. Indeed the very possibility of such a machine, and its present non-existence, are signs of just how half-baked are our present attempts to assess the human impact on the planet.

Whether it will take 20 years to build the global geography machine, or 100 years, will depend more on willpower and ingenious software than on the hardware. If computer technology continues to advance at the high rates of recent decades, PCs with gigabyte memories and supercomputers in the terrabyte range will soon be commonplace. It remains for human geographers and other social scientists to create the gear-trains of models, from the village of the world, that can also mesh on any scale with the physical and biological models of the natural scientists.

There is a special opportunity for the GIS/LIS community to take the initiative in much of this very strenuous and multidisciplinary work, provided you become model-minded and global-minded rather quickly. If you don't do it, someone else will, and there are no prizes for coming second. The same challenge confronts academic geographers. I need only quote Inez Fung, a physicist working for NASA on global modeling. When she was explaining to me her multidisciplinary outlook, she said: "We have a new way of doing business where you have to talk several languages. The universities have not caught up."

The Cussedness Factor

Nearly 90 years ago the English author G.K. Chesterton was scathing about H.G. Wells and the other social prophets of his time. He described the evolution of human society as a game of "Cheat the prophet", played by ordinary people. He wrote:

> The players listen very carefully and respectfully to all that the clever men have to say about what is to happen in the next generation. The players than wait until all the clever men are dead, and bury them nicely. They then go and do something else.

This is one aspect of what I mean by the Cussedness Factor. From a purely technical point of view, the element of unpredictability looms larger in human affairs than in most large-sale natural phenomena, and makes all predictive models of the human element in the Earth systems highly conditional. At one of the prime centers for geographic modeling, the International Institute for Applied Systems Analysis located in Austria, the teams have recognized the flaw in models based on what they call Conventional Wisdom -- the supposition that the world will remain much as it is, with changes as in recent decades continuing smoothly.

The IIASA teams have gone into a science-fiction mode, to imagine surprises that could radically alter all their forecasts of population growth, agricultural development and energy use. For example, they speculate about an urban plague driving people to a decentralized, rural existence. In another scenario, they visualize India emerging as the world's leading economic nation. These are all "What if?" conjectures, but given all the surprises in human history they are just as valid as Conventional Wisdom, which must be seen as a rather implausible case of "What if?": "What if the surprise is that there are no surprises?"

Wars, including nowadays the still-present risk of nuclear war, can have huge effects on human affairs and the natural environment. Disease, which includes nasty surprises like AIDS, can slash populations. Conversely, improvements in health have led to a large increase in the numbers of old people. This was not properly predicted, because demographers and doctors did not realize that if people stay healthy till 70, far more will live to 90 plus than ever before.

Apart from these life-and-death issues, basic economic parameters such as energy use have been very poorly forecast in recent decades. The energy crises of the 1970s were not anticipated, nor were the crippling effects on the nuclear power industry of the sharp public reaction to the accidents at Three Mile Island and Chernobyl. Any energy experts should not have been taken by surprise by the crescendo of public and governmental concern in the 1980s about carbon dioxide and the greenhouse effect. Meteorologists have been discussing these since the 1960s.

Organized concern about the environment and the limits of the Earth was itself a surprising novelty, which blew away the confident expectations of the 1960s that exponential economic growth would continue indefinitely. And this exposes the formidable philosophical, practical, and I should say political problem of the self-referential nature of all pronouncements or models concerning human beings. There is always a large element of conscious choice in human affairs, which includes the choice of responding or not responding to predictions. If, from satellite data of vegetation of Africa, I forecast famine in Ethiopia, and if relief agencies take action to avert the famine, then my forecast is falsified. A narrow-minded physicists might write it off as a dud forecast. But I can claim that it was implicitly a "What if?" prediction, assuming no remedial action.

Self-reference becomes politically charged when one asks: What are we studying the Earth system for? Or, putting the question another way, who will own the global geography machine? A general pious answer is easy enough. As our species becomes better informed about the Earth system and human interactions with it, we shall hope to use our natural resources more wisely and try to avoid harming our planet. But who is doing what to whom, in all this?

Information is power, and that applies to remote sensing data, where the military, law-enforcement and commercial applications make some people nervous about Big Brother in the sky. Here the situation is aggravated by the high cost of Landsat and Spot images, which some governments and large corporations can afford, but independent researchers cannot. In typical remote-sensing labs, I find the professor and his students working repeatedly over the same few images, half of which date back to the good old days when Landsat came cheap. Here I'm speaking not just of Third-World countries like Indonesia or Kenya, but of leading universities in the United States and Europe.

You need the resolution of Landsat and Spot to do serious human geography and land–cover studies from space. Governments cannot be said to care very much about the human factors in global change unless they hurriedly rethink their marketing policies for space images. It is not a coincidence that the most rapid progress in research discoveries and in applications have come with the NOAA satellite data, which are free.

GIS and LIS as much as any other form of information are potential power, as the census–dodgers who evaded even your splendid TIGER system seem to have sensed, however wrongheadedly. We had a spurious fall in population in the U.K. too, in this year's census, because of a hated poll tax. When that can happen even in scientific democracies, I have to wonder if we know the world's population to within even 20 per cent of the true figure.

By definition the statisticians know little about the activities of fraudsters, hoodlums, and smugglers which can be significant fractions of a country's gross national product. And governments tell lies. The meteorologists need not fear that the weather stations will fake the barometric pressure at Beijing or Baghdad, but governments may have powerful domestic or foreign–policy reasons to falsify quite basic data in human geography such as population change, agricultural yields, or coal production. Social scientists need to make more use of error estimates, as physicists and biologists do routinely, and not pretend that their head counts or acreages are accurate to the last digit.

I would also suggest that you abandon the term "ground truth", with its implication that data gathered at ground level are more accurate than satellite data. When the European Community began using remote sensing to monitor crop production it found, as expected, that some producers were faking their figures in a subsidies scam. But the researchers also found that some official enumerators, sent out by governments to identify crops field by field, headed instead for the nearest tavern and invented their data over a glass of wine.

All that is at the data–assimilation end of the global geography machine. What about the uses to be made of its output? There could be the same issue of privileged access due to high cost, as with remote sensing data. But that is less inevitable than the temptation to use the global geography machine as an aid to bossing people around. People in the Third World already mutter about ecocolonialism, meaning that the rich northern countries are using environmental issues as yet another excuse to meddle in their affairs. Global–change scientists only confirm these fears when they speak glibly about their hopes of managing the Earth's environment.

Wisdom about caring for the planet must begin with a dose of wisdom in dealing with the world's people. The worst pollution blights and environmental mismanagement, including the destruction of the Aral Sea, have occurred in totalitarian countries where managers have been most free to ride roughshod over the hopes and fears of local populations. The spectacular collapse of Leninism is, in my opinion, an occasion for reflecting on all forms of authority, in which I would include even well–meaning expert advice. The most dire statistics out of Africa are of course those concerning famine and AIDS, but to me it seems almost as shocking that 100,000 foreign experts are loose in that unhappy continent, giving contradictory advice.

The upheavals in what we used to call the Soviet bloc have attracted the headlines, but the Western democracies have discredited themselves too, in the eyes of the world's poor, by an abrupt switch in policies. Remember that just twenty years ago the tropical rain forest was called the jungle, and anyone who had not cleared it and let in the light was considered backward or even lazy. The World Bank and other agencies drove roads through the jungle to promote farming, forestry, mining, dam–building and so on. Then, during the past ten years, the compass of opinion swung through 180 degrees, and people like the Brazilians found themselves pilloried as the ravening destroyers of the rain forest. There are similar stories about rangelands, fisheries, river management, and so on, with the injunctions changing from "Develop!" to "Conserve!"

I say nothing for or against either development or conservation, but only note that the drastic change in Western expert opinion has left the governments and peoples of the Third World bewildered and angry. It is high time that we in the West gave up our bad habit of telling other people how to live. It would be wiser to focus our environmental zeal on our own local problems such as toxic waste disposal, our disproportionate contribution to global carbon dioxide, and the widespread damage to the forests of North America and Europe by air pollution. I shall be surprised if there is not a vigorous backlash from the Third World, directed at us, at next year's UN Conference on Environment and Development in Rio de Janeiro.

I commend to your attention a research project under consideration for the international Human Dimension program, which the psychologists are promoting. They are another profession, besides the geographers, with a foot in both the natural-science and social-science camps. The psychologists have joined forces with the anthropologists and with the pollsters of the World Association for Public Opinion Research, in proposing an international study of attitudes to the environment. The aims are to understand the mental models of the environment possessed by ordinary people, as individuals and cultures, and to see how values, motives, and information from neighbors or the media influence these models, and to examine the behavioral consequences that flow from these models and so shape the local landscapes.

At first sight all this is a long way from the typical concerns of GIS, and from its fruitful interplay with remote sensing. Satellites cannot see into the human mind. Or can they? In a certain practical sense they do, when they observe the landscapes created by distinctive human cultures. Geographers have been surprised by how clearly national borders stand out in some Landsat images: for example in abrupt transitions from farming to ranching along stretches of the U.S.-Canadian border, from well-watered fields to parched fields at the U.S.-Mexican border, from small fields to large fields at the Cold War West European/East European border, and then back form large fields to small fields at the Soviet-Chinese border. As people learn to read the satellite images more skillfully, I suspect they will see more subtle but revealing signs of cultures and cultural change in patterns of roads, settlements and recreational zones. Be that as it may, the proposal in the Human Dimensions program to push world-wide research on attitudes to the environment seems to me to go to the heart of the political issue.

The fate of the world, environmentally speaking, is in the hands of ordinary people. The human impact on the environment is mainly determined by the daily decisions of billions of individuals, who I call the true watchkeepers of Spaceship Earth. In cities, the impact is often indirect, as people choose what they buy and how they travel, but in the rural areas the impact is much more direct, as the inhabitants decide on land use and the season's crops. Especially in marginal areas of the Third World, people make life-and-death bargains with nature, such that if their environmental judgements are faulty they will see their children die. They have good reason to be cussed, when foreign experts tell them first one thing and then another.

But we in the rich countries still have a monopoly of the science, the satellites, and the supercomputer. Why spend billions of dollars on global-change research if the improved scientific knowledge cannot somehow enhance the wisdom of the entire human species, and influence those billions of personal decisions by people rich and poor? Here is the answer to the question of who should own the global geography machine. Everyone, everywhere, from presidents to peasants.

This is not as farfetched as it seems. We are halfway there already, with the only part of the global geography machine that is fully up and running -- the meteorological part. The bargain works as follows. I am on a small island in the Pacific, with a modest weather station. You are sitting by a supercomputer in Washington, D.C. or England or Japan, churning out global weather forecasts. You want my weather observations, as part of your primary data, so I feed them free of charge into the global telecommunications system of the World Weather Watch. In return you send me, free of charge, copies of your forecasts and your satellite images. I combine these with data from nearby islands to prepare my national weather forecast. I reserve the rights to disbelieve your computer forecast if I see storm clouds on the horizon that don't square with it, but usually I am grateful for your output. Then I issue my forecast free of charge, via the media, to the general public.

It doesn't matter that your equipment costs a million times more than mine. I command a piece of the Earth's atmosphere, from which you badly need data. It is far cheaper for you to send me copies of material that you are generating for your own purposes, than to come and set up your own weather station on my island. The net result is that, for all practical purposes, I am in effect part-owner of your supercomputer.

This approach can be generalized for the global geography machine. Consider a peasant in Bolivia, say, who commands a small part of the Earth's land surface. What he does with that land, what crops he grows, how much energy he uses, what sprays and fertilizers he uses, how he gets his produce to market -- all these things are primary data for any comprehensive model of the Earth system and the human impact. Now recall such principles as data-protection and freedom of information, and the right of individuals to inspect any files concerning their affairs. More positively, consider the end purpose of the global geography machine.

The more enlightened technical-assistance agencies now see their role as ensuring that people in the Third World have access to the latest scientific information and ideas. Where appropriate they help villages and local agencies to acquire tools such as PCs, GIS packages, satellite receivers, and the like. But they leave the decision-making about how to strike a balance between development and conservation to those who will prosper or starve as a result of the decisions--the local people themselves.

In the future, such services of information and technique will be available from the global geography machine. As in meteorology, there will be gray areas where companies or communities require special data sets or computations, for which some charge may be necessary. This may apply for example to explorations of scenarios in a "What if?" mode. But the basic data and routine computations of the global geography machine should, in my opinion, be freely available to all who contribute data to it, actively or passively--which means every last member of the crew of Spaceship Earth.

I do not underestimate the political implications of what I am saying. Indeed, I liken the new powers to observe the Earth from the space and model it by computer, to Galileo turning a telescope on the heavens. By contradicting medieval astronomy, Galileo's observations helped to undermine the secular authority of the Church, and to ease Western civilization back on the old Greek track of freedom of thought, scientific enquiry, and democracy. Direct observation of the Earth may easily have at least as great an impact. For a start, governments will find it much more difficult to tell lies. To the extent that information is power, its diffusion to anyone who wants it will give a further boost to people in power. The images, data, and models will also influence all our attitudes to questions of wealth and poverty, race relations, and the sharing of the Earth's resources, and help to change definitions and aims -- in another self-referential feedback in the system.

Take the idea of environmental quality. A clear, unsentimental definition is already long overdue. The geologic record shows that Nature is indifferent to how warm or cold the planet is, or how many species survive. The only objective test for the quality of the global environment is the quality of human life. The interpretation can be as broad as one wishes, and can include wildlife as a necessary asset, and pollution as an avoidable nuisance. But the overriding biological fact is that human life expectancy is about 15 years less in the world's poorer countries than in the rich ones. Human welfare and environmental wellbeing are not competing aims. They are identical.

I see you all as privileged craftsmen at the start of a great adventure of the human spirit. You are helping to fashion the ships for a rediscovery of the Earth and its peoples. It will be hard work. There will be storms and navigational blunders, but when the surviving ships come home they may be carrying the greatest treasure of all time. I mean the knowledge needed to abolish poverty in the world without harming the Earth. Bon voyage.

Developing Thematic Maps From Point Sampling

Using Thiessen Polygon Analysis

Robert C. Maggio
David W. Long

Remote Sensing/GIS Laboratory
Department of Forest Science
Texas Agricultural Experiment Station
The Texas A&M University System
College Station, TX 77845-2135

ABSTRACT

Federal agencies such as the USDA - Soil Conservation Service routinely conduct nationwide resource inventories to monitor natural resource use. These inventories are conducted using various sampling procedures. From the resulting databases, extrapolations are made to characterize the entire state or country. In most cases, tabular outputs depicting trends and summaries are generated. One such database developed for the Natural Resources Inventory (NRI) by the Soil Conservation Service is geo-referenced and contains over 30 landuse attributes describing landuse. The NRI database containing over 69,000 point samples for the 254 counties in Texas is used to develop thematic maps for various attributes using thiessen polygon analysis. Construction of thiessen polygons and their limitations are described.

INTRODUCTION

Government agencies have been gathering and storing large quantities of natural resource inventory data for years. These inventories are conducted to monitor changes in landuse as well as estimating production of agricultural products. From these inventories, decisions are made regarding federally supported projects aimed at maximizing the use of our natural resources, but at the same time, protecting them for future generations. Much of the resource inventory information is in the public domain and can be readily acquired. These data bases can be extremely valuable if the data is easily accessible and can be converted to a more usable form.

The U.S. Department of Agriculture - Soil Conservation Service (SCS) National Resources Inventory (NRI) is one such valuable resource data base. It contains information regarding soils, plant cover, land use, wetlands designation, slope, tillage practices, range condition, and conservation practices at hundreds of thousands of points throughout the United States. Summaries, means, and trends can presently be calculated from this data set, but there is no manner in which to evaluate it spatially other than developing these figures by county, district, or state.

National Resources Inventory

The NRI began in 1977 as a result of the Rural Development Act of 1972 (Public Law 92-419) which mandated the SCS to conduct nation wide resource inventories every five years and report the results. The inventory was repeated in 1982 and 1987 with several changes in procedures and quantity of data collected. The next inventory is scheduled for 1992 continuing the 5 year cycle.

The 1982 inventory was the most comprehensive study of our nation's resources to date with hundreds of thousands of sampling units measured across the entire United States. A database was designed for use in analyzing the inventory data and can it be queried through the SCS NRI Information System (USDA SCS 1987 Users Manual). For the 1987 NRI, the sampling percentage was heavily reduced since the emphasis was to relate trends in resource conditions between 1982 and 1987, rather than perform a complete enumeration. Approximately one third of the sample points were used in 1987, but each sample point used was one that was also measured in the 1982 inventory.

The sample points used in the NRI were derived using a two stage stratified sampling procedure. The county was the base region for sampling purposes, which were then stratified geographically to make sampling more efficient. First stage sample units (Primary Sampling Units - PSU) were randomly placed in the strata thereby providing sampling intensities commensurate with percentage landcover by land cover class in the county. At each location of a PSU, a 160 square block was located on an aerial photograph or map by the SCS District Conservationists for the area being sampled. For each PSU, information such as:

county name,
Major Land Resource Area (MLRA),
presence of farmsteads,
critically eroding areas,
water bodies,
and windbreaks are recorded.

The Secondary Sampling Units (SSUs) within each PSU were randomly placed, but were dispersed within the PSU. The location of each SSU was determined by randomly choosing coordinates that would locate the SSU by applying offsets, in feet, from the southwest corner of the square, 160 acre PSU. At each SSU, the following information was gathered:

Ownership (private, state, federal, etc.),
Soil mapping unit,
Degree of erosion ,
Irrigation,
Proximity to flood-prone area,
Landuse/landcover,
Cropping history,

Universal Soil Loss Equation (USLE) data,
Wind Erosion Data,
Wetlands,
Riparian Data,
Wildlife Habitat, and
Forest Type.

The State of Texas, containing 254 counties, has 23,000 PSU locations in the database with three SSUs in each PSU for a total of 69,000 data collection points. The PSUs were marked on air photos and 7.5 minute quadrangle maps for field location in the inventory process. Each of the 23,000 PSUs were transferred to United States Department of the Interior - Geological Survey 7.5 minute quadrangle maps with the three SSUs located within the PSU. These PSUs and SSUs were then digitized into a seamless GIS database using ESRI's ArcInfo. Figure 1 shows the position of PSUs and the numbering scheme used to reference each sampling unit in Bell county, Texas - SSUs are not shown.

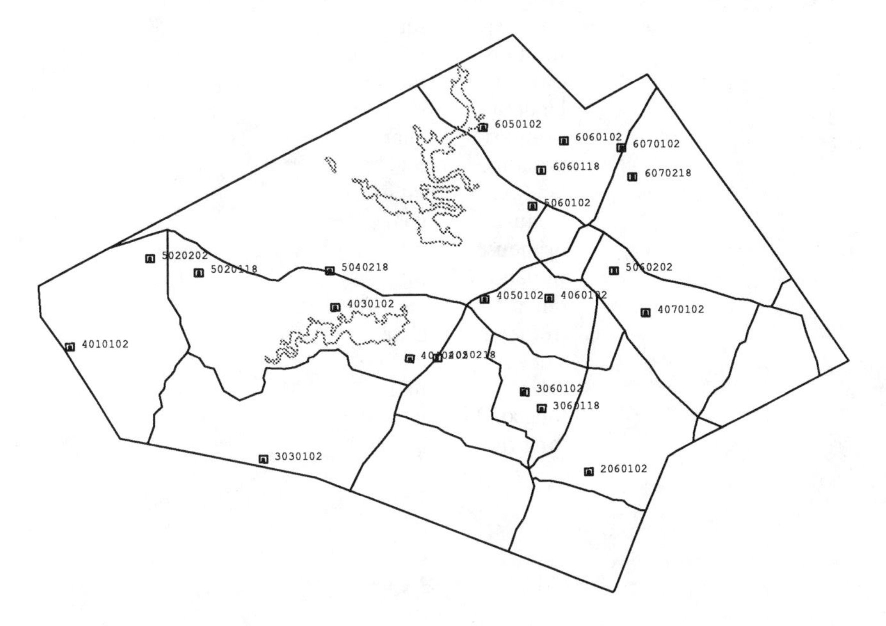

Figure 1. PSU locations and identification codes for Bell County, Texas.

NRI Database

The database created for the NRI consists of a collection of tables describing various attributes of soil, hydrology, cover, and land use at each data collection point. Several tables link codes to names such as FIPS code to county name and land class code to land class name. However, most of the tables provide data collected at each SSU. For example, the Cover Table contains information on percent woody canopy cover by species and the Practices Table contains information on conservation practices being used at the sampling point.

Most of the data collected at each site are found in the Trends Tables, a table developed from the raw PSU database depicting changes from one inventory period to another. There is one Trends Table for each inventory year: Trends82 and Trends87. Trends92 will be created after the 1992 inventory. The schema for the Trends82 table is shown in Table 1. The first 4 parameters (FIPS, PSU, Point, Year) are used to develop a unique key for linking this tabular data to the Texas PSU database containing the digitized location of each PSU and SSU.

Table 1. Trends82 table schema.

Name	Type	Key#
fips	int	1
psu	char	2
point	int	3
year	int	4
ifact	int	
rotation	int	
owner	int	
landclss	char	
primfarm	char	
broaduse	int	
landuse	int	
pasture	char	
primeuse	int	
wetland	int	
tilltyp	int	
tretneed	char	
sloplen	int	
usle	float	
rangcond	int	
rangtrnd	int	

THEMATIC MAP CREATION

In order to develop thematic maps from a database that is geo-referenced by points, techniques must be applied to take the field position of a point and extrapolate it to a polygon that covers the areal extent that the point represented according to the statistical sampling procedures used in the sampling design. There are several procedures available for accomplishing this task, but in this research thiessen polygon analysis was chosen (Boots, 1987; Boots, 1974; McCulloch, 1980).

The creation of thematic maps from the NRI data was accomplished by a three step process. First, thiessen polygons are produced in a GIS using the coordinates of the SSU points. Second, the attribute data from the NRI database are joined to the thiessen polygon coverage. Third, adjacent polygons with the same attribute have their common boundaries dissolved.

Thiessen Polygons

Producing thiessen polygons from a map of points is central in the automation of the thematic map making process (Figure 2).

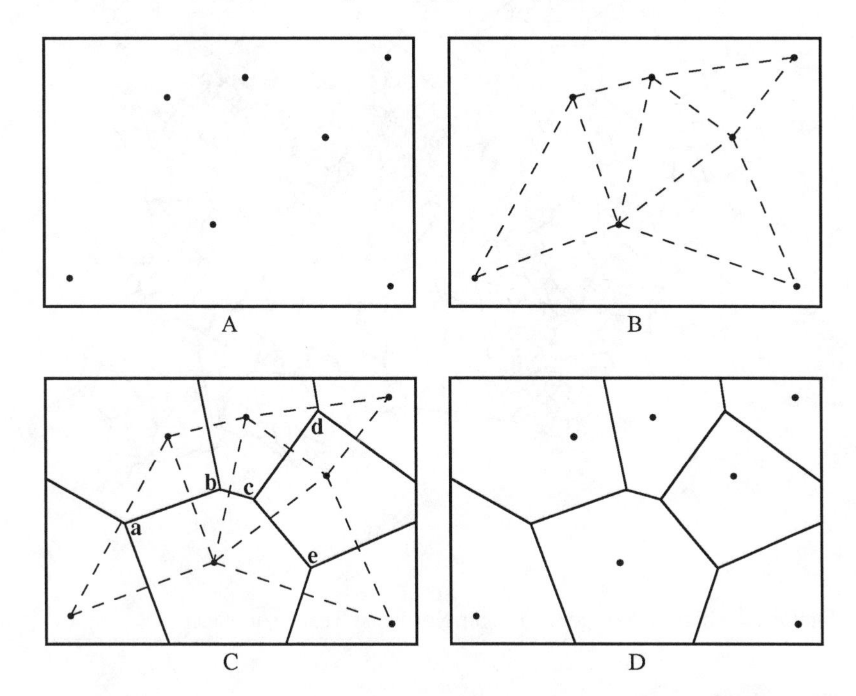

Figure 2. Steps in producing thiessen polygons from a point coverage: 2A) Point coverage. 2B) Points joined by lines forming triangles. 2C) Perpendicular bisectors of the lines are added and they meet in each triangle. 2D) Original lines are removed leaving thiessen polygons.

Thiessen polygon processing divides a map into regions using point data for division criteria. Each region is delineated in such a manner that any point in the region is closer to the defining point than to any other point in the data base. Polygons are constructed in a three step process. First the points in the coverage are joined to their neighbors by lines creating triangles (Figure 2B). Then the perpendicular bisector of each line is created (Figure 2C).

The three bisectors of each triangle meet at a point inside the triangle (Figure 2C, label points A through E). Third, the original lines are removed leaving a polygon around each point such that every location inside a polygon is closer to the originating point than any other point in the coverage Figure 2D).

Figure 3, representing an eight county area, shows the result of constructing thiessen polygons around these points in Northeast Texas. This map can be used to produce a thematic map by querying polygons that have the selected attributes.

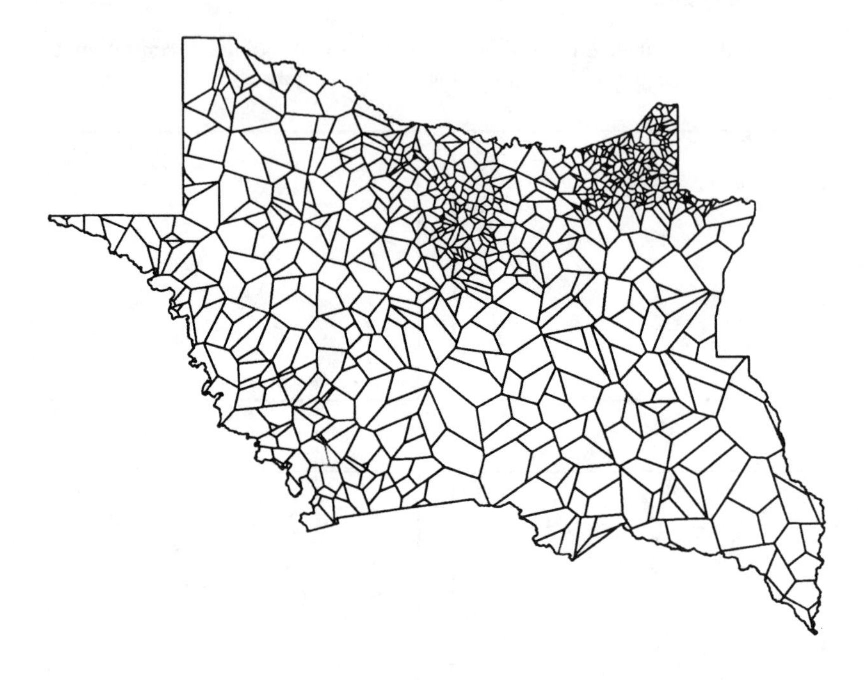

Figure 3. Thiessen polygons of eight Northeast Texas counties.

Thematic Maps

In order to link the SCS NRI attribute database to the GIS thiessen polygon coverage, a key common to both databases is required to link the two data types. The Trends82 table uses four variables as keys. The first variable, FIPS Code, identifies the state and county. The second variable identifies the PSU number. The third variable identifies the SSU point within the PSU. The fourth variable identifies the inventory year. The GIS requires that a single variable be identified as the key for linkage. In order to make the NRI database, Trends82, compatible with the GIS, a new variable was produced by concatenating the first four variables of the table. Once the thiessen polygons are created and each polygon has the SCS NRI data tied to it, a thematic map for a single attribute can be developed. Figure 3 shows the thiessen polygons without the SSU seed points.

Figure 4 depicts the distribution of the same attribute as Figure 3, except adjacent polygons with the same attributes have been merged to form larger polygons. Once this process is complete, classes within the variable are depicted by various shading patterns. Figure 5 shows the distribution of the Broaduse attribute from the SCS NRI database.

Each polygon in the thiessen map has only one attribute from the NRI database. Regardless of the attribute tied to the polygon, the thiessen polygon, remains the same - the seed point does not change, only the attribute data associated with the polygon. It is apparent in Figure 3, there are areas where the density of polygons is much greater than in others. The density is a function of the number of SSUs available for polygon processing. In the initial NRI inventory, a greater sampling intensity was selected for areas that contained greater diversity in landcover and a lower sampling intensity for areas of low diversity. If the number and size of the thiessen polygons can be compared to resolution on the map, the resulting map would have higher resolution in some areas than in others.

Since the size and configuration of each polygon generated by this process is dependent on the number and distribution of the SSUs, the resulting map is not one of the distribution of the variable being depicted. It merely provides a spatial approximation of the attribute data. Obviously, the greater the density of sampling points, the closer to the true spatial distribution the map becomes. But, if we revisit the objective for the SCS NRI, we find that it is to monitor change

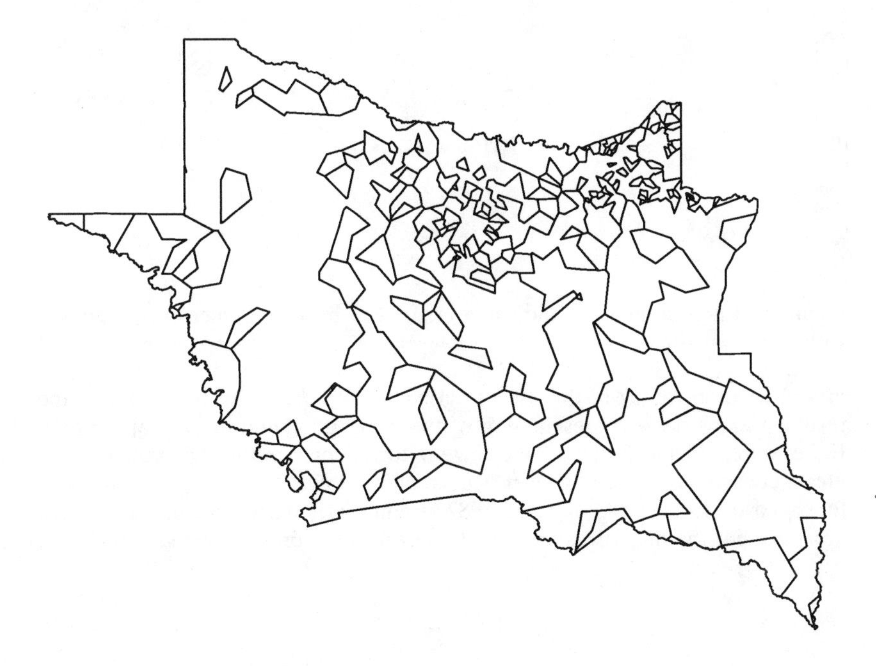

Figure 4. Thiessen polygon map after processing to aggregate polygons with like attributes.

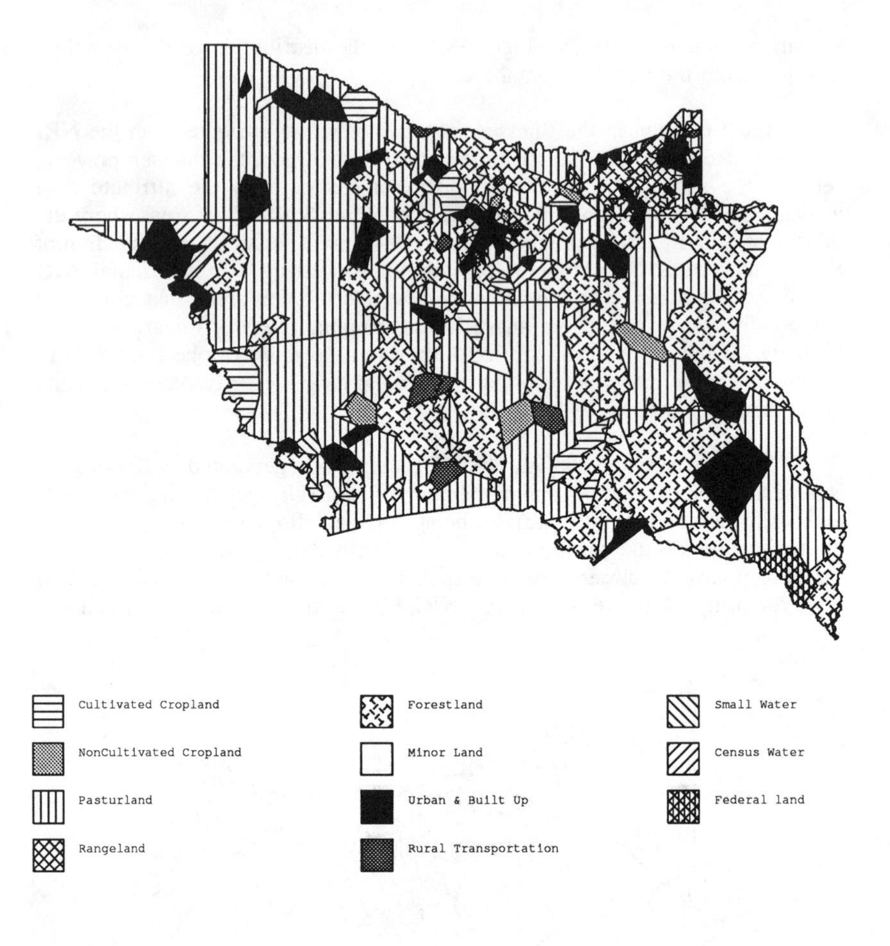

Figure 5. Landuse map of study area created by dissolving thiessen polygons with similar attributes.

in and develop trends in the use of agricultural lands. With this in mind, the resulting maps are a representation of the distribution and not a map thereof. For example, Figure 6 shows the change in forest cover in the eight county study area between 1982 and 1987. It was created by comparing the attributes for forest cover from the 1982 and 1987 inventories. Actual acreages are not accurate, but the map gives an indication of distribution of forest gain or loss.

DISCUSSION

Interpolating data taken from a point to produce an attribute for a polygon or region is quite dangerous, so this procedure or use of the resulting maps must be pursued with caution. The type of attribute involved is critical in allowing this interpolation to be safe. Attributes for which sampling techniques will give a

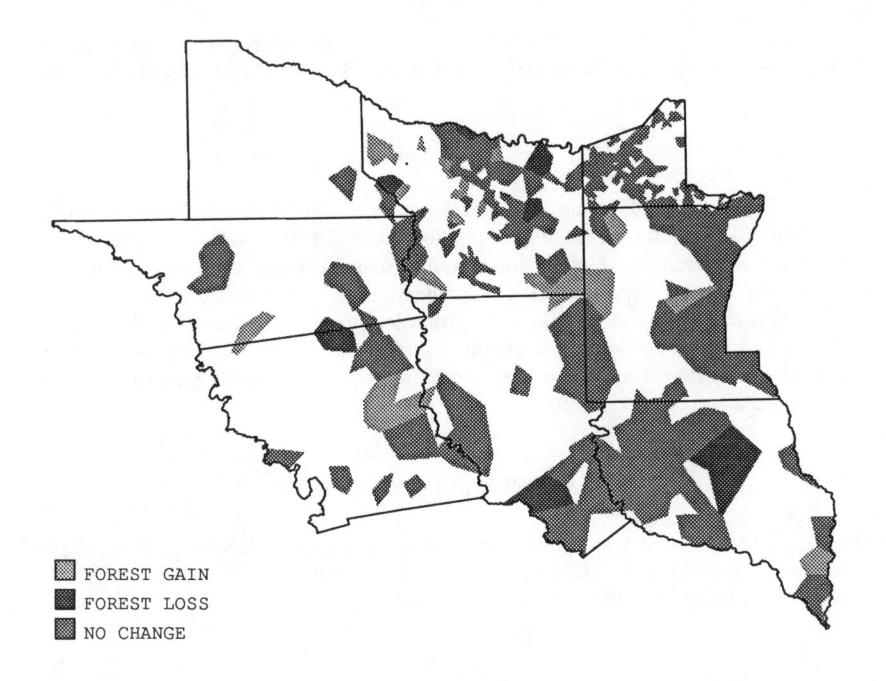

Figure 6. Forest cover change in eight Texas counties created by thiessen polygons using forest cover attributes for 1982 and 1987.

good indication of the entire region are better than attributes that are specific to a small area. For instance, percent tree cover works well because a series of point samples gives an indication of the regional tree cover. However, land ownership is a poor attribute to use for this procedure because the thiessen polygons produced have nothing to do with land ownership boundaries, and the subsequent dissolving of the polygons to make a thematic map will result in little relationship to reality.

The results of this procedure are not completely accurate because the polygon shapes are solely determined by the distance sample points are away from each other. The landscape for which the polygons and their attributes represent has nothing to do with the polygon shapes. If an area is lightly sampled, the polygons will be large. These polygons will have a greater chance of misrepresenting the true condition. However, if there are a large number of sample points, the thiessen polygons will be small in comparison to the entire map and the filtering procedures will produce an estimate of the actual condition from the random samples.

The research presented in this paper concentrates on the procedures used. Future work will involve comparisons of vegetation maps derived from thiessen polygon analysis for the entire state of Texas to ones developed through the use of both Advanced Very High Resolution Radiometer (AVHRR) data and a vegetation map developed from mosaiced Landsat MSS scenes. Resolution

resulting from thiessen polygon analysis for the entire state of Texas is comparable to the one kilometer resolution data derived from the AVHRR data.

CONCLUSION

The process of developing thiessen polygons and attaching NRI point data to them can greatly enhance the possible uses for the NRI database. This database is extremely valuable for resource managers and by tying it to a GIS, maps can be developed to facilitate interpretation and analysis of the data. However, caution must be taken in extrapolating point data to describe areas. Yet it can be done with a relatively high certainty of correct representation provided a sufficient number of sample points are used and the type of attribute data used is carefully chosen.

ACKNOWLEDGEMENTS

This research was funded through a Cooperative Research Agreement between the USDA - Soil Conservation Service, Texas State Office and the Texas Agricultural Experiment Station.

REFERENCES

Boots, B.N. 1974. Delaunay triangles: an alternative approach to point pattern analysis. In: Proceedings, Association of American Geographier 6, 26-29.

Boots, B.N. 1980. Weighting thiessen polygons. Economic Geography 56, 248-259.

McCullagh, M. J. and C. G. Ross. 1980 Delaunay triangulation of a random data set for isarithmic mapping. The Cartographic Journal 17(2) 93-99.

U.S.D.A. - Soil Conservation Service. 1986. National Resources Inventory 1987: Instructions for collecting sample data. Statistical Laboratory, Iowa State University. 40 pp.

AN EXPERT SYSTEM AND GIS APPROACH TO PREDICTING CHANGES IN THE UPLAND VEGETATION OF SCOTLAND

David Miller and Jane Morrice
Macaulay Land Use Research Institute,
Craigiebuckler,
Aberdeen,
Scotland.
AB9 2QJ

ABSTRACT

An expert system shell is used to combine environmental data expressed spatially, with knowledge on vegetation succession, to predict where changes will occur in the boundaries of vegetation types in upland Scotland. The philosophy behind the system and its design are outlined in the paper, with reference to one vegetation type - Bracken (*Pteridium aquilinum*) - as an example. Questions of data scale, resolution, content and reliability are tested in the course of the derivation of empirical models. Contextual measures of the distribution of Bracken with respect to the ecological environment are used within the assessment of likely changes in its boundaries. A more comprehensive implementation is described with the range of vegetation types extended to a larger sample of those found in upland Scotland, about which there exists some knowledge of likely changes in species composition through time. A test of the system using historical aerial photography and satellite imagery has demonstrated some of its strengths and weaknesses.

INTRODUCTION

Mapping the vegetation of upland Scotland can be done most rapidly and repeatedly using remotely sensed data. Interpretation, both manual and computer assisted, can be improved with *a priori* knowledge of environmental factors associated with different vegetation types. Environmental monitoring uses a wide knowledge-base in an increasingly multi-disciplinary field of research. Studies of the interactions between factors being monitored demands decisions on levels of detail to be considered in precision and accuracy, whether in underlying processes or geographical descriptions. The expert system design aims at permitting the user access to knowledge of botany, ecology, mapping, image processing and spatial data analysis for an informed and informative answer to their query.

Knowledge about semi-natural vegetation distribution and succession is a mixture of site specific studies at large scale, observations at smaller scales and general impressions at any scale. This provides a subject area where testing existing hypotheses and formulation of new ones can be explored at varieties of scales and accuracies with different types of models. One vegetation type - Bracken (*Pteridium aquilinum*) - has been used an example for designing the system, taken in the context of predicting where vegetation changes may occur.

Land Cover of Scotland

In response to concern about the rate of change of land use and the lack of complete data on the land cover and land use resource in Scotland the Scottish Office Environment Department commissioned a project to map the land cover of Scotland at 1:25 000 scale from aerial photographs. The Macaulay Institute was contracted to interpret and digitize the land cover information.

The project adopts a census approach and is planned around identification of 126 land cover classes interpretable from the photography. The minimum area of a semi-natural vegetation class interpreted as a single class is 10 hectares. Identifiable classes smaller than that will be combined as mosaics and coded according to the areal dominance of the component classes.

An alternative approach to mapping wide areas is based on use of satellite imagery, an approach which was rejected because of the unreliability of the availability of cloud free satellite imagery and some doubts over its ability to resolve the detail required for the land cover census (Harrison *et al.,* 1989). However, future updating of the dataset may utilise satellite imagery for those classes which it can adequately resolve, or target surveys towards where changes may have occurred.

Modelling approaches

Three types of model are used within this GIS application:

1. Geographical - describing pattern in datasets with respect to location;
2. Inductive - deriving relationships between datasets;
3. Knowledge-based - applying relationships developed external to the GIS and applying them to the geographic database

Following the principles of many knowledge-based systems or rule-based approaches to problem solving, Campbell and Roelofs (1984) outline the basis for artificial intelligence approaches to remotely sensed image analysis. Desachy *et al.* (1986), Erickson and Likens (1984) and Gilmore (1985) all suggest bases for using existing knowledge of a topic to improve or assist in geographical problem solving. These authors code characteristics of image processing and thematic map interpretation into a knowledge base. The probability levels describing class, context and presence are managed within the knowledge base.

MAPPING BRACKEN DISTRIBUTION

The land cover dataset, based upon air photographic interpretation, includes a category for bracken. The interpretation threshold used is that there should be greater than 50% ground cover of bracken. However, this vegetation type is likely to occur in areas less than 10 hectares, thus be less than the nominal limit for separate identification and will be represented as a component of a mosaic. The application of a knowledge-based approach to mapping upland vegetation with the incorporation of satellite imagery was developed using bracken (*Pteridium aquilinum*) as an example.

Experimental sites

Ten experimental sites were chosen around Scotland to represent a range of conditions in the physical environmental and their land uses. Observations were carried out in a hierachy of scales on vegetation types, principally on bracken at the larger scales.

1. Land survey of small areas (0.5 km^2), high accuracy (± 1cm standard deviation), high frequency (annually) for quantifying rates of change of bracken;
2. Photogrammetric plotting of larger areas (2-3 km^2), moderate accuracy (± 4m standard deviation), low frequency (approximately, 20 year intervals);
3. Satellite classification (400 km^2), low accuracy (± 25m), low frequency (10 years).

The hierarchy for each site was complete where large scale observations were made, but five sites comprised only satellite image classifications.

The soils, topographic and selected land use (forestry, built-up areas, and arable/improved grassland) data sets formed a digital databank for analysis with respect to satellite imagery of the area and the land cover map. A series of masks, derived maps and Landsat MSS and Thematic Mapper satellite imagery were used to define the hill land subject to bracken infestation.

A maximum-likelihood classification of the bracken distribution, low scrub and bent-fescue grasses was run using independent air photographic interpretation of selected land cover types for training data. The classification was split into two categories of greater than 50% or less than and equal to a 50% likelihood level of bracken presence. These two categories were then tabulated against the land cover digital dataset to provide estimates of variation within the land cover dataset as measured by classified satellite imagery.

EXPERT SYSTEM IMPLEMENTATION

Development of the expert system for mapping upland vegetation is using a 'shell' system (called SBS) developed by the Medical Research Council/Clinical and Population Cytogenetics Unit. The shell has been designed as a structured means for communication between "expert" routines with a BLACKBOARD - Engel *et al.*, (1990) and Tailor (1988) - (where information is passed between EXPERTs), following a set of priorities, with a view to solving a particular GOAL (for example, **Predict where bracken will spread**). A schematic diagram of the system appears in figure 1.

The USER INTERFACE is via keyboard and terminal display. The programme displays the goal being dealt with currently on the goal list and the EXPERT attempting to deal with it. The user specifies the problem, which is entered on the BLACKBOARD. The system SCHEDULER checks the available EXPERTs and each one compares its own goals with that of the problem. Those EXPERTs which have goals relevant to the current problem are then assigned to its solution according to specified priorities. They tackle the problem by adding new goals to the blackboard which in turn require further EXPERTs to be called on for a solution. Each EXPERT is comprised of "context's", which are groups of the rules which the EXPERT may initiate, similar in some form of detail or objective. Their priorities and rules can be modified by the user during programme execution.

A much simplified example of the run illustrated in Figure 1 follows the 'question and answer' form of communication between the EXPERTS and the Goal list in the BLACKBOARD.

SYSTEM EXPERTISE

The example above ignores the clarification the system may request (for example, *What geographical pattern has the current bracken presence*). The following section explains how such expertise can be incorporated into the system. The expertise of a system reflects the breadth of topics and details of rule content. Structures are provided for expressing the logic of achieving a set goal and the prerequisite goals to be achieved.

The expert content of the system is currently represented in the EXPERTs as exemplified in Figure 1. In the expert *neighbour* , the goal is to find the likelihood of there being a particular neighbour of bracken; the prerequisite is that bracken data exists; the rule is to read an adjacency matrix from a file.

A combination of the goal, prerequistites, priorities and alternatives which may be started form the 'expert' approach. Elements of the expert content can be dynamically altered during a run of the system. For example, a higher priority will be set for checking land use, in the context of bracken spread, where the location is less than 100 metres altitude and was previously improved grassland.

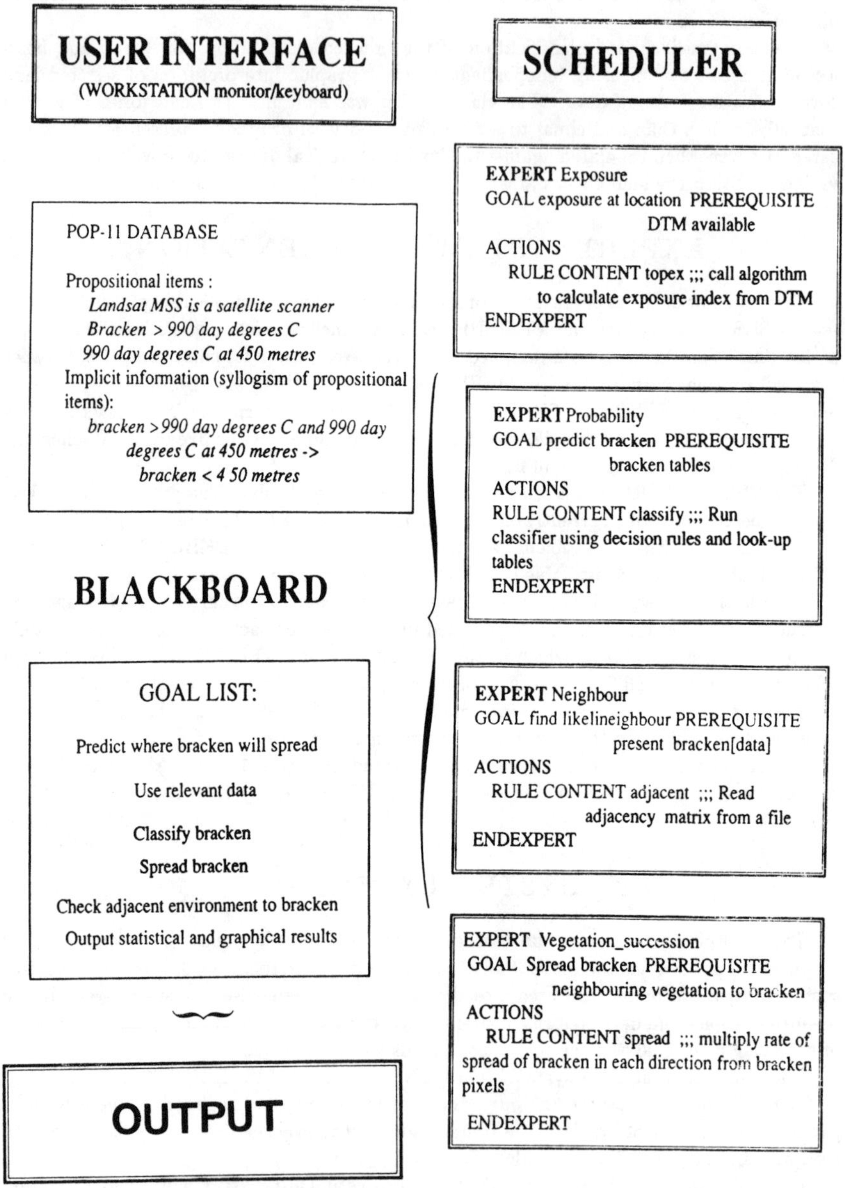

USER INTERFACE
(WORKSTATION monitor/keyboard)

SCHEDULER

POP-11 DATABASE

Propositional items :
Landsat MSS is a satellite scanner
Bracken > 990 day degrees C
990 day degrees C at 450 metres
Implicit information (syllogism of propositional items):
bracken >990 day degrees C and 990 day degrees C at 450 metres ->
bracken < 4 50 metres

BLACKBOARD

GOAL LIST:

Predict where bracken will spread

Use relevant data

Classify bracken

Spread bracken

Check adjacent environment to bracken

Output statistical and graphical results

OUTPUT

EXPERT Exposure
GOAL exposure at location PREREQUISITE
DTM available
ACTIONS
RULE CONTENT topex ;;; call algorithm to calculate exposure index from DTM
ENDEXPERT

EXPERT Probability
GOAL predict bracken PREREQUISITE
bracken tables
ACTIONS
RULE CONTENT classify ;;; Run classifier using decision rules and look-up tables
ENDEXPERT

EXPERT Neighbour
GOAL find likelineighbour PREREQUISITE
present bracken[data]
ACTIONS
RULE CONTENT adjacent ;;; Read adjacency matrix from a file
ENDEXPERT

EXPERT Vegetation_succession
GOAL Spread bracken PREREQUISITE
neighbouring vegetation to bracken
ACTIONS
RULE CONTENT spread ;;; multiply rate of spread of bracken in each direction from bracken pixels
ENDEXPERT

Figure 1: Schematic diagram of the SBS expert system shell with example system contents (Based on Baldock, *et al.*, 1987)

Knowledge base

In the development of the vegetation mapping expert system, using bracken as a specific land cover type, the methods of mapping bracken and analyses of the results are coded into EXPERTs, RULES and PROCEDURES using **POPLOG** and **POP-11** (Barrett *et al.*, 1986).

The knowledge base refers to ecological data on bracken presence: when, where and where not, and why not. *When* relates to the growth cycle of the plant; *where* was considered as the environmental niches providing a likely habitat for bracken; and *where* not as those environmental niches it will definitely not inhabit. Land use and bracken ecology determine the *why not* to expect for bracken presence.

Information types and database

The system executes its goals by means of 'matching' information between database entries and current problem details. This requires the information to have a defined structure and type. The library for knowledge within the system, accessible to all EXPERTs, is the POP-11 database. This is a mechanism which contains a list of data, where an item of data held may in itself be a list. Simple statements such as those in Figure 1 form the basis of the database content. Three forms of information are utilised within the system. These are listed in Figure 1 as explicit information, porpositional items and implicit information. Accompanying this information is some estimate of confidence in the relationship.

"Frame" representation of knowledge

Certain spatial datasets have information associated with their polygon units which need not be re-expressed in the form of multiple re-coded copies of the same dataset. Use of a GIS which interfaces with a relational database (for example Arc/info or Pamap) can utilise that facility for the polygon attributes. Within this system a structure called a FRAME is used as an alternative to the relationalo database. The example below is a simple FRAME with details of soil types. Analysis of the distribution of bracken on soil types is represented and the likelihood of bracken being associated with a particular soil type can be stored. Storage takes place in an attribute of the FRAME known as a SLOT (for example: wetness or % bracken). The CONFIDENCE attributed to the FRAME (for example, 'Brownforest' below) is the confidence level from the calculation of bracken presence on each soil type at 2 standard deviations. This will be used for calculating reliability of further analyses.

```
FRAME soils; ;;; Frame name

CONFIDENCE = 100 ;;; Confidence level of soils data

COMPONENTS = [Brownforest [...]; Peat [...]; .....]; ;;; Subframes of major soil-subgroups

ENDFRAME;

FRAME Brownforest; ;;; Frame with details of Brown forest soils

SLOTS Bracken = ; ;;; Default value for Bracken is the likelihood of its presence.

CONFIDENCE = 95; ;;; Confidence level of data on Brown forest soils

ENDFRAME;
```

Knowledge representation

Examples of the representation of the knowledge currently within the system are in the following forms:

1. Embedded in the EXPERT definitions and contents. In designing the individual EXPERTs, the procedure for mapping bracken has been formalised in a 'top-down approach'. At each level certain assumptions are made to allow the system to complete a run, although the degree of reliability has been set low where further work is required.

2. Information held in a database which is available for all EXPERTs. This is the POP-11 database indicated in figure 1.

3. Preprocessing of the test raster datasets is output as as intermediate files. For example, adjacent land cover types from the vegetation map for the test area have been summarised in a matrix. Knowledge of how to determine adjacency remains with the user while the current EXPERTs simply interpret the file.

4. The "Confidence" level associated with each FRAME (dealt with in the next section) which either reflects confidence in a statistical sense or an estimate of the importance of the qualitative information held in that FRAME.

BOTANICAL EXPERTISE

A knowledge of Botany or Plant Ecology relating to the nature of distributions of vegetation types, competition at the boundaries and vegetation succession is required to translate the distribution of vegetation types into more than just a map. Miles (1988) presents the results of studies into patterns of vegetation succession between vegetation types in upland Scotland. Two of the aspects vegetation succession described are :

1. Characteristics promoting and inhibiting successional change in semi-natural vegetation.

2. Magnitude of changes.

To the later aspect can be added the measurements obtained from the land survey and photogrammetric plotting from the experimental sites described in section 3.

Pteridium aquilinum

Botanical and ecological data about bracken have been abstracted from Miles (1988), McVean and Radcliffe (1962) and Page (1989). A summary of some of their records is as follows.

- Establishment of prothali is rare.

- Main spread is by rhyzomes.

- An invading front may advance up to 0.5 metres per annum and in "ideal" conditions up to 1.5 metres per annum.

- Many existing transitions to *Pteridium* dominated stands are from existing, low-density frond populations rather by invasion.

- Bracken does not grow much above 1500ft. (457 metres), suggesting that this indicates an intolerence to cold or exposure.

These "statements" include quantatative and qualatative information. Both (a form of) the statements and a reliability level are encoded into a FRAME which contains "expertise" about bracken. Similar FRAMES exist for selected grassland, moorland, forestry and agricultural classes.

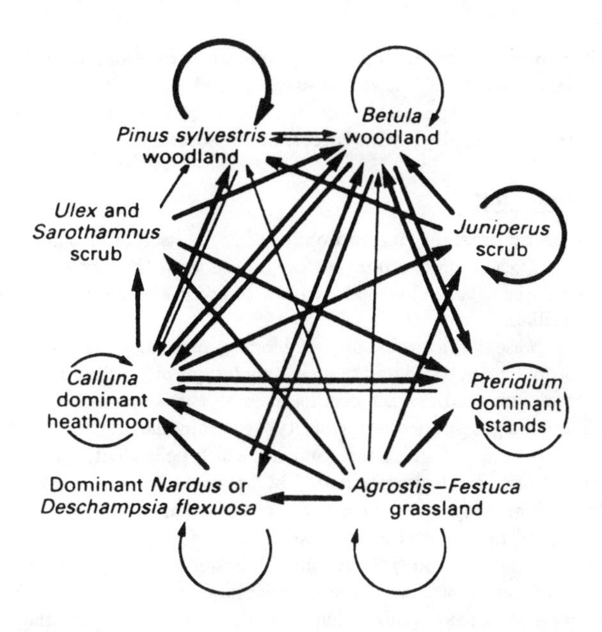

Figure 2: Successional transitions in the Scottish uplands between eight vegetation types (from, Miles (1988))

PREDICTING BRACKEN SPREAD

Application of the expert system has been made in several test sites in Scotland. The example of predicting bracken spread follows the order of GOALS in Figure 1. In response to the first GOAL of "Relevent data" Selected hypotheses were tested, using control with the system, relating to ecological or geographical limitations on bracken presence. The GOALS were phrased so as to invoke checks on data availability; data validity - scale and resolution for the task (Birnie and Miller, 1987); arithmetic calculations for assessing boundary errors in spatial datasets, (Burrough, 1986 and Brunsden *et al* ., 1990); and the derivation of Geographical models from available data.

Hypotheses relating to bracken distribution

An example of *guided* tests of hypotheses was that of upper altitudinal levels of grwoth of bracken.

The two probability levels of bracken presence were cross-tabulated with respect to digital terrain data at each experimental site. The presence of bracken increased to a maximum between 50 metres and 150 metres above sea level before tailing off to zero at an altitude of about 430 metres. An explaination of why this should be a limit was initiated by a new GOAL looking for data on exposure and temperature.

Meteorological records were used in association with the digital height image to produce a temperature image. This showed bracken presence tailing off at a mean annual minimum air temperature of about 3.3 degrees C. On average, the calculated accumulated temperature in day degrees C above 5.6 degrees C was around 990 day degrees C at the highest altitudes that bracken occurred.

Analysis of the DTM in each of the eight cardinal directions of the compass from each pixel in the dataset produces an expression of relative exposure called the TOPEX value. Small values (near to 0) relate to high exposure and high values (over 100) relate to sheltered locations. Bracken presence correlated well with sheltered exposure levels (Topex > 30).

This analysis provides examples of a geographical model from a DTM, inductive modelling based upon a sample of bracken presence and absence with respect to altitude and exposure and a rule-based approach where the 450 metre limit can be used as a surrogate for the other two datasets, albeit with caveats on its accuracy.

Probability mapping

The GOAL "Classify bracken" has two options: 1. a discrete classifier, 2. a maximum liklihood classifier. Where no thematic surface data is used the second option has a higher priority. However, the sotware has been written to accomodate thematic map data within the liklihood classifier.

In order to represent the probability of bracken presence, the terrain data sets were assigned prior probabilities based on the cross-tabulations of bracken presence and the environmental factors. This used bracken occurrence with respect to aspect, slope and altitude, after masking and weighting for the terrain types within the test area. The tabulation of bracken to landscape unit, of which the dominant soil type is identified, used the two likelihood categories of bracken. Units dominated by brown forest soils and humus-iron podzols have most bracken growing upon them (over 1,700 hectares) and the largest proportions of bracken compared to the total area of each group of map units (6.6% and 7.8%). The core areas with greater than 50% likelihood of bracken presence reflect the same pattern. Map units with non-calcareous gleys have the next largest proportion with bracken growing, although a larger area of peaty podzols had bracken present. However, the areas of bracken on peaty podzol map units could be confined to soils other than the dominant peaty podzol component. Peaty gley had the smallest area of bracken present - about 50 hectares.

The comparisons between bracken presence and the selected environmental factors in the area databank were used to assist classification in an enlarged area. A maximum-likelihood classification algorithm was modified (Kittler, 1983) to incorporate non-continuous data sets like soil types. Existing algorithms based on mean vectors and calculating equi-probability levels imply that any interpolated values in an image have some meaning. However, in a digital representation of a soil map, no meaning can be attributed to intermediate values. The facility to alter the prior probability of each class was also incorporated (Strahler, 1980). The measured preferences of bracken for certain topographic or soil conditions were used as weighted variables within the classification.

The preliminary prediction is being checked against aerial photographic evidence from 1987 and appears to be only 45% correct. The resolution of the databanks may be too small to identify changes in bracken presence over an eleven year period when compared to 1:24 000 scale photography. Therefore the predictions of *where* bracken spread might have occurred may be more successful than the estimate of the total *area* of bracken spread.

Bracken spread

The GOAL "Spread bracken" calls further GOALs before completion of the task. Three experts are called:

1. Bracken heterogeneity measured by the number of bracken polygons in an area minus the number of 'holes' in those polygons. This is calculated at different probability levels to give a measure of the susceptability of bracken spread according to the rules in the POP-11 database. This is called the *Euler number* .

2. Adjacent environment to that occupied by bracken. The altisude, slopes, aspect, soil types, exposure levels and land cover types adjacent to existing bracken are summarised and each bracken pixel is assigned a score of "suitability" of its neighbouring environment to bracken spread.

3. Vegetation succession model (Miles, 1988) which allocates likelihoods of spread according to the neighbouring vegetation type (Figure 2).

Output results

The results GOAL is either graphical output or statistical output. The tests run so far have shown accuracies of up to 70% in predicting changes in bracken where there was a bracken presence before. The best results were obtained in an Army training estate, possibly because there is least, conventional, human interference in land use (such as additional forestry or bracken clearance).

Poorer results of 29% success were obtained in areas where data resolution, was coarsest and available data was restricted to surrogates (for example the altitudinal mask of 450 metres). The system reports what data is used and a confidence score based upon the inputs used and error tolerances associated with them.

Comparison with the land cover dataset suggests the best results will be where the interpreted polygons are mosaics with a bracken component and satellite data is in good agreement and those polygons.

DISCUSSION

Knowledge of the accuracy of the information required for decision making is important within a system designed to incorporate a range of data types and reliabilities. The degree to which generalisation can be undertaken whether or not the data is available or appropriate is the over-riding guide the system will have to work to.

The system will require improved error train analysis and backward-chaining to permit the appropriate rules to be followed to match the desired accuracy and form of output. The extent to which dataset appropriateness can be assessed will then be dependent upon *a priori* understanding of the nature of the digital data available and predictions of errors at each stage in stepping through the goal list, until a final output and reliability estimate are acheivable. The principal limitation of the current system is the lack of the back-tracking to select the best levels of generalisation for the desired level of reliability.

A more mature system must state when the desired result is not possible with the resources available to it. This philosophy extends to the utilisation of any digital dataset within a GIS context. The question is whether or not the an expert system approach can best be used in the handling of, often disperate, forms of data to add informed caveats to the results obtained.

ACKNOWLEDGEMENTS

The authors would like to thank the Medical Research Council for the use of the SBS expert system shell and Ian Craw, Alistair Law and Matthew Wells for their adivce in implementing the system. Acknowledgement is also due to the Scottish Office Environment Department for use of their digital land cover data.

REFERENCES

Baldock, R A, Ireland, J and Towers, S J (1987), *SBS User Guide.* Medical Research Council and Population Cytogenetics Unit, Edinburgh, internal report.

Barrett, R, Ramsay, A and Sloman, A (1986), *POP-11: a practical language for artificial intelligence.* Ellis Horwood Limited, Chichester.

Birnie, R V and Miller, D R (1987), *Lessons from the bracken survey of Scotland: the development of an objective methodology for applying remote sensing techniques to countryside mapping.* In: Proceedings of Monitoring Countryside Change, Silsoe.

Brunsdon C, Carver S, Charlton M and Openshaw S (1990), *A review of methods for handling error propogation in GIS,* In: Proceedings, 1st European Conference on GIS, Amsterdam, 106-116.

Burrough, P A (1986), *Principles of geographical information systems for land resource assessment* , monographs on soil and resources survey No. 12, Oxford University Press, Oxford.

Campbell, W J and Roelofs, L H (1984), *Artificial intelligence applications concepts for the remote sensing and earth science community.* In: Proceedings of the Pecora IX Symposium, pp. 232-242.

Desachy, J, Castan, S and Fisse, G (1986), *Introduction of thematician knowledge in remote sensing imagery interpretation.* In: Proceedings of Mapping from modern imagery, RSS/ISPRS conference, Edinburgh 1986, pp. 418-439.

Engel, B A, Beasley, D B and Barret, J R (1990), *Integrating expert systems with conventional problem solving techniques using blackboards.* Computers and Electronics in Agriculture, **4,** 287-301.

Erikson, W K and Likens, W C (1984), *An application of expert systems technology to remotely- sensed image analysis.* In: Proceedings of Pecora IX Symposium, pp. 258-276.

Gilmore, J F (1985), *Artificial intelligence in image processing.* SPIE, 528, pp. 112-201.

Harrison, A R, Dunn, R and White, J C (1989), *A statistical and graphical examination of monitoring landscape change data.* Report to Department of the Environment, research contract PECD 7/2/47.

Kittler, J (1983), *Image Processing for Remote sensing,* Philosophical Transactions of the Royal Society of London, A 309, 323 - 335.

McVean, D N and Ratcliffe D A (1962), *Plant communities of the Scottish Highlands: a study of the Scottish mountain, moorland and forest vegetation.* Monographs of the Nature Conservancy, No. 1, HMSO.

Miles, J (1988), *Vegetation and soil change in the uplands* , In: Ecological change in the uplands, (Ed. M B Usher and D B A Thompson) Special publications series of the British Ecological Society, No. 7.

Page, C N (1989), *Taxonomic evaluation of the fern genus* Pteridium *and its active evolutionary state* , in Bracken'89: bracken biology and management, (Eds. J A Thomson and R T Smith), Australian Institute of Agricultural Science occasional publication No. 40. 23 - 34.

Strahler, A H (1980), *The use of prior probabilities in maximum likelihood classification of remotely sensed data.* Remote Sensing of Environment, 10, pp. 135-163.

Tailor, A (1988), *MXA - a blackboard expert system shell* , in Blackboard Systems, (Eds. R Engelmore and T Morgan), Addison Wesley, 315 - 334.

Towers, S J (1987), *Frames as data structures for SBS* , Medical Research Council and Population Cytogenetics Unit, Edinburgh, internal report.

GIS TOOLS FOR CONTAMINATED SITE REMEDIATION: THE BUNKER HILL SUPERFUND SITE

Daniel D. Moreno
Luke A. Heyerdahl
Dames & Moore
1125 Seventeenth Street, Suite 1200
Denver, CO 80202-2027
(303) 294-9100

ABSTRACT

The potential role of GIS in the Environmental Protection Agency's (EPA) massive Superfund program is illustrated by an ongoing Remedial Investigation and Feasibility Study (RI/FS) at the Bunker Hill site in northern Idaho. A GIS-based analysis of existing vegetation cover, soils properties, contamination, and physiography has yielded a Revegetation Plan for the site. The plan identifies those areas having the highest potential for successful revegetation.

INTRODUCTION

The EPA's massive Superfund program, designed to clean up the nation's most severely contaminated hazardous waste sites, is being assisted nationwide through the application of Geographic Information Systems (GIS). An illustrative example is the Bunker Hill Superfund site, located in the South Fork Coeur d'Alene River Valley in the northern Idaho panhandle near Kellogg (Figure 1). The project area (Figure 2) is approximately 7 miles long and 3 miles wide. The Bunker Ltd. smelter complex covers about 360 acres near Kellogg, and includes a lead smelter, zinc plant, impoundment areas, and mining facilities; these operations were discontinued in 1981.

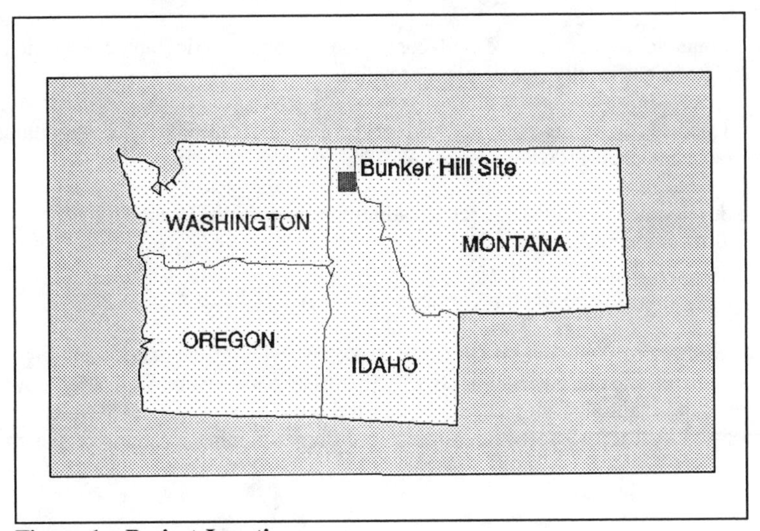

Figure 1. Project Location

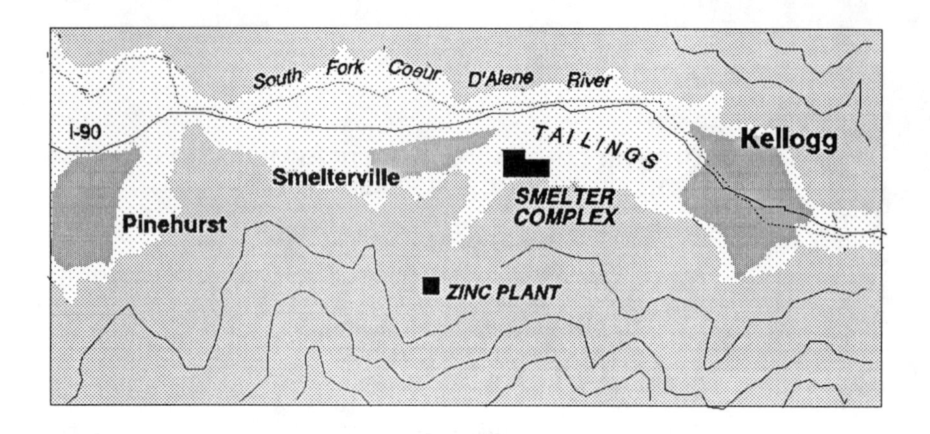

Figure 2. General Site Plan

The site has been extensively mined for lead, zinc, and silver ores for the past 100 years, resulting in degradation of ground and surface water from tailings and metal processing wastes. In addition, airborne emissions and deposition from lead smelting operations have seriously degraded natural vegetation on the surrounding hillsides. The major ecological change on the site has been the replacement of closed coniferous forests by invasive open scrub and woodland communities.

GIS ANALYSIS OBJECTIVES

The overall purpose of the three year RI/FS at the Bunker Hill site is to examine contaminant distributions in order to assess health risks and environmental threats. A further objective -- the focus of the GIS analysis -- was the preparation of a revegetation plan for the valley. Revegetation priorities were established for those unpopulated and undeveloped areas that met the following general criteria:

1. Need: areas that have experienced serious degradation of natural vegetation, as indicated by an analysis of existing vegetation cover and vigor.

2. Limitations: areas where soil contamination and physiographic conditions would not significantly impair plant growth.

3. Opportunity: areas having soil properties and physiographic conditions favorable for plant growth.

The development of a general revegetation plan is graphically shown by the GIS data flow diagram on Figure 3.

VEGETATION COVER ANALYSIS

The need for revegetation on the site is expressed by both the density and vigor, or health, of existing vegetation cover. Historical and existing vegetation cover were mapped through an analysis of Landsat Thematic Mapper (TM) imagery, processed by ERDAS software. Fourteen different training locations on the site

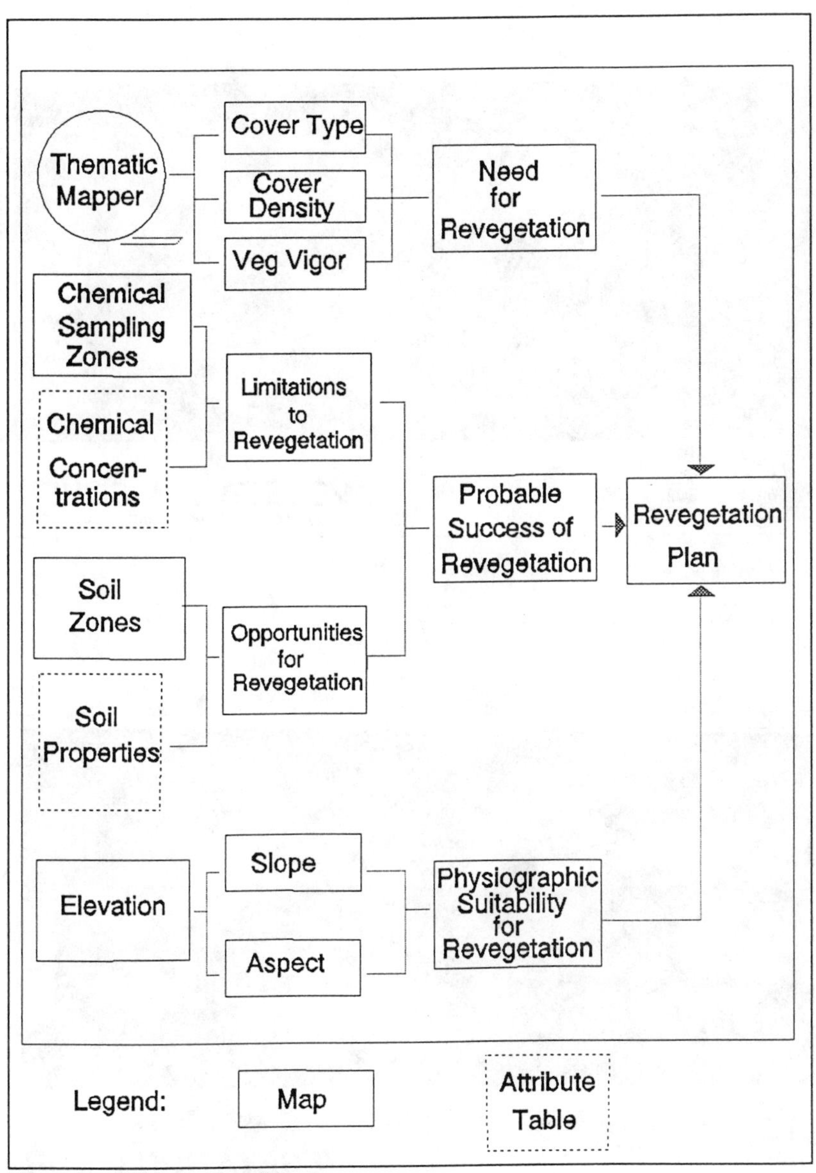

Figure 3. GIS Data Flow Diagram

were selected for field sampling and calibration of the imagery to observed cover density. Vegetation cover (Figure 4) was expressed as a density in percent. Vegetation vigor (Figure 5), was qualitatively classified into four categories ranging from very vigorous to very feeble.

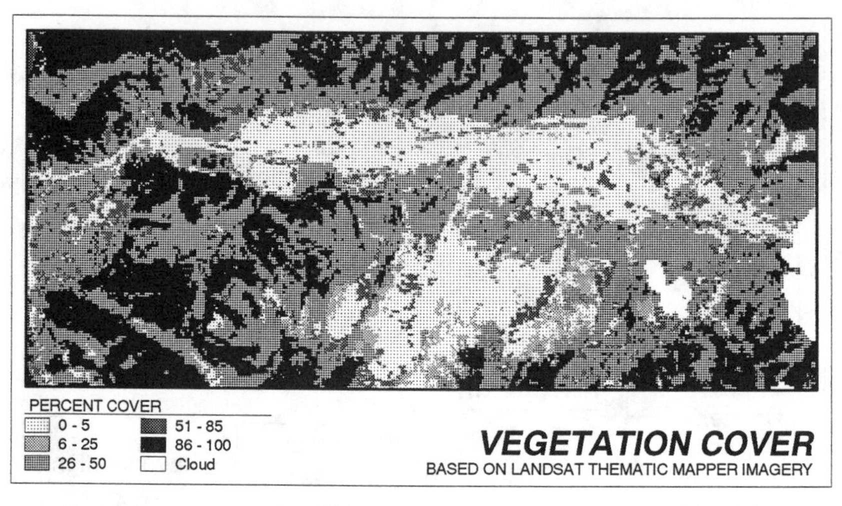

Figure 4.

Figure 5.

The areas showing the lowest percent cover and the most feeble vegetation are those most in need of revegetation (except most of the valley floor, which is urbanized). A large area in the south-central portion of the study area, in the vicinity of the smelter complex and zinc plant, shows particular damage.

LIMITATIONS AND OPPORTUNITIES

The vegetation cover and vigor maps help to identify candidate areas for revegetation. It was also necessary, however, to assess the physical conditions of those areas in order to assess probability for successful revegetation.

Chemical Concentrations

A major limitation to successful revegetation is the presence of arsenic and heavy metals which are toxic to plants. The project area was divided into several contaminant sampling zones (Figure 6) for which extensive shallow soil sampling was performed. An associated attribute table contained concentrations of arsenic, lead, zinc, mercury, and other potentially toxic elements. A second table contained the concentration thresholds for these elements that are considered toxic to the plant life found in the region. An analysis of sampled concentrations, with respect to toxicity thresholds associated with the sampled elements, yielded a map depicting limitations to revegetation (Figure 7).

Figure 6.

Physiography

Topographic conditions exert considerable influence on plant growth. Slope steepness and aspect, together with elevation, are principal agents in defining plant

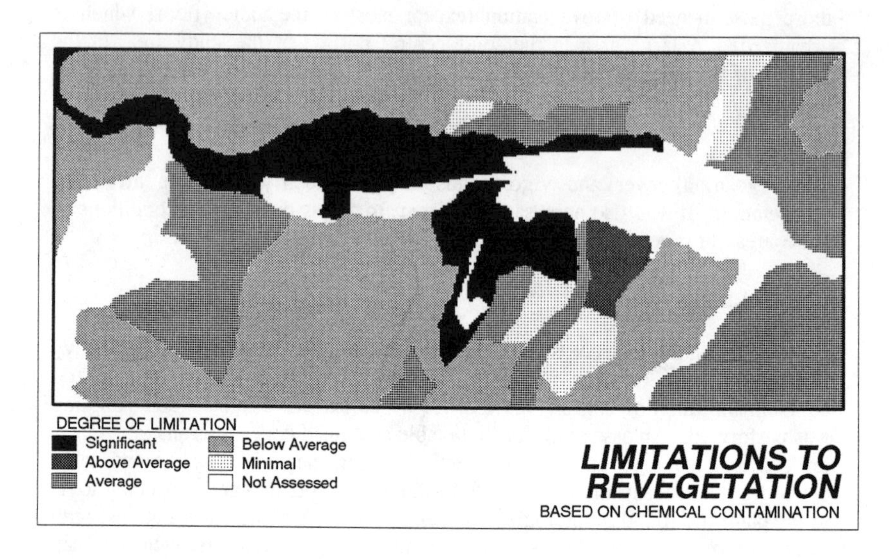

Figure 7.

community distributions. The topography map (Figure 8) clearly shows the well-defined valley floor and the dissected slopes within the project area.

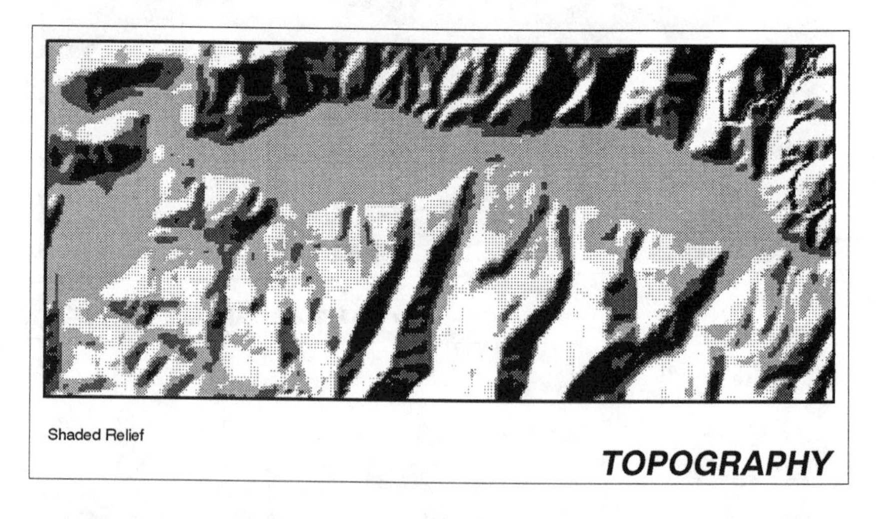

Figure 8.

Slope steepness affects growing conditions because it is one of the principal factors that control site moisture regimes. Steeper slopes generally contain less moisture because of increased rates of precipitation runoff. Four categories of slope were reclassified into relative levels of revegetation suitability (Figure 9).

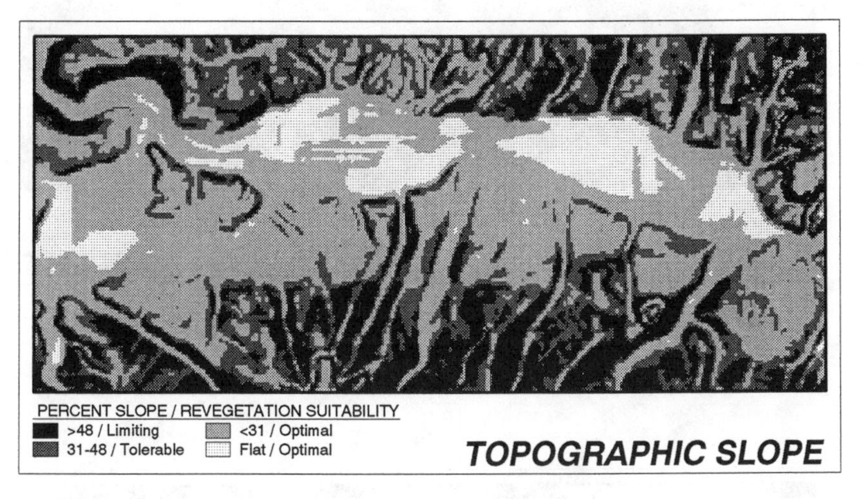

Figure 9.

Growing conditions are also affected by topographic aspect. Slopes exposed to the southeast, south, and southwest are generally drier than those with northeast, north, and northwest exposures because of higher solar evaporation rates. Thresholds were developed on this basis, whereby southeast to southwest aspects were equated to the most limiting plant growth conditions, while northern exposures were considered most favorable (Figure 10).

Soil Properties

Certain soil properties, such as air, water, nutrients, and stability, are necessary for plant growth. Soil zones having relatively uniform soil properties were mapped (Figure 11) and linked to an attribute table containing various parameters that are considered by the Soil Conservation Service to affect growth. These factors include organic matter content, acidity (pH), erosion factor K, bulk density, and water-holding capacity. An analysis of these factors yielded a map depicting opportunities for revegetation based on favorable soil conditions (Figure 12).

REVEGETATION PLAN

A composite map was prepared (Figure 13) to show the probable success for revegetation, based on the distribution of chemical limitations, soil properties, and physiographic conditions previously discussed. The composite model applied an

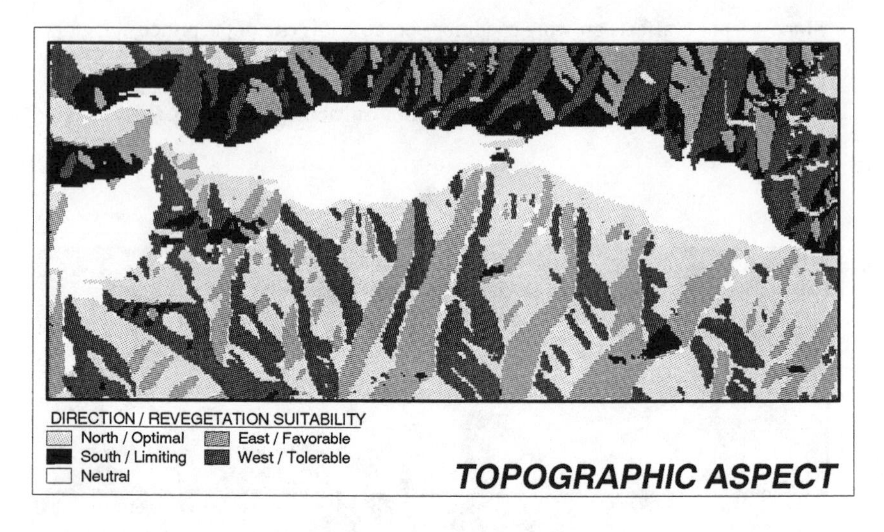

Figure 10.

Figure 11.

equal weighting to the three factors. The chemical influence is an inhibitor to plant growth beyond certain tolerance levels. The soil influence represents the natural capability of the site to support plant growth. The physiographic influence represents the site's natural range of soil moisture that is affected by runoff and

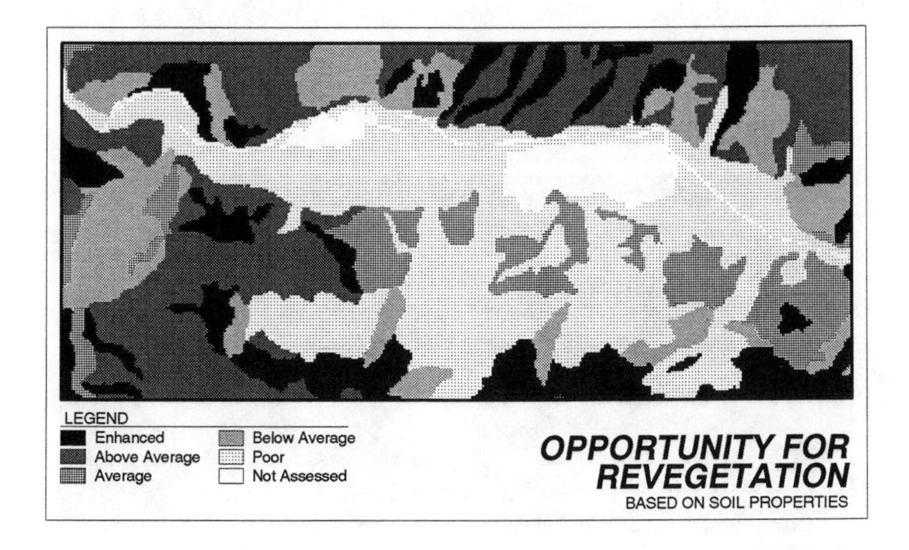

Figure 12.

evaporation. Positive concurrence in all three contributors to growth indicates a high confidence of probable success in revegetation. The absence of any one of the three creates a less favorable environment where special revegetation practices or monitoring may be indicated.

The general Revegetation Plan (Figure 14) shows the distribution of areas targeted for revegetation. Priorities for revegetation were based on areas that are most in need of treatment and show indications of success.

The GIS analysis has formed a solid basis for formation of a Revegetation Plan for the Bunker Hill site, which contains specific recommendations, including which species to plant, when planting should occur, and appropriate engineering and design specifications.

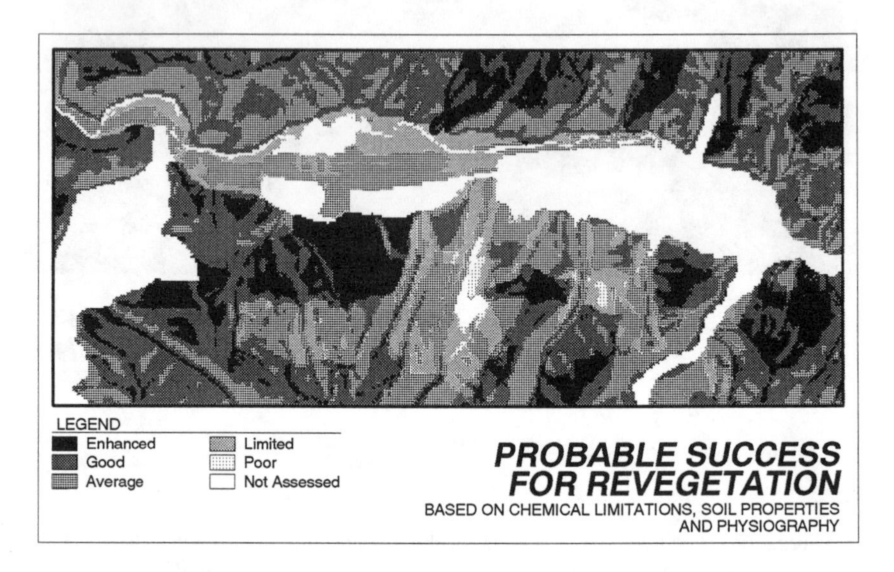
LEGEND
Enhanced
Good
Average
Limited
Poor
Not Assessed
PROBABLE SUCCESS FOR REVEGETATION
BASED ON CHEMICAL LIMITATIONS, SOIL PROPERTIES AND PHYSIOGRAPHY

Figure 13.

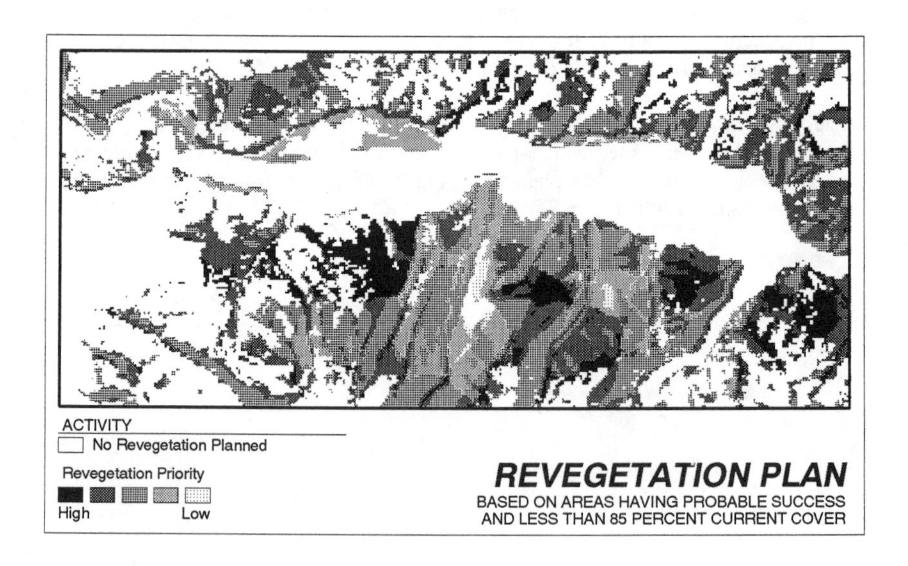
ACTIVITY
No Revegetation Planned
Revegetation Priority
High
Low
REVEGETATION PLAN
BASED ON AREAS HAVING PROBABLE SUCCESS AND LESS THAN 85 PERCENT CURRENT COVER

Figure 14.

James F. Hines, GIS Coordinator
Tallahassee-Leon County GIS
301 South Monroe St. Room 108
Tallahassee, FL 32301
(904) 488-2818

CANNONS TO THE LEFT, CANNONS TO THE RIGHT: PITFALLS OF A MULTI-PARTICIPANT GIS IMPLEMENTATION FOR A LOCAL GOVERNMENT

Abstract: System selection and conversion RFP's are the fun and easy part of a GIS. The real difficulties are obtaining funding, forging Interlocal Agreements, and solving internal organizational and people issues. Funding is the initial hurtle and a GIS consultant can be of great benefit in convincing policy makers to commit money to an expensive and long term project. Achieving an Interlocal Agreement between all the participants can prove even more difficult than the funding issue. All of the best arguments (economies of scale, cooperation, "sing from the same sheet of music"), carry little weight against years of mistrust and contention between City and County staff and policy makers. Add constitutional (elected) officials like the Property Appraiser, and Electric, Gas, and Water Utilities, and the internal political problems can seem insurmountable.

This paper will concentrate upon the funding, planning, and internal organizational problems encountered in the implementation of a multi-participant Geographic Information System in a rapidly growing mid-size urban setting of 200,000 population. The Tallahassee-Leon County project will have completed planimetric, topographic, and tax parcel base mapping and be operational by the Summer of 1992. This implementation experience should be of benefit to local governments about to ride forth into the GIS "valley of death". (with apologies to Tennyson).

BACKGROUND:

Planning for the local GIS was initiated in 1987. The impetus for the GIS effort was an attempt by local leaders to provide County and City departments with the tools necessary to deal with the concurrency aspect of Florida's Comprehensive Growth Management Plan which was passed by the State legislature in 1985. This legislation forces local governments to control growth by requiring that the infrastructure (roads, schools, parks, sewers, etc.) to support growth be in place or else be built concurrently with the development project. The details of the task of measuring, allocating, tracking, and enforcing concurrency were not complete, but there was almost universal accord that a GIS was the only viable solution. In 1988, FREAC (Florida Resources and Environmental Analysis Center) was selected to complete a Needs Analysis Cost/Benefit Study. A Coordinator was recruited to: manage the study, convince the policy makers that a GIS was feasible, procure funding, formulate an Interlocal Agreement for a multi-participant GIS, and develop an RFP for a system. The Interlocal Agreement required 18 months and many iterations but resulted in a tri-party agreement with Leon County, the City of Tallahassee, and the Leon County Property Appraiser as the principal participants. Up front costs were to be shared, with the City funding 50%, the County 40%, and the Property Appraiser 10%. While the agreement was being finalized a Request For Proposals for a system was developed. The Interlocal Agreement and the system contract (GeoVision software on a Digital hardware platform) were approved by both City and County Commissions in May 1990. The Interlocal Agreement specified and established two committees to oversee the GIS. An Executive Committee of three members represented by a City Commissioner, a County Commissioner, and the Property Appraiser would make policy and champion the project before their respective commissions. A Steering Committee represented by the three participants plus the GIS staff was set up to oversee technical issues and to implement the project.

INTRODUCTION:

With the maturing of the GIS industry and the expertise available from GIS Consultants, justification studies, system selection, and conversion RFP's have become a complicated yet straight forward exercise. There are a number of software packages which will provide a functional GIS for local governments. The task is to select the software which best fits a city or county's needs. The success or failure of a project will not be determined by the choice of software, but by the funding, organizational, and personnel involved. The real difficulties are:

1) obtaining long term funding,
2) determining what portions of the consulting and conversion work will be contracted out,
3) developing Interlocal Agreements,
4) overcoming organizational and personality turf wars,
5) attracting, training, and retaining quality staff,
6) gaining the time necessary to complete the project.

FUNDING:

Without adequate long term funding which extends through the procurement, startup, and conversion phases a GIS project can be destined for failure. A Bond issue is the best choice since the money is provided on the front end, can not be used for other purposes, and can be positively arbitraged (if allowed by tax laws) at an interest rate equal to the interest being paid on the bonds. The funds can then be expended as needed during the project life cycle.

Bond issues insulate the project from economic down-turns which could leave the GIS with a purchased system and no funding available for conversion services. Unfortunately bond issues must usually go before the voters for approval, and with current economic conditions the prospects of approval are remote.

The lease-purchase approach is a good choice for procuring the system hardware and software, and was the option taken by this project. Leasing has several advantages:

1) the cost is spread over four or five years which avoids large up front expenditures (attractive to policy makers),
2) no outlay of funds is required until equipment is installed and accepted to your specifications,
3) flexible terms can be negotiated so that obsolete hardware can be upgraded or replaced after two or three years and added to the lease
4) once contracted the lease payments are typically paid from debt service accounts over a four or five year period, and the task of procuring CIP funding each fiscal year is avoided.

An outright purchase is probably the least attractive due to the short life cycle of hardware (18 to 24 months). With the tremendous increases in processing speed for workstations (1000 MIPS by 1995 is speculated) a project could be locked in to obsolete hardware after only two years with no prospects for buying new equipment. Trends in hardware price/performance and maintenance cost are improving for the consumer so it is wise to delay purchases until equipment is absolutely needed. Next year the same dollars may buy a workstation with double the performance.

PLANNING:

Consultants can be of great benefit in planning and justifying a GIS, but be wary of Feasibility and Cost Benefit Studies. Feasibility studies should be renamed GIS Justification Studies. Technology has made GIS feasible for even small municipalities. That the Feasibility Study will recommend a GIS implementation is almost a foregone conclusion, but a study is still an excellent idea. A good feasibility study, done by professionals, provides concrete evidence which is crucial in convincing policy makers to invest substantial tax dollars. The study can be very beneficial by educating the potential users, creating support for the project, and by providing solid implementation strategies and "ball park" cost estimations for developing the various components of the system.

Cost Benefit Studies are usually based upon assumptions and methodologies which can be easily challenged by GIS opponents. It is very difficult to quantify factors such as, "staff time used to seek and verify information". To argue that a GIS will save 75% of current staff resources used for these tasks can foster the notion that a GIS will allow large reductions in staff. This will probably not be the case since the GIS will open up opportunities for public services that were not even conceivable prior to GIS and any staff that might have been eliminated will be needed to perform the addition work that additional capabilities will generate. It is more realistic to argue that with current trends, and without a GIS, in five years a city or county might need 100 additional staff to provide the same current levels of services. With a GIS, a staff increase of 50 positions may provide twice the level of service at a point five years in the future (an increase of 50 staff positions, but 50 less than would have been required without a GIS). This approach will also be more comfortable to existing staff who may view GIS as a threat to their job security. Strive to be as realistic as possible in selling the GIS but still create enthusiasm for the project.

The trend by State and Federal governments to shift more of the tax burden to local governments is creating a financial crisis for cities and counties. At the same time that funding levels are decreasing the State and Federal governments are mandating that local governments comply with more and more stringent regulatory mandates such as Comprehensive Growth Management Planning. The long range cost savings and technical capabilities of GIS technology will become more critical as municipalities struggle to provide more services with less staff and money.

Cost Benefit studies should be conservative in their estimates of pay-back times, staff reductions, and cost savings. Optimistic projections can create unreal expectations. Policy makers can be mislead with rosy pictures of short implementation and pay-back timeframes, (3-5 year payback means in addition to the 4-6 years for planning, implementation and data conversion). Decision makers must have an understanding of the complexity, cost, and time required for a project, or the GIS director will be faced with the old question "The hardware was delivered last month, where are the maps?"

Some provisions must be made to produce interim mapping and analysis products while the database conversion project is underway. Use of less accurate data from state and federal sources such as Census Tiger files, Digital Line Graph, Soil data, etc., can enable the GIS to show progress and buy the time needed to complete the conversion of the very accurate base map.

A multi-participant project prevents the proliferation of incompatible systems (GIS software systems seem to be created for maximum incompatibility by some grand design). It is theoretically possible to create a successful GIS project when three or four of the major participants have procured several different software systems running on different platforms and different operating systems, but the problems will be legion. Be wary of vendors who propose to develop translators between various systems. Good translators often do not bring across 100% of the data correctly, and poor translators can create nightmares. When either of the software systems is updated or revised (every six months is not unusual) the translator usually must be rewritten and tested. A typical city or county will not have the time or the in-house programming staff to continually rewrite and test translators. If translators are required they should be supported and updated by the vendor. Implementing a GIS is a monumental task even in the best of circumstances. The added burden of a multi-vendor system can be a very difficult obstacle to overcome.

ORGANIZATION:

The first priority of a project should be to achieve an interlocal agreement between the principal players. Because it "makes good sense" and is obviously the best way to proceed does not mean the task will be easy. All of the arguments (economies of scale, cooperation, "sing from the same sheet of music"), carry little weight balanced against years of mistrust and contention between City and County governments. Add constitutional (elected) officials like the Property Appraiser, and Electric, Gas, and Water Utilities, and the internal problems can be formidable. The value of a good Interlocal Agreement is in establishing the means for up front funding of the project as well as other issues such as: database standards, policies, procedures, shared costs, individual costs, funding, mechanisms for taking in new participants, and establishing committees to manage and oversee the project.

The most important objective after the system is up and running is to complete a base map with accuracy standards which satisfy the engineers, planners, and everyone in between. **The base mapping, on which all other mapping depends, should be at the best accuracy affordable.** Very little compromise can be allowed on accuracy issues without seriously limiting the utility of the end product. Disagreements about accuracy requirements will occur among the participants. Engineers will demand that the conversion be at 1" to 20' foot scales, which will be cost prohibitive, while planners can live with 1" to 800'/2000' scale mapping. The popular compromise is in the 1" to 100'/200' range.

Until the base map is completed departments cannot begin to build their layers of information using the base map as a reference. In this phase of converting the data it is better to narrow the scope to a few key departments (Property Appraiser, Engineering, etc.). If all the departments receive workstations in the beginning there is no data yet converted and the equipment cannot be fully utilized. The cost benefit ratio for hardware and workstations is such that it is wiser to purchase workstations and training "just in time", as they are needed. By delaying procurement more power and processor speed can be obtained for less relative cost.

PERSONNEL AND TRAINING:

The GIS staff consists of a Coordinator and a Systems Manager Analyst who are jointly funded by the GIS. In addition there are three Database Analyst/Administrators dedicated to the project who represent the City, the Electric Department, and Leon County. The GIS field is growing so rapidly that there is a severe shortage of experienced people. The manager or coordinator is usually faced with a choice between hiring Geographers who know little of computer science, or computer people who have no knowledge of mapping. Much training will be required in either case. If two positions are open hire a geographer and a database person, train them together, and their skills can compliment each other. The trend may be toward filling more of these positions with geographers: 1) they can usually be hired at lower salaries, 2) they typically have some computer mapping and GIS skills fresh out of college, and 3) GIS software is becoming more user friendly so that sophisticated core programming is not required or else is beyond the skills of an Analyst which a local government can afford to hire or retain (i.e. less need for computer analysts and more need for super users/mappers).

Once they are trained and experienced GIS personnel become prime targets for other GIS sites. Loss of a key person can easily set a project back 6 to 12 months. The director should prepare to wage constant war with the Administration and the Personnel Department to keep the wages and benefits as competitive as possible.

Within the Tallahassee-Leon County project extensive training of the GIS technical staff has been completed and is ongoing, and will always be ongoing. Software training is underway in three areas: GeoVision, ORACLE, and Digital (VAX and ULTRIX operating systems). Most of the training has been done in Tallahassee with formal and self-paced courses. Training of workstation operators from individual city and county departments is being done on-site. It is more economical to set up a training room with several workstations, fly in a vendor training expert, and train 6-8 operators at once. Sending large numbers of people out of town for training can be very expensive and is difficult to "sell" to Commissioners and Administrators. The departmental workstation operators, selected from existing staff, will be in charge of updating their "layer" of information in the system.

The RFP for the tax parcel mapping is near completion and will also include soil mapping and the creation of a road centerline node database. The tax parcel boundaries will be captured using a combination of GPS (Global Positioning System) points, platted subdivisions, and a best fit to the planimetric data (aerial photography). The unique parcel identification number will be attached to the parcel boundaries with a centroid. Using the parcel ID as a key to match the property appraiser's database records the GIS database will be bulk loaded by merging the vendor supplied files (ID, parcel lines, X-Y coordinates), with the Appraisers master file (ID, owner name, value, etc.).

POSITIVES:

Negotiating the Interlocal Agreement-

Acquiring state-of-the-art hardware and software after completing a thorough selection and evaluation process-

Arranging to lease-purchase the system at a reasonable rate, plus having great flexibility within the lease to replace old equipment with new as needed-

An agreement with CENTEL to provide FDDI fiber optic links between the three major GIS nodes (Local Area Networks). Original plans to install our own fiber network were shelved when CENTEL offered a very attractive deal if the GIS would serve as the initial site (in the USA) for this service. The network was installed on time and has functioned well.

Cost has been contained by the in-house development of RFP's and annual reports, which are usually farmed out to a consultant. Aerial photography and mapping contracts have been under budget.

No loss of trained staff (probably due as much to a sour national economic picture as to local government salary structures). It is sound policy to hire the very best people available and treat them as professionals.

Champions: It is vital to have policy makers at the Commission level who have the foresight to spend money now on a GIS which will save money in the future and provide better services to the citizens.

NEGATIVES:

Consultants: Make more use of consultants for writing and reviewing RFP's and Annual Reports. These tasks require such a large portion of the coordinator's time that it adversely impacts managing and planning the project. It is also a mistake to let the consultant do everything, since 1) it will be very expensive, and 2) it is necessary to get one's hands dirty in order to know what is involved in a GIS (the consultant will not always be there to hold your hand).

Training: When money is tight, travel and training funds become prime targets for budget and administrative people. The media has conditioned the public to view any travel by local government officials and staff as a "boondoggle", and travel money is becoming very difficult to obtain. This is tragic since a lack of training can doom a project. It is sheer folly to spend several million on equipment and conversion services and then place the project at risk by "saving" a few thousand on training and travel costs.

Participants with different priorities and agendas: It is difficult to balance the desires of participants when one has ample funding and little time while another has more time than money. Saving money by slightly extending the completion date of a contract will be viewed as a success by one participant and as a failure by another.

Internal politics and turf wars: Even a "perfect" Interlocal Agreement will leave some individuals on both sides with the feeling that the other side received more advantages or has more control. Every organization has a few "empire builders" who are determined to control a project like a GIS. The project coordinator must be prepared for periodic attacks from any and every side. These assaults can usually be deflected by reason, compromise, and diplomacy. If the project is supported at the upper echelons by policy makers who understand the complexity and the time required to complete an undertaking of this magnitude then the project can survive these internal conflicts.

BIOGRAPHY:

JAMES F. HINES
Tallahassee - Leon County GIS COORDINATOR

BA Geography/History, University of Tennessee, Knoxville
MS Geography-Computer Mapping, University of Tennessee

Eight years as a Graduate Assistant/Associate, and Senior Programmer Analyst with the University of Tennessee, working with the OAK RIDGE NATIONAL LABORATORY, Geographics Group (from the mid 1970's). Wide range of experience from digitizing, database administration, and mapping and analysis for a number of projects for Federal Agencies.

Two years as the Database Administrator for the Knoxville-Knox County GIS. Implemented an INTERGRAPH system, and managed a $ 2.5 million dollar data conversion project.

Three years as the Coordinator for the Tallahassee-Leon County Geographic Information System.

SPATIAL ANALYSIS WITH GIS: PROBLEMS AND PROSPECTS

Michael F. Goodchild

National Center for Geographic Information and Analysis

University of California
Santa Barbara, California 93106

ABSTRACT

In principle, GIS provides an ideal platform for supporting a wide range of analyses using geographic data. In practice, the linkage of GIS with analysis is impeded by numerous factors, and has been more successful in some fields than in others. This paper looks at the problems of integrating GIS with analysis in general, and at the prospects for greater integration in the future.

The paper presents a broadly based classification of methods of spatial analysis. Efficient support of any class requires that the appropriate data model be recognized by the GIS. In some cases, methods of analysis are written for continuous space, without explicit discretization, and thus cannot be implemented intact. Many methods of spatial analysis require a data model that abstracts space to a simple matrix of interactions between objects; the paper discusses the implications of this class for the design of GIS, and argues for the development of simple 'hooks'. In other cases, the data model required to support analysis includes features such as time, or the vertical dimension, that are not commonly available in current GIS.

BACKGROUND: GIS AND SPATIAL DATA ANALYSIS

There seems to be widespread agreement in the GIS community on two simple propositions: that as a technology, GIS has the potential to support many different types of analysis; and that this potential has not yet been realized. This theme is reflected in Openshaw's oft-quoted comment:

> "Such systems are basically concerned with describing the Earth's surface rather than analysing it. Or if you prefer, traditional 19th-century geography reinvented and clothed in 20th-century digital technology." (Openshaw, 1987 p. 431)

It is also recurrent in the collection of papers on GIS edited by Worrall (1990).

The potential to support analysis is reflected in many discussions of GIS:

> "Geographic information systems evolved as a means of assembling and analyzing diverse spatial data." (Star and Estes, 1990 p. 14)

> "...the Geographic Information System...is as significant

to spatial analysis as the inventions of the microscope and telescope were to science, the computer to economics, and the printing press to information dissemination." (Department of the Environment, 1987 p. 8)

But from a strictly pragmatic viewpoint, the reality of GIS today might be summed up in the following:

> A database containing a discrete representation of geographical reality in the form of static, two-dimensional geometric objects and associated attributes, with a functionality largely limited to primitive geometric operations to create new objects or to compute relationships between objects, and to simple query and summary descriptions.

GIS clearly needs stronger analysis and modeling capabilities if it is to meet its potential as a tool.

The case for Spatial Data Analysis (SDA) rests on the argument that explanation, understanding and insight can come from seeing data in their spatial context. There seem to be at least four separate arguments for this spatial perspective. First, space can provide a simple and useful indexing scheme. An archaeologist, for example, might record the locations of artifacts as they are unearthed at a dig simply in order to index them for later access. Geographic coordinates provide a kind of hashing code, on the assumption that two artifacts are unlikely to be unearthed at exactly the same location. A map provides a simple means of displaying the index, and finding artifacts given the human eye's extraordinary power to digest two-dimensional information. Spatial indexing is not likely to lead directly to insight, but it is a useful tool for handling large amounts of data.

Second, the spatial perspective allows easy access to information on the relative locations of objects and events, and proximity can indeed suggest insight. The Snow map showing the clustering of cholera victims in London during an outbreak of the disease in 1854 led directly to explanation, in the form of drinking water from a polluted well, and to an effective remedy for the outbreak (Gilbert, 1958). Although the explanation was immediately apparent from the map, it would have been virtually impossible to arrive at the same insight in any other way.

Third, a spatial perspective allows events of different types to be linked, in a process formalized in GIS as overlay. The fact that an event occurs in proximity to other events or objects can be very suggestive. For example, any environmental abnormality in the vicinity of a cluster of cancer cases is immediately suspect, because an individual is clearly more vulnerable to the local environment than to distant ones.

Finally, the distance between events or objects is often an important factor in interactions between them. In the physical sciences, distance can be a cause in itself, as in the inverse square laws of gravitation or electromagnetics. In the social sciences it is more likely to be a surrogate for information (we are more likely to know about nearby places than about distant ones) or for contact (we know more people locally) or time spent traveling (we prefer local shops to distant ones, all other things being equal) (Gatrell, 1983).

Spatial Data Analysis is a set of techniques devised to support a spatial perspective on data. To distinguish it from other forms of analysis, it might be defined as a set of techniques whose results are dependent on the locations of the objects or events being analyzed, requiring access to both the locations and the attributes of objects (Goodchild, 1987). Its techniques range from simple measures of the dispersion of a set of points to complex statistical tests of whether a set of points could have been generated by specific random processes (Ripley, 1981; Getis and Boots, 1978).

SDA provides a set of objective techniques to replace and augment subjective intuition. Unfortunately, while the spatial perspective can be very powerful as a source of insight, it can also be highly misleading. Ancient cultures found endless images in the random patterns in the night sky. More recently, there are many examples in the literature of false inferences drawn from apparent spatial patterns that later turn out to be no different from the outcomes of random processes. Haggett and Chorley (1969), for example, found that the average number of edges in the Brazilian administrative boundary network is very close to 6, and concluded that this supported the contentions of Christaller's Central Place Theory. But the bubbles in a polystyrene coffee cup also have very close to six edges on average, and analysis later showed that this is a necessary outcome of a theorem of Euler applying to any boundary network (Getis and Boots, 1978). Only this kind of objective analysis can determine whether visual cancer clusters are in fact abnormalities, or simply random events.

GIS needs SDA if it is to reach the potential implied by many of its definers and proponents, of a general-purpose tool for delivering a spatial perspective on data in a digital environment. SDA needs GIS if it is take advantage of the capabilities in GIS for data input, editing, display and mapping, and to be readily accessible to a broad user community.

PROGRESS TO DATE

If the arguments for linking GIS and SDA are so strong, why has so little progress been made to date? First, developments in the GIS industry largely reflect the demands of the GIS marketplace, which has been dominated for the past decade by applications in resource management, infrastructure and facilities management, and land information, where GIS tends to be used more for simple record-keeping and query than for analysis. Although a small minority of companies have stressed analysis in their development and marketing, SDA has had little impact on the GIS mainstream. SDA is also of greater interest in academic and scientific applications of GIS than in local government or the private sector, where GIS budgets and expenditures tend to have been much higher.

Second, despite its promises, SDA remains a comparatively obscure field. There are few books or reviews, and there is no easy way of organizing or codifying SDA. Notable exceptions include the now classic collection edited by Berry and Marble (1968), Unwin (1981), Upton and Fingleton (1985) and the recent book by Haining (1990). There are no widely accessible courses in SDA, and there is some concern that the introduction of GIS into many university programs may in fact have diverted resources from existing courses in SDA (Heywood, 1990). In the long term, linkage with GIS may lead to greater awareness of SDA and greater availability of courses and texts, but in the short term there is a distinct shortage of knowledge, experience and training on the SDA side.

Third, many techniques of SDA were developed in the 1960s and 1970s when GIS was still in its infancy, and cartography a technology of pen and paper. Early efforts to implement SDA in a computational environment had to rely on source code programming, notably in Fortran (for examples see Baxter, 1976; MacDougall, 1976). Although the 1970s saw the emergence of integrated statistical packages like SAS and SPSS, with no explicit support for coordinates or spatial objects except for mapping and display, in practice these provided the most readily available basis for implementation of SDA until the recent interest in GIS. In addition, the statistical techniques on which much of SDA is based are in many cases explicitly non-spatial, making assumptions about the lack of spatial dependence which fly directly in the face of the spatial perspective. Spatial autocorrelation has often been treated as a problem to be removed (Odland, 1988), rather than as an inescapable property of almost all spatial data.

It is only in the past few years, with the development of GIS, that a realignment of SDA with other explicitly spatial technologies like cartography, GIS and remote sensing has finally begun. But much SDA remains strongly linked to the aspatial environment of statistical packages like MINITAB and SAS (Griffith, 1988).

Fourth, many techniques of SDA are complex and difficult, requiring a very different approach from the intuitive, synthetic view often promoted for GIS. A glance through the pages of a journal like Geographical Analysis is sufficient to strike despair into the hearts of most GIS enthusiasts. As an academic specialty with little immediate connection with the world of practical application, SDA might be accused of emphasizing mathematical sophistication at the expense of practicality. Simple, intuitive techniques for exploring data in a spatial context have often been ignored in the search for elegant formulations. Analysis, particularly when intuitive, is often associated with the inductive approach to quantitative geography that fell out of favor in the 1960s, giving way to modeling and deduction. The move to mathematical modeling has also been one form of response to the critique of positivism launched by the social theorists in the 1970s (Gregory, 1978). This chain of thought suggests that the key to integrating GIS and SDA may lie in an emphasis, at least initially, on the more intuitive, exploratory techniques in the SDA toolkit.

THE ROLE OF DATA MODELS

A GIS database captures real geographic variation in the form of a finite number of discrete, digital objects. Because geographical variation is fundamentally continuous and infinitely complex, this process of capturing reality must involve abstraction, generalization or approximation. The rules by which the objects and their relationships are defined is termed a data model (Tsichritzis and Lochovsky, 1977). The variety of data models used in GIS is one of its complications and at the same time one of its strengths.

Reviews of GIS data models have been published by Peuquet (1984) and Goodchild (1991). Data models take two broad forms, depending on whether reality is perceived as an empty space populated by objects, or as a set of layers or fields, each defining the spatial variation of one variable. In very broad terms, the former view is more relevant to analysis and modeling in the social sciences, where discrete entities are conceived as interacting over space, and the latter is more relevant to the environmental and physical sciences, but exceptions abound.

Objects are normally modeled as points (P), lines (L) or areas (A), after appropriate generalization of form, for example by representing a city as a zero-dimensional point. Fields are modeled in GIS in at least six ways:

- a raster of cells, each defining the average value of the field within the cell (e.g. a remote sensing scene) (R1F);
- a raster of regularly spaced point samples (e.g. a digital elevation model) (R2F);
- a set of non-overlapping, space-exhausting polygons, each defining a class (e.g. a soil or vegetation cover map) (AF);
- a set of irregularly spaced point samples (e.g. a weather map) (PF);
- a set of digitized isolines (e.g. a contour map) (LF);
- a set of non-overlapping, space-exhausting triangles, each assumed to approximate elevations within the triangle with a simple plane (the triangulated irregular network or TIN model) (TF).

Data models define how geographic variation is represented, but also determine the set of processes and analyses that can be carried out. For example, it would be appropriate to use a set of point samples representing a field of atmospheric temperature (the PF model above) to interpolate a contour map or create an oblique view, using the attributes of each point to determine the elevation of the interpolated surface. But it would be meaningless to perform the same operation on a set of point objects (the P model) representing cities with attributes of population. Despite this, the two models may be stored identically in the GIS and the user may be unaware that a potentially meaningless operation is being performed.

Data models provide a logical and useful way of organizing the functionality of a GIS. They may also provide a framework for discussion of methods of SDA, since these are in principle extensions of basic GIS functionality. However many methods of SDA treat the issue of data modeling as a matter of implementation, rather than as an intrinsic property of the method of analysis. For example, suppose that a hydrological analysis requires the determination of ground slope, as an important factor in soil erosion. Slope is a well-defined property of any continuously differentiable mathematical surface. But it is not well-defined everywhere on the real landscape, which is characterized by frequent breaks of slope, and it is not defined independently of data model in any discrete representation of the land surface. In the TIN model, for example, it is well-defined and constant within triangles but indeterminate on their edges, and curvature is everywhere zero or indeterminate. In a contour model of the same real surface, slope must be inferred by some additional, as yet undefined process of interpolation.

In summary, an additional problem facing any effort to integrate GIS with SDA is that data modeling must be explicit in any use of GIS, but is often left undefined in SDA. Use of GIS forces the analyst or modeler to confront the issue of discretization directly.

DATA MODELS FOR SPATIAL DATA ANALYSIS

Although a wide range of data models are currently found in various GIS products, the range required to support a full array of spatial data analysis is much larger. Techniques of SDA can be arranged into several broad groupings depending on the underlying data model that is assumed in each one:

- points: techniques used to analyze an undifferentiated set of points, e.g. point pattern analysis (Getis and Boots, 1978);

- spatial objects with attributes: techniques that analyze an attribute matrix, and reduce space to a square matrix of spatial relationships between pairs of objects, e.g. measures of adjacency or proximity;

- networks of links and nodes: a range of techniques for analyzing networks in transportation and hydrology, based on attributes of network links and nodes;

- spatial interaction models: models of the interaction between pairs of objects, based on an analysis of the characteristics of origin objects, destination objects, and the spatial separation between them;

- raster techniques: methods of analysis based on the

representation of continuous layers as rasters of cells, and supported by the so-called raster GISs (a codification of this class has been developed by Tomlin, 1990).

Of these, the second and fourth require a matrix of relationships between objects that is missing in most currently supported GIS data models. Beyond these simple methods lie all of those models and techniques of analysis that require access to time or to the third spatial dimension. In other words, the current range of GIS products is far from adequate for supporting a full range of methods of spatial analysis.

THE NATURE OF A LINKAGE

One might define three different levels of linkage between GIS and SDA. Full integration would mean a common functionality accessible through a common interface, with associated conceptual structures. This seems unlikely to emerge given the nature of the GIS software industry and the unbounded nature of SDA. Close coupling would mean the recognition of common, high level structures in both systems so that information passed out of GIS and processed in SDA could be remerged without difficulty. For example, close coupling would require that the identities of objects be preserved when passed between GIS and SDA packages, so that if the order of objects changed, their identities would not be confused. To realize such close coupling and preservation of high level structures, the SDA system would have to know all of the data models in use in the GIS, which means in effect that it could not be a standard statistical package. Finally, loose coupling would mean that high level structures would be lost on transfer, and would thus have to be rebuilt on an ad hoc basis. This is the form of coupling that characterizes the relationship between many current GISs and other components of software federations, such as the statistical packages. Data must be transfered largely in the form of flat ASCII tables.

Practically, it seems that close coupling offers the most realistic alternative for improving on the current situation, which is characterized by loose coupling. For this to be achieved, however, there will have to be a much wider recognition of the role played by data models in GIS, and by the need to make discretization explicit in all aspects of spatial data analysis.

ACKNOWLEDGMENT

The National Center for Geographic Information and Analysis is supported by the National Science Foundation, Grant SES 88-10917.

REFERENCES

Baxter, R.S. 1976, Computer and Statistical Techniques for Planners, Methuen, London.

Berry, B.J.L. and Marble, D.F. 1968, Spatial Analysis: A Reader in Statistical Geography, Prentice-Hall, Englewood Cliffs, NJ.

Department of the Environment 1987, Handling Geographic Information. Report to the Secretary of State for the Environment of the Committee of Enquiry into the Handling of Geographic Information, HMSO, London.

Gatrell, A.C. 1983, Distance and Space: A Geographical Perspective, Oxford University Press, New York.

Getis, A. and Boots, B. 1978, Models of Spatial Processes, Cambridge University Press, London.

Gilbert, E.W. 1958, Pioneering maps of health and disease in England. Geographical Journal, Vol. 124, pp. 172-183.

Goodchild, M.F. 1987, Towards an enumeration and classification of GIS functions. Proceedings, IGIS 87: The Research Agenda, NASA, Washington, DC, Vol. II, pp. 67-77.

Goodchild, M.F. 1991, Geographical data modeling. Computers and Geosciences (in press).

Gregory, D. 1978, Ideology, Science and Human Geography, Hutchinson, London.

Griffith, D.A. 1988, Estimating spatial autoregressive model parameters with commercial statistical packages. Geographical Analysis, Vol. 20, pp. 176-186.

Haggett, P. and Chorley, R.J. 1969, Network Analysis in Geography, Edward Arnold, London.

Haining, R.P. 1990, Spatial Data Analysis in the Social and Environmental Sciences, Cambridge University Press, New York.

Heywood, I. 1990, Geographic information systems in the social sciences. Environment and Planning A, Vol. 22(7), pp. 849-854.

MacDougall, E.B. 1976, Computer Programming for Spatial Problems, Wiley, New York.

Odland, J. 1988, Spatial Autocorrelation, Sage, Newbury Park, CA.

Openshaw, S. 1987, Guest editorial: an automated geographical analysis system. Environment and Planning A, Vol. 19, pp. 431-436.

Peuquet, D.J. 1984, A conceptual framework and comparison of spatial data models. Cartographica, Vol. 21, pp. 66-113.

Ripley, B.D. 1981, Spatial Statistics, Wiley, New York.

Star, J.L. and Estes J.E. 1990, Geographic Information Systems: An Introduction, Prentice-Hall, Englewood Cliffs, NJ.

Tomlin, C.D. 1990, GIS and Cartographic Modeling, Prentice-Hall, Englewood Cliffs, NJ.

Tsichritzis, T.C. and Lochovsky, F.H. 1977, Data Base Management Systems, Academic Press, New York.

Unwin, D.J. 1981, Introductory Spatial Analysis, Methuen, London.

Upton, G.J.G. and Fingleton, B. 1985, Spatial Data Analysis by Example, Wiley, New York.

Worrall, L. 1990, Geographic Information Systems: Developments and Applications, Belhaven Press, London.

LINKING GIS WITH PREDICTIVE MODELS: CASE STUDY IN A SOUTHERN WISCONSIN OAK FOREST

Philip L. Polzer
Brauna J. Hartzell
Randolph H. Wynne
Paul M. Harris
Mark D. MacKenzie

Environmental Remote Sensing Center
University of Wisconsin - Madison
1225 West Dayton Street
Madison, WI 53706
(608) 262-1585

ABSTRACT

Geographic Information Systems offer the researcher an interface between spatial data and predictive models. Analysis of model results can be a complicated and time consuming task. Researchers can make use of GIS analytical capabilities as well as visualize the data before and after the model is run. This provides the researcher with a powerful tool for assessing model results. Further, it may enable the researcher to ask questions that previously were unable to be addressed.

This paper explores linking a GIS to a predictive forest succession model for Noe Woods, an oak forest in southern Wisconsin. The procedures and mechanisms for building this link are described as well as techniques for visualizing the model output. The link enabled spatial analysis and enhanced visualization that were previously beyond the capabilities of the model. The GIS relational database was effective in managing model input and output. The protocol established is germane to any number of models over a wide range of applications.

INTRODUCTION

Predictive mathematical models have long been important in the computer simulation of forest growth. Among the first complete dynamic models of forest growth were JABOWA (Botkin et al., 1972) and FORET (Shugart and West, 1977). These models and their progeny utilize both biotic parameters (e.g. maximum tree age, height, and diameter) as well as abiotic parameters (e.g. climatic variables, soil conditions, and solar insolation) in predicting ecosystem response. As Botkin et al. (1972) indicated, dynamic models serve both as a repository of integrated knowledge about a system as well as a relatively rapid and effective means by which hypotheses about an ecosystem can be tested.

Succession involves the change in natural systems and the understanding of the causes and direction of such change (Shugart, 1984). Forest succession models are often used to further enhance understanding of the processes underlying this change. In addition, succession operates over timescales of hundreds of years, making modeling an essential tool in fully comprehending the process of forest development.

The current generation of succession models, while quite advanced in terms of the incorporation of the sum of knowledge about forest ecosystems, are not well developed with respect to horizontal position in the landscape. Nonetheless, the ecological community has recognized that landscape pattern and process are intertwined (O'Neill et al., 1988). The use of GIS to represent and analyze data in a spatial context has thus become widespread in research focused on advancing the fundamental understanding of ecological systems and dynamics (Lillesand et al., 1989; Johnston, 1989).

The area of interest in this paper focuses on Noe Woods, a southern Wisconsin oak forest in the University of Wisconsin Arboretum, located in Madison. The Arboretum was established in 1934 in order to represent presettlement plant communities in Wisconsin and the upper Midwest through the process of ecological restoration. Today the Arboretum is recognized as a leader and pioneer in the development and techniques for restoring and managing ecological communities and has given birth to a new discipline, restoration ecology.

Noe Woods represents a relatively undisturbed forest community. Set aside as a minimal-management area, Noe Woods has had no large scale species introductions and only judicious removal of exotic species (Jordan, 1981). Data detailing the dynamics of tree species in two hectares of Noe Woods have been collected since 1956. These data include the diameter at breast height (dbh), basal area, species, and location of individual trees with a dbh greater than 10 cm (McCune and Cottam, 1985). These data represent a unique body of observations about forest dynamics due to their completeness, long-term nature, and spatial context. These traits make the data amenable to incorporation in a forest succession model. The intrinsic spatial representation of the data make them ideal for inclusion in a relational GIS serving as a centralized data archive and analytical tool.

RATIONALE AND SIGNIFICANCE

By using a GIS as the interface between the data and model, researchers can not only make use of GIS database management and analytical capabilities, but also visualize the spatial component of the data before and after the model is run. This provides researchers with a powerful tool in assessing model results. In brief, linking the model to the GIS provides a mechanism by which a non-geographic model can be made horizontally spatial.

A link between a GIS and a model has applications not only to forest succession models but to any number of models. Development of procedures and mechanisms for building this link using a general forest model and the Noe Woods data provides other researchers the basis on which to build links to other models.

PROJECT OBJECTIVES

Two objectives pertinent to this link were addressed:

1) Explore the feasibility of linking a GIS to a forest succession model.
2) Investigate the application of the linked model to Noe Woods.

These objectives can be restated in the form of questions that this project has studied. Can a link be forged effectively? How can the GIS enhance the analysis of model results?

METHODS

Several steps were necessary in order to answer the questions posed above. These steps, though, should be viewed as parts of a whole. They come together to form the link between the model and the GIS. Though presented in a sequential fashion, the steps are in fact interdependent. This link can be thought of in a circular fashion, starting with the GIS, moving through the data extraction process to the running of the model, followed by the conversion of results back to a GIS format (Figure 1). These steps are as follows:

1) Database development and model parameterization
2) Construction of the data link to the general forest model
3) Model automation and execution
4) Results of analysis and visualization

The forest model used was ZELIG (Urban, 1990). This model was chosen because it is conceptually easy to understand, its predecessors have been validated in a number of geographic contexts, and the source code was available. It was also amenable to the incorporation of horizontal data. ZELIG is a forest succession model which simulates the establishment, annual diameter growth, and mortality of individual trees on a model plot corresponding to the zone of influence of a canopy-dominant tree. This plot size is determined by the model as a function of sun angle (latitude dependent) and the typical canopy height of the stand. The plot size or grid cell for the Noe Woods simulation was thus determined to be 200 m^2 (14.1 x 14.1 m) upon parameterization. Implementation requires placement of parameter estimates (Table 1) for local weather, site characteristics, and tree species growth parameters into the DRIVER file for the model run. Initial forest cover is written to the INDATA file (Table 1) for each plot. As noted by Burrough et al. (1988), the necessary model parameters and input data to a GIS-linked model should be supplied by the GIS. Therefore, the database design process consisted of the concurrent development of the GIS database with the model parameterization.

As Urban (1990) notes, a single plot depicts only one possible "trajectory" of forest dynamics due to the stochastic nature of the simulator. The model was therefore run 50 times for each of the 100 plots with the aggregation representing the average "trajectory" of the forest over time.

Figure 1. The GIS - ZELIG Link

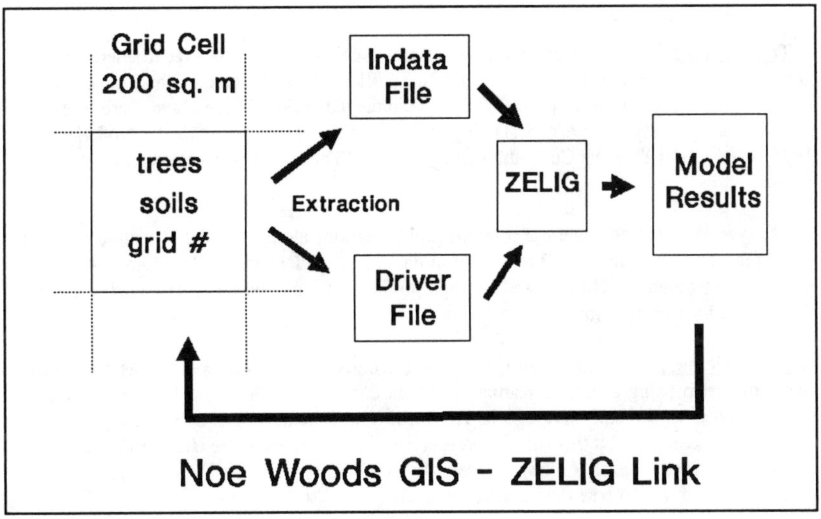

Database development and model parameterization

PC Arc/Info (version 3.4D) was the GIS software used. This was complemented by a variety of other software packages. These software packages were implemented in an IBM compatible 386-based computer environment.

The first step in this phase of the project was to collect and enter the data which would be the basis for the GIS used in this project. Primary data coverages were the tree information (McCune and Cottam, 1985) and the Huddleston (1971) soil survey. A secondary coverage, consisting of a 10 x 10 square grid covering the area for which tree data were collected, was created to provide the spatial basis of the model.

Table 1. Model Parameters

DRIVER file	**INDATA file**
Mean monthly precipitation and temperature	Plot number
Site characteristics (latitude, longitude, elevation)	Number of trees
Soil characteristics (wilting point, field capacity, soil fertility)	Species data (code, dbh, height)
Species growth parameters (maximum height, diameter, age, shade, and drought tolerance seeding rate)	

Tree Coverage. Information on tree parameters such as stem count, tree diameter, basal area, and species exists for a 34 year period. In addition, all trees greater than 10 cm dbh were mapped in reference to a permanently monumented grid. These data were measured at approximately five year intervals starting in 1956 (exact years: 1956, 1961, 1968, 1971, 1973, 1978, 1983, and 1989) (McCune and Cottam, 1985). These data were converted to the Arc/Info format.

The Noe Woods tree coverage contains point locations along with the measured basal area of each tree. In order for ZELIG to use these data for a model run, the basal area was converted to dbh. Tree height is also required for input to the model and was calculated using the equation formulated by Ker and Smith (1955).

Soil Coverage. A detailed soil map of Noe Woods (Huddleston, 1971) was scanned from a hard copy map using a flatbed scanner and then converted to Arc/Info format. Soil attributes were attached to the soil coverage in Arc/Info. The critical values of wilting point and field capacity, not collected by this survey, were derived by comparing the Huddleston data to similar soil types in Holtan et al. (1968). Seven different soil types were identified for the region of Noe Woods for which tree data exist (Table 2).

Grid Coverage. As noted previously, the ZELIG model requires a plot size of 200 m^2 in the Noe Woods implementation. This results in 100 plots for the 2 ha of Noe Woods for which tree data exist. A 10 x 10 grid with a 200 m^2 cell size was thus generated in Arc/Info to cover this area. For each grid cell, the dominant soil type and all trees located in it were determined by a GIS overlay process. All of the coverages were georeferenced to State Plane Coordinates using seven control points.

Table 2. Noe Woods Soil Types (Huddleston, 1971)

Soil Number	Name and Description	Extent
1	Dresden loam, very fine sand substratum variant	13%
3	St. Charles silt loam	32%
4	McHenry silt loam, deep variant	19%
5	McHenry silt loam	22%
6	St. Charles silt loam, heavy Bt variant	2%
7	Sisson sandy loam	8%
10	St. Charles silt loam, overwash variant	2%

Overlay Operations. Three coverages (trees, soils, and grid) were overlaid to create one coverage which contained the tree point data and the dominant soil type for each grid in the Noe Woods study area. Information was extracted from this coverage to drive the model. It also allowed the model results to be related to the GIS through the grid number.

Model Modifications. The original ZELIG model was designed to run on a mainframe computer. The amount of memory it used had to be reduced to run on a personal computer. This step involved altering the values in the ZELIG source code which affected the allocation of memory. The value for the maximum number of trees per plot (MT) was changed from 1000 to 250 and the value for the maximum height of the foliage profile (MH) was changed from 100 to 75 m.

In order to provide model output data in a form that could be exported to dBASE or other software packages (i.e., ASCII format), the source code of the model was changed to produce three new files. The first contains species data for each year of the model run, the number of each species in 10 cm dbh size classes, basal area, density, and relative importance value. The second contains aggregate stand data on the plot such as basal area and dbh. The third file was a variation of the second which facilitated the production of area percentage graphs.

Model Parameterization. Species parameters (i.e., maximum age, maximum height, maximum diameter, growth parameter, limits of growing degree days, shade tolerance, drought tolerance, nutrient stress tolerance, and relative reproductive (seeding) rate) are also required for each species used in the model. There are twelve tree species in Noe Woods (Table 3). The model documentation provided the parameters for most species in Noe Woods. The remaining species were paramaterized using values obtained from Fowells (1965), Preston (1976), and Solomon et al. (1984).

Other DRIVER file parameters include the mean monthly temperature and precipitation with their respective standard deviations for the study site. Monthly averages for both of these parameters dating from prior to 1900 were obtained from the Wisconsin State Climatologist for Truax Field in Madison and monthly means and standard deviations were calculated.

Table 3. Noe Woods Tree Species

Common Name	Scientific Name	Abbreviation
Red Maple	*Acer rubrum*	ACRU
Butternut	*Juglans cinerea*	JUCI
Bigtooth Aspen	*Populus grandidentata*	POGR
Cherry	*Prunus serotina*	PRSE
White Oak	*Quercus alba*	QUAL
Burr Oak	*Quercus macrocarpa*	QUMC
Black Oak	*Quercus velutina*	QUVE
Black Locust	*Robinia pseudoacacia*	ROPS
Basswood	*Tilia americana*	TIAM
Elm	*Ulmus americana*	ULAM
Hackberry	*Celtis occidentalis*	CEOC

Construction of the data link to the general forest model

The second phase of this project was the construction of a data link from the GIS to the forest model (Figure 1). This involved the conversion of the necessary data to a format compatible with ZELIG, running the model, obtaining the results, and reconverting them to a format acceptable for the GIS in order to spatially analyze and visualize the results. The use of PC Arc/Info 3.4D made dBASE programming logical for the construction of the data link.

In forging this link, two factors had to be considered: how to extract tree data and store them in an INDATA file and how to make available the dominant soil type data in the extraction process. To address the first issue, a dBASE command file was written to access the tree data from one or all of the grid cells. In order to obtain an average trajectory of forest dynamics, the tree data extracted for one grid cell had to be repeated for the appropriate number of plots (50) over which the model would be run. In running the model over the study area, the extraction of tree data resulted in 100 INDATA files, one for each grid cell. The second factor to be considered was a method to make available in the model automation process the dominant soil type for each grid cell. As a result of the overlay operation previously described, the dominant soil type number was stored as an attribute of each grid cell.

For each of the three output file types, 100 data files were produced, one for each grid cell. The output results for each file type were agglomerated over the 100 grid cells using a dBASE command file. The command file produced three dBASE files which could be related to the GIS through the grid cell number. These files represent the results for the Noe Woods study area at the species level and stand level for one model run.

Model automation and execution

Through the use of DOS batch files and dBASE command files, the link was partially automated. Maintaining 100 INDATA files was a problem in and of itself along with the problem of running the model for each of these tree data files. As an organizing method, therefore, the name of the INDATA files incorporated the grid cell number from which they were extracted together with the dominant soil type. These could be organized into various subdirectories representing each of the seven soil types. Seven DRIVER files were created with the appropriate values of wilting point and field capacity for each dominant soil type and were stored in the respective subdirectories, together with the model program. DOS batch files were created to run the model for each of the INDATA files in a subdirectory and a controlling batch file was written to direct the running of the individual batch files. This process allowed the use of one batch file which executed a complete model run over all INDATA files.

Baseline data, as represented by the 1956 tree data, were extracted from the GIS and the model was run to simulate changes over a 200 year time period. Model runs were completed for two different input parameters, one with all species and one with Black Locust (ROPS) and Red Maple (ACRU) removed, this having been a management practice.

RESULTS

The final step in this project was the analytical phase and the development of techniques for visualizing the results. Figures 2 and 3 show the results of the 200 year runs. These two figures represent traditional methods for visualizing results from a forest succession model. The percent of total basal area is plotted for five species categories: the four dominant species and one for the remaining species. From these figures the effects on the modeled composition of Noe Woods of removing Black Locust and Red Maple from the model run can be seen. With all species present, Red Maple is the most dominant species. However, when Black Locust and Red Maple are not considered, the model produces a simulated oak dominated forest which seems to better reflect the expected composition of Noe Woods.

Figures 2 and 3 show the uncertain nature of model performance in the first 40 years. There is a precipitous drop in the major species in the early years before they return to dominance later in the simulation. This drop is possibly due to a combination of factors. The lack of seedling and sapling data in the McCune and Cottam (1985) study required to appropriately initalize the plot resulted in the model randomly seeding a number of minor species during the initial phase of model simulation. These minor species are eventually suppressed due to competition as the major species return to dominance.

Figure 2. Percent Basal Area for Unmanaged Model Run

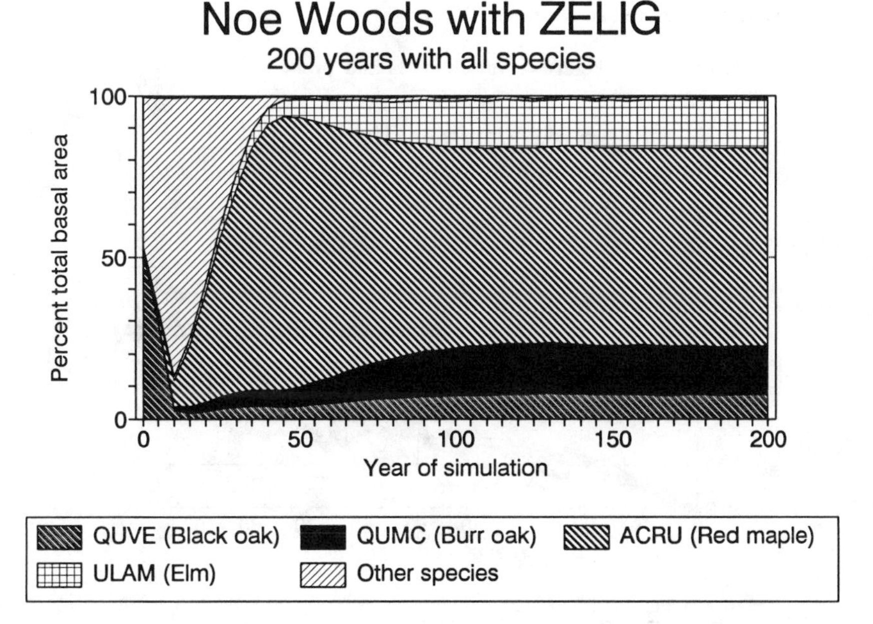

Figure 3. Percent Basal Area for Managed Model Run

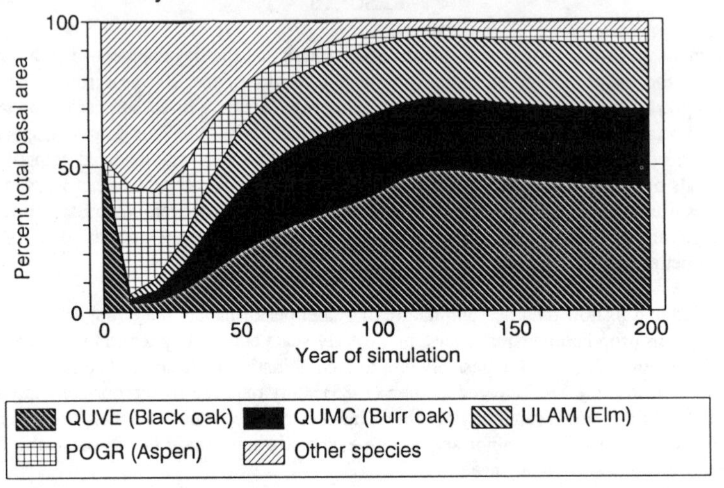

Figure 4 shows three model runs with three of the soil types described by Huddleston (1971). This illustrates how soil differences affect a model run. Due to the stochastic nature of the model this graph should be treated with caution in terms of characterizing the effects of soil type on forest productivity, as only one run with the same random number generation was made using each of the soil types. However, this has the advantage of making major "events", such as the death of a large tree, more easily noticed.

Figure 4. Variability in Soil Types

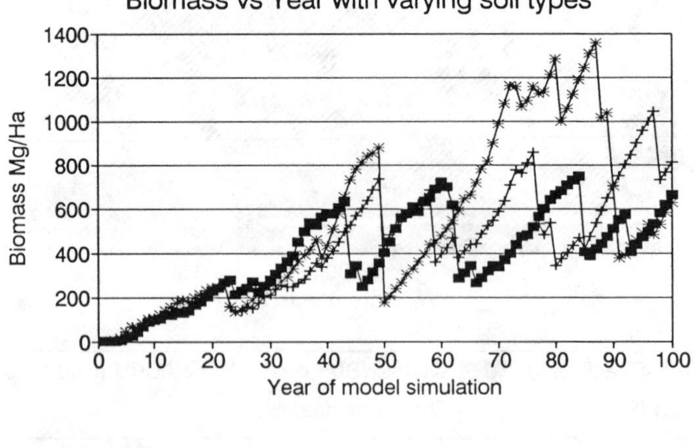

One of the advantages of linking the GIS to the model is the utilization of the graphical capabilities of the GIS which allows the model results to be visualized spatially. The GIS was used to visualize and analyze the results of the four model runs. Figure 5 shows a series of 40 year time slices of the composition of the simulated forest. The dominant species for each grid cell was determined on the basis of basal area and each grid cell was coded based on the dominant species. This provides a method for analyzing the progression of forest succession spatially and temporally in terms of dominant species composition.

Figure 5. Change in Dominant Species for Managed Model Run

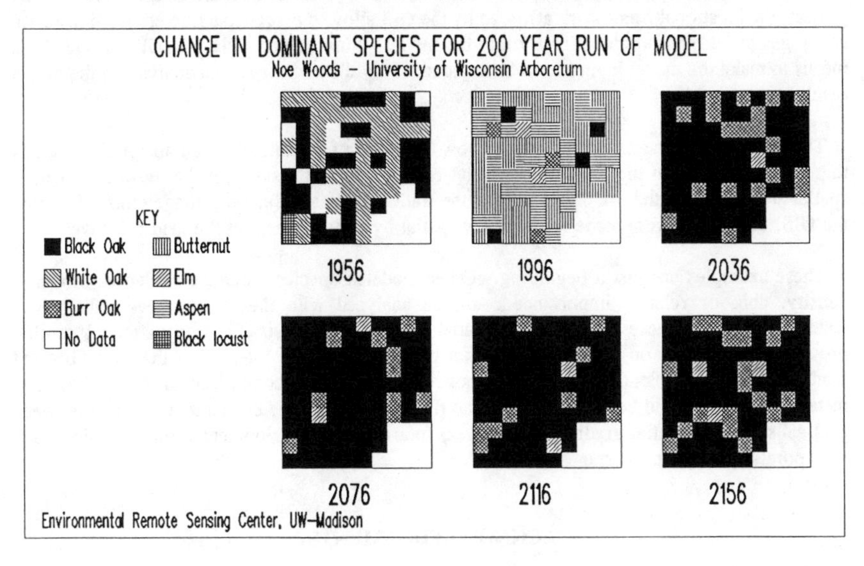

SUMMARY AND CONCLUSIONS

This research demonstrates that, not only can a model be linked to a GIS, but that the link provides considerable advantages to the user. The advantages include the management of the input and output data, the additional spatial analysis capabilities, and the visualization of the model results. The development of the link, while not a trivial task, was achieved using simple DOS batch files and dBASE command files.

A GIS integrates spatial data and their characteristics via use of a relational database. In this project, the locational framework afforded by the GIS allowed diverse parameters from different coverages to be incorporated directly into the model. Most importantly, the GIS provided the means to make the model input and output spatial. This allowed for spatial analysis of the model results.

The techniques presented in this paper show the range of visualization and analytical methods that can be utilized in the analysis of model results. Model results can be viewed in a non-spatial arena through the use of graphs and the traditional basal area percentage plots. Through the GIS, the results were made horizontally spatial by relating them at the grid cell level.

These examples are just a beginning. Other modeled species parameters such as biomass, density, dbh, or relative importance could be analyzed with these techniques. Together, techniques such as these allow for the spatial and non-spatial visualization of model results, providing more tools on which analysis can proceed. Further integration between GIS and models could be developed to provide a more complex spatial component to the model. For instance, the GIS could be used to model the flow materials (i.e., nutrients, moisture) between grid cells along a spatial gradient. However, future model development should consider GIS incorporation in the initial stages.

ACKNOWLEDGEMENTS

The research presented in this paper grew out of the Institute for Environmental Studies 1990-91 Environmental Monitoring Practicum at the University of Wisconsin (Environmental Monitoring Practicum, 1991). We would like to thank Drs. Frank Scarpace and Evelyn Howell for their guidance in this effort. Dr. Greg Armstrong and the Arboretum staff are appreciated for their support in this project. We would also like to thank Dr. Dean Urban for providing the ZELIG source code.

BIBLIOGRAPHY

Botkin, D.B., J.F. Janak, and J.R. Wallis. 1972. Some ecological consequences of a computer model of forest growth. Journal of Ecology **60**: 849-873.

Burrough, P.A., W. van Deursen, and G. Heuvelink. 1988. Linking Spatial Process Models and GIS: a marriage of convenience or a blossoming partnership? Proceedings, GIS/LIS '88. San Antonio, Texas.

Environmental Monitoring Practicum. 1991. REGIS: Restoration Ecology Geographic Information System. Demonstration Pilot Projects for the University of Wisconsin - Madison Arboretum. Institute for Environmental Studies, University of Wisconsin, Madison, WI.

Fowells, H.A. 1965. Silvics of Forest Trees of the United States. Agriculture Handbook Number 271. Washington, United States Department of Agriculture, 762 pp.

Holtan, H.N., C.B. England, G.P. Lawless, and G.A. Schumaker. 1968. Moisture-Tension Data for Selected Soils on Experimental Watersheds. Agricultural Research Service Publication Number 41-144, Washington, United States Department of Agriculture, 608 pp.

Huddleston, J.H. 1971. Characteristics and distribution of soils in the Noe Woods, University of Wisconsin Arboretum, Lake Wingra Basin. International Biological Program Deciduous Forest Biome Memo Report #71-42.

Johnston, C.A. 1989. Ecological Research Applications of Geographic Information Systems. Proceedings, GIS/LIS '89. Orlando, Florida.

Jordan, W.R. 1981. The Arboretum. University of Wisconsin - Madison Arboretum.

Ker, J.W. and J.H.G. Smith. 1955. Advantages of the Parabolic Expression of Height-diameter Relationships. Forest Chronicles **31**: 235-46.

Lillesand, T.M., M.D. MacKenzie, J.R. VandeCastle, and J.J. Magnuson. 1989. Incorporating Remote Sensing and GIS Technology in Long-Term and Large-Scale Ecological Research. Proceedings, GIS/LIS '89. Orlando, Florida.

McCune, B. and G. Cottam. 1985. The Successional Status of a Southern Wisconsin Oak Woods. Ecology **66**(4): 1270-1278.

O'Neill, R.V, J.R. Krummel, R.H. Gardner, G. Sugihara, B. Jackson, D.L. DeAngelis, B.T. Milne, M.G. Turner, B. Zygmunt, S.W. Christensen, V.H. Dale, and R.L. Graham. 1988. Indices of landscape pattern. Landscape Ecology **2**: 63-69.

Preston, R.J. 1976. North American Trees, third edition. Ames, Iowa, Iowa State University Press, 398 pp.

Shugart, H.H. 1984. A Theory of Forest Dynamics: The Ecological Implications of Forest Succession Models. New York, Springer-Verlag, 278 pp.

Shugart, H.H. and D.C. West. 1977. Development of an Appalachian Forest Succession Model and its Application to Assessment of the Impact of the Chestnut Blight. Journal of Environmental Management **5**: 161-179.

Solomon, A.M., M.L. Tharp, D.C. West, G.E. Taylor, J.W. Webb, and J.L. Trimble. 1984. Response of Unmanaged Forests to CO_2-Induced Climate Change: Available Information, Initial Tests, and Data Requirements. Springfield, National Technical Information Service, 93 pp.

Urban, D.L. 1990. A versatile model to simulate forest pattern. A user's guide to ZELIG version 1.0. Charlottesville, Department of Environmental Sciences, University of Virginia, 108 pp.

DESIGN OF BUFFER ZONES FOR CONSERVATION AREAS AND A PROTOTYPE SPATIAL DECISION SUPPORT SYSTEM (SDSS)

Zhao Yang
Management Information System
City of West Palm Beach
P.O. Box 3366 200 2nd Street
West Palm Beach, FL 33402
(407) 659-8019

David M. Sharpe
Department of Geography
Southern Illinois University
4525 Faner Hall
Carbondale, IL 62901
(618) 536-3375

INTRODUCTION

The 'boundary area' issue and our approach

The issue of landuse conflict around the legal boundaries of conservation areas has significant effects on the species survive within the conservation areas.

Many researchers have been involved in this "boundary area" issue from various perspectives like economics (Machlis and Tichnell, 1987), law (Hiscock, 1986), sociology (Kellert, 1986; Lemons, 1987; Turner and Brooke, 1988), landscape ecology (Newmark, 1985; Schonewald-Cox and Bayless, 1986; Ambrose, 1987; Schonewald-Cox, 1988), and population ecology (Shaffer, 1981; Newmark, 1985). This research has prepared the ground to reach a powerful multi-disciplinary synthesis in the buffer zone design issue.

This research looked at this design problem from management per se. We consider that a buffer zone should be the product of a certain land management strategy adopted by land use managers on the conservation core area and its surrounding land, which can fulfill their conservation goal at the least economic cost because it is always the most acceptable solution to the general public. The buffer zone boundary and its spatial configuration are derived from such a management strategy and will change with it. If we can find out an optimal land use pattern to fulfill a certain conservation goal, the 'biotic boundary' in a management sense will be the spatial edge of such a land use configuration. In other words, a buffer zone design depends upon the conservation goals set by decision makers.

The following empirical research questions address this problem:

1) What effect does the initial spatial ecosystem configuration and their evolution history in a conservation core area and surrounding areas have on the setting management goals by decision-makers?

2) Is the management on the buffer zone a separate issue from the management of the conservation core area?

3) In what way could we response to the pattern change on the ecosystem configuration between core area and buffer zone during the planning period?

In order to answer above questions, a methodological question is:

4) How can a spatial management alternative range which could result in different spatial-temporal landuse

patterns required by a given management goal in the area be defined for a decision maker? The Spatial Decision Support System is the answer for this question.

The study area and indicator species

The study area chosen for this research is one of the cores and its surrounding areas for an international biosphere reserve located in the Land Between the Lakes (LBL) of Tennessee Valley Authority (TVA).

Land Between The Lakes has a unique and outstanding feature: 170,000 acres of forest land in one continuous ownership bounded on three sides by water. LBL is also designated as a National Recreation Area, indicating recreation has high priority. Hunting is also important. In addition, forest management is practiced in order to manage wildlife habitat, and for forest products. This mix of land use demands on LBL makes it necessary to coordinate land uses as much as possible, and to minimize conflicts inherent in the several uses, especially conflicts between timber production and development of wildlife habitat. Several core conservation areas have been identified in LBL. This research was conducted in one core area, Working Area 39, and its surroundings, Working Areas 38, 40, 41, 44. The study area is located on the Rushing Creek, Ky.-Tenn. 7.5 minute quadrangle map. The total study area is 44 square kilometers. Each Working Area is divided into a number of stands, each 10-50 ha in size. There are 343 stands in the study area. Most of the landscape is a series of narrow ridges with moderate to steep slopes and narrow valleys. The soils are gravelly, infertile, and generally not well suited to crops.

Among wide variety of wildlife species, the pileated woodpecker (Dryocopus pileatus) was given a special consideration because their habitat requirements fulfill the majority species' important habitat requirements in the area. The pileated woodpecker inhabits both coniferous and deciduous forests, but is restricted to areas containing mature, dense, productive stands (Bock and Lepthien 1975). The critical components of pileated woodpecker habitat are large snags, large trees, diseased trees, dense forest stands, and high snag densities (Bull 1975).

Spatial Decision Support System

Because of its complex, dynamic ecological and social-economic context with numerous qualitative variables, buffer zone design is a spatial multi-objective land management problem with a high degree of structure. The solution for such a problem requires a sophisticated technique. The recent development in Geographic Information Systems (Burrough, 1986) and Decision Support Systems (Densham and Rushton, 1988) opens a new and more efficient approach Spatial Decision Support Systems (SDSS) to explore this emerging spatial decision sub-field.

Although the difference in implementation of the system, the key design idea should be the same for any types of SDSSs: to develop a highly-interactive, computer-based system that integrates aggregated spatial choice models and spatial interaction model (Fotheringham and O'Kelly, 1989) to solve spatial-oriented, semi- or ill-structured management problems. Due to different missions, each type of SDSS should have a unique system structure. The main goal of the SDSS for the buffer zone design is to meet the needs of decision makers in dealing with this complex spatial-temporal management issue . Therefore, the structure of the SDSS for the problem should be unique.

RESEARCH PROCEDURES

In order to demonstrate the effects of spatial-temporal pattern of ecosystems on the buffer zone design, information is required on (1) initial spatial pattern of vegetation; (2) the potential change on the initial spatial pattern of vegetation; (3) ecological and economic evaluations on such spatial-temporal pattern change.

Base Map Digitizing and Study Resolution

The basic map scale for this research is 1:20,000 in order to cope with the raw data map scale, i.e., forest inventory maps which are based on 1:20,000 aerial photographs. A USGS 7.5 minute topographic map (1:24,000) - Rushing Creek, KY.-TENN. was used as the base map in this research. The boundaries of total 660 forest stands in this map sheet were digitized using pc-ARC/INFO. An attribute file for 343 forest stands which are composed of the research area (including 5 LBL work areas) was created as the basic relational data base for this research.

Vegetation modeling

This research has partially relied on the extensive empirical modeling of ('ground truth') vegetation from existing forest inventory data in the study area. The direct gradient analysis method was used in this study (Gauch, 1982). 150 field samples are analyzed to produce direct gradient 'models' that express each forest stand's vegetation structure and floristics as a function of other, easily obtained coverage in SDSS (such as elevation, topographic position, aspect, bedrock types, etc.).

This gradient-modeling procedure is utilized in conjunction with the digitized maps, often to provide floristic, structural and diversity details not available from the maps alone. The direct gradient analysis in the study is based on bedrock type which is highly related to top soil type, and elevation.

According to field observation, 480-500 feet elevation line in the area has a significant ecological meaning. All of the range top is above such elevation, underlain by Qtg gravel deposits, and is covered by Chestnut and Post Oak dominated community. The slopes in the study area are underlain by limestone deposited in different geological time. The slopes above 480-500 feet elevation are covered by Stable White Oak community which has stable species composition. The slopes below that line are covered by Successional White Oak Community which is dominated by white oak and sugar maple.

Spatial and Non-spatial Variables

A. Local Ecological Variables

In order to cope with the ecological evaluation of spatial pattern, most variables in this modeling process are typical forest inventory indices which can be used as variables in Habitat Suitability Index (HSI) model for the indicator species, in this research, Pileated Woodpecker. Additionally, some ecological variables have to be derived from such inventory data.

V1: basal area (square feet/acre);
V2: number of trees > 51 cm (20") dbh/0.4 ha (1.0 a);
V3: Number of trees > 5.08 cm (2") dbh/0.4 ha(1.0 a);
V4: snag density (number of snags/0.4 ha (1.0 acre));
V5: all stems average dbh (inches);
V6: average dbh of trees > 51 cm (20 inches) dbh;
V7: distance from water source;
V8: distance from the legal boundary of core area;

The values for variable V1 - V6 were obtained from the forest inventory data (1985-1990) provided by the Division of Forestry, TVA. The values for V7 and V8 were obtained by a series of spatial analyses using ARC/INFO.

B. Economic Variables

V9: board feet/acre;
V10: cords/acre;
V11: distance from main roads.

The values for V9 and V10 were obtained from TVA forest inventory data (1985-1990). The values for V11 were the results of a series of spatial analyses using ARC/INFO.

The distance of each stand from water source (rivers), which is a habitat requirement for nesting pileated woodpeckers (Conner et al. 1975). The stands in the study area were classified as less than, or greater than, 150 meters from water.

The final results of the above process are 11 ARC/INFO coverages which represent the initial spatial configuration of those important variables.

The procedures of modeling pattern change over time

(1) The simulation program

The current vegetation of each stand in the study area was assessed using the most recent available forest inventory data (1985-1990) from the Division of Forestry, TVA.

The simulation program used in this research is TWIGS which was developed by the North Central Forest Experiment Station, U.S.D.A., Forest Service. TWIGS is a stand growth-and-yield simulation program with management and economic components developed for use in the North Central United States. The program used in this research is the model for Central States.

(2) Variable Simulation and Temporal Database

A temporal relational database structure was created for this study area. There are two time indices in this temporal database. The real time, (or Langran's 'world time', Lum's 'logical time' 1990) in this temporal database was actual forest stand age. It was used to trace the real forest succession events and the database time, TWIGS-age, was used to trace the history of the forest succession stages. The structure of this temporal data base is designed as two relations linked with database time index: TWIGS-Age. Through running TWIGS, historical features of forest stands at certain TIWGS-Age, from open land to mature forest, are summarized into three temporal relation tables (TWIGS_AGE table) with TWIGS-Age as the primary key. Each table represents one of three vegetation communities in the study area. The values in these three relation tables are tested with previous research or field investigations during the summer and fall of 1990. Another temporal relational table (basic data base table) which includes the current forest succession data is created with TVA forest inventory data (1985-1990) and data updated by field investigation conducted during summer and fall of 1990. The 11 non-spatial and spatial variables identified above are used as attributes in this relation table with TWIGS-Age as the primary key.

In order to estimate the TWIGS age for each forest stand in the study area, a procedure written with relational algebra expresses was developed to locate the corresponding TWIGS-age for the tuples in basic data base table.

(3) Temporal Data Base Establishment

Once calibrated and tested, TWIGS-AGE Table was used in TWIGS to project the stand characteristics for a number of stand age classes, i.e., TWIGS-AGE, for each of the three community types. The TWIGS-AGE was divided into three classes, i.e., 40 - 60, 60 - 80 and 100 - 135 years old. The stand characteristics of each age class were entered into TWIGS-AGE Table. The entries in the TWIGS-AGE Table were compared with the current stand characteristics of each stand in the study area and a TWIGS-Age that corresponds to current stand structure was assigned to the stand. The whole process was done automatically in SDSS.

In this way, a current TWIGS-Age distribution in the area was revealed. Future stand structure of each stand in the study area was forecasted by referring its future age, e.g., 20 years into the future, to the entries at the next higher TWIGS-Age class in TWIGS-AGE Table. The detailed relation expressions used in SDSS are not described here.

Based on TWIGS-AGE Table and Basic Data Table, a series of relational tables about changes in forest succession stages were developed according a relational algebra expressions.

Ecological evaluation of the ecosystem spatial pattern

A consistent ecological rating system using TWIGS-Age was developed according to the forest community history in this particular study area. Except several remote stands, the forest communities in the area are fairly young with estimated age around 40 -60 years old. Consistent harvesting timbers gives the low basal area inventory in most stands.

Eight ecological variables (V1 - V8) were used to

evaluate the stands ecologically. Considering spatial ecological factors, the rating system has three-level structure: the first level has two categories: core and buffer. The real ecological meaning of the distance from the core area does exist. The second level has two categories: stands with distance within or beyond 150 meters from the rivers. Ecological preference was given to those stands within 150 meters distance from the rivers because they provide nesting sites for the pileated woodpecker. The third level uses other six ecological variables to classify the stands ecologically According to Conner and Adkisson's (1976) research, the best nesting sites have the mature forest with basal area more than 118 square feet/acre and density more than 192 stems/acre.

According to such ecological evaluation categories, the most ecologically favorable habitats (the first class habitats) are those stands that their basal area is more than 118 square feet/acre, their stem density is more than 192 stems/acre, their locations are within 150 meters from river and within core area.

Economic evaluation of certain spatial-temporal alternative

Given initial spatial pattern of ecosystems and its change in the study area, SDSS provides the capability for managers to evaluate the economic cost and possible way to reduce it by choosing suitable spatial management alternatives.

The value of three economic variables: V9 (board feet/acre); V10 (cords/acre); V11 (distance from major roads) were used as economic indices to evaluate the spatial-temporal pattern.

The estimation procedures for a given spatial management alternative are as follows:

1. To summarize the total values for V9 (current and projected board feet volume of sawtimber)) and V10 (current and projected pulpwood volume for poletimber) for a chosen spatial management alternative during a planning period.
2. To estimate the total transportation cost according to V11 - the distance from roads. Because the concrete figure for such cost was not available, the absolute distance from the major roads was used as a general economic indicator in this study.

The Spatial Decision Support System Components

In order to provide above functions for the SDSS at less cost, a micro-computer based system configuration was adopted. The configuration of hardware and software for the SDSS project currently is still evolving. pc-ARC/INFO software produced by ESRI, the basis of the GIS is installed on a Gateway 2000 386-microcomputer (Corolado). The simulation program, TWIGS, is installed on the same machine. SQL DS, which is used to perform relational operations in the SDSS system, is installed on a IBM 3390 mainframe which is connected to the 386- microcomputer with modem at speed of 9600 baud. The peripherals include a Calcomp 9600 digitizing tablet, a HP painjet color printer which is used to produce output maps from pc-ARC/INFO and reports from TWIGS.

The center-piece of the software system is pc-ARC/INFO. The forest stands inventory data provided by Forestry Division, TVA were downloaded to the 386-microcomputer. The processes which support the decision making for the buffer zone design require that data can flow between different software components at a high speed. The best way to increase such speed is to download data to the same computer on which the major software programs are residing.

Six PASCAL procedures were written as interfaces between SQL DS, TWIGS and pc-ARC/INFO in order to allow that data and results flow between different softwares.

EMPIRICAL RESULTS AND ANALYSIS

Scenarios used in the simulation of stands development

The values in the modified HSI variables were simulated using SDSS for three twenty-year periods (1990-2010, 2010-2030, 2030-2050). Two management strategies have been run for this analysis. The management goal of Management Strategy No.1 is to prohibit all the timber production in the entire area during the management cycle regardless of high economic cost. The management goal of Management Strategy No.2 is to develop the first class habitats in this ecologically poor area in the early stage of the planning cycle until the area of the first class habitats reach 50% of the entire area and to maintain the area of the first class habitats at this level afterwards. This strategy has four components: 1) no timber harvest in the core area throughout the planning period in order to allow development of class 1 habitat stands in the core area; 2) during the period 1990 to 2010, allow stand development throughout the surrounding area in order to build up the inventory of class 1 stands; 3) in 2010, harvest selected class 2 stands in the buffer zone that are greater than 1400 meters from the core area, emphasizing stands that will not be needed to supplement the class 1 stands; 4) in 2030, begin to harvest class 1 stands that are greater than 1400 meters from the core area boundary in order to reduce the total area (including the core area and buffer area) of class 1 stands toward the 50 percent target and allow for the harvesting of high quality timber products.

The spatial-temporal simulation results

The development of stands in each of three community types was simulated over twenty-year growth cycles using TWIGS. Each TWIGS-Age class has lower and upper bounds for stand density, basal area, average dbh of all stems and of stems > 51.8 cm, and timber volumes expressed in board feet (international 1/4 inch rule). Snag density and dbh are also considered. However, in this simulation, mortality for each TWIGS-Age was not recorded. The simulation of standing timber shows decreasing stand density with age, but increasing basal area and average dbh, and the development of large trees (>51.8 cm dbh) as the stand approaches a TWIGS-Age of 100 years.

The economic value of the stands, expressed as board-feet of sawtimber, increases as TWIGS-Age increases. Sawtimber volumes increase dramatically as stands mature, e.g., from 7400 board-feet per Ha at the beginning of the 80-100 year TWIGS-Age to over 36,000 board-feet at a TWIGS-Age of 135 years in successional white oak stands. Consequently, the value of stands increases with age in two mutually exclusive dimensions, as an economic timber resource and as habitat for species such as pileated woodpecker that favor mature forest stands.

Simulation of stand development for all stands in the study area in the absence of timber harvesting suggests how rapidly the stands will approach maturity and become Class 1 habitat for pileated woodpeckers. Currently, 2.5 percent of the area, out of a total of 1100 Ha, are Class 1. All of which is located outside the core zone. Within 20 years, in 2010, the area of Class 1 habitat will increase to 8160 Ha, a small proportion of which will be in the core area. However, the current (1990) age of the stands is such that many of the stands that have the potential for becoming Class 1 habitat will do so in the following management cycle. By 2030, 71 percent of the area would be in Class 1 habitat. The areas that have not attained Class 1 status are in valley bottoms, many of which are currently managed as open land, and some upland areas where stands develop Class 1 characteristics slowly. The trend continues for the following management cycle, after which 78.6 percent of the study area is Class 1 habitat for pileated woodpeckers.

This simulation suggests that significant areas of Class 1 habitat would develop over the next four decades, and that the economic value of the stands would increases sharply, as well.

If in this exercise we limit the need for Class 1 habitat to 50 percent, it is clear that there is an opportunity to combine protection of the habitat qualities of the core area with timber harvesting in the buffer zone. Spatially, the stands within core area are not treated as designated conservation area, i.e., the configuration of biosphere reserve is not created by this management strategy. Simulation of the second management strategy demonstrates one possible result of managing the area to develop Class 1 stands (and thus mature stands) to 50 percent of the study area and to develop core and buffer configuration for this biosphere reserve.

Management strategy No. 2 involves allowing Class 1 stands to develop until the desired area is achieved. Timber harvesting would not occur until 2010, when Class 2 stands would be harvested. Harvesting of Class 1 stands outside core area would begin in 2030 when the management goal of 50 percent of the study area in Class 1 stands would have been achieved. Timber harvesting would be restricted to outer portions of the buffer zone, greater than 1400 m from the core area boundary in this scenario. The impact of this management strategy is to reduce the area of Class 1 habitat in 2030 relative to the no-cut option. In 2030, there would be 2552 Ha of Class 1 habitat, which is 58 percent of the study area. Thus, harvesting has reduced the area of Class 1 habitat by 20 percent, nearly 600 Ha comparing with No.1 strategy. If the inner buffer zone, defined as buffer zone area within 1400 m of the core area, were allowed to be narrower, e.g., 1000 m, a greater area of Class 1 and Class 2 stands would have been eligible for harvesting, and the area in Class 1 habitat could have approached the 50 percent goal more closely. The impact of harvesting on the spatial distribution of Class 1 stands is shown by comparison of the two maps in 2030 . The core area undergoes the same trends in both options because timber harvesting is not allowed there. The inner buffer zone undergoes similar development, as well, because no harvesting has been allowed there during two management cycles. The stands in the outer buffer zone, however, are harvested as they reach maturity, and do not develop into Class 1 stands.

No.2 strategy provides timber for the area in entire planning period. Only spatial conflict is most of cutting locations are far away from the main road because the harvesting are only allowed in outer portion of the buffer zone.

DISCUSSION AND IMPLICATIONS

Management goals and landscape spatial-temporal patterns

The first conclusion of this study is that the landscape spatial configuration and its evolutionary stages are the first consideration of determining management goals. For instance, landscape configuration in this study area has following features: 1. The entire landscape, including core area and its surroundings, is an ecologically poor area. There is no Class 1 habitat in the core area and very few of them in the surrounding area. 2. The changing landform and soil types provide a gradient of different types of vegetation communities across the landscape. From the bottom of the valley to the range top, there is a gradient of open field, Successional White Oak Community, Stable White Oak Community and Post-Chestnut Oak Community. 3. Such landscape configuration provides different growth speed for different vegetation communities. Among them, Successional White Oak Community has the fastest speed and Post-Chestnut Oak Community has the lowest one.

Such landscape spatial-temporal configuration provides a starting point for managers to set up the management goal for the core and buffer areas. Politically, there is very insignificant conflicts between conservation and timber production at this time because of low timber quality of the forest in landscape. However, the low ecological quality also limited the conservation goal for the area. This is a moderate goal which was based on the unique characteristics of this area. If we set up the conservation goal to be that when 50% of the area reaches Class 2 Habitats, the timber cutting begins, the ultimate conservation and economic return will be much less than the current goal, although the cutting cycle will be much fast (about 10 years).

The interactive nature of management for core and buffer area

This analysis has shown how a buffer zone can be managed to meet management goals that relate to the core area. In fact, attainment of management objectives for a core area requires that a management plan for a buffer zone be developed so that the habitat in the buffer zone supplements that of the core. Due to the economic consideration, we can not set this entire area under no timber production option which would be to costly. Carefully managing core and buffer area will make the conservation goal more reachable because of the economic return provided by timber production.

Management of core area and buffer zone is an ongoing and dynamic process because of the interactive nature of management of core and buffer area. Manager must be able to respond to changes in the spatial structure of habitat in both core area and buffer zone. These changes may result from trends in stand species composition and size classes that are a consequence of stand conditions at the start of the management cycle. Management also must be able to respond to exigencies, e.g., unforeseen disturbances such as fire and windthrow, or changing timber demand.

The other aspect of this on-going process is that the succession look-up table has to be adjusted according to more field study and more advanced survey techniques. SDSS has to be run again and again in order to keep up with new ecological study in order to adjust the projection of stands development.

The spatial aspect of this dynamic feature provides a dynamic configuration in this biosphere reserve area. In this example, at the beginning of the planning period, a large buffer zone was set aside in order to give ecologically poor core area time to develop Class 1 Habitats. After 40 years restriction on timber cutting on both core area and buffer area, core area has fully developed (Class 1 Habitats has occupied 100% of core area) and 58% of study area wiil be covered by Class 1 habitats. Spatially, buffer zone boundary (the boundary of Class 1 Habitats in this case) has expanded 2500 meters from the legal boundary of core area. It has exceeded the original conservation goal for the entire area. Therefore, timber cutting has been introduced to maintain original conservation goal and to reduce the economic cost.

Considering 'spatial contagion' nature of the ecosystem configuration, the cutting will start from the outer area of the buffer zone, in this case, from Class 1 Habitats (stands) which are 1400 meters away from the legal boundary of the core area. The result of such cutting makes the buffer zone around the core area shrieked spatially. We can image that as long as the conservation goal is constant (e.g., the area of Class 1 Habitats or certain population size), the interactive nature of management for core and buffer area will make buffer zone boundary move outwards or inwards periodically. This nature will make the cooperation between different agencies who has the ownership to the buffer area more dynamic.

Although in this study, the entire area is under a single ownership, the buffer zone design issue in real world will equire intensive cooperation among different owners in the surrounding area.

Finally, the management decisions involved are too complicated to be resolved without a sophisticated database and analytic capability when any significant area is to be managed. A SDSS, a GIS coupled with a stand simulator such as TWIGS, to organize databases and provide a structure for exploring management options, is needed.

REFERENCES

Bull, E.L., 1975. Habitat utilization of the pileated woodpecker. Blue Mountains, Oregon, M.S. Thesis, Oregon State Univ., Corvallis. 58pp

Conner, R.N. and C.S. Adkisson, 1976. Discriminant function analysis: A possible aid in determining the impact of forest management on woodpecker nesting habitat, For. Sci. 22(2): 122-127

Dasmann, R.F., 1988. Biosphere Reserves, Buffers, and Boundaries. BioScience, 38, 487-489

Densham, P.J. and Rushton, G., 1988. Decision Support System for Locational Planning, In Behavioural Modelling in Geography

and Planning, edited by R.G. Golledge and H. Timmermans (London:Croom-Helm), pp.56-90
Gauch, H.G., 1982. Multivariate Analysis in Community Ecology, Cambridge University Press, Cambridge, London
Hiscock, J.W., 1986. Protecting National Park System Buffer Zones: Existing, Proposed and Suggested Authority. J. Energy Law Policy, 7, 35-94
Newmark, W.D., 1985. Legal and Biotic Boundaries of Western North American National Parks: A Problem of Congruence. Biol. Consv., 33, 197-208
Schonewald-Cox, C.M. & J.W. Bayless, 1986. The Boundary Model: A Geographical Analysis of Design and Conservation of Nature Reserves. Biol. Conserv., 38, 305-322
Schonewald-Cox, C.M., 1988. Boundaries in the Protection of Nature Reserves. BioScience, 38, 480-486
Schroeder, R.L., 1983. Habitat Suitability Index Models: Pileated Woodpecker, FWS/OBS-82/10.39. U.S. Dept. of the Interior, Fish and Wildlife Service. Washington, DC 20240

MODELLING SPATIAL PATTERNS OF DIGITAL ELEVATION ERRORS FOR DRAINAGE NETWORK ANALYSIS

Jay Lee and Peter K. Snyder
Department of Geography
The University of Georgia
Athens, Georgia 30602

INTRODUCTION

In recent decades, digital elevation models have been developed and provided to users for performing a wide variety of terrain analyses. The digital elevation model (DEM) can take the form of a regular matrix of elevations in which elevations are spaced evenly apart in two orthogonal directions. Published DEMs are usually supplied with reports of accuracy levels as a reference for users. For example, the accuracy of the DEMs as stated by the USGS is not greater than 7.0 meters Root Mean Squared Error (RMSE) (USGS 1987). But how realistic or accurate is it to perform terrain analysis functions from a DEM with a stated accuracy of 7.0 meters?

This paper examines how errors in a DEM impact on the accuracy of an extracted drainage network by simulating a controlled magnitude and spatial pattern of errors onto the surface of the DEM. By examining how the number of drainage network cells classified varies as a function of the standard deviation and spatial autocorrelation of the errors simulated onto the surface of the DEM, the overall impact of the errors on extraction of the drainage network can be inferred.

ALGORITHMS FOR EXTRACTING DRAINAGE NETWORKS

Earlier research on automated extraction of the drainage network was pioneered by Peucker and Douglas (1975) who analyzed topographic features of grid cells according to the patterns of elevation changes with neighbor cells. Later on, Mark (1983), O'Callaghan and Mark (1984), Band (1986), Morris and Heerdegen (1988), Jenson and Domingue (1988), and Hutchinson (1989) developed various algorithms for extracting drainage network based on raster DEMs. These algorithms performed to different degrees of success but all encountered problems with single cell pits in flat areas due to high signal-to-noise ratio (O'Callaghan and Mark 1984). In terms of the width of the channels, it is not until Tribe's work (1990) that most algorithms extracted drainage networks with width of one cell. Tribe's approach employed both elevation changes as well as slopes between grid cells to derive more realistic drainage networks. In a similar fashion, we devised a set of procedures using IDRISI, a raster based GIS, to determine which grid cells are part of drainage networks.

DATA AND METHODOLOGY

We used a USGS DEM which covers the area of Herbert Domain, TN and is located in the Appalachian Plateau Province. The spatial resolution of this DEM is 30 meters and covers an area of 9 km^2. The minimum elevation in the DEM is 474.6 meters and the maximum elevation is 569.0 meters.

Our procedures for assessing the magnitude and spatial patterns of elevation errors can be divided into three main stages: (1) pre-processing of the DEM, (2) running the error simulations, and (3) extraction of the drainage network.

The first stage, pre-processing of the DEM, involved converting the data to IDRISI file format, extracting a sub-image from the original DEM, and applying a low-pass filter twice to remove single-cell pits in the flat areas of the DEM. The first step extracted a sub-image of 100 X 100 cells from the original 7.5 minute quadrangle DEM. Conversion of the file from a grid structure to a sequential structure, as required by IDRISI, was the second step in the pre-processing of the DEM. The size of the sub-image was chosen because the simulation program is limited to a size not larger than 100 X 100 cells. The WINDOW program in IDRISI was used to extract the sub-image and asked for the top-left and bottom-right coordinates of the original DEM image. The third step in the pre-processing stage required that the image be filtered twice with a low-pass filter. A low-pass or smoothing filter was used to remove the pits or errors inherent in the creation of the DEM.

The second stage of the methodology is to simulate different magnitudes and spatial patterns of the errors onto the DEM surface. The simulation program[1] applies errors with a user-specified standard deviation (i.e., magnitude) and spatial autocorrelation (i.e., spatial pattern) of the errors. The process of generating the errors onto the DEM surface was developed by Goodchild (1980) and was adopted in Fisher (1990). First, for each cell in the DEM, a random number was generated with a mean of zero and a standard deviation defined by the user. Then, the spatial autocorrelation of the error matrix was calculated and stored for later comparison. Next, two cells were randomly identified, swapped, and the new spatial autocorrelation of the errors was calculated. If the new spatial autocorrelation was closer to the user-defined spatial autocorrelation, then the swap was retained, otherwise the two cells were swapped back and two other cells were tried. This process continued until the spatial autocorrelation of the errors minus the user-defined spatial autocorrelation was within a certain threshold. Last, the error matrix was added to the DEM and the simulation was complete. In the results presented here,

---

[1] The authors wish to thank Dr. Peter Fisher of Kent State University for his generosity in making his simulation program available for this study.

twenty simulations were run per test. In total, 320 runs were performed.

The third stage extracted the drainage network from the DEM through the use of several IDRISI programs. First, an examination of the original DEM and the elevation ranges was made to determine at what elevations the drainage channels are located at. Since the DEM of this area is located in a physiographic region characterized by flat plateaus and deeply incised river channels (Figure 1a), it is rather simple to assume that the drainage network can be defined by a certain elevation range. Therefore, the elevation range of the drainage network was found to occur between 0 meters and 505 meters. The elevation image was then reclassed, using RECLASS in IDRISI, from 0 meters to 505 meters as equal to 1, and 506 meters and up as equal to 0. Next, a slope image was created from the DEM image using the SURFACE program and reclassed as well from 0 degrees to 10 degrees as 1, and 11 degrees and up as equal to 0. The ranges for the slope map were chosen because it is assumed that the channel bottoms are flat areas (i.e., slope under 11 degrees). The third step was to overlay the reclassed elevation and slope maps using the MULTIPLY option in IDRISI's OVERLAY. Therefore, all cells where there was both a slope less than 11 degrees and an elevation between 0 meters and 505 meters was classified as a drainage network cell. The final step is to use the AREA program in IDRISI to count the number of cells classified as drainage network cells for analysis in the results.

RESULTS

In order for comparisons to be made between the simulations performed and the original DEM, the original DEM needed to be processed *via* the same IDRISI procedures for extracting the drainage network with the same criteria for RECLASS. This process was first carried out for the original DEM and the number of drainage network cells classified was found to be 1,137 out of 10,000 cells. Figure 1b shows the classified drainage cells without any error applied. The extracted drainage network seems to agree well with those displayed on the topographic map (Figure 1a). The procedures were then repeated for every simulated DEM. Two of the simulated results are shown in Figures 1c and 1d in which some of the drainage cells are mis-classified due to the added errors.

The results of the simulations with varying spatial autocorrelation values and a fixed standard deviation of 7.0 meters showed that the number of drainage cells classified increases as a function of the spatial autocorrelation of the errors. The results are listed in Figure 2a.

A linear regression model was applied to examine the relationship between drainage cells and spatial autocorrelation (Figure 2b). The regression line was found to be:

$$Y = 523.086 + 381.697X$$

where X represents the spatial autocorrelation, i.e., Moran's I (see

Goodchild 1986) and Y represents the number of drainage network cells classified. The r-square value for this model is 0.775.

A logarithmic transformation was applied to the total number of drainage network cells classified for all twenty simulation runs. The resulting linear model was found to be:

$$Y = 2.733 + 0.228X$$

where X represents the spatial autocorrelation and Y represents the logarithmic values of total number of drainage network cells classified. This model has an r-square of 0.814 which indicates a fairly strong relationship.

Upon examination of Figure 2b, the number of drainage network cells classified increases with an increase in the spatial autocorrelation of the errors. Therefore, the results of the extracted drainage network will be more reliable if the errors were more spatially autocorrelated (i.e., similar errors being located closer to each other). Alternatively, errors with a random distribution tend to have less of an effect on the reliability of the resulting drainage networks. Another observation should be pointed out here is that Figure 2b seems to suggest that there is a dramatic increase in the number of drainage network cells classified beyond $I = 0.5$.

The reason that the number of drainage network cells classified increases with an increase in spatial autocorrelation of the errors is thought to be a result of the aggregation of errors such that the likelihood that the errors will affect those cells making up the drainage network is lessened as the errors become more aggregated (i.e., higher spatial autocorrelation). Figure 2a lists the percentage of cells not classified as part of the drainage network (i.e., errors of omission). These values are achieved by taking the average number of cells classified and dividing it by the total number of cells classified in the original DEM image. This value (as a percentage) is then subtracted from 100 percent to give the total percentage of cells not classified. The average percent omission refers to the least number of cells that are not classified because some cells are mis-classified as part of the drainage network when in fact they are not. Figure 2c shows how with an increase in the spatial autocorrelation of the errors the number of cells not classified decreases to a low of 15.1% at $I = 0.9$.

The second set of simulations varied the standard deviation while holding the spatial autocorrelation constant at 0.0. The results of these simulations are listed in Figures 3a. Figure 3b shows a strong linear relationship between the average number of cells classified and the standard deviation of the errors. As the standard deviation of the errors increases, the number of drainage network cells classified decreases. A linear model was generate from this data and the resulting model is:

$$Y = 1241.750 - 94.227X$$

where X represents the standard deviation and Y represents the number of drainage network cells classified. The resulting r-square value is 0.986. Therefore, there appears to be a strong negative relationship between the magnitude of the errors in a random pattern and the average number of drainage network cells classified.

It is obvious that as the amount or magnitude of errors decreases, the more accurate or similar the DEM will be to the original DEM regardless of the spatial patterns of the errors. For instance, errors of one meter in standard deviation applied to the DEM are such a small amount that regardless of the spatial pattern, the DEM more closely approximates the original DEM. As the magnitude of the errors increases, the more the simulated DEM will change from the original DEM. Figure 3a lists the average percentages of drainage network cells not classified as a function of standard deviation. Figure 3c depicts a graph of this linear relationship. As the standard deviation increases from 1.0 meter up to 7.0 meters, the average percentage of drainage network cells not classified increases from 1.3% up to 49.2%. Although the simulated surface still underestimates the true number of drainage network cells, the smaller magnitude (i.e., standard deviation) affects the extraction procedures less.

CONCLUSIONS

Because of the time involved in manually mapping the drainage network from a topographic map, the DEM has been seen as an excellent new data source for automating this once tedious and specialized process. But how can one be sure that the results obtained from any of the algorithms for the extraction of the drainage network are accurate? The USGS published their DEMs with a stated RMSE of 7.0 meters, but how do those errors affect the extraction of the drainage network? This paper attempted to answer these questions by demonstrating how the accuracy of a DEM impacts a procedure for the extraction of a drainage network *via* simulating errors under controlled circumstances. By comparing the results of the simulated surfaces with that of the original, a better understanding of how the magnitude and spatial pattern of the errors affect the extraction procedure can be determined.

The results from the simulations performed in this paper have demonstrated that there is a strong quantifiable impact that the accuracy of a DEM has in extracting drainage networks. The impact of errors on the reliability of extracted drainage networks from DEMs has been substantiated throughout the study.

An understanding of how the accuracy of the procedure for the extraction of drainage networks changes as a function of the magnitude and spatial pattern of the error can give new insight into the prediction of how accurate a resulting drainage network will be for different levels of accuracy of a DEM. This ability to describe the level of accuracy of a particular terrain analysis function for various DEM accuracies may be valuable to organizations wishing to set standards for terrain analysis and

GIS operations. By determining at what level of accuracy of the results are required, the accuracy of the required data source (i.e., DEM) can be determined. This would alleviate the need for trial and error tests to be performed on DEMs with different accuracies.

REFERENCE

Band, L. E. 1986. Topographic Partition of Watersheds with Digital Elevation Models. Water Resources Research, **22(1)**:15-24.

Eastman, R. 1990. IDRISI:A Grid-Based Geographic Analysis System. Worcester, Mass.: Clark University.

Fisher, P. F. in press. 1990. First Experiments in Viewshed Uncertainty: The Accuracy of the Viewshed Area. Photogrammetric Engineering and Remote Sensing (draft).

Goodchild, M. F. 1980. Algorithm 9: Simulation of Autocorrelation for Aggregate Data. Environment and Planning A, **12**: 1073-1081.

Goodchild, M. F. 1986. Spatial Autocorrelation. **CATMOG 47.** Norwich, UK: Geo Books.

Hutchinson, M. F. 1989. A New Procedure for Gridding Elevation and Stream Line Data with Automatic Removal of Spurious Pits. Journal of Hydrology, **106**:211-232.

Jenson, S. K. and Domingue, J. O. 1988. Extracting Topographic Structure from Digital Elevation Data for Geographic Information System Analysis. Photogrammetric Engineering and Remote Sensing, **54(11)**:1593-1600.

Mark, D. M. 1983. Automated Detection of Drainage Networks from Digital Elevation Models. Proceedings of Auto-Carto 6, **Vol 2,** Ottowa, Ontario, Canada, pp. 288-298.

Morris, D. G. and Heerdegen, R. G. 1988. Automatically Derived Catchment Boundaries and Channel Networks and their Hydrological Applications. Geomorphology, **1**:131-141.

O'Callaghan, J. F. and Mark, D. M. 1984. The Extraction of Drainage Networks from Digital Elevation Data. Computer Vision, Graphics, and Image Processing, **28**:323-344.

Peucker, T. K. and Douglas, D. H. 1975. Detection of Surface-Specific Points by Local Parallel Processing of Discrete Terrain Elevation Data. Computer Graphics and Image Processing, **4**:375-387.

Tribe, A. 1990. Towards the Automated Recognition of Landforms (Valley Heads) from Digital Elevation Models. Proceedings of the 4th.

International Symposium on Spatial Data Handling, **Vol 1**, Zurich, Switzerland, pp. 45-52.

U.S. Geological Survey. 1987. Digital Elevation Models: U.S. Geological Survey Data User's Guide 5. Reston, VA: U.S. Geological Survey.

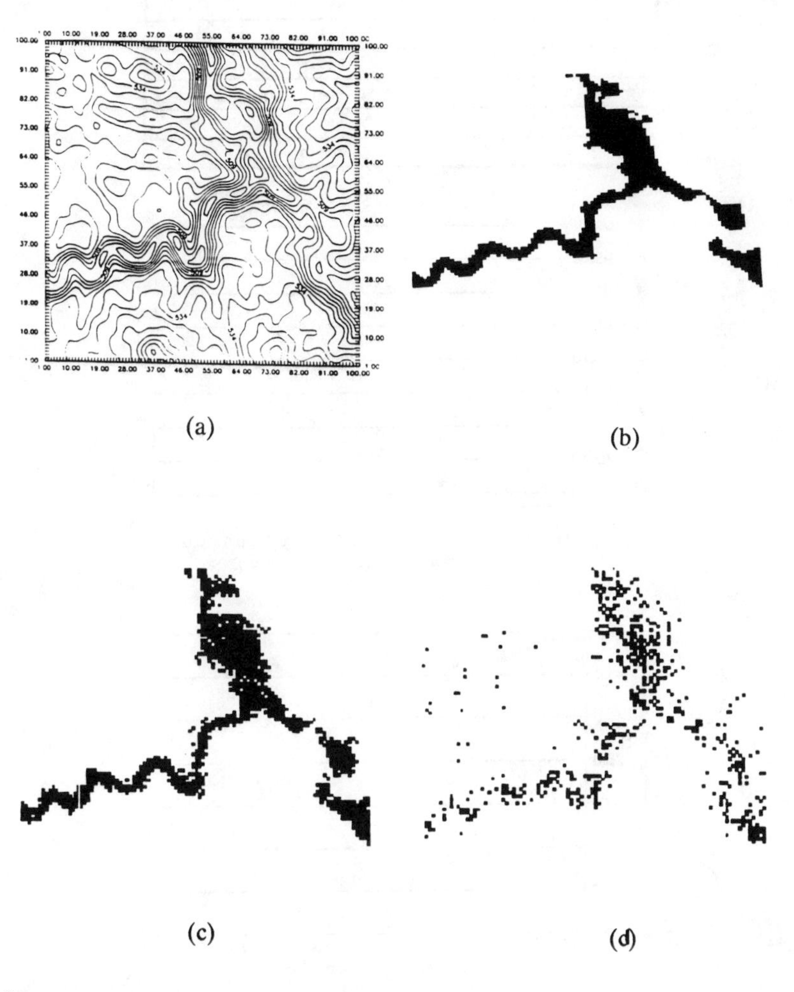

Figure 1: (a) Contour Map of Herbert Domain, TN
(b) Drainage Cells without Simulated Errors
(c) Drainage Cells with Errors of Spatial Correlation of 0.0 and Standard Deviation of 2.0 meters
(d) Drainage Cells with Errors of Spatial Correlation of 0.2 and Standard Deivation of 7.0 meters

(a)

TEST #	FILE SEED	TARGET S.A.	S.D.	AVG. NO. CELLS CLASSIFIED	AVG. OMISSION
1	A	0.9	7	965.5	15.1%
2	Q	0.8	7	832.1	26.8%
3	F	0.7	7	760.0	33.2%
4	P	0.6	7	701.4	38.3%
5	E	0.5	7	663.4	41.7%
6	O	0.4	7	648.8	42.9%
7	D	0.3	7	613.6	46.0%
8	N	0.2	7	604.7	46.8%
9	C	0.1	7	581.4	48.9%
10	B	0.0	7	577.8	49.2%

(b)

Avg. No. Drainage Cells Classified
1000.0
950.0
900.0
850.0
800.0
750.0
700.0
650.0
600.0
550.0
0.0 0.1 0.2 0.3 0.4 0.5 0.6 0.7 0.8 0.9 1.0
Spatial Autocorrelation (Moran's I)

(c)

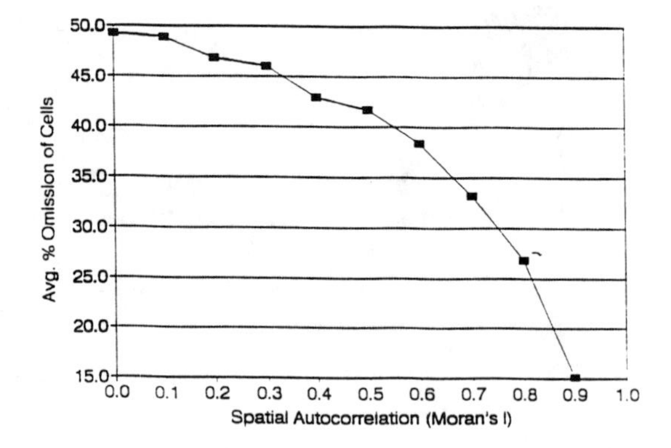

Figure 2:
(a) Results of Simulations Performed with Varying Spatial Autocorrelation Values and a Fixed Standard Deviation of 7.0 Meters
(b) Relationship between Average Number of Drainage Cells Classified and Spatial Autocorrelation
(c) Relationship between Average Percentages of Omission of Cells and Spatial Autocorrelation

(a)

TEST #	FILE SEED	TARGET S.A.	S.D.	AVG. NO. CELLS CLASSIFIED	AVG. OMISSION
10	B	0.0	7	577.8	49.2%
11	M	0.0	6	667.7	41.3%
12	H	0.0	5	766.2	32.6%
13	L	0.0	4	876.0	23.0%
14	I	0.0	3	981.4	13.7%
15	K	0.0	2	1062.1	6.6%
16	J	0.0	1	1122.1	1.3%

(b)

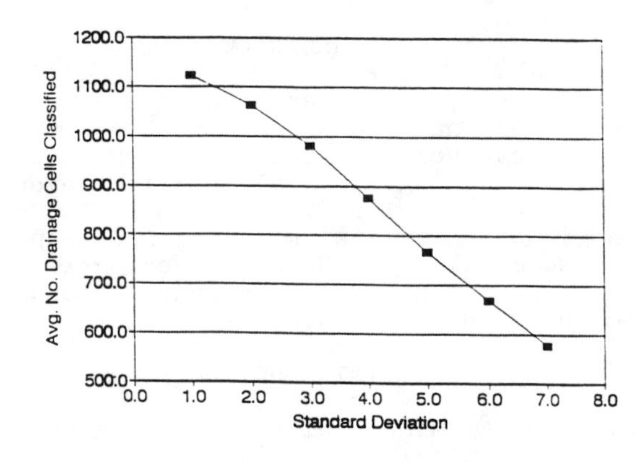

(c)

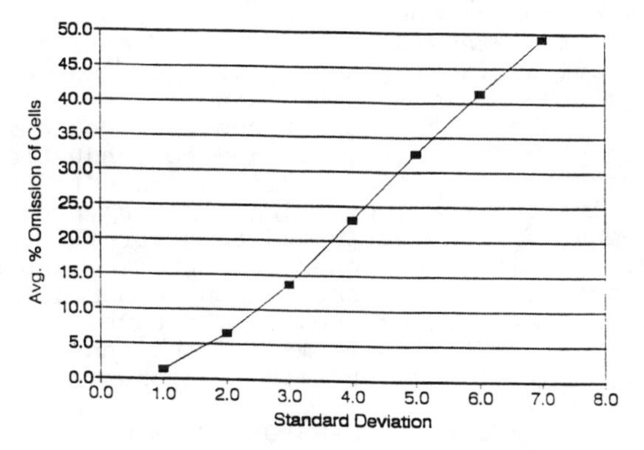

Figure 3 (a) Results of Simulations Performed with Varying Standard Deviation Values and a Fixed Spatial Autocorrelation of 0.0
(b) Relationship between Average Number of Drainage Cells Classified and Standard Deviation
(c) Relationship between Average Percentages of Omission of Cells and Standard Deviation

BUILDING A REGIONAL TRANSPORTATION MODEL WITH GIS SOFTWARE

William Barrett
Transportation Planner, Traffic Forecasts Section,
Minnesota Department of Transportation
395 John Ireland Boulevard, Room 820
St. Paul, MN 55155
(612) 296-1600   FAX # (612) 297-3160

ABSTRACT

Mn/DOT and the Metropolitan Council are currently updating their Regional Transportation Planning Model of the Seven-County Metropolitan Area. The regional model update consists of conducting a Travel Behavior Inventory, redefining the region's Traffic Assignment Zones and rebuilding the Regional Road Network file. This paper introduces the work flow model which Mn/DOT is using to integrate the several different data sources of the Regional Model update into one planning system.

INTRODUCTION

The seven-county Minneapolis/St. Paul area is the major population and employment center in Minnesota. The Twin Cities are home to about 2.5 million people and over 1.3 million of the state's employment. The existing Regional Transportation Planning Model has been basically unchanged since the late 1960s when it was first implemented. Mn/DOT, together with the Metropolitan Council's transportation planning staff, decided to conduct a complete Travel Behavior Inventory (TBI) in order to update, adjust and redesign the transportation planning process and its related models. The estimated cost for this is about $1.8 million. A selection process was used to choose a consultant team to conduct the TBI and other data-gathering activities. Mn/DOT has the task of completely redefining the region's Traffic Assignment Zones and the underlying Regional Road Network for both 1990 and the future.

REGIONAL TRANSPORTATION MODELS

What is a regional transportation planning model and why do we have one? A regional transportation planning model (RTPM) is actually a set of models that are used to inventory and then forecast a region's population, employment, income, housing and the demand for automobile and transit in a region.

To build a RTPM, the region is divided into smaller geographical units called Traffic Assignment Zones (TAZs). A TAZ is used in the modeling process to produce or attract vehicle/transit trips. For example, a TAZ in the central business district or a major retail shopping center will attract trips based on equations associated with the type of employment in the TAZ. On the other hand, a TAZ in the suburbs will produce vehicle/transit trips based on equations associated with population, income, housing and automobile availability. These productions and attractions are assigned to a computer road network which uses the attributes associated with that network such as speed, capacity and volume to find a path from the origin to the destination TAZ.

In the Minneapolis/St. Paul area regional model, there are 1165 TAZs and 35 external stations. An external station is a point at the edge of the region where automobile/transit traffic enters the model area but no socioeconomic data is available to model these trips. Instead, roadside interviews are conducted in which the vehicle is pulled over to the side of the road and the driver interviewed to determine the origin, destination and purpose of the trip. In modeling, the purpose of the trip is important because it affects trip behavior. For example, shopping trips which include food-clothing-miscellaneous tend to be much shorter trips than the work trip. To model these differences requires that separate categories be used in the modeling process.

In summary, a regional model is actually a set of social, economic and behavioral models which route vehicle/transit traffic between production/attraction TAZs, by purpose, via a computerized road network that attempts to simulate real world driving conditions.

BENEFITS OF REGIONAL MODELS

The RTPM of the Minneapolis/St. Paul area is used for many purposes and by a diverse group of individuals and organizations. The primary purpose of the RTPM is to forecast the demand for travel 20 years into the future to allow Mn/DOT to plan road upgrades and improvements for the highway system. In doing these forecasts, Mn/DOT and the Metropolitan Council analyze different scenarios of highway networks and socio-economic growth forecasts. Scenario testing is probably the most common benefit of regional planning.

The Metropolitan Council uses the RTPM as a guide when it works with the cities and counties in the region to determine the location and future growth of population and employment. The Regional Transit Board uses the RTPM to analyze and predict transit use in the region. Some cities and counties use the regional model to build their own model. A city or county planner can analyze the impact of different scenarios on their own highway system. The Pollution Control Agency has used our forecasts to analyze vehicle miles of travel (VMT) to estimate the amount of pollution that will occur in the future. Consultants and developers use the regional model as input to site location analysis and other analyses like highway corridor planning.

CREATING THE REGIONAL TAZ SYSTEM

It was understood from the beginning that our update of the RTPM would coincide with the 1990 Census. The first step was to create a map of the 1990 census tracts so we could evaluate a TAZ system based on the census tracts. The census tract map was created using Intergraph's PC-Microstation product on an IBM Model 80 PC with a 386 processor. The base maps are the 1:24000 street series maps of the seven-county area digitized by Mn/DOT's Cartographic Section. In effect, we borrowed the digitized line work to create our own 1:24000 1990 census tract map. The census tracts were overlaid onto the 50-street series maps and plotted at the 1:24000 scale.

This procedure quickly pointed out that census tracts could not be used as TAZs. Major roads including the Interstate Highway System (I-35, I-394 and I-94 in the Twin Cities) split many census tracts right down the middle. Census staff assured us that they could give us socio-economic data by TAZ as long as we did not cut census blocks into parts. As a result, we decided to use census block boundaries as our lowest common denominator and to include the Regional Road Network, geographical boundaries, political boundaries and land use considerations into determining the TAZ system.

Mn/DOT then modified the existing census tract map into TAZ90, a regional TAZ system with 1165 internal zones and 35 external stations. The 1990 TAZ system differs significantly from the 1970 system. In general, the interior of the metropolitan area lost TAZs. For example, Minneapolis dropped from 225 to 120 TAZs and St. Paul showed similar losses. These losses were recouped by suburban communities like Eden Prairie which increased from 7 to 35 TAZs.

To evaluate the regional TAZ system, another 50-series of maps was plotted and the Metropolitan Council, the seven counties and all the larger cities were invited to review their respective zone systems. Numerous changes were suggested and adopted at this time. Finally, on April 1, 1990 the first version of TAZ90 was plotted and distributed to everyone involved in the RTPM update.

THE GRAPHICS ENVIRONMENT AND BUILDING THE REGIONAL ROAD NETWORK

With the zone system in place, Mn/DOT turned its attention to building NET90, the computer map and link card set that simulates the 1990 road network that people can drive on in the Twin Cities area.

Once again we turned to the cartographic street series maps and wrote computer programs in EDG (Edit Design Graphics) to strip the line work representing the Interstate and State Highways, County State Aid, and County Roads and merge them into one file representing the seven-county area Regional Road Network (RRN).

Another 50-series of plots was created and the volume of vehicle traffic on the road, a minimum of 1,000 cars per day, was used to eliminate or add roads to the RRN. Our original programming did not bring city streets into the RRN. Many city streets carry volumes high enough for them to be included in the RRN and many highways in the rural areas do not carry enough volume to be included in the RRN. A manual review of the RRN was used to add roads to and subtract roads from the RRN based on volume of traffic per day. Also, the centroids were manually added and centroid connectors were digitized to connect the centroids to the RRN.

THE INTRODUCTION OF GIS SOFTWARE

We introduced GIS software into our system at this time. The software chosen was Intergraph's MGE (Modular GIS Environment), MGA (MGE GIS Analyst), and IRAS32 (Raster Data) software modules to run on an InterPro 6040 workstation with ORACLE as the relational database. We chose ORACLE because Intergraph had announced it will support ORACLE and raster data on the PC in the future and we wanted to continue to use the IBM model 80's we had purchased in 1988.

The MGE module had an application we had tested successfully, the make intersection program. This program automatically creates intersections of all the line work. But this also created two problems: one, we had intersections of line work that were not really intersections, for example, bridges; and two, transportation planning software allows only one vector to represent all the attributes occurring between two intersections. Mn/DOT did not digitize the roads from intersection to intersection, consequently more than one vector could occur between intersections.

To solve the first problem, the bridge access problem, required a manual phase where we reviewed the RRN and added freeway ramps and coded the network to represent the actual movements that were possible at each intersection. All of the coding by design was then set aside on what we call "protected levels", levels in the computer design file that allowed us to automate the solution to the second problem -- having one and only one vector from intersection to intersection.

The second problem was solved by writing a computer program in UCM language. The UCM language is a MACRO programming language that is provided with the Intergraph software. Essentially, the program looks at each vector in the file to determine if one end of the vector is at an intersection, that is, if more than two vectors exist at an endpoint. When an endpoint at an intersection is found, the other endpoint of the vector is examined to see if more than one vector exists at that point. If so, this one vector represents an intersection to intersection movement and the next element in the file is examined. But, if only one other vector is found at the endpoint opposite the intersection of vectors, the vertices are written to the database and the program continues until an intersection of vectors is found. Once an intersection is found, the multiple vectors are rewritten to the file as one vector using the vertices written to the database. The program continues until all vectors in the file have been examined. No manual intervention is necessary.

The program is written this way to solve the problem of which direction the vectors were originally digitized. By starting at the endpoint that we know is at an intersection we know which direction the vertices should be written to the file. The GIS software makes this possible after the make intersection program runs. The vectors share a common coordinate as an endpoint.

In the graphics file at this point, each vector represented an intersection to intersection movement. Special coding by design existed where intersections of line work were not intersections or where access needed to be limited. Also, the centroids were numbered and connected to the RRN. All we were missing were numbers at the unique intersections. Another program was written in the UCM language to do just that -- number intersections. Essentially, the program reads each vector in the graphics file, and looks at its endpoints for a number; if no number exists, the program increments by one and places the number at the intersection. This continues until all elements in the file have been examined.

Once the automatic numbering program is finished, the work in the graphics environment is completed. The centroids are in and connected to the RRN, and all vectors represent one and only one segment of a road between intersections. All the intersections have a unique node number assigned to them.

THE DATABASE

The work flow model uses ORACLE, a relational database, in partnership with the MGE/MGA software. ORACLE uses a SCHEMA to know which tables to associate with the maps in the RTPM project. The maps are TAZ90 and NET90. The three tables are LINKDAT, NODEDAT and TAZDAT. LINKDAT is associated with the NET90 and holds the road attribute information such as a_node, b_node, distance, speed and capacity. NODEDAT is associated with NET90 and holds the node numbers and their X, Y coordinates. TAZDAT is associated with TAZ90 and holds the population, housing, employment and behavioral data (from the TBI) for the TAZs. In terms of GIS, LINKDAT is the line table, NODEDAT is the point table, and TAZDAT is the polygon table. Lines, points and polygons are the three primary elements in a GIS environment.

The MGE software provided a straightforward way to develop the link, node and TAZ information from the graphic elements of the graphics file and store it in the ORACLE tables. In MGE, you make features out of graphic elements and tie each feature to a record in an ORACLE table. Our features are centroids, centroid connectors, interstates, river bridges, state highways, county roads, county state aid highways, ramps, city streets, external connectors, nodes, external stations, TAZ boundaries and TAZ centroids. Each feature points to one of the three tables, LINKDAT, NODEDAT or TAZDAT.

Once the graphic element is a feature and has a blank database record attached to it our custom UCM programs create the data from the graphic elements. For example, NODE.UCM examines each centroid and node number in the file and writes its X, Y coordinate and node number into NODEDAT. LINK.UCM examines each line feature (interstate, state highway, etc.), finds the numbers at its endpoint to make a_node, b_node, calculates its distance and writes all this information into the LINKDAT table.

In this sequence, the NODEDAT table is populated automatically; no other data is needed because we have the node number and its X, Y coordinate. On the other hand, the LINKDAT table has some information programmed into it (a_node, b_node and distance) but is missing other data like speed, capacity and number of lanes. We can populate the rest of each record in the table using both automatic input and manual input. For example, we use automatic input for the centroid connectors information because the speed, capacity and other information is the same for all connectors. But, the interstate links are built manually. For the manual input, we made a 50-series of plots with the node numbers on the map. The database records are transferred from ORACLE to RBASE, a different database package, which runs on our Local Area Network (LAN). Then anyone with a LAN connection can use the plots to populate the other attribute information in LINKDAT, such as one-way roads, Average Annual Daily Traffic (AADT), facility type, speed, capacity, etc. The TAZDAT table will be populated automatically once the Metropolitan Council, Department of Jobs and Training and the consultants finish their work on the TBI.

All of the previous discussion is about developing the attribute information for the links, nodes and TAZs that the transportation planning package needs to model vehicle/transit traffic. But we had to develop programs to move the results of the transportation planning packages PLANPAC, TRANPLAN and EMME2 back into ORACLE. To accomplish this, Mn/DOT personnel have written FORTRAN programs which reformat the standard output files of PLANPAC, etc., into the a_node, b_node format that ORACLE requires. These programs allow us to import the results of the modeling process into the GIS software.

THE PILOT PROJECTS

Three pilot projects have been developed to test and define the various software packages MGE/MGA, TRANPLAN, PLANPAC, EMME2 and our own UCM and FORTRAN programs. In our first pilot project, we developed TAZs and a road network just south of the Minneapolis CBD. Using all of our programs and MGE/MGA we populated 19 TAZs and 161 link cards with real data. The primary purpose of this work was to enable TRANPLAN to rebuild the pilot project network using its NEDS (Network Editing and Display System) software. TRANPLAN successfully rebuilt our network and this enabled us to test minimum time paths with NEDS.

But we also wanted to test other TRANPLAN models such as the gravity model and the trip generation/trip distribution models. We used our Waseca Bypass Study to accomplish these tests. During the summer of 1990, Mn/DOT personnel collected speed, capacity and routing data in the Waseca, Minnesota area. In addition, we conducted a license plate study and interviewed business owners. The purpose of all of this data collection was to model a north and south bypass of the area.

Using the 1:100,000 USGS quad maps, the custom programs we had written and MGE/MGA, we recreated a Waseca network and trip table in about two days. We then successfully tested the other TRANPLAN models.

The most important pilot project is Anoka County. Anoka County, part of the RTPM, has 124 TAZs, six external stations and 885 links in the RRN. Anoka County is important because we used it to write FORTRAN programs which pass information between PLANPAC/TRANPLAN and the GIS software. For example, PLANPAC/TRANPLAN have standard reports for path building. A path is the set of links used to travel from a production to an attraction zone. But the standard output file of PLANPAC/TRANPLAN is not readable by ORACLE. We have successfully written FORTRAN programs which reformat the standard output into a_node, b_node pairings along with path information that ORACLE can read. This allows us to display the path set in the MGE/MGA environment. (It should be noted that the creators of TRANPLAN, the Urban Analysis Group, are changing their software to allow for ASCII file transfers. This will eliminate the need to write special FORTRAN programs to move information between TRANPLAN and the GIS software.)

SPATIAL ANALYSIS AND THE IRAS32 SOFTWARE

The pilot project just south of Minneapolis is being used to evaluate the use of aerial photography as a tool in transportation planning. The IRAS32 software allows aerial photographs to be scanned into a raster data file. This raster file can then be registered to a state plane or other coordinate system. This allows the analyst to overlay the aerial photograph, a raster data set, onto the vector data set of the MGE/MGA software. For example, by doing a query in our topological pilot project the analyst determines that zone 11 is void of population. The analyst quickly locates this area, attaches the aerial photograph and reviews the results of the query. A plot of the result can also be used to review the query.

In another test of the MGE/MGA/IRAS32 products, the analyst may want to conduct a corridor analysis of Interstate 35W. Since the vector data are topologically structured, TAZs that are adjacent to I-35W (or one mile away) can be easily found. Again, a query is made, and the aerial photograph is attached to the area for the analyst to review the contents of the TAZs which are adjacent to Interstate 35W.

As part of the RTPM update, Markhurd Corporation did a seven-county fly-by which consists of 231 aerial photographs. We will scan these photographs and register them to the state plane coordinate system so that we can use them to check our other data sources for accuracy.

CONCLUSIONS

Several conclusions are possible at this time. Arguably, a work flow model based on GIS software is too complex for the average person to understand. Computer programming has been used extensively to piece the system together. This is a problem because considerable training is required to understand and use the system at its full capacity.

A potential problem is money. This is an expensive system. Mn/DOT and the Metropolitan Council will spend $1.8 million on the TBI. The MGE/MGA/IRAS32 software, riding on the 6040 InterPro workstation with ORACLE, is approximately $60,000 of the total budget.

On the positive side, the work flow model uses cartographic maps as its base and it will always be compatible with these base maps. This allows the system to communicate with others who do not understand the planning process. The ability to communicate through a recognized cartographic product is a definite strength.

Also, introducing spatial analysis into the transportation planning process is a major improvement in our ability to analyze regional systems. This automation of a heretofore manual process has the potential to radically improve our understanding of the transportation planning process. This is a major strength of this system.

The RTPM update integrates many sources of information into one GIS system. This makes the data more accurate because the integrity of the data has been systematically analyzed and constructed all through the process. Overall, this improves the model's output, making it an essential benefit to all users of the system.

Finally, this system delivers what it was originally designed to do: rebuild the RTPM. Ultimately, the RTPM will provide better planning information to other sections of Mn/DOT, other government agencies and local governments in the Minneapolis/St. Paul area.

REFERENCES

(1) **Calibration and Adjustment of System Planning Models,** Dane Ismart, Federal Highway Administration, December, 1990.

(2) **Characteristics of Urban Transportation Demand,** U.S. Department of Transportation, UMTA, July, 1988.

(3) **Calibrating and Testing a Gravity Model for Any Size Urban Area,** U.S. Department of Transportation, Federal Highway Administration, August, 1983.

(4) **Programming with User Commands,** Mach N. Dinh-Vu, Pen and Brush Publishers, Melbourne, Australia, 1990.

(5) **Microstation GIS Environment (MGE) Reference Manual,** Intergraph Corporation, Huntsville, AL, 1989.

(6) **Microstation Analyst (MGA) Reference Manual,** Intergraph Corporation, Huntsville, AL, 1989.

(7) **Microstation 32 Reference Guide,** Intergraph Corporation, Huntsville, AL, 1989.

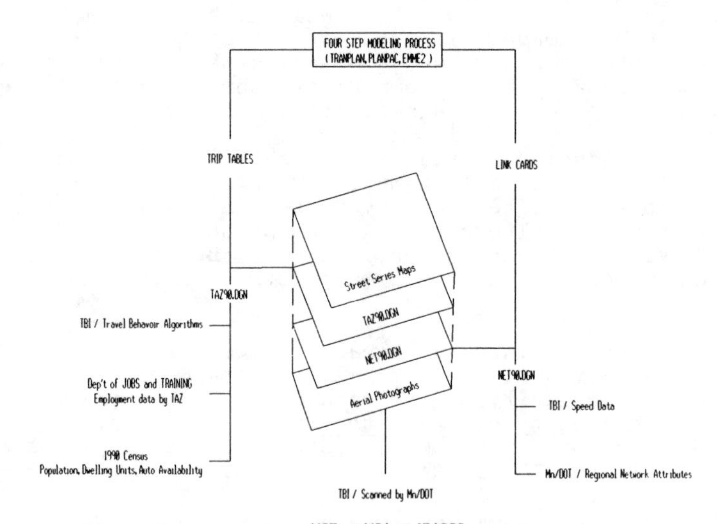

MGE / MGA / IRAS32

WORK FLOW MODEL

FOR REGIONAL TRANSPORTATION MODEL UPDATE

MINNESOTA DEPARTMENT OF TRANSPORTATION

Appendix A

IDENTIFYING POTENTIAL SITES FOR DEVELOPMENT OF HOSPITAL FACILITIES UTILIZING GEOGRAPHIC INFORMATION SYSTEMS TECHNOLOGY

Allen P. Marks, Ph.D.
Department of Finance and Real Estate
Kogod College of Business Administration
The American University
Washington, D.C. 20016

ABSTRACT

Determination of hospital bed need and optimal site location can be viewed as a special case of retail market analysis. The problem facing both health planners and hospital providers is to determine the appropriate number and location of hospital beds and services.

A geographic information system model was developed in order to determine the need for additional hospital facilities within the State of Maryland. The model was based on the following criteria: facility size, distance to other facilities and metropolitan areas, percent of the population over age 65, existing land use, site availability and percent slope, and the availability of existing infrastructure such as water and sewer.

A weighted Voronoi technique, based on hospital bed size, was used to determine the service area for existing hospitals (supply). Potential mapping was used to determine the future need for hospital facilities (demand). Once the gap between demand and supply was identified, potential sites were evaluated based on land use characteristics and the availability of infrastructure.

REAL ESTATE ACQUISITION DECISIONS WITH GIS: RANKING PROPERTY FOR PURCHASE WITH AN EXAMPLE FROM FLORIDA'S ST. JOHNS WATER MANAGEMENT DISTRICT

by

J. William McCartney, Ph.D.
Baskerville-Donovan
316 South Baylen Street, Suite 300
PO Box 13370, Pensacola Florida 32591
(Phone: 904-438-9661. Fax: 904-433-6761)

and

Grant Ian Thrall, Ph.D.
Homer Hoyt Institute
and
University of Florida
3121 Turlington Hall, Gainesville Florida 32611
(Phone: 904-392-0494. Fax: 904-392-3584)

ABSTRACT

Making a decision on property acquisition is complex when only one decision maker is involved. When more than one decision maker must approve a purchase, land acquisition becomes a complex compromise between interest groups and strongly held preferences. For a GIS to be successful in assisting which property to acquire, a procedure to generate the required compromise must be built into the design of the GIS analysis. A general method that accomplishes this is presented; it uses the Delphi strategy of consensus building. Criteria is created for ranking land for the purpose of acquisition. Land use data from Florida's St. Johns Water Management District is used for the illustration.

INTRODUCTION

An appropriate decision strategy built into the design of a GIS can improve real estate decisions of private corporations where a consensus among management or stockholders must be obtained. Examples of firms in the private sector that would benefit from a decision strategy linked to a GIS include rail roads, retail chains and franchises, and land investment and land development corporations. The same procedure can improve real estate decisions made by the public sector as well. Examples of public programs that can benefit from formally identifying a decision strategy and operationalizing that strategy in a GIS includes local governments acquiring land for a comprehensive public open-space and parks system, and state governments acquiring land to meet some environmental or water management goals.

The Delphi procedure is used to derive criteria and their weights for the evaluation of land to be purchased. The criteria and weights are then programmed into a GIS which in turn is operationalized on an appropriate data base; a choropleth map developed by composite scoring is used to display the consensus order in which lands should be acquired. Twelve Delphi-generated land selection criteria are defined and mapped; the lower Oklawaha River Basin of Florida's St. Johns Water Management District is used as the study area.

From its acquisition from Spain in 1821 until the early 1970s, the policy of the State of Florida was that the waters of the state were a hindrance to its optimum growth and development. Human modification of the state's hydrologic character was an acceptable activity if not a desirable practice. The 150-year period from 1821 to 1971 was filled with a multitude of schemes and projects to structurally alter the natural hydrologic conditions of the state.

In his October 1980 address to the fifth annual meeting of the Northwest Florida Water Management District, Governor Bob Graham stated that "Floridians can no longer afford the all-out dig it, drain it approach to development. We've played all the tricks on Mother Nature that we can get away with." The next spring the Florida legislature, at the governor's direction, enacted the nation's first comprehensive non-structural water management program -- Save Our Rivers.

Save Our Rivers (SOR) established the Water Management Lands Trust Fund (Section 373.59, Florida Statutes) and provided an initial $35,000,000 annually to the state's five water management districts for land purchases. The 1990 appropriation for land purchases exceeds $50,000,000. To date the districts have acquired over 500,000 acres of land; this figure is expected to increase to over 1,000,000 by the turn of the century.

Specific acquisition priorities for the annual purchase of water resource-related lands is the responsibility of each district's governing board. Each board was charged with formulating its own land acquisition goals, objectives, and strategies, including land selection criteria and methodologies. As the SOR program becomes more established and as its visibility as a nationally significant environmental land protection program is increased, not only is a district called upon to justify the selection of its purchases, it must also justify why other lands are not selected.

The following will present an objective computer-assisted decision strategy (CADS) to identify land acquisition priorities. Development of the CADS was assisted by the St. Johns Water Management District. Water Management Districts throughout Florida are considering adopting the CADS presented here, with some modifications to fit their location conditions; however, it should be emphasized that as of this writing, no Water Management District has adopted our approach to prioritizing land for acquisition.

BACKGROUND

A variety of techniques have been used for the evaluation and ranking of land. Related to the methodology presented here is (1) the **manual overlay** technique; (2) **matrix analysis**; (3) **GIS/CAM**; and (4) **GIS/MCS**.

The **manual overlay** technique uses a base map and other geographically referenced or keyed overlays (i.e., land use, drainage, and slope). Thrall (1979) employed an overlay technique to denote inequities in property appraisal techniques.

Matrix analysis has a list of criteria on one axis and values of occurrences on the other. It has been used to value, rate or rank land areas. Thrall, et al. (1988) proposed a ranking scheme utilizing matrix analysis to prioritize land for acquisition by public agencies. This study uses a basic component value, component weight to establish relative importance, and a computer-assisted algorithm to establish rankings; their objective process for green-space acquisition in the City of Gainesville and Alachua County, Florida, relies upon fifteen weighted criteria. The result is an orderly and rational technique for public real estate acquisition. Other matrix analysis studies includes that by Dyer et al. (1982). These efforts require the prior identification of parcels to be evaluated. The preselection process thus allows for the exclusion of sites that may, once evaluated, score high among alternatives. Also, Hepner (1984) notes that the lack of a scaling procedure in the matrix analyses results in valuing some data (criteria) beyond their true importance.

The **GIS** approach can be partitioned into two methods: (a) computer aided mapping (CAM) and multioverlay composite scoring (MCS). **GIS/CAM** is automated cartographic techniques that allows for general analysis and evaluation similar to any static map. Applications of CAM are decreasing (Dangermond, 1988). CAM lacks the sophistication of GIS. **GIS/MCS** allows for an entire area to be evaluated thereby eliminating the potential problem of not identifying all appropriate parcels. Marks et al. (1991) use GIS to improve the retail site selection decision. The literature also provides several examples of GIS suitability/capability assessments. Dangermond and Freedman (1984) cite planning applications for environmental suitability analysis. Lyle and Stutz (1983) develop methods for suitability scaling for land-use decisions which are influenced by environmental concerns.

The methodology presented below integrates the above techniques and concepts. The methodology introduces a criteria development methodology into a GIS design and analysis. The resulting methodology is applied to the problem of water resource management; the problem can also be viewed as a public real estate issue. The methodology is an extension of the earlier work of Thrall et al. (1988), and their Delphi criteria identification for land acquisition. Unlike the work of Thrall et al. (1988) where individual parcels are submitted for consideration, GIS technology allows all lands in the database to be evaluated and ranked. The database can in theory include all parcels in the land acquisition universe.

APPLICATION OF THE DELPHI PROCESS TO DEVELOP LAND ACQUISITION CRITERIA

A general discussion of the criteria for the selection of participants for a Delphi, and photographs of the actual questionnaires mailed to participants of this Delphi presented here, is in Thrall and McCartney (1991). For a more detailed account see McCartney (1991). The Delphi was comprised of three rounds of questionnaires. There was an 85% response rate to the questionnaires. The membership of the final Delphi round can be divided into eight categories: public environmental agencies - 10 members; private conservation association - 9 members; private citizens - 9 members; water managers - 6 members; other government - 5 members; university faculty - 3 members; economic development agencies - 2 members.

The first round of the Delphi was mailed on June 8, 1990. The materials consisted of a cover statement outlining the instructions and background, and a form requesting that the participants list the ten (or less) most appropriate (and mappable) criteria which should influence the selection of water management lands. The results of the first round were tabulated, a second questionnaire and cover letter were prepared for Round Two.

The second round of the Delphi was mailed on July 5, 1990. Its questionnaire listed 28 criteria and basic justifying or defining support statements. The members were requested to select ten of the criteria with which they most agreed, and to modify any of the defining statements. The result was that the number of criteria was reduced from 28 to 15 as consensus began to be developed within the committee.

The third round of the Delphi was mailed on July 27, 1990. The questionnaire requested that the committee: (1) list in order, up to ten, their preference for water management and land acquisition criteria; (2) to rate their feeling of relative importance of each criteria to each other by assigning a weight from 1 to 5 to each preferred criteria; and (3) to strike the defining statements with which the committee member did not agree.

Upon the return of the third questionnaire, the results were tabulated to obtain the 10 most preferred criterion and average recommended weighting for each. Because of the way in which the third questionnaire was phrased, the lower the weighting-score the higher the ranking. All individual criteria not being selected by a committee member were assigned a value of 11 for the final tabulation. After scoring the questionnaires, it was determined that three criteria were very closely concentrated at the tenth (lowest) position. This required a decision to either use the highest ranking nine criteria or to acknowledge a three-way "tie" for tenth place and utilize twelve. After conferring with the program coordinator at the St. John's Water Management District, the decision was made to use twelve criteria for the identification of lands.

The criteria and their defining statements were developed and ranked by the Delphi committee as follows:

1. Groundwater Protection/Recharge

Total Score: 123; recommended weight: 3.2

* Protection of water wellfields, sinkholes, and areas affecting public supplies.
* Acquisition of future wellfield sites.
* Areas of high recharge to potable aquifers.
* Areas of high recharge to surface aquifers.
* Any lands needed to insure protection of present and future ground water supplies.

2. Buffers

Total Score: 126; recommended weight: 2.9

* Shoreline laterals, protective corridors around lakes and along streams.
* Barrier islands.
* Uplands around springs, wetlands, lakes, streams.
* Scenic natural zones along water courses.
* Upland shorelines to prevent development.

3. Wetlands

Total Score: 158; recommended weight: 2.2

* Coastal marshes.
* Mature hardwood swamps.
* Fresh water marshes.
* Uplands buffers along wetlands.
* U.S.F.W.S. Natural Wetlands inventory areas.

4. Surface Water Quality

Total Score: 162; recommended weight: 2.4

* Purchase of areas to buffer and filter waters from agriculture, cattle, mining, landfills, or forestry operations.
* Areas acquired to naturally detain and filter storm waters.
* Areas of high site sensitivity.

5. Flood Control and Water Storage

Total Score: 176; recommended weight: 2.0

* River floodways.
* River, stream, lake, 20 year floodplains.
* River, stream, lake, 100 year floodplains.

* Areas for structural water movements.
* Reduction of flood damages.
* Maintenance of natural properties of floodplains.
* Areas of water storage.

6. Habitat Protection

Total Score: 191; recommended weight: 2.5

* Protection of habitat for rare, endangered or threatened species of plants or animals.
* Habitat protection for all species not only water related.
* Areas appropriate for wildlife sanctuaries.

7. Estuarine Dependent Lands

Total Score: 193; recommended weight: 2.5

* Areas which affect the water quality of estuaries.
* Areas which affect the nutrients and productivity of estuaries.
* Protection of all the estuarine system.
* Protection of naturally productive estuaries.

8. Unique Water Features

Total Score: 205; recommended weight: 2.0

* Protection of wetlands and uplands around first-magnitude springs.
* Protection/preservation of river rises, karst windows, geologically rare lakes.
* Purchase presently developed unique hydrologic features to restore to natural or protected state.

9. Plan Implementation and Specially Designated Areas

Total Score: 209; recommended weight: 2.0

* Acquire areas identified needing protection in SWIM plans.
* Purchase of areas recommended in local government comprehensive plans.
* Areas identified in other state and regional plans.
* Areas of uplands and wetlands along waters having special designations such as OFW, Aquatic Preserves, executive order areas, Areas of Critical State Concern, Works of the District and Class I waters.

10. Headwaters

Total Score: 235; recommended weight: 1.1

* Wetlands and uplands around stream headwaters.
* Wetlands and uplands around springs.

11. Connectedness

Total Score: 241; recommended weight: 1.5

* Lands in close proximity to other public lands.
* Wildlife corridors are legitimate concerns and should be acquired.

* Acquisition of entire ecological system.
* Needed to join public land holdings.

12. Biodiversity

Total Score: 244; recommended weight: 1.6
* Areas of high or unique mix of plant and/or animal species.
* Pristine natural communities.
* Unique habitat to specific area.

The three criteria that did not score sufficiently to be included are development pressure, restoration, and recreation. It should also be emphasized that the criteria and their defining statements are solely the product of the Delphi process and no modifications have been made by the authors or others. It should also be emphasized that the process was designed to produce geographically-based selection criteria and that administrative, financial, and political criteria - such as cost, size, distribution, degree of interest - while important to the actual purchase, were not intended to be utilized for the identification of lands for acquisition.

THE STUDY AREA: ST. JOHNS RIVER WATER MANAGEMENT DISTRICT'S OKLAWAHA RIVER BASIN

The Director of the St. Johns River Water Management District's land acquisition program selected the study area based on the operational and planning needs of the district. Program Director John Hankinson selected a 375 square mile area in the lower Oklawaha River Basin in Putnam and Marion Counties which lie in northeastern Florida, southwest of Jacksonville.

The study area, shown in Figure 1, lies in Northeastern Florida, Southwest of Jacksonville, Florida. The study area includes about 35 miles of the Oklawaha River and its flood plain, the northeastern area of Ocala National Forest; and three significant lakes (Lake Delaney, Lake Kerr and Mud Lake). It also includes a number of minor Oklawaha tributaries and spring runs. The area also includes the eastern works of the halted Cross Florida Barge Canal, encompassing the Eureka lock and dam, Rodman dam and its reservoir Rodman pool (Lake Oklawaha) and the canal and lock connecting the pool with the St. Johns River.

The district recommended the study area for several reasons. The area is close to its headquarters in Palatka. The area had an adequate data base installed in ARC/INFO. The district had previously purchased Caravelle Ranch which was located in the study area and the district wanted to "test" their intuitive selection to the evaluation procedure proposed by this study. The overriding reason was the inclusion of the Cross Florida Barge Canal. If the canal becomes completely deauthorized, then the federal government would have to make a determination as to the disposition of the canal land. This research could provide some of the evidence as to how the lands should be disposed.

The research was performed on a Sun workstation using ARC/INFO GIS software. The district employs nine people in its GIS section. With the exception of one prior study on groundwater recharge, the district had used the capabilities of GIS solely for computer mapping and for storing spatial data.

Figure 1. Land Acquisition Ranking

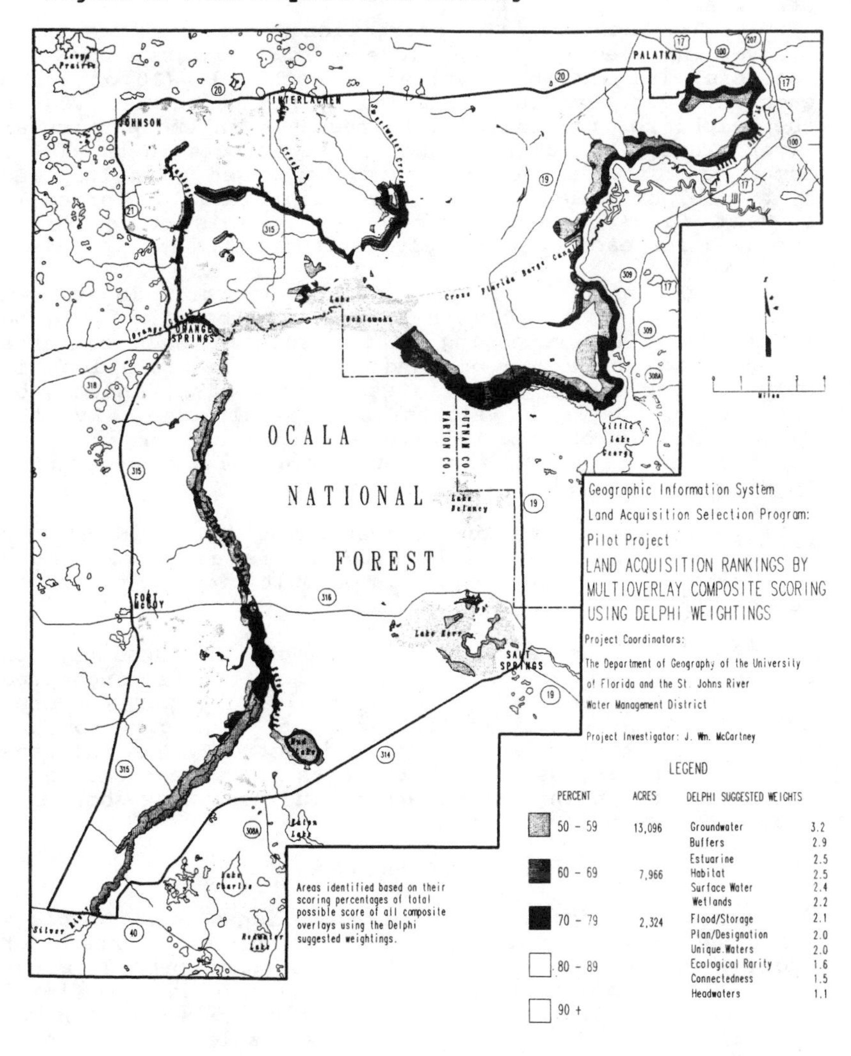

APPLICATION OF THE GIS TO THE LAND ACQUISITION CRITERIA

STEP 1: A detailed presentation as to how each of the twelve criteria was programmed into the GIS is reported in McCartney (1991). Essentially, a map is generated for each criteria. The map displays whether or not the criteria is present at a specific location. The data was not uniformly of sufficient quality to code each site by a continuous scoring procedure that would indicate the degree to which the criteria was present. Instead a dichotomous approach was followed. The map layer was divided into four acre cells; the cell is given an attribute score of ten if any evidence for the criteria existing there was noted, zero otherwise.

STEP 2: The spatial information from the resulting twelve maps became twelve layers in the subsequent GIS analysis. The attribute value for each cell (10 or 0) in each layer in turn is multiplied by the weight derived in the Delphi for the criteria represented in the particular layer. Thus, the value of the cell attribute in the layer named "1. Groundwater Protection/Recharge" is multiplied by the Delphi weight of 3.2; the attribute value for cells in that layer become either 0 or 32. This procedure is repeated for each of the twelve criteria layers.

Each of the resulting Delphi weighted criteria layers can be examined individually in STEP 2. The benefit would be an increased understanding of the role of each criteria (layer) and how a map of land acquisition would look if only the particular criteria were used. Also, instead of the Delphi derived weights being multiplied by the dichotomous (0 or 10) in STEP 2, a variety or range of weights can be used; the result would be a geographic sensitivity analysis.

STEP 3: The values of corresponding cells in STEP 2 among the twelve layers of criteria are added to form a composite score. The result is a multioverlay composite score. This is shown in Figure 1.

Figure 1 is the spatial display of the assembly of criteria scored by the Delphi committee's suggested weightings. By the Delphi weighting, 2,324 acres scored over 70, shown as 70% in the legend of Figure 1. No four acre cell scored over 80% in this study. The Oklawaha River floodplain, especially the area below Lake Oklawaha, Mud Lake, and spring run creeks in the northwest show high scores for priority of acquisition.

CONCLUSION

The research reported on here establishes a methodology for the identification and ranking of lands for acquisition by the public sector. The same procedure can be used in the private sector as a highly beneficial management tool. In this application of the general method, twelve Delphi-generated land selection criteria are defined and mapped as individual GIS layers. The procedure introduces objectivity to the identification of lands through standardization of evaluation criteria and the employment of those criteria equally across the landscape. The result is the production of a composite map identifying lands for potential acquisition.

LITERATURE CITED

Berry, B.J.L., and E. Teicholz. (1983). Computer Graphics and Environmental Planning. Englewood Cliffs, NJ: Prentice Hall.

Dangermond, J. and C. Freedman. (1984). Findings regarding a conceptual model of a municipal data base and implications for software design. Geographic Information Systems for Resource Management: A Compendium, ed. W.J. Ripple, pp. 100-124. Falls Church, VA: American Society for Photogrammetry and Remote Sensing.

Dyer, A.G., B. Smit, M. Brklacich, and R.S. Rodd. (1982). Land evaluation model (LEM): concept, design and application. Computers, Environment and Urban Systems, 7, 367-376.

Hepner, G.F. (1984). Use of value functions as possible suitability scaling procedure in automatic composite mapping. Professional Geographer, 36, 468-472.

Lyle, J., and F.P. Stutz. (1983). Computerized land suitability mapping. Cartographic Journal, 20 (1), 39-50.

Marks, A., G. Thrall and M. Arno. (1991). Using a GIS to perform retail market analysis: a hospital case study. Paper presented at the Annual Spring Session of the Weimer School for Advanced Studies in Real Estate and Land Economics, Homer Hoyt Institute, North Palm Beach, FL.

McCartney, J.W. (1991). A Procedure For Ranking Lands For Public Acquisition With AN Example From Florida's Save Our Rivers Program, Ph.D. Dissertation, University of Florida.

Thrall, G. (1979). A geographic criterion for identifying property tax inequity. Professional Geographer, 31 (3), 278-283.

Thrall, G.I., B. Swanson, and D. Nozzi. (1988). Green-Space acquisition and ranking program (GARP): A computer-assisted decision strategy. Computers, Environment and Urban Systems, 12, 161-184.

Thrall, G.I., and J. W. McCartney. 1991. Keeping the garbage out: Using the Delphi method for GIS criteria. Geo Info Systems, 1 (1), 46-52.

CONFESSIONS OF A GIS PROJECT MANAGER.

Michael Cully
Project Coordinator, St. Lucie County, Florida
Automated Mapping and Geographic Information System Project
2300 Virginia Avenue
Fort Pierce, Florida 34982
(407) 468-1746

ABSTRACT

The local government, multi-agency and jurisdictional GIS Project Manager faces formidable technical and political issues. The technology is new. GIS projects are multi-disciplinary and each is different. The typical manager has limited experience in all phases of the project. The project manager's best source of information is the experience of other projects. GIS presentations typically tout the great triumphs of technology not the failures. This presentation features the wrong turns of GIS projects in the hope that others may avoid making similar mistakes.

INTRODUCTION

The local government, multi-agency and jurisdictional GIS Project Manager faces formidable technical and political issues. GIS technology and projects are new. GIS projects are multi-disciplinary. Each GIS project environment is different. The typical manager has very limited direct experience in all phases of the project. The Project Manager's best approach to survival is to gather specific information about the experience of other project managers.

GIS presentations tend to shout hosannas for the great triumphs of technology. Project managers hear all the good news about successful projects. If there is GIS heaven, then there must be GIS hell. But rarely do presenters open their kimonos to reveal the unsavory, not so successful, details of GIS projects. For the project manager, stories of "How we got our butt kicked and survived" are as important as rosy pictures of sweet success.

The perspective of this presentation will be to feature the miss-steps and wrong turns experienced by an evolving county-wide, multi-agency and jurisdictional GIS project and other similar projects. Along the way, I hope to offer some good advice gleaned from many different sources on how to avoid making similar mistakes.

Confession is good for the soul, they say. The spirit of this presentation is to help other GIS project managers slog out of the swamp of GIS implementation to the promised land of GIS heaven without gettin' lost, fallin' down, or steppin' in somethin'.

WHO DO WHAT ? - PROJECT ORGANIZATION

Our project is basically organized as a joint project between the St. Lucie County Property Appraiser, Community Development Department, and Automated Services. The Property Appraiser is an elected constitutional officer in Florida. The Community Development Department falls under the Board of County Commissioners. Automated Services is a centralized data processing and technical support organization serving all the county government elected officials and governed by a board of governors consisting of those officials.

The Property Appraiser wants to automate the maintenance of his appraisal maps and as a future benefit develop GIS applications for the appraisal function. The initial thrust is "maps in the drawer". Community Development wants applications to support planning, zoning, etc. Automated Services is essentially a technical support resource for these organizations and is the channel for funding for the project.

But there is no "GIS Czar". No person or organization is charged with pushing GIS from concept to completion and armed with the full support of an all empowering authority. There is no central GIS organization with all the GIS funding and resources gathered into it.

The project functions through the continued cooperation of the participants. The Property Appraiser's mapping department is working on the map conversions and Community Development is completing a street center line map as a base map for application development. Progress is being made. And this project probably would not be in existence in any other form.

I do not believe that this is the optimal organizational form for GIS projects. Though I do not advocate "Grand Design" large scale projects (more on that later), I believe that GIS base map and application development should be separated from map maintenance and application usage.

When the press of "normal" business hits the Property Appraiser's mappers or the Community Development Planning technicians, these people will, quite naturally, allocate their time to handling that work. GIS creation is a long term project. The payoffs of GIS to the folks on the ground facing the everyday onslaught of work are unclear and in the distant future. The old saying that it's hard to remember that you intend to drain the swamp when you're up to your rear end in alligators, applies here.

GIS development must be the only job of the people who are assigned to the project. If they have any other duties then the GIS project will suffer. I advocate a separate joint development team empowered to "get it done." Otherwise you are inviting delays that will result in, at best, lost confidence in project and, at worst, funding cuts.

YOU CAN'T TELL THE PLAYERS WITHOUT A PROGRAM - PERSONNEL

In case you missed this point somewhere along the road, GIS projects take a long time from start to finish. Our project planning is based on the premise that we will acquire GIS technical knowledge by training existing staff. This assumes that we have existing competent staff, and - here's the big one - that we can retain them for the life of the project.

As our project evolved, three critical players emerged. These individuals become more important to the project than their relative positions within their own organizations would indicate. Regardless of what your favorite vendor says, GIS systems have not achieved the level of "user friendliness" associated with typical personal computer applications. The users have to work hard at understanding and using these systems to the fullest.

Not every cadastral mapper or planning technician has the background, desire, or gluttony for punishment to master GIS software and technology. The ones who become "super users" are the brain trust of the project. They are the people who know how to handle the problems in dealing with the application systems.

Technical support people are no exception. Programmers who can handle mainframe COBOL applications are lost when they have to master system administration of networked workstations and application development in a graphical environment. At the county or municipal level, finding and keeping programmers who are good at GIS applications and support is difficult.

The people who emerge as these key players become very critical to the success of the project. I am a firm believer in the principle that no one is irreplaceable but the replacement costs vary from individual to individual. For the key players the replacement costs are much higher in time and money.

In our project these people get the funding support for advanced training because they have demonstrated great return on the investment. We have clustered more and more knowledge and assumed responsibility in these individuals. We base our future planning on the assets of their knowledge. The timely project success depends on these people.

So what happens when one of these key people decides to make a career change? In our project the technical support programmer's oft stated career goal was totally GIS oriented. A couple weeks after completing some expensive GIS training, she informed us that she was leaving the computer field for the medical field.

So what happens is you begin a search for a new programmer and hope you can find someone who has some applicable, similar experience. Then you start the training process all over again. In the meantime, you make do with stop gap measures. The execution of your plan goes into a holding pattern.

After this scenario, you vow never again to put all your eggs in on basket, etc. and rely on one person. But the fact of life on the county/municipal level is that you do not have the resources to achieve complete personnel redundancy. You will, in fact, repeat the same cycle and again find yourself with a good many of your eggs in someone's basket.

Every project is going to lose people at some point. The key is to get the project to critical mass before these people start to leave. The longer a project is in the development stages the more exposed the project is to delays caused by key players leaving. Therefore, I believe that you should place great importance on rapidly transferring GIS technology out of the base map and application development phase into the users hands.

When you get the technology out into the users hands, you widen the GIS system knowledge base throughout the organization. More people who can become super users are exposed to the technology. The best example of empowerment of users is the personal computer. St. Lucie County Automated Services supports over 300 personal computer users. I am constantly amazed at the variety of applications our users put their PC's to work on. I believe GIS technology can be just as enabling as the PC.

People problems extend beyond the individuals directly involved in GIS implementation. In the initial stages to our project, we have intentionally limited the number of participants. We believe that this gives the project flexibility in determining the project direction. We have talked to other potential participants to brief them on what we are doing and to offer cooperation as we have products they can use.

In several cases, the managers were interested in what we were doing. They recognized the logic of sharing a base map and other information. No formal agreements were reached but their GIS oriented plans included cooperation with our project. In the course of two months, one of these supporters received an new job offer he could not refuse and another died suddenly. Just as suddenly all plans of cooperation were cast asunder.

Even when you have agreements and funding participation, you can find yourself at the mercy of other people. One organization joined with us to build a street center line map with an address location facility. In order for this organization to use the map, we had to test if the address ranges, street names, etc. were correct. The test we had used previously was to try to locate all the telephone service addresses in our 911 dispatch system on the map. We were locating these in the high eighty percent range. This was pretty good for this stage of the project.

We wanted to concentrate on providing the other organization with the best "hit rate" we could. They needed the center line map sooner than we did and had a smaller number of addresses to locate. We asked for an address file of the addresses they wanted to locate. We also asked that they standardize the addresses to postal service conventions and our street names file. We then could concentrate on correcting the errors in our map that directly affected the other organization's use of the map.

The individuals coordinating the project agreed to this and we waited. When we did get the file several months later, no effort had been made to clean up the addresses to conform to the standards. The MIS director of the other organiza- tion passed this off as "we don't need to do that". In order for us to get the project done, we edited their addresses to conform with the standards, ran the tests, and corrected our map.

Remember that the more participants in a project, the greater the opportunities for delays caused by non-cooperation. Neither the other coordinators nor I could force this individual to provide us with the information we wanted in the format we needed. He simply chose not to cooperate and the project suffered.

THE MYTH OF THE GRAND DESIGN - PROTOTYPING

"The Grand Design" is a phase which the GAO used to describe some very large and overly ambitious federal computer systems projects. These projects were characterized by the attempt to include everything for everybody. The GAO's position was that huge amounts of money were spent on systems which had no chance of solving the smallest user need in the near future.

GIS projects sometimes fit this description. There are numerous examples of large, all encompassing projects which promise to integrate the world with GIS. In many ways, "Grand Design" GIS projects present more opportunities for failure than the information systems the GAO criticized. The technologies that must be integrated to produce a seamless base map of some specified accuracy add enough complexity to qualify. Adding multiple agencies and political jurisdictions completes the qualifications.

Viewing these projects from the most pessimistic perspective, every additional user organization wants everything formatted in its own unique way for free. Endless committee meetings are required to bring the user requirements back to reality. The user organizations generally do not have the internal technical expertise to understand abstract GIS or computer systems design issues. They have simplistic views of the complexity of these systems and do not know the hard dollar investment required for what they think is are simple requirements.

For an average sized county or municipality to launch into a "Grand Design" scale project with the normal in-house expertise is the surest road to GIS hell. Every local government situation is different. What works for your neighbor may not work for you. The best way to approach GIS is to start with a well defined, limited goal, prototype or pilot project.

The prototype approach allows you to see what works for your project and the principal participants. Instead of trying to be everything to everybody, try being very useful to somebody. To do this in GIS requires that you build a base map for a small area. This tests the methods you propose to create the base. It points out the possible shortcomings and areas of concern like a proper control network (more on that later). And it gives you something to demonstrate to your potential users what is possible and, just as important, what is difficult to accomplish without great added expense.

The pilot project educates the principal decision makers on the technology. Do not underestimate the necessity of educating these folks. I have lost count of the number of times I have heard comments of higher level managers like "I didn't realize this stuff was so complicated". You must make the managers aware that implementing a GIS system is not just "pushing a button". The development of a prototype GIS system helps you educate managers in the context of an functioning system.

You can then use your prototype to investigate the feasibility of specific applications. The potential users can be shown concrete, working solutions on a small scale instead of "it'll be just like this only different". The users expectations can be molded in line with what is available for them. The hype of GIS has raised expectations of non-technical users far above what the GIS industry can deliver in a reasonable time frame and for a reasonable amount of money.

One last point on pilots or prototyping. When you are designing a prototype project, concentrate on providing solutions to a real world, ground level user problem. Make the system work for the people who will use it. No GIS system will succeed if the people who must use it, do not want to use it.

A fellow once told me, "Data processing people have the wrong perspective. They persist in trying to build management information systems instead of building information management systems based on what the worker needs to help get the work done." Make your GIS an information management system. The rest will follow.

WHERE DO IT GO? - THE CONTROL NETWORK ISSUE

"Ah yes, it's a fine half section map, Freddie Fast Mapper, but where do it go in the seamless county wide base map we're a-buildin'?" We have good quality "paper" appraisal maps. They were drawn, redrawn, and maintained under the same dedicated supervision for over 20 years. They were compiled from best quality source materials available and edge matched to adjoining sections. They have consistent drafting standards. They look pretty, too. They are fine maps indeed.

As we are converting these maps to digital format and fitting them together into a seamless county wide base map, we are finding some inconsistencies. We are finding some holes in our survey files. We don't know where it go!

When we started investigating the GIS field several years ago, we discovered two basic approaches to base map creation. The first approach involved extensive and expensive surveying and photogrammetry work. We heard cost figures of a half millon dollars before the first map was converted for counties of the same size. We knew we could not expect to spend money like that. If that was what it cost, we could not afford it.

Further investigation revealed that these projects were launched by counties with a large investment in public utilities. These counties could justify the large expenditures in relation to the accuracy required to support utility facility management and expansion. We did not have county owned public utilities. We looked for other projects more like us.

We looked at our situation. We had good quality maps. We did not need extreme accuracy to support the Property Appraiser's appraisal map function or to support applications intended for Community Development. So we went to the Board of County Commissioners and presented our findings. If the Commissioners did not anticipate entry into public utilities, we recommended that we use our existing maps and good survey files to create our base map. We could adjust the maps as better information become available. The Commission concurred. We went to RFP.

Behold the clear vision of hindsight. We did not understand the implications of this decision. Yes, we can produce wonderful half section maps and maintain them with relative ease. We can fit sections together in a limited fashion. But we are experiencing problems fitting them together in large areas of the county. We have "maps in the drawer", an automated mapping system, but we do not have a GIS map base.

Do not ignore the requirement for a framework to hang your base map on. There are a number of good ways to establish that framework. GPS is becoming more affordable. There are new techniques available through analytical photogrammetry. We hope to leverage state supplied aerial photography using these techniques. We are waiting for a clear winter day to get that photography.

SOME RECOMMENDATIONS

Limit the Initial Participants

Start your project with a limited number of initial participants. Avoid "The Grand Design". This allows the maximum flexibility to changing the direction of the project as reality is revealed to you.

Establish a Control Network Plan

You must have a control network to create a seamless base map. The density of this network will be determined by the final map accuracy required. Remember that you can start with an overall framework and then densify within that network in the future.

Concentrate on producing useful applications.

Decide on a small number of useful applications for the GIS. Develop these applications and get them in the users hands. Successful applications will drive demand for additional GIS development. An over ambitious development plan that does not produce useful applications for a long time will stagnate your project.

Use a street center line base map.

A good way to approach the initial applications is with a street center line base map with an address location facility. This map can be completed by a third party vendor as a startup product. Consider using USGS quad sheets as a base for this map. This is a popular base and there may be some graphic data available from other sources. Register your center line map to the control network. As the parcel level map becomes available you can combine the two maps and the applications developed for both.

Prototype

Use prototyping to demonstrate the feasibility of the proposed mapping procedures and each application. Testing your plans on a small area will give you good information about difficulty and costs. Continue to build on your prototype. Additional applications can be developed using the prototype. You can demonstrate the functionality of new applications to the users. Remember that you must provide something that is useful to your GIS users – not what you think is useful to them.

Jump start applications development.

Consider using your GIS vendor or third party conversion vendor to build some of the initial applications. As part of your prototype project, select an application that has commonality with other applications. For example, locating a list of addresses and producing a pin map is common to a variety of applications. Using an outside vendor for this first application, you get "instant GIS" and a sample of how to develop other similar applications.

Establish realistic expectations for cadastral base map creation.

Parcel level base maps do not happen overnight. Even with large amounts of money available for third party conversion, the cycle of initial conversion, checking, corrections, and checking can extend for several years. Make the decision makers aware of the steps involved and the time conversion will require. Remember that full time, in–house, quality control people are necessary with third party conversion.

Get useful applications to the user base.

The rapid transfer of applications to the user base is very important to maintaining momentum in a GIS project. If you can provide useful products the demand for additional applications will drive your project. If you do not get useful products into your users hands early in the project you will be faced with selling GIS to the organization again and again.

Lastly..

One of our vendor's technical support people is fond of ending our frequent conversations on "how you do this and what to do about that" with the same comment. I know he means the best when he says this but I have always found it humorous. So after discussing various ways to "bet your GIS career", I offer his standard parting words to you. Good Luck !

KEEPING THE (GIS) PROJECT ALIVE

Randal L. Krejcarek, P.E.
Gwinnett County Department of Transportation
75 Langley Drive
Lawrenceville, GA 30245

ABSTRACT:

Budget cuts, budget cuts and more budget cuts. This all too common cry from tax payers and politicians is heard at all levels of government, from federal down to the local level. The idea of selling GIS, and its overwhelming price tag, to elected officials is not a routine exercise. In 1988 Gwinnett County, Georgia Board of Commissioners authorized the development of a comprehensive Geographic Information System. Over the past two years the road to implementation of this GIS project has seen many highs and lows. When massive budget cuts came in late 1988, this same Board considered discontinuing the GIS project. Persistence does pay off though. How does a multi-participant GIS project manage to survive when elected officials are looking to cut costly long term projects? This paper will examine how Gwinnett County utilized a combination of participant involvement, creative financing and other important issues noted by the Board of Commissioners in order to "keep the (GIS) project alive". As a result of these efforts a GIS hardware/software solution was presented to, and approved by, the Board of Commissioners, in Dec 1990.

INTRODUCTION

Gwinnett County, Georgia experienced phenomenal growth in the 1980's. From 1980 to 1989 there was an increase of approximately 163,492 persons, or 98%. (Gwinnett County Planning and Development, 1990) Figure 1 shows the yearly estimated population from 1980 through 1989.

Along with being one of the fastest growing counties in the nation, Gwinnett County also experienced the problems associated with such rapid growth. Some of these problems were:

- a rapid rate of parcel splits (approximately 10,000/year maximum at peak growth)
- the inability to produce a timely and up-to-date county map
- numerous databases created that entirely or partially duplicate other existing data bases
- increasing difficultly for various departments to be able to communicate efficiently with each other

FIGURE 1

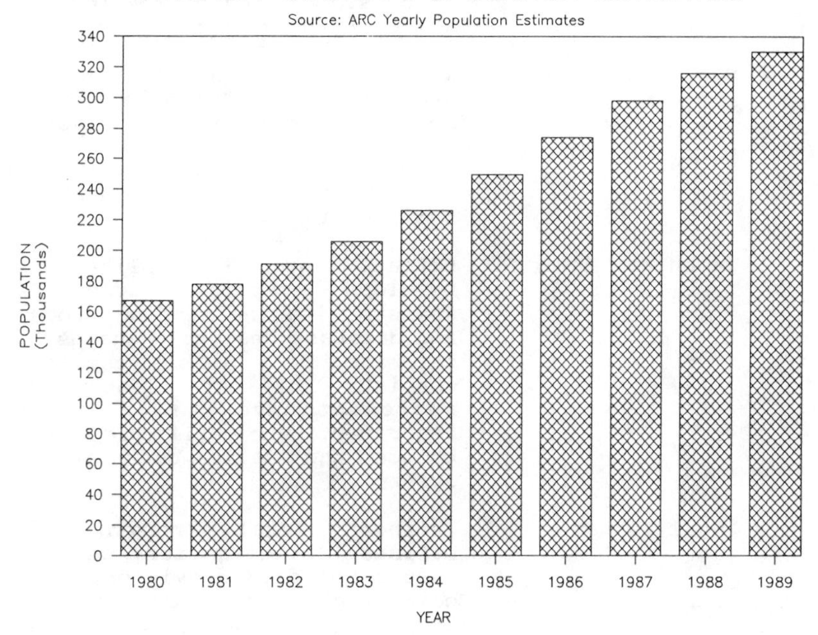

As a result of this tremendous growth the Gwinnett County, Georgia Board of Commissioners (Board) identified two areas in need of immediate attention: first, the various Departments required up-to-date geographical information in order to effectively address daily decisions and second, more effective communication between departments addressing critical geographical issues. The various Departments were also finding it difficult to keep maps current due to the labor intensive tasks of manually updating numerous, and often redundant, maps.

THE GIS ROAD BEGINS

In early 1988 the Board authorized the development of a comprehensive Geographic Information System (GIS). This authorization was broad in scope but included three important aspects: 1) active participation of each Department within the county, 2) the hiring of a GIS project manager and 3) development of a comprehensive implementation plan.

Four committees were formed to deal with the various aspects of the project. The committees and their functions are as follows:

Executive	Approves strategic planning and budgetary activities for the project and is directly involved in critical decision making process. This committee includes all seven Department Heads.
Management Review	Involved in strategic planning and assisting in annual budget requests. Coordinates activities within the County. This committee consists of at least one representative from each Department.
Technical Review	Defines needed applications. Develops mapping standards. Assists with the mapping process and database design. Manages the database. This committee consists of at least one representative from each Department who will ultimately become the daily user of the system.
Legal Issues	Defines legal requirements. Defines policies for security issues and data provision. This committee is a subcommittee of the Management Review Committee.

THE EARLY STAGES

The most costly aspect of any GIS is data conversion. Gwinnett County project was not atypical. The county has approximately 129,000 parcels covering 430 square miles. Initial cost estimates for data conversion ranged from $4 to $10 million, depending on accuracy requirements and what geographical information would be converted.

The management review committee decided to leave several options available for the "what geographical information would be converted" question. This was done mainly because the exact cost of converting each feature was not known.

The accuracy question was handled differently. Development within Gwinnett County has progress from the southwestern part of the county to the northeastern part of the county. In 1988 there was a distinct line dividing the more densely populated area in the southwesterly part of the county with the less populated area in the northeasterly part. Therefore this line represented the logical dividing line for use of 1" = 100'and 1" = 200' conversion. As development progresses to the northwesterly part of the county this part of the GIS database would be enhanced to 1" = 100'.

These decisions were followed by contracts with two consultants to help the County put together a request for proposals (RFP). The county hired both consultants on a part-time, or as needed basis, mainly to help put together the conversion RFP, review the conversion RFP, and help put together the hardware/software RFP. This arrangement worked very well for the county.

To help reduce conversion costs the County decided to use in-house staff to mark existing utilities (sanitary manholes, water valves and fire hydrants). This decision saved the county over $300,000. This decision also required coordination between the county and the firm chosen to perform the conversion work. The painting of utilities took approximately five months to complete.

THE FIRST SETBACK

Gwinnett County's fiscal year is from January 1 to December 31. Budgets are initially prepared in July of the year preceding their expenditure. Therefore in July 1988 a budget was put together to allow the county to select a conversion vendor and begin some of the initial conversion work (aerial photography, setting global position stations (GPS) and setting vertical control monuments).

Late 1988 the Board faced an unusual situation, population growth within the County dropped off drastically. This meant that projected revenues for 1989 would not be as high as expected. The Board immediately started to look for programs to reduce current spending on or eliminate. GIS was one of the first programs examined by the Board.

This change in the Board's support of GIS caught most of the people on the Management Review and Executive Committees by surprised. The conversion vendor had already been selected. Plans were also being made to complete the aerial photography in early 1989.

The Management Review Committee immediately put together the information collected for the benefit/cost analysis. This information was presented to the Board in late 1988. During this presentation all management review and technical review committee members were present to show their support of the project. Specific tasks, which would benefit from GIS, from each department were targeted in the presentation. This was done to demonstrate the applications of GIS to all Departments.

Table 1 shows estimated annual tangible benefits to various departments as a result of GIS. (Gwinnett County GIS Project, 1991) Numerous intangible benefits, such as: creating inventory databases (which currently do not exist), reapportioning of maintenance districts based on service request density, rapid and accurate response to construction field adjustments and providing the County with excellent horizontal and vertical ground control for future development, were also noted.

The Board agreed that the benefits, both tangible and intangible, of GIS far out weighed the costs of implementation. As a result the Board decided to retain funding for the GIS project. The (GIS) project was kept alive.

TABLE 1

GWINNETT COUNTY GEOGRAPHICAL INFORMATION SYSTEM ESTIMATED ANNUAL BENEFITS	
DEPARTMENT	ESTIMATED ANNUAL BENEFITS
PUBLIC UTILITIES	$240,000
TRANSPORTATION	150,000
PLANNING/DEVELOPMENT	60,000
HUMAN SERVICES	25,000
TOTAL	$475,000

THE PROJECT CONTINUES

The first set back did not delay the project, in fact the conversion vendor was selected and aerial photography completed on time in the spring of 1989.

The next order of business was to define the pilot area. The criteria used was simple; find an area with a good mix of land use, proposed utility construction projects, proposed road construction projects and highly recognizable by everyone.

About this same time the Board decided that some of the road design work currently farmed out to consultants should be completed in-house. This decision, whether the board knew it or not, gave the GIS project a well needed shot in the arm.

The Transportation Department put together a proposal which included increasing staff and purchasing a computer aided design and drafting system (CADD). The Board approved this proposal and a CADD RFP was carefully put together. This RFP included a section on GIS considerations. The reason for including GIS considerations was to let all CADD vendors know that the County has a long range plan to install a GIS. The system chosen for CADD had to be either a proven GIS or GIS compatible system. In late 1989 Transportation Department did purchase a CADD system which met all CADD and GIS requirements.

THE SELLING CONTINUES

When it came to preparing the 1990 budget the Management Review Committee decided to switch the funding of GIS from the general fund to the capital improvement program (CIP) budget. This was an important change because the project was entering the most costly phase, data conversion, and a long range commitment from the Board was critical.

The decision to fund GIS from the CIP budget was not an easy decision. CIP's are usually "a plan that outlines the improvements to capital assets that an agency intends to implement during a multi-year period, given the current projections of available funds for that time". (Special Report 54, APWA, 1988) The key terms are "capital assets", generally capital assets are physical items, such as roads, wastewater treatment plants, etc. The idea of funding data conversion costs and software purchases under CIP were new to the Board. Tying these costs to the idea of being able to spot and track physical (geographical) features throughout the entire county was the link used to sell GIS under the CIP budget.

The project so far has accomplished the following tasks:

- 80 GPS monuments set and shot
- 240 vertical benchmarks set
- marked existing sanitary manholes, water valves and fire hydrants for aerial photography
- aerial photography completed

Table 2 shows the proposed CIP task breakdown for completing data conversion and purchasing hardware/software. This indicates that the County will be paying for data conversion through 1995. It is important to point out that through a special financing arrangement with the conversion vendor the county will receive all data by the end of 1993, but distribute the payments through 1995. Table 3 indicates the projected costs to accomplish these tasks.

In order to show the Boards that all Departments were committed to GIS, a funding scheme was set up to access funds from various accounts in different Departments. Other Departments, such as Transportation, committed an unspecified amount of funds, if needed.

CURRENT STATUS

The 1992 proposed CIP will be submitted to the Board in November 1991. The 1992 CIP is broken down into two parts: Prior Years Expenditures and 1992 - 1996 Expenditures. To date the County has obligated approximately $1.343 million to GIS. Table 4 shows the proposed financing method and Table 5 shows projected costs for this same time period.

TABLE 2

CAPITAL IMPROVEMENTS BUDGET GEOGRAPHICAL INFORMATION SYSTEM 1990 - 1995 BREAKDOWN OF TASKS PERFORMED BY YEAR						
	1990	1991	1992	1993	1994	1995
ANALYTICAL CONTROL	X					
VERTICAL CONTROL	X					
PLANIMETRICS		X	X			
TOPOGRAPHY		X	X			
CADASTRAL		X	X	X		
HARDWARE/SOFTWARE		X	X	X	X	X
MISCELLANEOUS	X	X	X	X	X	X

TABLE 3

CAPITAL IMPROVEMENTS BUDGET GEOGRAPHICAL INFORMATION SYSTEM 1990 - 1996 BREAKDOWN OF TOTAL EXPENSES BY TASK	
ANALYTICAL CONTROL	$ 230,000
COMPLETE VERTICAL CONTROL	25,000
PLANIMETRICS	2,400,000
TOPOGRAPHY	1,200,000
CADASTRAL	2,654,000
HARDWARE/SOFTWARE	1,136,000
MISCELLANEOUS	310,000
TOTAL	$7,955,000

A RFP for hardware/software was sent out in mid 1990. A vendor was chosen in November 1990 and the Board approved a hardware/software contract worth approximately $485,000 in December 1990. The amount budgeted in 1991 to purchase hardware/software was only $200,000. The special financing agreement set up with the conversion vendor allowed the County to "borrow" $300,000 from the 1991 Professional Services account (see Table 5) to pay for the hardware/software purchase made in 1990-91. The $300,000 (in $100,000 increments) will be moved from hardware/software account for 1992, 93 and 94 back into Professional Services for each year.

This "borrowing" of monies allowed the County to purchase adequate hardware/software in 1990-91 to handle the expected quality control work load, which is due to start arriving in late 1991.

TABLE 4

CAPITAL IMPROVEMENTS BUDGET GEOGRAPHICAL INFORMATION SYSTEM 1992 - 1996 PROPOSED FINANCING METHOD (thousands $)							
	PRIOR YEARS	1992	1993	1994	1995	1996	TOTAL
GENERAL	1,537	851	840	601	601	450	4,880
SEWER/WATER	889	382	382	383	383		2,419
FIRE	110	55	55	54	54		328
RECR.	110	55	55	54	54		328
TOTALS	2,646	1,343	1,332	1,092	1,092	450	7,955

NOTES: GENERAL = General CIP Fund
SEWER/WATER = Public Utilities Renewal and Extension Fund
FIRE = Fire Fund
RECR. = Recreation Fund

TABLE 5

CAPITAL IMPROVEMENTS BUDGET GEOGRAPHICAL INFORMATION SYSTEM 1992 - 1996 PROPOSED EXPENDITURES (thousands $)							
	PRIOR YEARS	1992	1993	1994	1995	1996	TOTAL
PROF SERV	2,050	1,043	1,042	942	942	350	6,369
EQUIP	436	200	200	100	100	100	1,136
ADMIN	60	40	40				140
MISC	100	60	50	50	50		300
TOTALS	2,646	1,343	1,332	1,092	1,092	450	7,955

CONCLUSION

The Gwinnett County GIS experience proves that persistence and unity do pay off in a multi-participant GIS. If the County had not been unified in selling GIS the Board would have found themselves facing the same geographically related problems they faced back in 1988.

The idea of selling GIS, and its overwhelming price tag, to elected officials is not a routine exercise and is often neglected by technically oriented staff. The success of Gwinnett County's GIS project is the result of the unique approach Gwinnett took in implementing this project and not neglecting the constant need to sell GIS to the Board.

REFERENCES

American Public Works Association, Good Practices in Public Works, (Special Report 54, 1988), p. 13.

Gwinnett County Department of Planning and Development, 1989 Population/Race/Households/Housing Estimates, (Gwinnett County Planning Division, 1990) p.11.

Gwinnett County Department of Administrative Services, Gwinnett County GIS Cost Benefit Analysis Update, (Gwinnett County 1991)

Gwinnett County Department of Administrative Services, Proposed 1992 Capital Budget, (Gwinnett County 1991)

INTEGRATING GIS AND ENVIRONMENTAL MODELING AT GLOBAL SCALES

Michael F. Goodchild

National Center for Geographic Information and Analysis

University of California
Santa Barbara, California 93106

ABSTRACT

GIS has significant potential to support environmental modeling. This paper explores the issues involved in integrating the two fields, and develops the particular context of global-scale research. GIS offers a range of interesting data models, and the proliferation of data models within the current generation of GIS products is seen as both an advantage and a disadvantage to integration. The functionality of GIS is not currently well adapted to environmental modeling, and there is a need for a more carefully defined interface.

INTRODUCTION

In its broadest sense, the term 'Geographic Information System' refers to any digital information system whose records are somehow geographically referenced. However this very general definition conveys little sense of the nature of a GIS, or of its applications, and would be of little help to someone interested in the potential of GIS for supporting global scale research. In terms of its functions, a GIS is a system for input, storage, analysis and output of geographical data, and it is generally accepted that of those, analysis is the most important. Very generally, a GIS may be described as a system for support of geographically based decisions, or a 'Spatial Decision Support System' (Cowen 1988). Again, these terms are likely of little help in evaluating the usefulness of GIS as an enabling technology for global scale environmental research. GIS finds uses in management of geographically distributed facilities, analysis and modeling of geographical data, manipulation of information for making maps, and management of natural resources - in other words, its development as a technology likely has had little to do to date with environmental modeling as such.

Like any information system, a GIS combines a database with a set of procedures or algorithms that operate on the database. Because of the geographical nature of the data, the input and output subsystems must be unusually elaborate, and must rely on specialized graphics hardware such as plotters, digitizers and scanners. Historically, the development of GIS has been to some extent constrained by the availability of suitable specialized hardware. At the same time the database itself must be structured to handle the complications of geographical data. Data modeling, or the process by which the real world is measured and captured in discrete database records, is particularly difficult for geographical data and has been the subject of much research and development effort.

Finally, the design of efficient algorithms for standard geographical operations has also proven to be a major challenge, exacerbated by the very large volume of much geographical data, particularly imagery.

GIS has often been seen as a valuable tool for scientific research - as an 'enabling technology' for a wide range of disciplines (for example see Zubrow, Allen and Green 1990 for a discussion of GIS applications in archaeology). We use the term 'spatial analysis' to describe a set of techniques for analyzing geographic data - techniques whose results are not invariant under changes in the locations of the observations being analyzed. Under this definition many models and techniques of analysis are not spatial - changes in the locations of observations will not normally affect the outcome of a regression analysis. Thus while the statistical packages (e.g. S, SPSS, SAS) exist to support a wide range of statistical analyses, GIS can be seen as existing to support spatial analysis. In essence, a GIS provides a geographical perspective on information.

The GIS industry is currently enjoying a period of dramatic expansion, and growth rates of over 20% are often reported. However it is clear that the reason for much of this growth has little to do with the application of GIS as a scientific tool, or to environmental modeling in particular. While numerous universities have developed GIS courses (Morgan 1987) and invested in GIS hardware and software, sales to governments, utilities, the military and resource-based corporations for information management vastly exceed sales for scientific research. In recent years much development effort in the GIS software industry has gone into information management-related capabilities, and relatively little into spatial analysis and modeling.

The purpose of this paper is to offer a series of reflections on the current state of GIS applications in environmental modeling, with particular reference to global scales. It looks in detail at the assumptions that would lie behind an enthusiastic endorsement of GIS. The first section discusses the vital issue of data modeling, compares current GIS data models, and asks whether current thinking on GIS data models can inform the modeling of environmental processes. The second section looks at GIS functionality, and at the functional requirements of environmental modeling, and asks what functions GIS should be expected to perform in this set of applications.

GIS DATA MODELS

Standard models

Many geographical distributions, such as those of soil variables, are inherently complex, revealing more information at higher spatial resolution apparently without limit (Mandelbrot 1982). Because a computer database is a finite, discrete store, it is necessary to sample, abstract, generalize or otherwise compress information. 'Geographical data modeling' is the process of discretization that converts complex geographical reality into a finite number of database records or 'objects'. Objects have geographical expression as points, lines or areas, and also possess descriptive attributes. For example, the process of sampling weather-related geographic variables such as atmospheric pressure at weather stations creates point objects and associated measured attributes.

GIS technology recognizes two distinct modeling problems, depending on the nature of the distributions being captured. When the distributions in reality are spatially continuous functions or 'fields', such as atmospheric pressure or soil class, the database objects are creations of the data modeling process. The set of objects representing the variation of a single variable are termed a 'layer', and the associated models are 'layer models'. However there are numerous instances where the database objects are defined a priori, rather than as part of the modeling process. The object 'Lake Ontario' is meaningful in itself, and has an identity that is independent of any discretization of a binary water/land variable over North America. We refer to these as 'object models'. In a layer model every location by definition has a single value of the relevant variable, whereas in the object model there would be no particular problem in allowing a location to be simultaneously occupied by more than one object. For example the 'Bay of Quinte' is also in 'Lake Ontario'. The term 'planar enforcement' is often used to reflect the fact that objects in a layer model may not overlap; planar enforcement clearly is not relevant to the object models.

A major difficulty arises in the case of the object models when a well-defined object has no equally well-defined location. For example, the spatial extent of Lake Ontario would most likely be defined by some notion of average elevation, but this is not helpful in deciding when Lake Ontario ends and the St Lawrence River begins. Many geographical objects have inherently fuzzy spatial extents. One common solution to this problem is to allow objects to have 'multiple representations' - spatial extents that vary with scale. A river, for example, might be a single line at scales smaller than 1:50,000, but a double line at larger scales. Both geometric and topological expression vary in this case as the object changes from line to area.

The layer models

The purpose of layer models is to represent the spatial variation of a single variable using a collection of discrete objects. A spatial database may contain many layers, each able in principle to return the value of one variable at any location (x,y) in response to a query. Because information is lost in modeling, the value returned may not agree with observation or with the result of a ground check, so accuracy is an important criterion in choosing between alternative data models. We define the accuracy of a layer as $E(z - z')^2$ where z is the true value of the variable, as determined by ground check, and z' is its estimated value returned from the database. Note that z may be inherently uncertain because of definition or repeated measurement problems.

Six layer models are in common use in GIS:

1. Irregular point sampling: the database contains a set of tuples <x,y,z> representing sampled values of the variable at a finite set of irregularly spaced locations (e.g. weather station data).

2. Regular point sampling: as (1) but with points regularly arrayed, normally on a square or rectangular grid (e.g. a Digital Elevation Model).

3. Contours: the database contains a set of lines, each consisting of an ordered set of <x,y> pairs, each line having an associated z value; the points in each set are assumed connected by straight lines (e.g. digitized contour data).

4. Polygons: the area is partitioned into a set of polygons, such that every location falls into exactly one polygon; each polygon has a value which is assumed to be that of the variable for all locations within the polygon; boundaries of polygons are described as ordered sets of <x,y> pairs (e.g. the soil map).

5. Cell grid: the area is partitioned into regular grid cells; the value attached to every cell is assumed to be the value of the variable for all locations within the cell (e.g. remotely sensed imagery).

6. Triangular net: the area is partitioned into irregular triangles; the value of the variable is specified at each triangle vertex, and assumed to vary linearly over the triangle (e.g. the Triangulated Irregular Network or TIN model of elevation).

Other possibilities, such as the triangular net (6) with non-linear variation within triangles (Akima 1978), have not received much attention in GIS to date.

Each of the six models can be visualized as generating a set of points, lines or areas in the database. Models (2) and (5) are commonly called 'raster' models, and (1), (3), (4) and (6) are 'vector' models (Peuquet 1984); storage structures for vector models must include coordinates, but in raster models locations can be implied by the sequence of objects. Models (3) and (6) are valid only for variables measured on continuous scales.

Models (4), (5) and (6) explicitly define the value of the variable at any location within the area covered. However this is not true of models (1), (2) and (3), which must be supplemented by some method of spatial interpolation before they can be used to respond to a general query about the value of z at some arbitrary location. For example, this is commonly done in the case of continuous-scaled variables in model (2) by fitting a plane to a small 2x2 or 3x3 neighborhood. However this need for a spatial interpolation procedure tends to confound attempts to generalize about the value of models (1), (2) and (3).

In practice, model (6) is reserved for elevation data, where its linear facets and breaks of slope along triangle edges fit well with many naturally occurring topographies (Mark 1979). It would make little sense as a means of representing other variables, such as atmospheric pressure, since curvature is either zero (within triangles) or undefined (on triangle edges) in the model. Models (2) and (4) are frequently confused in practice, since the distinction between point samples and area averages is often unimportant. Models (1) and (3) are commonly encountered because of the use of point sampling in data collection and the abundance of maps showing contours respectively, but are most often converted to models (2), (4), (5) or (6) for storage and analysis. The ability to convert between data models, using various algorithms, is a key requirement of GIS functionality.

The object models

Objects are modeled as points, lines or areas, and many implementations make no distinction in the database between object and layer models. Thus a set of lines may represent contours (layer model) or roads (object model), both consisting of ordered sets of <x,y> pairs and associated attributes, although the implications of intersection, for example, are very different in the two cases.

Object models are commonly used to represent man-made facilities. An underground pipe, for example, is more naturally represented as a linear object than as a value in a layer. Pipes can cross eachother in object models, whereas this would cause problems in a layer model. Most man-made facilities are well-defined, so the problems of fuzziness noted earlier are likely not important. Another common use of object models is in capturing features from maps.

Object models are also commonly used to capture aspects of human experience. The concept 'downtown' may be very important in building a database for vehicle routing or navigation, forcing the database designer to confront the issue of its representation as a geographical object. In an environmental context, McGranaghan (1989a,b) has shown how this issue is important in handling the geographical referrents used in herbarium records.

Finally, object models can be conceptualized as the outcome of simple scientific categorization. The piecewise approximation inherent in layer model (4) assigns locations to a set of discrete regions, in the geographical equivalent of the process of classification. In geomorphology, the first step in building an understanding of the processes that formed a given landscape is often the identification of 'landforms' or 'features', such as 'cirque' or 'drumlin'. Band (1986), among others, has devised algorithms for detecting such objects from other data. Mark (1989) has discussed the importance of categories in the GIS context, and there is growing interest in understanding the process of object definition and its effects. For most purposes, environmental data modeling is dominated by the layer view, and its concept of spatially continuous variables. It may even be possible to go so far as to state that object models have no place in the modeling of environmental processes. But the object view is clearly important, particularly in interpreting and reasoning about geographical distributions.

Network models

Both layer and object models have been presented here as models of two-dimensional variation. An important class of geographic information describes continuous variation over the one-dimensional space of a network embedded in two-dimensional space. For example, elevation, flow, width and other parameters vary continuously over a river network, and are not well represented as homogeneous attributes of reaches. Models (1), (2), (4), (5) and (6) can all be implemented in one-dimensional versions, but none are supported in this form in any current GIS.

Choosing data models

In principle, the choice of data model should be driven by an understanding of the phenomenon itself. For example, a TIN model will be an appropriate choice for representing topography if the earth's surface is accurately modeled by planar facets. Unfortunately other priorities also affect the choice of model. The process of data collection often imposes a discretization, the photographic image being a notable exception. The limitations of the database technology may impose a data model, as for example when a 'raster' GIS is used and the choices are therefore reduced to layer models (2) and (4), or when a 'vector' GIS is used and a cell grid must be represented as polygons. Finally convention can also be important, particularly in the use of certain data models to show geographic variation on maps. For example, digitized contours are used in spatial databases not because of any particular efficiency - in fact accuracy in a layer sense is particularly poor - but because of convention in topographic map-making.

Relationships

A digital store populated by spatial objects - points, lines and areas - would allow the user to display, edit or move objects, much as a computer-aided design (CAD) system. However spatial analysis relies heavily on interactions between objects, of three main forms:

- relationships between simple objects, used to define more complex objects (e.g. the relationships between the points forming a line);
- relationships between objects defined by their geometry (e.g. containment, adjacency, connectedness, proximity); and
- other relationships used in modeling and analysis.

Examples of the third category of relationships not determined by geometry alone include 'is upstream of' (connectedness would not be sufficient to establish direction of flow, and a sink and a spring may not be connected by any database object). In general, a variety of forms of interaction may exist between the objects in the database. In order to model these, it is important that the database implement the concept of an 'object pair', a virtual object which may have no geographical expression but may nevertheless have attributes such as distance, or volume of flow.

Recent trends in data modeling

Recently there has been much discussion in the GIS community over the value of 'object orientation', a generic term for a set of concepts that have emerged from theoretical computer science (see for example Egenhofer and Frank 1988a,b). Unfortunately the debate has been confused by the established usages of 'object' in GIS, both in the sense of 'spatial object' as a point, line or area entity in a database, and also 'object model' as defined here.

Three concepts seem particularly relevant. 'Identity' refers to the notion that an object can possess identity that is largely independent of its instantaneous expression, with obvious relevance to the independence of object identity and geographic expression in GIS. 'Encapsulation' refers to the notion that the operations that are possible on an object should be packaged with the object itself in the database, rather than stored or implemented independently. This has interesting implications for the modeling of distributed systems. Finally, 'inheritance' refers to the notion that an object can inherit properties of its parents, or perhaps its component parts. As a geographical example, the object 'airport' should have access to its component objects - runway, hangar, terminal - each of which is a spatial object in its own right.

Of the three concepts, inheritance seems the most clearly relevant, particularly in the context of complex objects, and in tracking the lineage of empirical data. It seems increasingly important in the litigious environment which surrounds many GIS applications to track the origins and quality of every data item.

Encapsulation seems to present the greatest problems for modeling using GIS. In a modeling context, the operations that are permissible on an object are defined by the model, and are therefore not necessarily treatable as independent attributes of the object. This issue seems particularly important in the context of the discussion of object orientation in location/allocation modeling by Armstrong, Densham and Bennett (1989). For example, one can rewrite the shortest path problem by treating each node in the network as a local processor, making it possible to encapsulate the operations of a node with the object itself. It is possible that this process of rewriting may lead to useful insights in other models as well.

A related debate is that over procedural and declarative languages: a user should be able to declare 'what' is required (declarative), and not have to specify 'how' it should be done (procedural). But are these largely distinguishable in a modeling context, and do they imply that the modeler should somehow surrender control of the modeling process to the programmer?

The role of data models in environmental analysis and modeling is clearly complex. Models written in continuous space, using differential equations, are independent of discretization. But for all practical purposes modeling requires the use of one or more of the data models described here. Perhaps the greatest advantage of GIS is its ability to handle multiple models, and to convert data between them.

FUNCTIONALITY FOR ENVIRONMENTAL MODELING

The statistical packages are integrated software systems for performing a wide variety of forms of analysis on data. By analogy, we might expect GIS to integrate all reasonable forms of spatial analysis. However this has not yet happened, for several reasons. First, while the analogy between the two systems may be valid, there are important differences. The statistical packages support only one basic data model - the table - with one class of records, whereas GIS must support a variety of models with many classes of objects and relationships between them. Much of the functionality of GIS must therefore be devoted to supporting basic housekeeping and transformation functions that would be trivial in the statistical packages.

Second, spatial databases tend to be large, and difficult and expensive to create. While many users of statistical packages input data directly from the keyboard, it is virtually impossible to do anything useful with a GIS without devoting major effort to database construction. Recently there have been significant improvements in this situation, with the development of improved scanner and editing technology.

Third, while there is a strong consensus on the basic elements of statistical analysis, the same is not as true of spatial analysis. The literature contains an enormous range of techniques (for examples see Serra 1982; Unwin 1981; Upton and Fingleton 1985), few of which could be regarded as standard.

Because of these issues and the diversity of data models, GIS has developed as a loose consortium, with little standardization. While ESRI's ARC/INFO and TYDAC's SPANS are among the most developed of the analytically-oriented packages, they represent very different approaches and architectures. Among the most essential features to support environmental modeling are:

- support for efficient methods of data input, including import from other digital systems;
- support for alternative data models, particularly layer models, and conversions between them using effective methods of spatial interpolation;

- ability to compute relationships between objects based on geometry (e.g. intersection, inclusion, adjacency), and to handle attributes of pairs of objects;

- ability to carry out a range of standard geometric operations, e.g. calculate area, perimeter length;

- ability to generate new objects on request, including objects created by simple geometric rules from existing objects, e.g. Voronoi polygons from points, buffer zones from lines;

- ability to assign new attributes to objects based on existing attributes and complex arithmetic and logical rules;

- support for transfer of data to and from analytic and modeling packages, e.g. statistical packages, simulation packages.

Because of the enormous range of possible forms of spatial analysis, it is clearly absurd to conceive of a GIS as a system to integrate all techniques, in contrast to the statistical packages. The last requirement above proposes that GIS should handle only the basic data input, transformation, management and manipulation functions, leaving more specific and complex modeling to loosely coupled packages. Whereas the statistical packages are viable because they present all statistical techniques in one consistent, readily accessible format, GIS is viable for environmental modeling because it provides the underlying support for handling geographical data, and the 'hooks' needed to move data to and from modeling packages.

The argument that GIS is a technological tool for the support of science is widely accepted, and reflected in applications from archaeology to epidemiology. Geography provides a very powerful way of organizing and exploring data, but the map has lagged far behind the table and graph because early generations of scientific computing tools made it so difficult to handle. GIS has finally provided the breakthrough, although it remains far from perfect. If we were to draw an analogy between GIS and statistical software, which began to emerge in the 1960s, then the current state of GIS development is probably equivalent to the state of the statistical packages around 1970. But GIS and statistics are ultimately very complementary sets of tools, both capable of supporting an enormous range of scientific inquiry.

To date, the major success of GIS has been in capturing and inventorying the features of the earth's surface, particularly as represented on maps, and in supporting simple queries. There has been much less success in making effective use of GIS's capabilities for more sophisticated analysis and modeling. It is hard to find examples of insights gained through the use of GIS, or discoveries made about the real world. In part this comment is unfair, because such insights would be next to impossible to document. In part the reason is commercial - the market for GIS as an information management tool is far larger than that for spatial analysis, and vendors have invested relatively little in developing and promoting analytic and modeling capabilities. And although GIS is a major improvement, it is still difficult to collect, display and analyze data in geographical perspective. Finally, Couclelis (1991) has made the point that the current generation of GIS concentrates on a static view of a space occupied by passive objects, and offers little in support of the analysis of dynamic interactions.

GIS is a rapidly developing technology for handling, analyzing and modeling geographic information. To those sciences that deal with geographic information it offers an integrated approach to data handling problems, which are often severe. The needs of global environmental modeling are best handled not by integrating all forms of geographic analysis into one GIS package, but by providing appropriate linkages and hooks to allow software components to act in a federation.

ACKNOWLEDGMENT

The National Center for Geographic Information and Analysis is supported by the National Science Foundation, Grant SES 88-10917.

REFERENCES

Akima, H. 1978, A method of bivariate interpolation and smooth surface fitting for irregularly spaced data points. Algorithm 526. ACM Transactions on Mathematical Software, Vol. 4, p. 148.

Armstrong, M.P., Densham, P.J. and Bennett, D.A. 1989, Object oriented locational analysis. Proceedings, GIS/LIS '89, ASPRS/ACSM, Bethesda, MD, p. 717.

Band, L.E. 1986, Topographic partition of watersheds with digital elevation data. Water Resources Research, Vol. 22, p. 15.

Couclelis, H. 1991, Requirements for a planning-relevant GIS: a spatial perspective. Papers in Regional Science Vol. 70(1) pp. 9-20.

Cowen, D.J. 1988, GIS versus CAD versus DBMS: what are the differences? Photogrammetric Engineering and Remote Sensing Vol. 54, p. 1551.

Egenhofer, M.J. and Frank, A.U. 1988a, Object-oriented databases: database requirements for GIS. Proceedings, IGIS: The Research Agenda, NASA, Washington DC, Vol. 2, p. 189.

Egenhofer, M.J. and Frank, A.U. 1988b, Designing object-oriented query languages for GIS: human interface aspects. Proceedings, Third International Symposium on Spatial Data Handling, p. 79.

Mandelbrot, B.B. 1982, The Fractal Geometry of Nature, Freeman, San Francisco.

Mark, D.M. 1979, Phenomenon-based data structuring and digital terrain modeling. GeoProcessing, Vol. 1, p. 27.

Mark, D.M. 1989, Cognitive image-schemata for geographic information: relation to user views and GIS interfaces. Proceedings, GIS/LIS '89, ASPRS/ACSM, Bethesda, MD, p. 551.

McGranaghan, M. 1989a, Incorporating bio-localities in a GIS. Proceedings, GIS/LIS '89, ASPRS/ACSM, Bethesda, MD.

McGranaghan, M. 1989b, Context-free recursive-descent parsing of location-descriptive text. Proceedings, AutoCarto 9, ASPRS/ACSM, Bethesda, MD, p. 580.

Morgan, J.M. 1987, Academic geographic information systems education: a commentary. Photogrammetric Engineering and Remote Sensing, Vol. 53, p. 1443.

Peuquet, D.J. 1984, A conceptual framework and comparison of spatial data models. Cartographica, Vol. 21(4), p. 66.

Serra, J.P. 1982, Image Analysis and Mathematical Morphology, Academic Press, New York.

Unwin, D.J. 1981, Introductory Spatial Analysis, Methuen, New York.

Upton, G.J.G. and Fingleton, B. 1985, Spatial Data Analysis by Example, Wiley, New York.

Zubrow, E., Allen, K. and Green, S. 1990, Interpreting Space: GIS in Archaeology and Anthropology, Taylor and Francis, London.

GLOBAL CHANGE OPPORTUNITIES FOR GIS: A Panel Discussion

Mapping, monitoring, and modeling are basic themes that appear throughout global change research programs. These themes are included within broad research objectives such as "documenting long-term changes in the earth system on a global scale, collecting data necessary for process studies and modeling, improving understanding of global change through focused studies of earth system processes, and developing integrated models within the earth system" (National Research Council Committee on Global Change, 1990, *Research Strategies for the U.S. Global Change Research Program*: National Academy Press, Washington, D.C.). Mapping, monitoring, and modeling are also fundamental to the science elements of the U.S. Global Change Research Program, such as climatic and hydrologic systems, biogeochemical cycles, ecosystem dynamics, and human interactions. Global change research requires a cross-disciplinary approach with emphasis on analysis and modeling at multiple time and space scales. Strong linkages among observational programs, process studies, and integrative modeling are necessary. In terms of land surface processes research, detailed spatial data are needed to support the development and application of advanced, quantitative land models. This need is in sharp contrast to data requirements of first generation models. Translation of scientific results, including integration of complex earth science data sets, into the policy formulation process is a challenge. These types of global change research activities, as well as requirements for advanced spatial data analysis, suggest an important role for geographic information systems (GIS).

This panel discussion describes current GIS applications, and explores potential opportunities for using GIS's in mapping, monitoring, and modeling activities of global change research. In terms of mapping, the U.S. Geological Survey has developed a strategy for large-area land characterization to help address the land data needs of the global science community. The current developmental status of a prototype land cover characteristics data base for the conterminous United States will be summarized. For this mapping, a GIS permits integration of diverse data sets that can provide tailored digital thematic map products to meet unique user requirements. Results of the National Center for Geographic and Information Analysis sponsored First International Symposium on Integrating GIS and Environmental Modeling (Boulder, Colorado, September 1991) demonstrate the role of GIS technology in global change process modeling. A GIS is also an essential tool for detecting and monitoring land surface change, as well as investigating biospheric processes and land-atmosphere interactions at multiple scales. The pilot projects of the Consortium for International Earth Science Information Network focus on the human dimensions of global change and illustrate the use of GIS technology for integrating physical and socioeconomic data to support policymaking.

The format for this panel discussion consists of four overview presentations on key topics followed by 30 minutes of open panel discussion. The presentation topics are:

Topic 1: A Strategy for Large-Area Land Characterization: The Conterminous U.S. Example

Donald T. Lauer
Acting Chief
EROS Data Center
U.S. Geological Survey
Sioux Falls, SD 57198

Topic 2: Integrating GIS and Environmental Modeling at Global Scales

Michael F. Goodchild
Co-Director, NCGIA
(National Center for Geographic Information and Analysis)
University of California-Santa Barbara
Santa Barbara, CA 93106

Topic 3: Using GIS for Land Surface Monitoring in Global Change Research

Louis T. Steyaert*
Research Physical Scientist
521 National Center
U.S. Geological Survey
Reston, VA 22092
(* Session Organizer and Chairperson)

Topic 4: Integrating Diverse Data Sets for Policymakers

William R. Kuhn
University of Michigan, and
CIESIN
(Consortium for International Earth Science Information Network)
2200 Bonisteel
Ann Arbor, MI 48109-2099

EMBEDDING QUALITY INTO COUNTY-WIDE DATA CONVERSION

Robert J. Garza
Timothy W. Foresman, Ph.D.
GIS Management Office
Clark County
Las Vegas, Nevada 89155

BIOGRAPHICAL SKETCH

Robert Garza is the Quality Assurance Supervisor for the Clark County GIS program. Tim Foresman is the GIS Manager for the Clark County program and sits on the Board of Directors for AM/FM INTERNATIONAL.

INTRODUCTION

Through careful planning and attention to detailed process control design, efficient and automated database construction can be performed with embedded quality control for a geographic information system. Otherwise, the integrity of the database and the fidelity of subsequent analysis degrades and demeans the investment of money and labor in the system or enterprise. Establishing the appropriate quality control for data conversion efforts requires formalization of data handling procedures, conversion techniques, documentation standards, and product specifications. The memorialization and documentation of these essential elements are necessary to implement a comprehensive program and ensure adherence to standards for quality control. While planning, implementation, and enforcement of these methodologies are major components of a quality assurance approach, group commitment and vigilant monitoring effort are required to fulfill the ultimate strategic goals prescribed. Experiences from the southern Nevada GIS network convey the realistic benefits from efforts at educating a constituency of committed GIS participants on the operating principles of quality assurance for development of a large, integrated system.

In any business or local government environment, support for the total quality concept must begin with top management, only then can it filter down to those key individuals responsible for daily operational control. Once all organizational levels have made this commitment to total quality, caution must be taken to prevent the waning of interest and nonconformance. Continual revision and feedback review of existing policies and procedures can effectively ensure improvement of methods and maintenance for desired system quality.

Clark County's commitment to improving, and increasing the degree of excellence for all networked agencies within the county, is demonstrated by the creation of a quality assurance program and rating system. As promoted through the GIS Management Office, these methodologies describe the ideas, principles, policies, and components that sustain the total quality concept mandated by the County Manager. The GIS Management Office assists each agency with development of their quality assurance program utilizing these methodologies as reference guidelines. One county agency, the Clark County Assessor's Office, is currently undergoing a complete conversion of parcel records for the greater Las Vegas Valley area. Associated with this major undertaking are a host of actions related to development of their quality control objectives and processes. These include outside consultation for design specifications, vendor qualifications and selection, and the pressures of keeping up with the area's tremendous population growth rate. The Assessor's Office is addressing these issues along with the quality planning, to create a data base that meets the baseline goals of most other agencies as well as its own.

DEMING'S PHILOSOPHY

The methods used by Clark County to implement quality control follow closely the philosophy of total quality and the improvement of quality espoused by Dr. W. Edward Deming. Fundamentals of Deming's philosophy have become general reference points for the education and administration of the GIS program. Tools and methodologies for the improvement of competitiveness and productivity, as adapted from Deming's business models, can just as easily be applied to municipal environments where efficiencies translate into improved public service (Deming, 1986). Quality control is designed to create a foundation of the basic elements to succeed, whether for a GIS enterprise or other professional discipline, by planning and embedding quality as part of a process. Full comprehension of how to capture the quality assurance philosophy can be difficult and painful to GIS administrators because unfortunately no blue-print exists to build the perfect spatial database.

Accuracy of spatial database elements has been researched and analyzed by many, and in almost every analysis the focus is on specifications and standards. Quality is often defined as conformance to specifications (Hawkes, 1986). In reality, one should not just settle for conformance of specifications, because specifications can easily vary, with time, from the target mark. Continuously reviewing the functional requirements of the database for the user community can help to keep focus on improving the product desired with adjustments made to processes and specifications when necessary. These dynamic adjustments can then be used to address the realities of process control based on the operational needs for both system administration and user participation.

Chances for success for multiparticipant systems can be enhanced if the development of the database incorporates the proper planning elements, training support, process control, and team commitment. To prosecute a successful data conversion effort, therefore, requires a structured approach (Gitlow et al, 1989). The following structure, based on Deming's philosophy, can assist embedding quality into the design process.

FOUR STAGES FOR QUALITY CONTROL

The planning stage represents the first important component of spatial database development. This first stage defines the difference between appropriate and inappropriate data. Deming suggests a simple process called the PDCA (Plan-Do-Check-Act) cycle to illustrate how to recognize potential problems and eventually provide improvements in a process, (Figure 1). Planning defines the collection of information regarding possible variables of a data handling process. Consumer (user) needs are gathered in what is known as Quality of Performance Studies. Based on this information an analysis is performed and reported in a second document known as a Quality of Design Studies. Quality of Conformance Studies are then created to describe how to specifically improve a process.

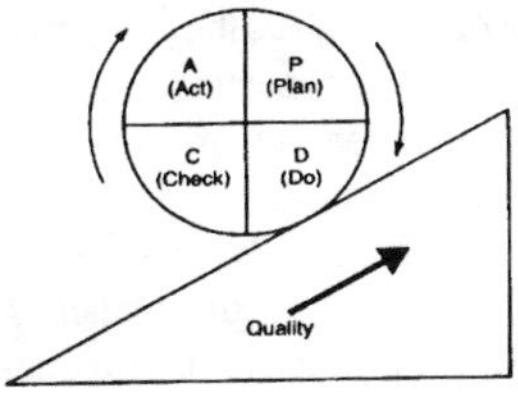

FIGURE 1
The Deming (PDCA) Cycle

Adopted from:
(Gitlow et at, 1989)

In familiar GIS terms the Quality of Performance Studies refer to the Requirements Analysis. These documents define specifications, standards and uses as outlined by the user. In the Quality of Design Studies a formal strategic implementation plan is documented based on variables identified and evaluated in the previous stage. The Quality of Conformance Studies are used to formally define those specifications outlined by the customer as documented in the Quality of Performance Studies. All studies are sequential and iterative so that as concrete plans are made they always improve the process. Constructive tools used in these studies include flow charts, PERT charts and system diagrams.

The second stage, Do, is placed into motion only when all planning tools have been formalized and the user and developer are confident that no misinterpretation or misunderstanding of data standards exist. A prototype is then scheduled to test the efficacy of the planning stage. A prototype should be treated as a trial of the plans set into motion. As an effective tool to assess quality in a project, prototypes can be seen by users as a demonstration of how the process works in an operational setting (Wachtel, 1990). Testing enables feedback from the user community thereby providing better design input for a system. All the plans laid out can fail miserably if this critical stage is skipped. Clark County has heartily endorsed the prototype approach for all GIS related projects. Prototypes enable timely adjustment for a variety of system implementation as well as thorough testing of QA schemes.

The third stage, Check, basically covers the monitoring aspects of the prototype. Findings made at this stage can be used to implement process improvements. Differences between user needs and project performance can be effectively evaluated and reduced by providing revised input for this cycle or again at the planning stage, or continue to improve the prototyping session. Formalization and documentation of all procedures is important for this on-going cycle of stages.

Prototyping reinforces the fact that we learn most from our mistakes. It is the hands-on, sometimes painful realities of life's experiences which make us better administrators and system designers. Alfred North Whitehead (1925) said it best "...knowledge does not keep any better than fish." We can readily see that it is the prototyping stage that truly prepares us for the actions of the operational setting.

The fourth stage, Act, poignantly demonstrates that planning for quality and practicing quality principles are separate struggles. Deming's philosophies for planning fall in line with his beliefs for creating process control - "End dependence on mass inspection." He argues that inspecting a product will not improve its creation process and that the inspection process itself is not error free (Deming, 1986).

There are two basic choices for quality control inspection (Gitlow et al, 1989). Inspect all or nothing, or base the quality of the lot on a statistical sampling thereof. In data conversion processes, how well defined the process control is determines which rule should be used. Having quality built into the process minimizes the amount of inspection required, thus reducing the cost associated with editing or inspecting (Pyzlek, 1991) Both manual and software check points must be installed at every stage of development. Since detailed flow charts and PERT diagrams are in place (plan stage) it is simply a matter of identifying unforeseen bottlenecks and addressing those problems accordingly. Flow charts of critical paths are an important element of process documentation. Flow charts help determine the logical and most efficient flow for the process. If a task position looks illogical on the white board it will likely form into a bottleneck on the production line. In most cases, the development team will be able to understand and then efficiently improve on the process by installing the proper quality control for each task. PERT diagrams can be used to display functional sequence of and which task cannot be started before the previous task has been completed and approved. The ideal situation is to refine each task so that the proper quality parameters are addressed each time. When all key variables are controlled within the process, questions of data quality will be minimized.

QUALITY CONTROL INTEGRATION DURING PARCEL CONVERSION

At Clark County the parcel layer of the Assessor's Office represents one of the most important layers of the GIS regional network. Legal parcels function as the GIS database foundation for many agencies in the region. The county decided to contract for the bulk conversion of its parcel mapping assets. A vendor, Smartscan of Boulder, Colorado, was selected for the conversion contract based partially on their application of in-house process control and partially on their proven experience with automated conversion techniques. With 275,000 parcels, growing at a rate of 1000 per month, to convert over a two year period, there was no question that automated techniques be applied to track all of the data captured. Attributes and vector data are encoded in a systematic process controlled by automated step-checking algorithms thereby lessening the chances for input error.

Parcel data conversion incorporates a series of steps beginning with the establishment of horizontal control, adjustments and merging of parcel subsets, recording of map line and symbology attributes, and final compilation of the cartographic and attribute files. Conformance standards were thoroughly reviewed in the request for proposals and throughout the initial planning cycle with the conversion vendor. The prototype incorporated five sections to test and adjust both the processes and the process checks between the vendor and the Assessor's cartographic staff.

Due to the fully automated structure of Smartscan's conversion processing, algorithms were created to track the spatial adjustment made during the horizontal control and parcel aggregation. Their quality control algorithms calculate the change in movement of the parcel centroid and the change in area as each parcel is referenced for encoding. With these embedded calculations, quality control can be automatically reviewed for subdivisions or parcels of differing quality for any specific range of values. Quality control calculations are automatically encoded into the attribute tables, (Figure 2).

QUALITY ASSURANCE RATING

The cornerstone of the Clark County GIS Quality Assurance (QA) Program is the QA rating system which is attached to every data element during the conversion process (Foresman and Fain, 1991). No where is the idea of truth in labeling more elegantly designed than in this methodology. The QA value is comprised of only two digits characterizing the source used to create the data and the cartographic scale of each original data source. By attributing accuracy assessments for each data element, Clark County reduces both the risks and misuses of a public domain data base, (Figure 3). In this manner liability is placed on the users of the data rather than on its creators.

This approach does not prevent the use of powerful GIS analytical techniques to manipulate and combine various data sets for multiparticipants. Significantly, it promotes the overt awareness and responsibility in performing GIS analysis by ensuring that all users are fully aware of the varying quality, both in scale and source, of the data they are manipulating. The bottom line is that the end result is essentially no better that the weakest link of data utilized. However, comprehension is greatly increased in the application of multiple inputs for decision making. This methodology along with a firm foundation for process conformance and control will retain quality for data conversion and maintenance in a dynamic database for many years to come.

CONCLUSION

Clark County has been fortunate in its quest for attaining high levels of quality control during the creation of the area-wide GIS network. The use of automated techniques to capture and embed quality control values has been facilitated by the support of an educated and committed, multiagency team. The combination of advanced automated process controls with a comprehensive appreciation of quality assurance fundamentals has enabled the county-wide GIS network to capture value added knowledge without incurring additional cost. This quality assurance approach will in turn lead to increased database valuation as the system grows, because exacting knowledge of historic data has always been the missing ingredient for may promising GIS studies. And, as Lord Acton reminds us, those who don't study the past (quantitatively) are destined to repeat its misfortunes.

QA VALUE CONSTRUCTION

Spatial accuracy value	Scale	Source value	Construction document
0	Reserved	0	Reserved
1	Control points	1	Reserved
2	Design drawings	2	Reserved
3	1:1,200	3	CCAO ARC/INFO COGO
4	1:2,400	4	WANG COGO
5	1:4,800	5	COGO subdivision traverse with reference to geodetic control, such as section corner
6	1:9,600	6	COGO subdivision traverse with no reference to geodetic control
7	1:24,000	7	Subdivision located by occupancy only
8	1:50,000	8	Subdivision placed by reference to surroundings
9	≥1:100,000	9	Other

PARCELS.PAT	Column number	Item name	Definition
Parcels quality assurance	37	PCLQA	2,2,1
Map source	47	MAPSOURCE	1,1,1
Change in centroid (feet)	48	CENTROID-CHG	3,3,1
Change in area (%)	51	AREA-CHG	2,2,1

FIGURE 2
Parcel QA Value Construction

Adopted from:
(Foresman and Fain, 1991).

Figure 3

Quality Assurance Tracking through GIS Analysis

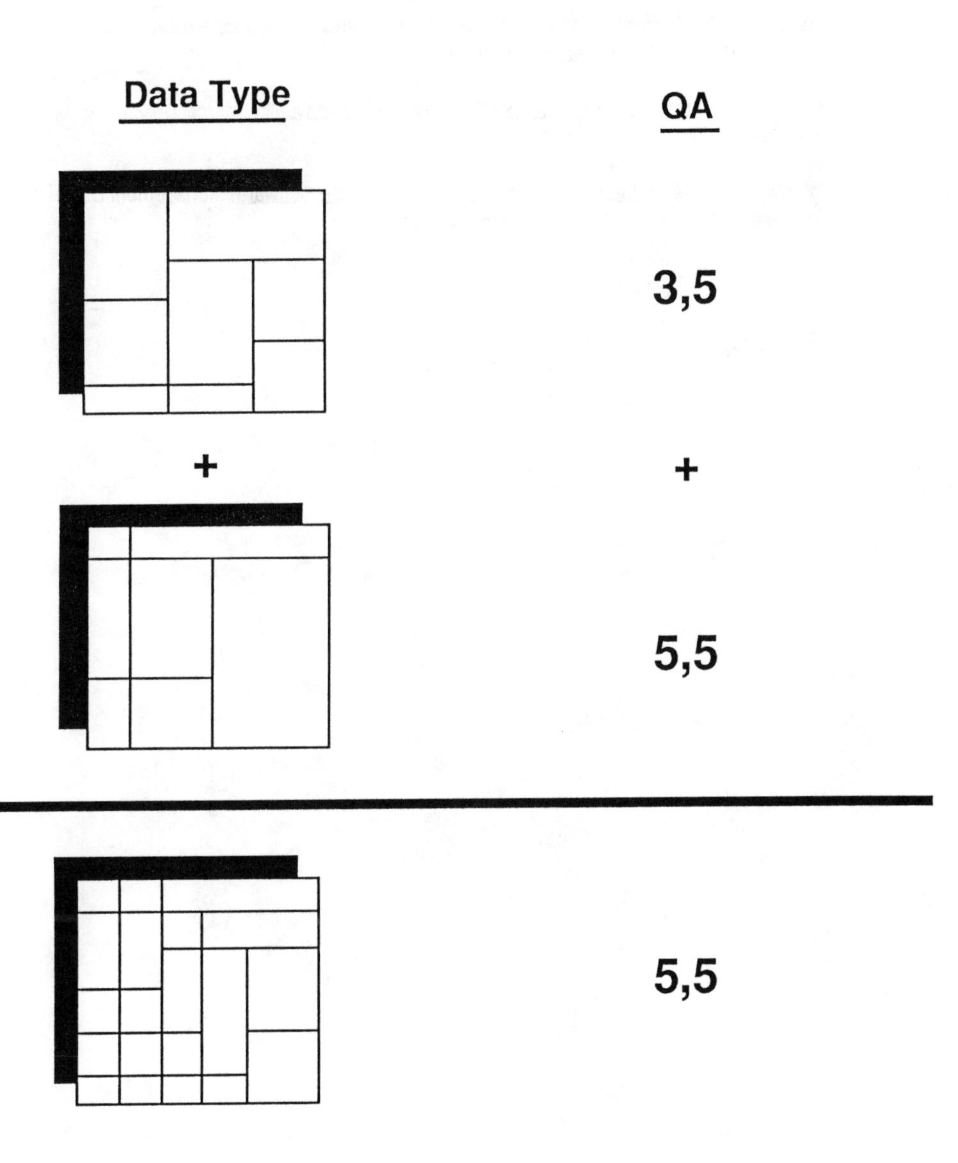

REFERENCES

Deming, W.E., 1986, Out of the Crisis, MIT Press, Cambridge MA.

Foresman, T. W., and Mickey A. Fain, 1991, Taking The Gamble Out of a Parcel Layer, GEOINFO SYSTEMS, Vol.1, No. 7, pp. 27-40.

Gitlow, H., Shelley Gitlow, Alan Oppenheim, Rosa Oppenheim, 1989, Tools and Methods for the Improvement of Quality, Irwin Inc., Homewood, IL.

Hawkes, A.G., Monitoring and Productivity in Conversion Projects, Proceedings of Keystone IX AM/FM Conference, (August 1986).

Pyzlek, T., 1991, What Every Manager Should Know About Quality, Marcel Dekker Inc., New York.

Wachtel, R.A., 1990, Structured Methodologies and CASE Software, Management Development Foundation, Ltd., Colorado Springs, CO.

Quality
Through Conversion Process Management

Mickey A. Fain
SMARTSCAN, Inc.
2344 Spruce Street
Boulder, Colorado 80302

ABSTRACT

The quality of the data destined for Geographic Information Systems continues to be a topic of debate and an area that requires serious scrutiny. Quality is first defined in a measurable manner and the costs of poor quality data are discussed. The way to achieve data quality is to view conversion as a process. Process management makes a project-by-project approach obsolete and is used to implement a "Zero Defect" philosophy. High quality throughout the process then becomes less expensive than a final quality review phase that requires re-routing and re-working after the fact. The process itself ends up assuming a great deal of the responsibility for the quality of the end product, in effect eliminating opportunities for human error.

WHAT IS QUALITY

To be meaningful within the GIS world, the term quality requires a measurable definition. It is not sufficient to equate quality to accuracy. Accuracy by whose measure? And what of all of the other important aspects of data acquisition, such as scheduling and format?

Quality in this context means meeting the user's reasonable expectations the first time around. It does not mean delivering a Mercedes Benz instead of a Yugo. In other words, it does not mean delivering greater accuracy than required, or exceeding specifications, but delivering data which *precisely meets* the specifications of the user. Quality, then can be clearly defined as meeting the user's expectations as expressed in reasonable and quantifiable terms as project specifications.

Thus, quality can incorporate not just accuracy, but scheduling, project reporting — every interaction and communication within the process, including the data. This definition of quality reflects the true way in which vendors are evaluated in an industry where meeting deadlines is nearly as important as meeting specifications.

WHAT IS THE COST OF POOR QUALITY?

Most people assume that higher quality implies higher cost. However, when this quantifiable definition of quality is used, as the quality goes up, the overall cost comes down. Does this mean that you pay less for a Mercedes Benz than for a Yugo?

Of course not. But if the conversion process does a better job of meeting user expectations the first time around, it is only logical that costs will go down. This is as true of an in-house operation as it is in a client/vendor relationship.

Let's take a look now at the typical data conversion process. Generally, the vendor (or the conversion department) does the work and implements a QC process that involves inspecting 100 percent of the work done. Errors are found and some percentage of the data is sent back to previous steps to be re-worked. The data loops around again to the QC process, where another 100 percent inspection is done. A valid attempt to catch all the errors has been made before the data goes to the user.

Then what happens? The user receives the data, and another 100 percent inspection takes place. Errors are often found during this inspection, and some data is returned for re-work. This data must again pass through the QC process, both on the conversion side and on the user side. If all goes well, the loop finally closes when fewer than "X" number of errors are found.

WHAT'S WRONG WITH THIS PICTURE?

The amount of time this process takes is the most obvious problem. However, let's go beyond efficiency for a moment. The real problem is that *it doesn't work.* If 100 percent inspection really worked, the user would not have to do it again. And we know it doesn't work because users frequently find errors in data that has already been inspected for errors and found to be "error-free." Often, in fact, the conversion QC process finds yet more errors when the data comes back from the user, and then the user finds different errors when the data is delivered once again — and on and on, ad nauseam.

Why does this occur? Inspecting spatial data is no easy task. Liken it to "finding Waldo" in a tri-county area, or proofreading the entire Encyclopedia Britannica without a spell-checker. Someone will always find another mistake, no matter how assiduous one is. Thus, if the goal is accurate data, this is *not* a good methodology.

There are further inefficiencies to contend with, however. All errors must be communicated by the user: lists of "Should be a 6 instead of a 7," and so on. And each item that requires re-work implies time, and thus expense. And, since conversion technicians are re-doing work they've already done, rather than moving on to the next data to be dealt with, it's easy to see how this can also impact the delivery schedule: 1 week to convert, 2 days to QC, 1 week to re-work, 1 day to inspect, 1 week for user inspection, 1 week for re-work, 2 days to inspect, and so on.

But the greatest cost of all cannot be measured in days, or weeks, or dollars. What about all the errors that were not caught? And what of the cost of the decisions based on those errors? Finally, consider the legal liability this implies, not to mention the lack of confidence such errors can generate regarding the quality of the entire database.

WHAT IS THE ANSWER?

Do it right the first time. And a "Zero-Defect" approach to the conversion process accomplishes this. The only way to deliver quality data is to design a process that ensures that it is created accurately the first time and does not rely on a 100 percent inspection process — which we know does not work.

Red Paint

Since spatial data is extremely complex, let's talk about process management in terms of red paint for just a few minutes.

One method of manufacturing red paint would be to hire 10 people, give them all the ingredients and put them to work matching a color card. The quality control would take place at the end of the process, when each can of paint would be compared to a master color card. Is each can close enough? Some cans would meet the standard and some would not.

The customer purchasing 8 gallons of red paint would also need to inspect each one to make sure they all appeared to match, perhaps opening as many as 15 cans before finding 8 that he or she considered close enough. This is because each of the 10 individuals hired to make the paint was using a slightly different process.

To correct this problem, the process must be clearly defined. The order in which each ingredient is added must be spelled out, the lighting must be standardized, the temperature must be specified, how long each can is stirred must be stated — every variable that might affect the quality of the paint must be taken into consideration and spelled out as part of a process. This would ensure that the exact same process is followed in the manufacture of every can of red paint.

Clearly then, the inspection process could change dramatically once the process is under control. Once there is reasonable confidence in the process, it would be possible to inspect every other can. Then perhaps only every fifth can would need to be opened. Then every tenth can. Ultimately, mathematical sampling formulas dictate the percentage of cans that need to be inspected.

BUT WHAT IF THERE ARE STILL ERRORS?

The paint might still change color slightly. Once the process has been controlled, however, there can only be two reasons for this and both are easily remedied.

1. It is possible that someone did not follow the process precisely. This can be prevented through training and discipline.

2. The second way in which this might happen is if a variable within the process was missed. Perhaps, over time, dirt has collected under the rims of the cans. The solution is obviously not to send the one bad can of paint back. It's not even sufficient to send the whole batch back. The process itself must be analyzed and changed to include the step "thoroughly wash cans before using," and everyone involved must be educated regarding the change made to the process.

THE ZERO-DEFECT STRATEGY

For process management to work, it is essential that every corporate or administrative level subscribe to a Zero-Defect strategy. Otherwise, it is possible to say, "Well, 1 out of 10 cans is not bad." Even if the goal of Zero-Defects is never achieved, it is necessary to always strive for this level of quality because it *forces* continual improvements to the process.

This effectively converts errors into opportunities to improve the process. As though this were not beneficial enough, consider the savings during the final QC process alone. No longer a 100 percent inspection ritual that takes place at the end of the paint manufacturing process, quality assurance is built into the process itself, and QC is limited to some small percentage of the total number of cans. The user's inspection process is also radically simplified (if not eliminated), and the number of rejected cans is minimal and always accompanied by a process improvement. Delivery times are met or exceeded, implying further savings for all concerned.

DATA APPLICATIONS

This process management approach can be applied within the complex environment of GIS as effectively as it can be used within the simple world of red paint. Moreover, it is equally useful in data conversion, data maintenance and data handling — virtually any aspect of GIS data management.

The first step in applying this approach is to divide the data process in question into phases. This makes it possible to look at smaller and smaller pieces of the process.

Then it is helpful to look at each small step of the process as having "vendor" and "user" responsibilities. Ask, "What does the user (the next step in the process) expect of this phase?" And each step then defines the specifications which the previous step (their "vendor") must meet before delivering the data to it. This is how QC is incorporated into every step of a process — by defining the specifications of every single step.

Next, ask each person in each step of the process, "What are the tasks within each step that, if completed, will produce Zero Defects?" Have them write these tasks down in order and disregard at first how long it takes to complete these tasks.

One of the issues that is sure to come up is, how detailed should this listing and description of tasks be? The answer is simple: Only detailed enough to produce the same result every time, regardless who is completing the task. Leave out all information that is not necessary to achieve this end. For example, it is not sufficient to state, "Stick the label on the can" (returning briefly to the red paint analogy), but it is not necessary to specify "Pick up label with right hand and..." Perhaps something along the lines of, "Align upper left hand corner 1/2" from the rim and..." will turn out to be sufficient. There is an element of trial and error to this stage in the implementation of process management.

Then, start up the process. Every time an error is detected in any step of the process, **STOP** the process (at SMARTSCAN, a horn was sounded throughout the production area). Analyze the error, change the process in such a way that it cannot occur again, document the change and inform the entire crew of the change.

Slapdash, makeshift changes are not sufficient. The changes made to the process may be simple, but they must address the true flaw in the process that caused the error.

For example, during the initial stages of the data conversion process it was necessary to transfer numbers from a hardcopy table to unique points on a hardcopy map. Initially, a technician found the point on the map and pencilled in the number next to it. Inevitably, errors were made — numbers were transposed, duplicated, misassigned. This is a difficult process to which to apply the Zero Defect philosophy. Rather than assign a "checker" to review every number pencilled in by the tech, however, which would have been the traditional approach that accepts correcting errors as sufficient rather than demanding accuracy in the first place, a simple but ingenious solution was found.

The tables were photocopied onto adhesive-backed paper and the numbers were cut apart. That provided exactly the same number of labels as points on the map. Numbers could not be transposed or duplicated, and if one happened to be misassigned, the mistake would be discovered before the last number was stuck to the map, because the incorrect number would be found in its place. The process actually forces the task to be done accurately the first time.

By using all of the technology available to us today (like adhesive-backed paper!), there are many innovative ways to accomplish this same end. And "continual process improvement" cannot help but benefit the industry as a whole.

CONCLUSION

Usually, when someone says, "From now on, we're going to concentrate on quality," the response is, "How can you afford to?" The perception has been that high quality is somehow more costly than poor quality. But if a poor quality product cannot be used and is unacceptable, making it necessary to ultimately deliver accuracy, poor quality actually becomes far more costly than producing a high quality product the first time around. It then is appropriate to ask, "How can we afford not to concentrate on quality."

By focusing on the process, striving for Zero-Defects and transforming every error into an opportunity to improve the process, it is possible to ensure consistent, across-the-board quality on every aspect of every project all of the time.

FURTHER READING

Crosby, Philip B., **Quality is Free**. New York: New American Library, 1979.

Berry, Thomas H., **Managing the Total Quality Transformation**. New York: McGraw-Hill, Inc., 1991.

"DMA DIGITAL PRODUCT SPECIFICATIONS AND STANDARDS"

Paul E. Frey
Defense Mapping Agency
8613 Lee Highway
Fairfax, Virginia 22031-2137

BIOGRAPHICAL SKETCH

Mr. Paul Frey is a Physical Scientist assigned to the DMA Plans and Requirements Directorate, Specifications and Standards Division and is responsible for activities concerning the development of Digital Product Specifications and Standards.

ABSTRACT

DMA is addressing major issues confronting digital product development and the user environment. The trend in major weapon systems is to fuse several products together using GISs. We are currently converting our product specifications and standards to Military Specifications (MIL-SPEC) and Military Standards (MIL-STD). A high degree of user education and training and the development of specifications and standards will be required in order to get the maximum benefit.

INTRODUCTION

The purpose of this paper is to present the scope of the DMA Standards Activity and information from a two phase study which DMA conducted for the past 18 months. DMA must understand both the customer environment and the major issues affecting the use of digital products. We know our users have different systems, requiring different products, and data sets at varying scales. The use of spatial data in Geographic Information Systems creates an environment where products with different intended uses will be displayed at the same time. User education, specifications and standards will maximize effective use of these products.

TRENDS

The trend in DoD weapons/systems is toward the capability of merging, or fusing, multiple data sets. DMA reviewed 289 systems, of which 154 have identified a requirement to display fused MC&G data sets to some extent.

Systems	DMA Products
107	2
13	3
34	4 or more

The degree of fit required for each product grouping varies among systems and the present intended uses for the data. For example, if you take the 1:50,000 scale ARC Digitized Raster Graphic (ADRG) (source map compiled in 1976) over the Nolanville (Ft. Hood), Texas area and Digital Feature Analysis Data (DFAD) Level I, 2nd Edition (Photo compiled) 1987 at 1:50,000 scale, when fused you will see as you might expect some inconsistencies in the alignment of features. This is exaggerated if you zoom in on the area. For the majority of users this can be tolerated, but for some analysts it may not be acceptable. DMA is considering coping with such inconsistencies by providing these users, through utility software, the means to adjust the control coefficients.

CURRENT ACTIVITY

DMA is managing a standards and specifications program which addresses: a) converting current product specifications and existing standards to Military Specifications (MIL-SPEC) and Military Standards (MIL-STD), b) major issues in the development and use of digital product specifications and standards.

DMA has identified ten major issues which must be resolved in the production and distribution of digital products. These are:

Develop standards and specifications for customer data exchange.

Formalize standards and specifications for existing digital products: video maps, Point Position Data Bases, ARC Digitized Raster Graphics, Digital Terrain Elevation Data, World Vector Shoreline, Tactical Terrain Data, Interim Terrain Data, and other unique products.

Address compression of digital data for distribution.

Conduct media analysis for all DMA computer readable products.

Development of MC&G utility software.

Reach final agreement with U.S. and international producers on standards and specifications for producer data.

Develop criteria for "make vs. buy" decisions on all computer readable products.

Address the packaging (mix) of the digital product.

Determine the most effective and efficient Digital Map Display Product.

Advanced Product Concepts and Prototype Digital Product Applications.

Each of these issues will be addressed in turn. We have approximately a six year schedule which actually began in January 1990. We completed Phase II of the study in May 1991. To resolve some issues and consequently develop standards may take longer than the estimated six years while others can be accomplished sooner.

One of the first considerations for DMA was to understand the user environment. DMA is the producer of large amounts of digital data contained in MC&G data bases. Using the DMA Digital Production System we will provide our DoD users consisting of the military services, theater commanders and other agencies, with product specific or core product information. At the national level DMA also supports the federal and civil community data exchange efforts in the development of the Spatial Data Transfer Standard (SDTS). We are currently working with the United States Geological Survey (USGS) to include a relational model, the Vector Product Format (VPF) as a Federal Profile. At the international level, DMA also supports, through the Digital Geographic Information Working Group (DGIWG), the Digital Geographic Information Exchange Standard (DIGEST) which will be used primarily by NATO Nations. DIGEST will also include VPF. We believe that at some time in the future most users of digital MC&G data will be in a Geographic Information System (GIS) environment where the need to fuse more than one product will be the rule rather than the exception.

MAJOR ISSUES

Develop standards and specifications for customer data exchange.

We believe that adoption of a single exchange standard per product form (Vector, Raster and Formatted Text) is the best technical and least costly solution to data exchange for the following reasons: a) Reduces the cost for development and maintenance of translators. b) Eliminates requirement for maintenance of multiple versions of the same product. c) Aids in the fusion of different products. d) Simplifies MC&G Utility Software development. This may not meet all customers requirements. The question is being asked of DMA, why are you developing a Vector Product Standard (VPS)? This is being developed as a customer oriented standard for product exchange, emphasizing performance, direct access and storage requirements. VPS is topological, relational and thematically layered and designed for single implementation. Single implementation means that one set of feature attribute codes are used for all products.

DMA must consider the organization of standards for ease of implementation. A generic standard would list referenced military standards which address format, tiling, digitizing conventions, and feature coding schemes and other national and international standards. Product specifications in the form of MIL-SPECS reference appropriate MIL-STDs, content specific information (including thematic and gazetteer information, coverage index, media, compression / decompression, etc.).

The Phase I study also looked at trends in Geographic Information Systems (GIS). GISs are capable of producing tailored and finished maps (including generalization, feature displacement, symbolization). Our study concluded that in the next five years, the GIS industry will adopt a common exchange format(s) (market driven) and will continue to support product specific formats. Relationally structured GISs will continue to dominate the market for at least the next five years. Within 10 years, technology will allow users, with a desktop networked workstation, to satisfy many of their own MC&G needs in-house assuming the availability of all necessary attendant DMA sources and services.

We are currently submitting draft MIL-SPECS for 21 products and 7 MIL-STDs. Since this is a new area for us, we are looking at the controls and the organization to handle new standards. This brings us to the second issue.

<u>Formalize standards and specifications for existing digital products.</u>

A Mapping, Charting & Geodesy Technology (MCGT) Area has been established within the Defense Standardization Program (DSP). A Departmental and Agency Standardization Office (DepSO) has been staffed to manage this area. We are converting current product specifications and standards to MIL-SPECs and MIL-STDs. We will establish a hierarchy of standards in the following manner:

ENVIRONMENT LEVEL	-Customer Workstations.
EXPLOITATION LEVEL	-MC&G Utility Software.
DATA DIRECTORY LEVEL	-Global Information (accuracy & currency info, spatial referencing scheme, indices, tiling scheme, etc).
PRODUCT LEVEL	-Product MIL-SPECS.
DATA DICTIONARY LEVEL	-Structure, Feature Coding.
FORMAT LEVEL	-Record & Field Labels and
Lengths,	Delimiters.
MEDIA LEVEL	-File & Volume Labels, Physical Characteristics.

Many users cannot store or handle the massive amount of data in their given areas. Therefore digital compression and decompression is required. Some users will be able to tolerate some loss of both resolution and/or data. Others will not. Whatever scheme is determined for a given user, maintenance may be required.

<u>Address compression of digital data for distribution</u>

The task was to determine which current compression/decompression techniques are best suited for MC&G digital products. Choose compression techniques which preserve the detail of the Product data set to the maximum extent possible. The metric integrity should be unaffected. Over 91 algorithms have been examined applicable to the five types of MC&G data. The following algorithms are candidates based on selection criteria and the assessment of quality effects and suitability:

DATA TYPE	ALGORITHM	RATIO(range)
Imagery	Subband Coding	8 - 32:1
Textual	Arithmetic Coding	3 - 4:1
Raster	Vector Quantization	8 - 10:1
Gridded	Arithmetic Coding	3 - 4:1
Vector	Shannon Coordinate Chains	8 - 33:1

A rigorous test by DMA will provide sufficient information to select specific algorithms.

We are aware of new technology and developments in the media area. We hope to ask ourselves all the right questions when it comes to the phase in/phase out criteria. Decisions will be made taking into account, among other things, the volume of each type. The next issue addresses transportable media.

Conduct media analysis for all DMA computer readable products.

The task was to assess magnetic, optical, magneto-optical, video and telecommunication off-the-shelf alternatives. Over 20 different media types were analyzed. The "media of choice" for producer and DMA customer exchange of low (i.e., <1 MB) to medium-volume (i.e., 1 to 500 MB) products:

MEDIA TYPE	PRODUCER EXCHANGE	DMA CUSTOMER EXCHANGE
9-track CCT	Yes	Yes
IBM 3480 cartridge	Yes	Yes
CD-ROM	Yes	Yes (preferred)
Floppy disk	No	Yes (low-volume)

Very Large Data Storage (VLDS), although a single-source media, should be considered the "media of choice" for exchange of high-volume (i.e., >500 MB) products (e.g., imagery). Based upon emerging technology trends and projected market trends, the 8mm cassette should replace VLDS for exchange of high-volume products. As rewriteable optical media mature (e.g., 5.25"/3.50" CD), those media will be of future value to DoD/DMA for allowing update of deployed MC&G products, and based on emerging trends, they may replace CD-ROM. Videodisc should continue to be used for analog data transport, and telecommunications can be a cost-efficient means of distribution under normal operating conditions.

Development of MC&G utility software.

Utility Software is standardized (tested, validated and configured) which allows users to perform the MC&G applications in a consistent and repeatable manner. It addresses user applications where the preservation of the quality and accuracy of MC&G digital data is necessary to the success of a mission. It supports: a) data importation, display and demonstration, b) datum transformation, coordinate conversion, projection/grid generation, c) scale transformation (generalization), d) compression and decompression, e) panel/merge operation. We do not want to be in the business of user application software, but we want to drive the user to accept utility software for standard proven technical solutions to problems.

Reach final agreement with U.S. and international producers on standards and specifications for producer data.

Agreement to support specific Producer Exchange Standards should be based on validated producer exchange requirements and identification of specific products/data sets. We are proceeding in the following manner: a) Support the NATO Digital Geographic Information Exchange Standard (DIGEST). Propose amendments to DIGEST that support DMA vector and raster exchange standards, b) Support the Federal/Civil Spatial Data Exchange Standard (SDTS). c) Gain acceptance of customer oriented standards (Vector Product Standard (VPS), Raster Product Standard (RPS), and Textual Product Standard (TPS)), for producer exchange with DMA.

Develop criteria for "make vs. buy" decisions on all computer readable products.

The scope and task of this issue was to develop a functional definition of operations for the reproduction and distribution of DMA machine-readable products deployable on physical media, considering existing capabilities and committed plans of the DMA Production Centers. Evaluate and recommend the most appropriate means of reproduction (i.e., in-house versus contracting) for each of the recommended customer exchange media, considering cost and qualitative criteria. The required number of media masters and replicates (i.e., total digital product distribution) is forecast for FY97, based on: the DMA FY91 Initialized Program, identified advanced weapons/systems, compression, media, and product packaging/survey, and statistical analysis of DMA/PR's Advanced Weapons and Systems Database (AWSDB). The following table projects the preferred media type, cumulative masters and replicates, and the mean replicates/master.

MEDIA TYPE	FY97 MASTERS	FY97 REPLICATES	FY97 REP.L/MSTR.
Floppy Disk	500	100,000	200
CD-ROM	1,500	14,800,000	10,200
8mm Cassette	1,700	1,500,000	900
Videodisc	2,600	1,200,000	500

The selection Criteria Evaluation of Reproduction Alternatives was based on qualitative and quantitative selection criteria considered for each of the available options within each of the media. The preferred method of reproduction for each medium is the option with the highest rating, and the overall ratings for each option (for each of the nine media) are summarized in the following table:

MEDIA TYPE	IN-HOUSE WRITE-ONE	IN-HOUSE REP.L	CONTRACTED REP.L
Preferred:			
Floppy Disk	B	A	C
CD-ROM	F	D	C
8mm Cassette	F	A	F
Videodisc	--	D	A
Other:			
9-track CCT	B	A	C
IBM 3480 Cart	B	A	C
VLDS	D	A	--
3.50" MOD	B	A	--
5.25" MOD	B	A	--

Note: A = very favorable, B = acceptable, C = cause for concern, D = not recommended, F = rejected

Address the Packaging (mix) of the Digital Product.

The scope of this issue was to develop a set of recommendations regarding the logical structuring (packaging) of DMA digital products on the DMA Customer Exchange Media to enhance the Customer's utility while not significantly impacting DMA costs. DMA would adopt a regional atlas packaging scheme aligned with the U&S Commands' Area of Responsibilities (AORs) for all digital products. Given the diversity in product combinations, package products in the following manner: One product line per medium (e.g., only DTED on the CD-ROM). Maximize product coverage on a single medium and the customer's flexibility is preserved, while filling the media to capacity in integer whole regions.

Determine the most effective and efficient Digital Map Display (DMD) Product.

The task will be to conduct multi-product comparison of DMD technologies: ARC Digitized Raster Graphics (ADRG), compressed ADRG, Navy compressed aeronautical chart, USA/ETL Electronic map display, UK Extended color code/ARC standard raster graphic, and Digital Chart of the World. The comparisons will be based on two scales: 1:50,000 TLM, 1:1,000,000 ONC, and accomplished in conjunction with analytical compression tests. The Warrior Support Center will be used for DMD product analysis.

Advanced Product Concepts

The last issue was to define the Products 2000 Plan, concept, and methodology and to proceed according to that plan. Products 2000 will be undertaken to initiate actions to gain insight into the customers' intended uses of digital MC&G data and working knowledge of the customers' work scenario. To initiate collection and analysis of emerging technologies, played against assessments of DMA requirements. DMA will focus on the future direction of DoD/INTELL customer systems development and private industry. DMA will develop capabilities to synthesize and assess evolving DoD/INTELL requirements (mid 1990's-2000) for new forms of MC&G machine readable products, features and uses of deployed environments. The plan is to develop a concept for a new family of products derived from core, machine readable product databases.

CONCLUSION

DMA is addressing major issues confronting digital product development and the user environment. A high degree of user education and training and the development of specifications and standards will be required in order to get the maximum benefit.

REFERENCE

DMA, May 1991, DMA Digital Products Study: Uses, Specifications and Related Standards, Vol. IA, IB, II, III.

ZOMAP: A KNOWLEDGE-BASED GEOGRAPHICAL INFORMATION SYSTEM FOR ZONING

Jun Chen
School of Urban and Regional Planning
University of Waterloo,
Waterloo, Canada, N2L 3G1

Peng Gong
Department of Surveying Engineering
The University of Calgary
Calgary, Alberta, T2N, 1N4

Ross T. Newkirk and Gary Davidson
School of Urban and Regional Planning
University of Waterloo,
Waterloo, Canada, N2L 3G1

ABSTRACT

A knowledge-based geographical information system, named ZOMAP, was developed to help planners for ZOning MAp Preparation. ZOMAP is composed of four major parts: a geographical information system (GIS), a knowledge base, an inference engine, and a controller. While the GIS is used to store maps and data needed for zoning map preparation and to conduct data manipulation, the knowledge base stores expertise and heuristic knowledge obtained from various sources. The inference engine searches knowledge and matches rules from the knowledge base to generate a zoning designation. The controller is used to link each separated parts and to control operation of the entire system. ZOMAP was developed using ARC/INFO GIS software package and C programming language on a MICRO VAX II under VMS operating environment. It runs on PCs under DOS and Sun Workstations under UNIX as well. ZOMAP was tested in a rural area located in Township of West Wawanosh, County of Huron, Ontario, Canada.

INTRODUCTION

Urban and regional planning is a multidimensional and multidisciplinary activity embracing social, economic, political, and technical factors. Conventional mathematical modelling techniques, such as Operation Research (OR) which mostly deals with numerical analysis of data, lack the capability of incorporating heuristic or qualitative knowledge of planners into problem solving. Advanced artificial intelligence (AI) in combination with geographical information system technique (GIS) hold promise in this area. The objective of the study is to apply the Knowledge Base System (KBS) and GIS techniques to zoning by-law preparation which is one of the fields of planning, where heuristic knowledge and human expert experience dominate, and a large volume of spatial data is required.

Zoning

Zoning has been defined as the planning instrument that deals with the land uses and the physical form of development on individual parcels of privately owned land (Hodge, 1987). According to the Ontario Planning

Act, 1983 a zoning by-law "defines the uses permitted in specific locations within a municipality and the specific development standards relating to those uses" (Ontario Ministry of Municipal Affairs, 1983).

Generally speaking, a zoning by-law is a composite of two major parts: text and maps. The text part states exactly what land uses and provisions are provided in a municipality. It includes: "(1) where buildings or structures may be located; (2) types of building height, side yard dimensions and setback from the street" (Ministry of Municipal Affairs, 1985). Zoning maps are used to show graphically the zone designation for each piece of land in the whole municipality. Usually, once the official plan has been adopted, the provisions for the zoning by-law are set out. In a county or region, a zoning provision which is suitable for one municipality can be used in other municipalities with a little variance which ensures the accordance to their official plans. However, zoning maps should be produced independently to indicate the locations and zone designations for each individual municipality.

Knowledge-Based Geographical Information System

A knowledge-based geographical information system (KBGIS) is a technique integrating GIS and KBS techniques. It is composed of two major components. The first component is a spatial database management system (DBMS) that organizes the properties of spatial objects by their locations. The second component is a knowledge base that organizes domain knowledge and expertise.

Geographical information systems. GISs are computer-based technology devoted to the storage, management, analysis, and display and output of spatial data (Burrough, 1986). Spatial data refers to geographically-referenced data which can be represented using three basic topological components: points, lines, and polygons.

In a KBGIS, the major role of a GIS plays is to provide and manage spatial data. It offers better organization of the data and adds the dimension of geo-referencing to the data. Spatial data typically handled by a GIS include the locational information which represents the location of the spatial object and attribute which refers to the characteristics of a geographical object. Spatial data is usually stored in one of two formats: raster and vector format. In raster data format, data are arranged as arrays of grid cells. Digital images are most suitable to be stored in this format. The vector representation of a spatial object is based on the attempt to represent the object as exactly as possible. It is particularly suitable for representing phenomenological data.

Knowledge-based system. A KBS consists of two major parts: a knowledge base and an inference engine. The knowledge base is a collection of specific and general facts and descriptions of relations between them. Knowledge structure and representation are the key steps in the development of a knowledge base (Charniak & McDermott, 1985). They are tools for illustrating what experts know. One of the most commonly used methods of the knowledge representation is the production rule system. A production rule system consists of a currently perceived state or context (IF-component), the goals of the individuals, an appropriate action (THEN-component) and a state the decision maker expects to reach if the action is taken. The expression of a rule can be represented as:

IF (condition1 & condition2 ... or conditionN1 or conditionN2)
THEN (action)

The condition is usually a conjunction of predicate that test properties about the current state. The primitive action is some simple action that changes the current state.

There are two major advantages using the production rule system. The first is that it is intuitive and native for most people and heuristic knowledge is represented easily. The second is that it is easy to modify the rules in the knowledge base and to keep a consistent rule-base structure because rules can be changed independently (Chandra & Goran, 1985).

In a KBS results are obtained by simply stating what knowledge is required, leaving them up to the control mechanism of the system to infer the results in the best way it can. For this reason this control mechanism is often called inference engine (Frank et al, 1987).

There are two reasoning techniques which can be used to search the knowledge base in a KBGIS: forward-chaining and backward-chaining (Frank et. al., 1987). The forward-chaining is a data-driven mode whereby the inference engine goes from known facts, releases all the rules whose premise have been checked, adds the facts supplied by working memory to obtain satisfactory results. Backword-chaining is goal driven, with the inference engine attempting to assemble sufficient facts in the working memory to prove hypothesis.

SYSTEM DESIGN AND DEVELOPMENT

ZOMAP System Structure

Based on the the concepts of a GIS and a KBS, a prototype system named ZOMAP has been developed. It was implemented for ZOning MAp Preparation with the ARC/INFO GIS software package and C programming language on a Micro VAX II computer operating under VMS. The structure of the system is illustrated in Figure 1. The entire system is composed of four parts: a GIS, a knowledge base, an inference engine, and a controller.

The GIS is used for data storage, maintenance, management, analysis, and output. The knowledge base is a storehouse where the knowledge needed for system operation are organized as rules. How to extract the knowledge or rules efficiently and precisely from the knowledge base is the task of the inference engine. The database, the knowledge base and the inference engine are independent parts in the system and can access each other through the controller.

Development of ZOMAP

The procedure of the development of ZOMAP involves four stages: (1) information collection and knowledge acquisition; (2) database development; (3) knowledge base construction; and (4) the development of the controller.

Information collection and knowledge acquisition. There are three kinds of information required for zoning: documents, maps and airphotos. The sources, content of the information and their uses in ZOMAP are illustrated in Table 1.

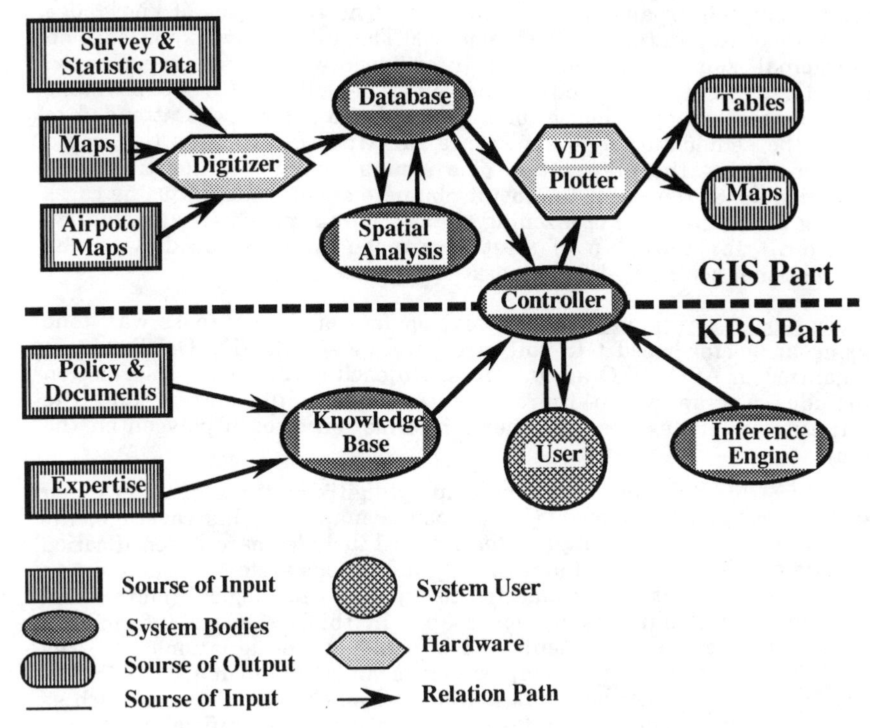

Figure 1. The system structure of ZOMAP

Class	Name	Contents Used	Use	Source
Document	County Official Plan	policy	KB	County
	WWP Secondary Plan	policy	KB	County
	WWP Bkgd study	data & maps	Ref./DB	County
	Tax roll	ownership	DB	WWTwp
		unit class	DB	
	Other zoning By-law	policy	Ref	County
	Unit code definitions	definitions	KB	WWTwp
Map	Assessment Map	maps, par. no.	DB	Asse. Off.
	Official Plan Map	map	DB	County
	wetland and related area map	map	DB	CA
	Air photo Map	map	DB	CA
Air Photo	Air Photo for WWT	everything	Ref	County

Notes: KB: Knowledge Base DB: Database
CA: Conservation Authority Ref: Reference
WWtwp: West Wawanosh Township

Table 1. Information sources and usage.

Knowledge acquisition is the gathering of information, decision, and relationships from any sources available. The techniques of knowledge acquisition used in this study include : (1) The authors' own knowledge on municipal zoning; (2) personal interviews with some experienced planners. This provided "hand on knowledge"; (3) open house attendance: An open house for a zoning by-law draft was attended to allow the senior author to experience the kinds of problems that could happen during the preparation of a zoning by-law; (4) questionnaire: Questionnaires were sent to consult planners about some confusing cases during the preparation of a zoning by-law; and (5) practice: A zoning by-law draft for township of west wawanosh was produced with the conventional zoning by-law preparation process.

Database development: The development of the database was done using an vector-based GIS software package --- ARC/INFO. Data are organized in ARC/INFO as coverages. In each coverage the information includes a map which represents topological information and the attributes which express the theme information of each polygon on the map.

The ZOMAP database was built up by digitizing three kinds of major original maps. They are: (1) the parcel map: In this coverage, the attribute information contained for a parcel includes parcel identification and its area, unit code and accessibility of drainage system if it is in urban area. All information is connected to a parcel's topological information; (2) the official plan designation map: In this coverage, information related to the map represents the official plan designations; (3) the wetland and related area map which is an airphoto map: With the topology map, the attribute information includes wetland classes, such as upland, wetland, engineering flood area, natural and artificial corridor.

An intermediate map coverage was then generated by overlaying the three original coverages. In this coverage, all information in the three original coverages is contained and will be used later as the condition part of production rules to allow the inference engine to derive proper zoning designation.

Knowledge base construction: As mentioned before, the production rule system or rule-based knowledge representation utilizes "if condition then conclusion" statements. Given a situation the inference engine seeks to satisfy the condition in the "if part" of the rule. A true condition results in a conclusion. The conclusion here is a zoning designation. For example:

IF Official plan designation = Agriculture
& Unit Code = FL & FRU
THEN Zone Designation = AG1

FL means farm land and FRU farm residential unit. AG1 is a zone type adopted in County of Huron, Ontario, Canada.

In order to organize knowledge more efficiently, a block structure of rules is used in the system. The official plan designation parameter is the most important one in determination of zoning designation. It indicates the system where to go to search and match the rules.

In ZOMAP, forward-chaining is used to conduct conclusion from known conditions. The general procedure can be described as following:

```
IF OPD (Official Plan Designation) = Agriculture
THEN
OPEN AG-BLOCK
Loop Start
SEARCH Rule-i
MATCH Rule-i
IF TURE
GET ZD (Zoning Designation)
ELSE
GO TO LOOP again
DESIGNATE "ERROR"
```

The construction of the knowledge base and inference engine is done using C programming language.

The development of the controller. The database, the knowledge base and the inference engine are all independent sub-systems in ZOMAP. A connector is developed to link each sub-system. The connector is the controller. It is a program written in C programming language. The roles of the controller plays in ZOMAP are:

- when the data file is ready, it starts up the knowledge base;
- starts up the inference engine;
- creates a new file to store the results; and
- writes the result into the file.

The file created by the controller contains two columns. One is the polygon numbers which refer to the polygons in the overlaid coverage. The other is zoning designation derived from ZOMAP. This file will be used as a data source and input to the attribute table of the overlaid coverage in the INFO module. ARC/INFO can then be used to display the final zoning result in colors or shaded patterns.

RESULTS AND RECOMMENDATIONS

Analysis results

ZOMAP system was tested in part of the Township of West Wawanosh, County of Huron, Ontario, Canada. The township of West Wawanosh is a rural community. Its primary development goal is to strengthen and preserve the agriculture economy. The result generated by ZOMAP (Figure 2) is compared with the one produced using conventional zoning map preparation method (Figure 3). Several differences can be observed:

- There are more land parcels in Figure 2;
- Some polygon boundaries in Figure 2 are shifted;
- There are several redundant arcs appearing in Figure 2;
- Some polygons are designated as "error" in Figure 2;

The reasons causing these problems are the quality of raw maps and the illogical attribute combinations formed during the overlay. In the conventional method, these can be done easily by planners because they can adjust the location of the polygon boundaries and reasoning the illogical attribute combinations. Also we can observe that all polygons which are not zoned in error in Figure 2 are consistence with those polygons in Figure 3. All polygons with error designation are caused by illogical attribute combinations. These combinations can be detected by the knowledge base. The error polygons can be redesignated easily by interactive editing.

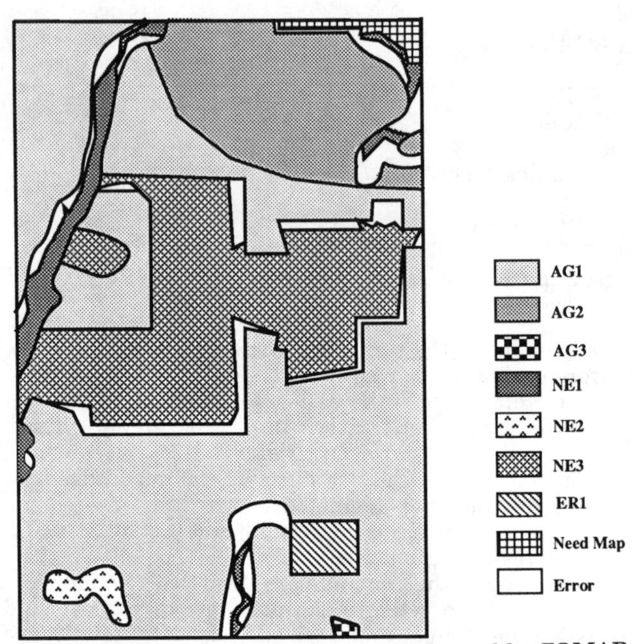

Figure 2. Zoning Map of the Study Area Prepared by ZOMAP

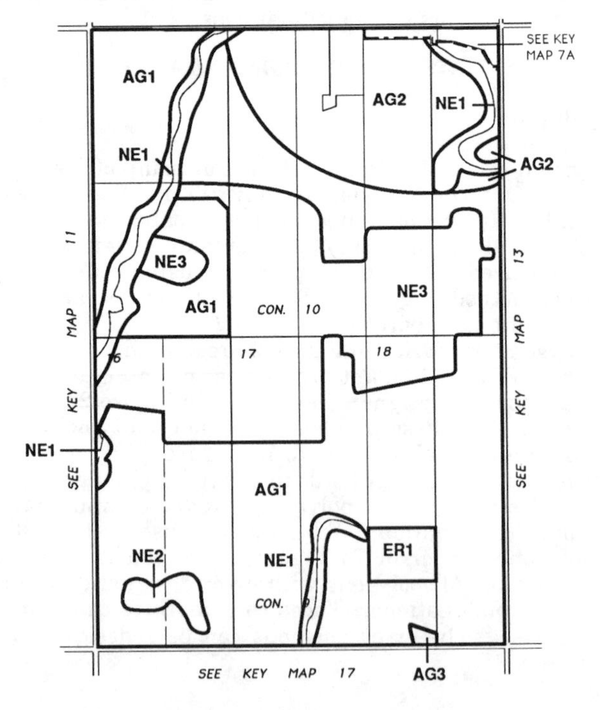

Figure 3: Zoning Map of the Study Area Prepared by Conventional Method

Recommendations: Even though some problems have been identified, it should be pointed out that the zoning map produced using the CAD software in planning department (such as Huron County) may not be as accurate as the one done by ZOMAP. This is because in the conventional process the boundary of each zone is traced by hand with the judgement of planners or technicians since no better and easier way can be used.

By reviewing the whole test process, some suggestions were made to improve the accuracy of data input and result: (1) to adopt a GIS at the very beginning; (2) to adopt a uniform coordinate system to digitize all maps; (3) to adopt a set of relatively accurate base maps as a reference to locate polygon boundaries on official plan maps, wetland and related maps and zoning maps; and (4) to use a uniform coding system which is suitable not only for knowledge base coding but also for municipal affairs, such as tax system and administration in a local municipality.

CONCLUSIONS

This research represents the first effort to build a knowledge-based geographical information system for zoning. A prototype KBGIS for zoning(ZOMAP) was developed by integrating ARC/INFO GIS software package and a knowledge-based system developed in C programming language on Micro VAX II under a VMS operating system. It can also run on Sun Workstations under UNIX and PCs under DOS as well. With the test of ZOMAP, we can conclude that the system is a cost-effective, operation-efficient, reliable, and useful system for zoning map preparation. This is especially true when the information provided is accurate.

REFERENCES

Burrough, P. A., 1986, Principles of Geographical Information Systems for Land Resources Assessment, Toronto: Oxford University Press

Chandra, N. and W. Goran, 1985, Steps Toward a Knowledge-based Geographical Data Analysis System, Geographical Information Systems in Government, pp.749-763

Charniak, E. and D. McDermott, 1985, Introduction to Artificial Intelligence, Massachusetts: Addison-Wesley

Frank, A. U., D. L. Hudson, & V. B. Robinson, 1987, Artificial Intelligence Tools for GIS, Proceedings: International Geographic Information System (IGIS) Agenda, Vol II, pp 257-271

Hodge, G., 1989, Planning Canadian Communities, Nelson Canada

Ontario Ministry of Municipal Affairs, Research and Special Projects Branch, 1985, An introduction to Community Planning, Toronto, Ontario.

Ontario Ministry of Municipal Affairs, 1983, Planning Act 1983

DYNAMIC STATISTICAL VISUALIZATION OF GEOGRAPHIC INFORMATION SYSTEMS

E. Bruce MacDougall, Professor
Department of Landscape Architecture and Regional Planning
University of Massachusetts
Amherst, MA 01003
(413) 545-6608
Internet: ebm@larp.umass.edu
Bitnet: BruceMa@UMass

ABSTRACT

This paper reports on the application of dynamic statistical visualization as a method to facilitate exploratory analysis of data typically stored in geographic information systems. It demonstrates how commercially available statistical packages may be used to explore GIS data stored in raster format, and presents results from an operational prototype program for data stored in vector format.

INTRODUCTION

One of the least developed capabilities of GIS technology is its use to explore or browse spatial data to form hypotheses about patterns and anomalies that may be significant for a particular problem. The GIS analyst typically proceeds from a set of a preconceived notions about how factors are related, with little preliminary study of the actual data other than to ensure that is acceptably error-free. There was an analogous situation in statistics, where until the last decade, data analysts generally proceeded by forming and testing hypotheses. More recently, however, the notion of exploratory data analysis (EDA) has been recognized as an important step in the analysis process. Further, it is now coupled with dynamic visualization on microcomputers so that it has become not only an important way of thinking, but also a powerful tool for studying data.

This paper reports on the application of dynamic statistical visualization to the exploration of geographical data. The principal problem posed by this application is the incorporation of a suitable map display that is updated as are other statistical views. For raster data, it is possible to use existing EDA programs, although the map display is somewhat contrived, and leaves something to be desired in terms of its clarity. However, after a simple transformation of raster data, the GIS analyst has access to many statistical views and reports. For vector data, the author has developed a prototype program that includes a display and those statistical views that seem most useful. Further, it incorporates features that are specifically intended for browsing spatial data such as histograms by both polygon count and area of the map occupied, and partitioned clustering that allows the influence of distance to be explicitly varied.

EXPLORATORY DATA ANALYSIS: TRENDS AND OUTLIERS

The idea of data exploration as a useful and important statistical approach is generally attributed to Tukey (1977), and has subsequently been elaborated and enhanced through developments of statistical visualization and dynamic graphics. The basic characteristic of exploratory data analysis that distinguishes it from conventional statistical approaches is its emphasis on exploration as compared with confirmation. It focuses on searching for and understanding trends and outliers (global patterns and local anomalies), rather than assist in fitting models that have been determined a priori.

In particular, EDA focuses on identifying "outliers," that is, cases that are unusual either because they represent special situations that require further examination ("rogues"), or because the data are in error ("blunders"). The basic set of EDA techniques are common statistical graphs such as bar charts, histograms and scatterplots, but also less well known graphs such as scatterplot matrices, influence and leverage plots, density ellipses, spin diagrams (including principal components), and cluster membership.

The effectiveness of these tools is greatly enhanced though multiple linked views and dynamic graphics. Data are viewed through two or more statistical graphs in which selected ("brushed") cases in one view are highlighted in all others, as well as in a data table (equivalent to a spreadsheet). Software vendors refer to this feature as "hot links." Subsequent study is done through isolating a selected individual case or subset of cases.

PREVIOUS WORK

The potential of EDA and dynamic statistical visualization for exploring geographic data has been recognized by a number of authors (Haslett, et al 1990; Monmonier 1990; Openshaw, et al 1990; Walker and Moore 1988; Wills, et al 1990). The work reported in this paper extends these earlier efforts in two ways. First is a demonstration of how recently available software may be applied to the exploration of raster data. Second is an extension of findings and thoughts about the application of these techniques to data stored in vector (polygon) format that was first described in MacDougall (1991).

EXPLORING RASTER DATA

There are a number of commercial statistical packages that are oriented towards exploratory data analysis. The author has used two of these with raster data on a Macintosh computer: JMP, from SAS Institute, and DataDesk 3.0, from Odesta Corporation. These and similar programs first require a conversion of the original grid (with implicit coordinates) to a table where each row represents a mesh point or cell in the grid, and with explicit coordinates. Figure 1 illustrates this process for two three by three raster data sets. For the research reported here, this conversion was done with a short program written in BASIC, but it could also be done with some cutting and pasting operations in a spreadsheet, or even within the applications by someone proficient with their data manipulation features.

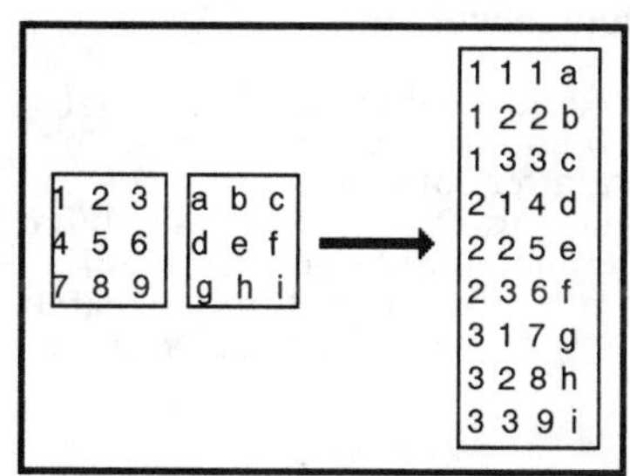

Figure 1. Conversion of two raster data sets to a table format with explicit row and column coordinates.

To generate a map as one of the statistical views, one uses the "spin" feature, which is intended to show relationships among three variables as a three-dimensional plot. The spin plot is oriented so that the y axis represents north-south coordinates, the x axis east-west coordinates, the z axis any other variable, and the viewpoint is directly along the z axis.

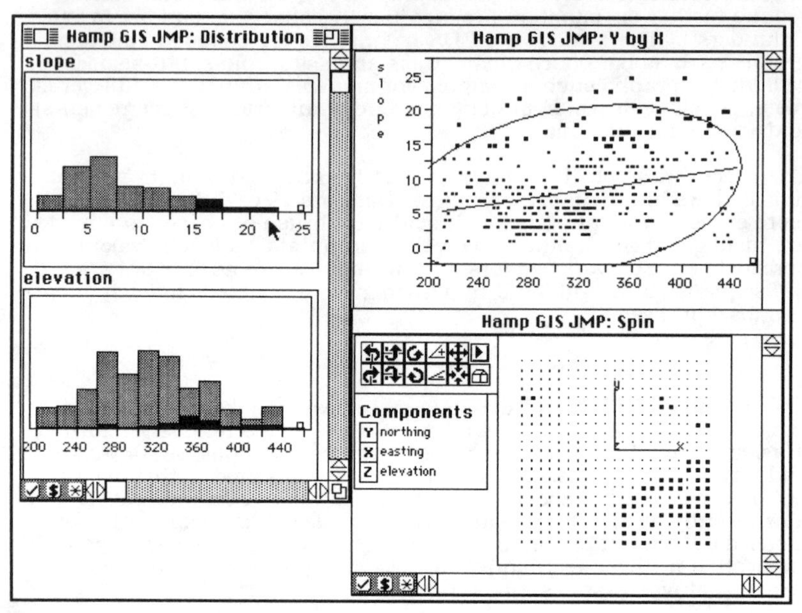

Figure 2. Example of statistical views of raster data.

Figure 2 is a screen dump that illustrates how JMP may be used to examine a raster data set, in this case a matrix of 21 rows by 21 columns for an area in the south part of Amherst, Massachusetts. The left window shows histograms for elevation and slope, and the upper right window a scatterplot relating slope and elevation. Included on the plot is the regression line from a model that predicts that slope is dependent on elevation, and a density ellipse that encloses 95 per cent of points if the relationship between the variables were bivariate normal. (Points outside this ellipse are therefore candidates for further examination as blunders or rogues.) The lower right window is the spin diagram. For this example, steep slopes were selected in the upper histogram, and immediately highlighted in all the other displays. Also available (but not shown in Figure 2) are comprehensive text reports that give descriptive statistics, and parameters of any model fitting.

The spin feature in these programs may also be used to simulate aerial views. The window on the right of Figure 3 shows the same map as in Figure 2, and the left is a view of this data after rotation so that the viewpoint is to the north, although still close to directly overhead. What is not obvious from these figures is the dynamic nature of interaction with the program. Changing a viewpoint is quickly done through "pitch, yaw and roll" options, or the entire plot may be set to spin indefinitely in any direction.

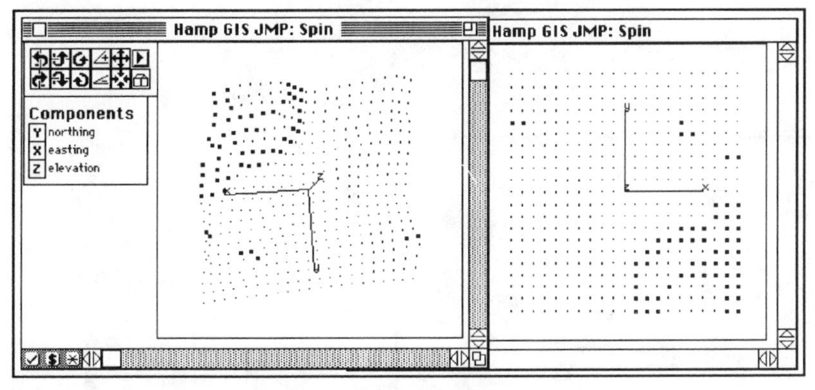

Figure 3. Spinning a plot of a raster data set.

A final and important comment on exploratory analysis of raster data sets with such statistical packages is that with larger (and more realistic) data sets than 21 rows by 21 columns, the sense of dynamic graphics is quickly lost even on faster microcomputers. For example, the author loaded a 200 by 200 grid into JMP on a Mac II accelerated to 50 MHz, and found that updating was agonizingly slow, enough that the experiment was abandoned.

A PROTOTYPE INTERFACE FOR VECTOR DATA

As useful as commercial programs might be for dynamic statistical visualization of raster data, they are of little value for exploring vector data, arguably the more common format of geographic information systems. As a result, the author developed an operational prototype on the Macintosh that implements dynamic graphics and those statistical views that appeared most immediately useful. The principal data used to test and apply this prototype were land cover maps for parts of Massachusetts originally coded in Arc-Info format. The variables used were the land cover type, and the size and shape of polygons, i.e., one categorical and two continuous variables.

Figure 4 shows output from this prototype for a small part of a data set used in training courses at the University of Massachusetts. In this case, the bar representing residential land use has been selected, and the corresponding views updated. Note that the bar charts and histograms are shown both by count (number of polygons) and area. At first this may seem to be a subtle and perhaps unnecessary distinction, but in many ways reflects one of the fundamental issues that separate the GIS and LIS communities in that the former is generally more interested in the total area of the map taken up by a particular category or class, and the latter is concerned with the total number of parcels in a category or class.

Considering this figure in a statistical sense, we see that there is a bimodal distribution of both size and shape. In fact, this occurs only because of the particular measures used for size and shape: size is the log of the polygon area, and shape is the log of the ratio of perimeter to area. One of the features of the prototype, and an important part of operations in exploratory data analysis, is the ability to re-express or transform variables by their logarithms or square roots (Velleman and Hoaglin, 1981, p. xv). Figure 5 shows the options available for re-expression, based on measures described by Forman and Godron (1986).

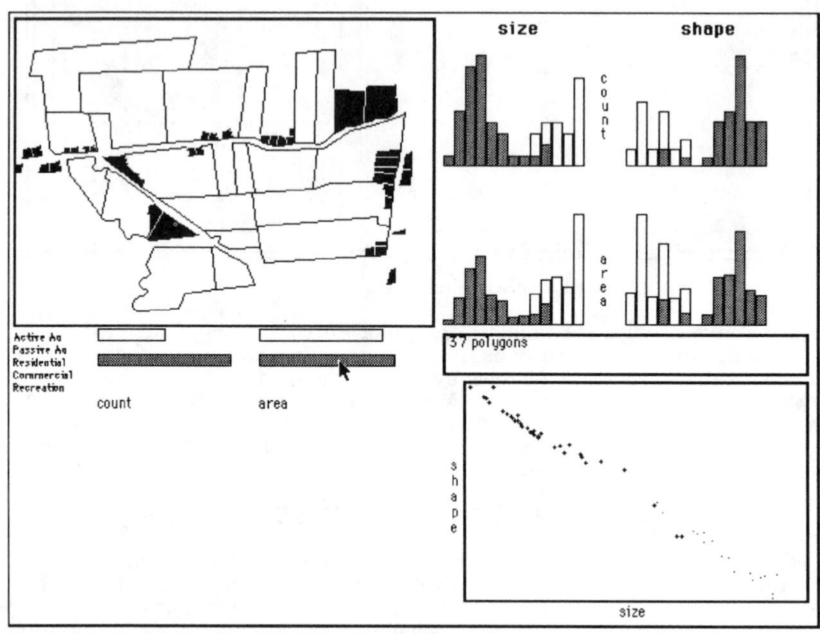

Figure 4. A screen display from the prototype.

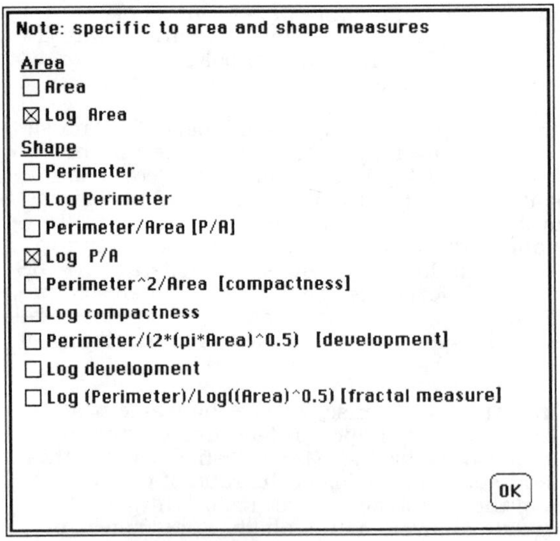

Note: specific to area and shape measures

Area

☐ Area

☒ Log Area

Shape

☐ Perimeter

☐ Log Perimeter

☐ Perimeter/Area [P/A]

☒ Log P/A

☐ Perimeter^2/Area [compactness]

☐ Log compactness

☐ Perimeter/(2*(pi*Area)^0.5) [development]

☐ Log development

☐ Log (Perimeter)/Log((Area)^0.5) [fractal measure]

OK

Figure 5. Re-expression options for polygon area and shape.

These alternative ways to express shape have proved to be quite useful in studying polygon data, particularly when several hundred polygons are involved. One typically starts by viewing area and perimeter untransformed, and then proceeds to various alternative measures in their original and then logarithmic form, at each stage selecting parts of the various views, and building up a sense of skewness or central tendency and the pattern of outliers. (Interestingly, slivers tend to be very obvious in scatterplot views as outliers.)

Another necessary part of the exploratory process is isolating particular sets of polygons for further study. For this purpose, the prototype provides a capability to restrict statistical views to a subset of polygons. Figure 6, which illustrates this feature, shows the subset of polygons representing industrial land use for part of Worcester, Massachusetts. (The line background has been hidden to better show the highlighted polygons.)

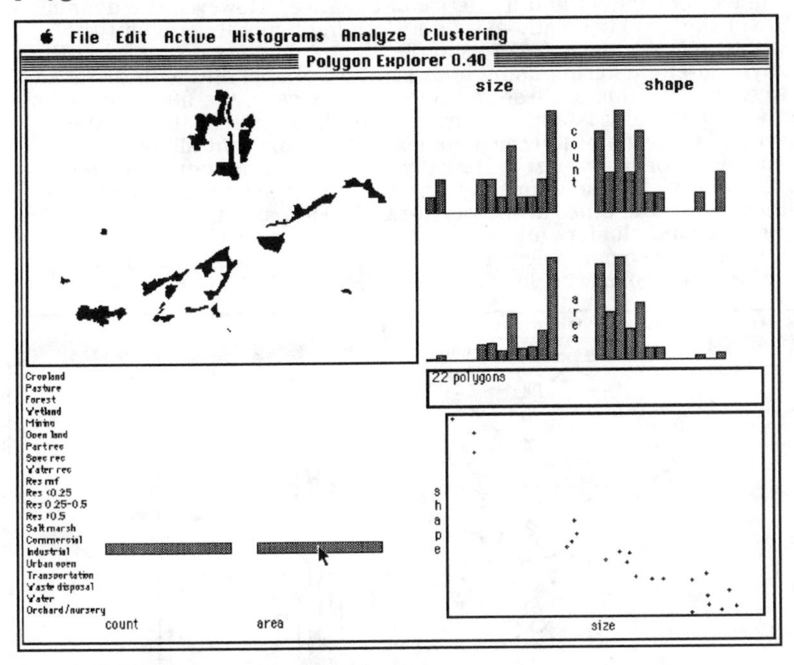

Figure 6. View of a subset of polygons.

CLUSTERING

In addition to bar charts, histograms, scatterplots and spin diagrams, there are a number of tools in exploratory data analysis that lend themselves to dynamic statistical visualization. The one chosen for investigation in this work was clustering, and specifically partitioned or non-hierarchical clustering. In the prototype, the user selects the variables to be used in the cluster calculation, and how distance between polygons should be involved (Figure 7).

The clustering calculation, based on an algorithm in Jain and Dubes (1988), first normalizes variable values and distances between polygon centroids so that they have a mean of zero and a standard deviation of 1. Distances are then multiplied by the weight chosen in the dialog box.

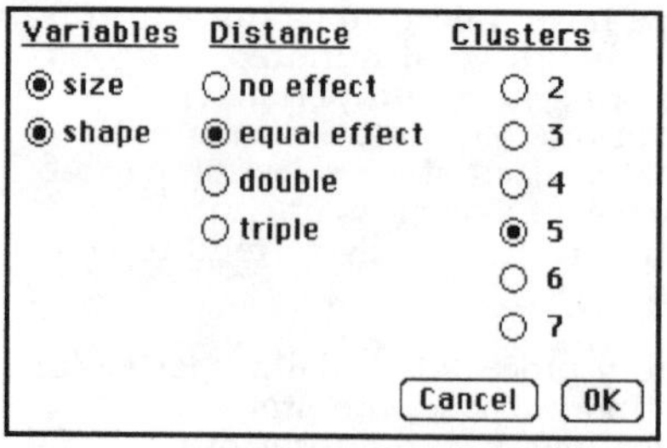

Variables Distance Clusters
size
shape
no effect
equal effect
double
triple
2
3
4
5
6
7
Cancel
OK

Figure 7. Options for clustering.

Exploratory analysis of clusters is considerably more challenging than analysis of observations, particularly given the capability to quickly vary the number of clusters and influence of distance. However, the dynamic nature of the display greatly assists in forming hypotheses about the difference among clusters. For example, Figure 8 shows output from the prototype after clustering using both size and shape, and with distance set to twice the value of these variables. In this case, the fifth cluster has been selected, and it is apparent from other displays that this cluster consists of polygons of intermediate size and shape, but all located in the east and northeast part of the map. The third and fourth clusters show a similar size and shape pattern, but one is located in the south-southeast, and the other in the north-northwest part of the map. The first and second clusters are polygons distributed evenly across the map, in one case small polygons with high shape values, and in the other, large polygons with low shape values.

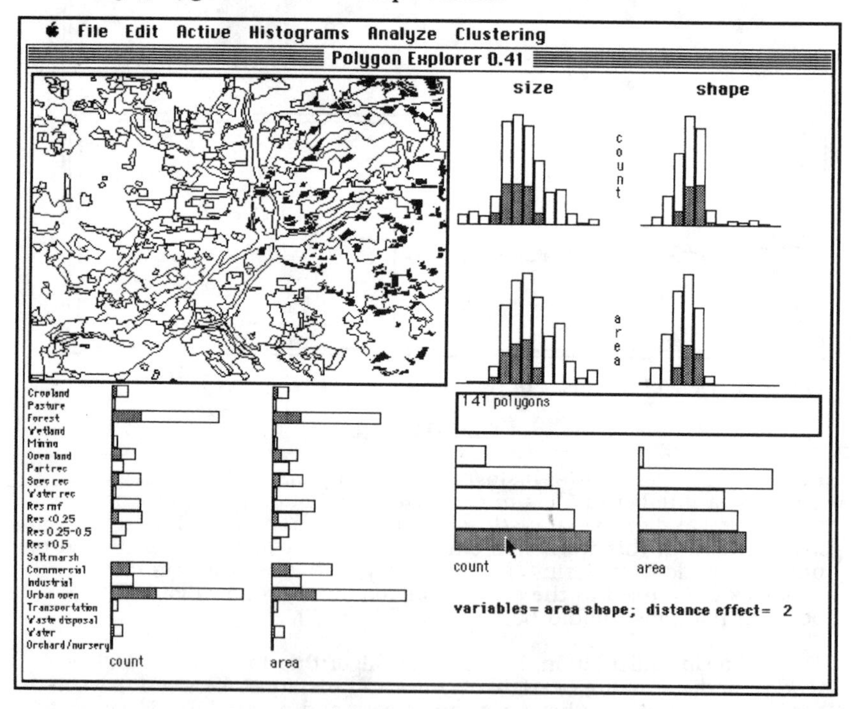

Figure 8. Screen display after clustering.

CONCLUSION

This study demonstrates that the application of the ideas of dynamic statistical visualization to the exploration of data stored in geographic information systems results in a set of novel and potentially powerful tools. These capabilities are readily available for the study of raster data sets, although the sense of dynamism is lost with more realistic data sets. The prototype for vector data described here indicates that there is considerable potential for exploring this data format. The extent to which such tools are useful and informative in exploratory data analysis remains to be considered, and is clearly dependent on the statistical knowledge of the GIS analyst.

ACKNOWLEDGEMENTS

Eun Hyung Kim provided important assistance in acquiring and converting the test data. This research was supported in part by the Massachusetts Agricultural Experiment Station.

REFERENCES

Forman, R.R.T., and Godron, M. 1986. Landscape Ecology. Wiley, New York.

Haslett, J., Wills, G., and Unwin, A. 1990, SPIDER- An Interactive Statistical Tool for the Analysis of Spatially Distributed Data: Int. J. Geographical Information Systems, Vol. 4, pp. 285-296.

Jain, A.K., and Dubes R.C. 1988. Algorithms for Clustering Data. Prentice-Hall, Englewood Cliffs, NJ.

MacDougall, E.B., 1991, A Prototype Interface for Exploratory Analysis of Geographic Data: Proceedings of the Eleventh Annual ESRI User Conference, Vol. 2, pp. 547-553.

Monmonier, M. 1990, Strategies for the Interactive Exploration of Geographic Correlation, Proceedings, Fourth International Symposium on Spatial Data Handling, pp. 512-521.

Openshaw, S., Cross, A. and Charlton, M. 1990, Building a Prototype Geographical Correlates Exploration Machine: Int. J. Geographical Information Systems Vol. 4, pp. 297-311.

Tukey, J.W. 1977, Exploratory Data Analysis. Addison-Wesley, Reading, Massachusetts.

Velleman, P.F., and Hoaglin, D.C. 1981, Applications, Basics and Computing of Exploratory Data Analysis. Duxbury Press (Wadsworth), Belmont, California.

Walker, P.A., and Moore, D.M. 1988, SIMPLE - An Inductive Modelling and Mapping Tool for Spatially-Oriented Data: Int. J. Geographical Information Systems Vol. 2, pp. 347-354.

Wills, G., Bradley, R., Haslett, J., and Unwin, A. 1990, Statistical Exploration of Spatial Data: Proceedings, Fourth International Symposium on Spatial Data Handling pp. 491-500.

MINIMUM-ERROR PROJECTION FOR GLOBAL VISUALIZATION

Piotr Laskowski

Intergraph Corporation
Huntsville, AL 35894-0001

ABSTRACT

A new minimum-error map projection is proposed to facilitate the visualization of global phenomena, such as displays of pollution or global-change data. The projection is specifically designed to achieve a balance of the three basic geometric characteristics of a world map by simultaneously minimizing the deviations from equal-area, conformal, and equidistant representations of the globe. Because of this "triple balance of errors," this projection could be termed Tri-Optimal. The Tri-Optimal projection has many positive characteristics: unavoidable distortions in polar regions are arranged to suggest the perspective view of the globe, simple polynomial mapping equations guarantee fast numerical processing, and projection coefficients maintain optimality for an arbitrary central meridian. The concept of a Tri-Optimal projection is briefly explained, and the complete set of mapping equations is given.

INTRODUCTION

In recent years there has been a growing need for the visualization of global phenomena in Earth related disciplines. To facilitate comparative and statistical analyses, the data of the global scope should be represented on a contiguous map projection which shows the entire suface of the Earth without interruptions or sharp corners and which has a minimum amount of distortion. In addition, the ideal map projection for global visualization should create the impression of the roundness of our planet. Numerical simplicity of mapping equations, which is important for the real time GIS transformations, displays, and animations, should be a desired feature.

A new minimum-error world map projection is presented which is specifically designed for the visualization of global phenomena. The Tri-Optimal projection was constructed with the unique concept of the "triple balance of errors," which achieves the best compromise among equal-area, conformal, and equidistant representations of the globe.

TRI-OPTIMAL PROJECTION OF THE WORLD

The concept of Tri-Optimal world projection uses the original combined measure of map distortions, discribed in detail in (Laskowski 1991). This new analytic error

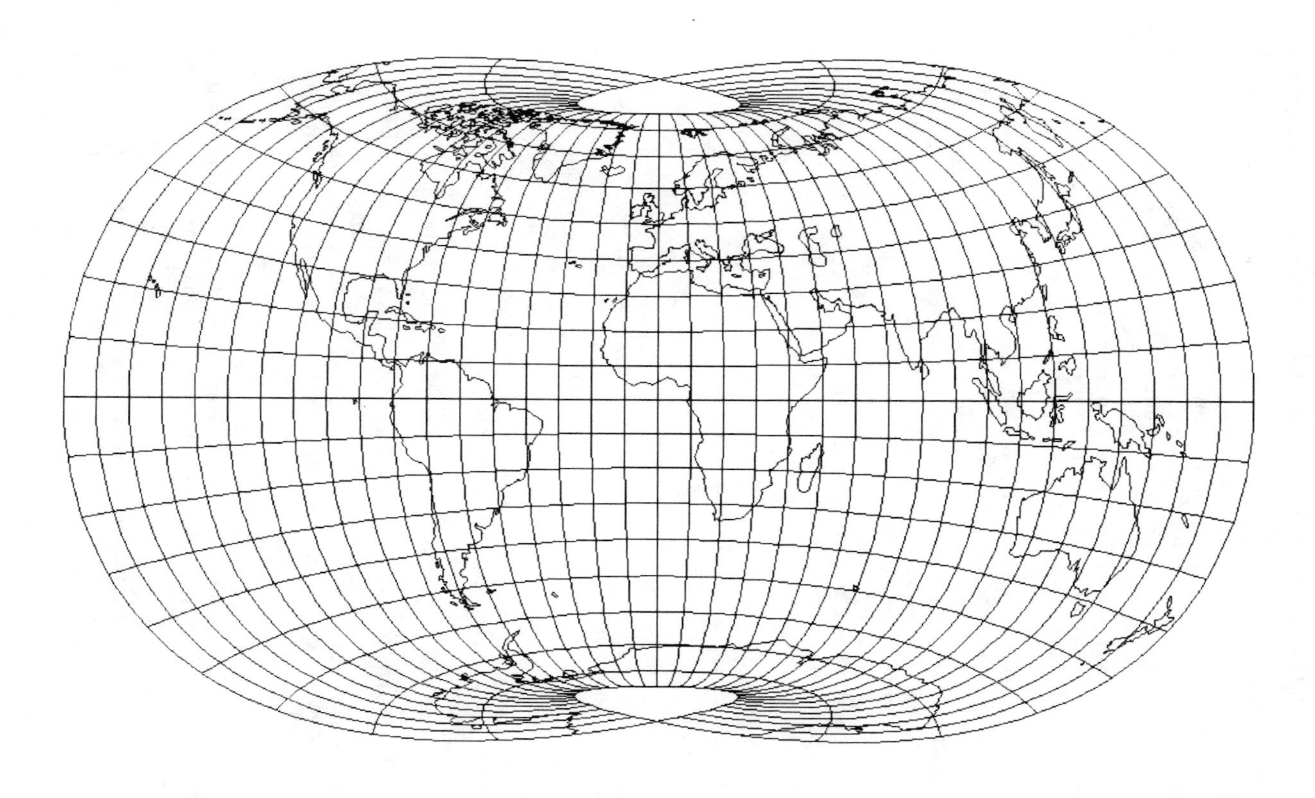

TRI-OPTIMAL PROJECTION OF THE WORLD

measure combines in its three components the three important map characteristics: departure from conformality, departure from equal-area, and departure from true distances. The first two components date back to Airy (1861) and measure the deviations of Tissot's error elipses from unit circles. The third component, first used by Gilbert (1974), represents the measure of errors in finite distances on the globe. All three components are evaluated numerically as mean-square errors averaged over 5000 points randomly distributed on the surface of the globe (Laskowski 1991). Finally, the combined error measure was minimized over the family of projections having the fifth-degree polynomials in longitude and latitude as mapping equations.

The optimal projection was constructed by numerical minimization of the combination measure. The procedure resulted in the best compromise among equal-area, conformal, and equidistant world projections, on which the combined mean square errors in areas, shapes, and finite distances have been simultaneously minimized.

The multidimensional minimization algorithm has solved for 10 optimal polynomial coeficiants, which are provided in Table 1.

Table 1. The coefficients of the Tri-Optimal minimum distortion polynomial projection.

$a_{10} = 0.975534$	$b_{01} = 1.00384$
$a_{12} = -0.119161$	$b_{21} = 0.0802894$
$a_{32} = -0.0143059$	$b_{03} = 0.0998909$
$a_{14} = -0.0547009$	$b_{41} = 0.000199025$
	$b_{23} = -0.0285500$
	$b_{05} = -0.0491032$

The mapping equations for the Tri-Optimal projection have the following polynomial form:

$$X= a_{10}\lambda + a_{12}\lambda\phi^2 + a_{32}\lambda^3\phi^2 + a_{14}\lambda\phi^4 \qquad (1a)$$

$$Y= b_{01}\phi + b_{21}\lambda^2\phi + b_{03}\phi^3 + b_{41}\lambda^4\phi + b_{23}\lambda^2\phi^3 + b_{05}\phi^5 \qquad (1b)$$

When used with the coefficients from Table 1, the above equations result in the Tri-Optimal world map projection, shown in Figure 1.

By design, the Tri-Optimal projection exhibits symmetry about the central meridian and equator (both represented by straight lines, with the equator evenly divided by the meridians).

In addition to being mathematically optimal (the minimum-error projection), the above representation has many other useful characteristics which make it well suited for general-purpose global visualizations, as well as for statistical comparisons and quantitative data evaluations.

The main features of the Tri-Optimal projection may now be listed.

- Minimum-distortion projection of the world.

- Triple balance of errors: best compromise among equal-area, conformal, and equidistant representations.

- Simple mapping equations.

- Optimal coefficients valid with arbitrary selection of central meridian.

- Whole globe represented as apple-like oval shape, without interruptions or sharp corners.

- Unavoidable distortions in polar regions are shaped to mimic the perspective view of the globe and to create an illusion of viewing simultaneously the front and the back side of our planet.

CONCLUSION

The properties of the Tri-Optimal world projection make it the best candidate for the base cartographic representation used with the global data visualizations. This novel representation keeps a well balanced mixture of geometric distortions at the absolute minimum, a feature important for accurate spatial analyses. The projection is well suited for general-purpose global visualizations, which represent the whole world without interruptions or sharp corners, but which enhance the perception of roundness of our planet. Inexpensive mapping equations make it ideal for animation of dynamic global phenomena.

Most importantly, the Tri-Optimal projection has the smallest possible geometric distortion among all projections which are representable by 5-th degree polynomials in longitude and latitude (and most projections are well representable as polynomials through the Taylor series expansion). Due to this analytic optimality, the new projection is especially well suited for statistical and quantitative data analysis.

REFERENCES

Airy, G.B. 1861, Explanation of a projection by balance of errors for maps applying to a very large extent of the Earth's surface; and comparison of this projection with other projections: London,Edinburgh, and Dublin Philosophical Magazine, series 4, v.22, no.149, pp.409-421.

Gilbert, E.N. 1974, Distortion in maps: SIAM Review, vol.16, no.1, pp.47-62

Laskowski, P. 1991, On a mixed local-global error measure for a minimum distortion projection: Technical Papers, 1991 ACSM-ASPRS conference, vol.2, pp.181-186.

LARGE-FORMAT HARDCOPY DISPLAY OF TM, DLG, AND DEM DATA IN THREE-DIMENSIONS

E. Lynn Usery
Jennifer S. Norton
Department of Geography
University of Wisconsin-Madison
Madison, Wisconsin 53706

Geographic information system (GIS) datasets containing volumetric information can be displayed in three dimensions. This paper presents methods for combining vector-based map data such as U.S. Geological Survey (USGS) digital line graphs (DLG's) with digital elevation model (DEM) terrain and Landsat Thematic Mapper (TM) digital satellite images to create large-format three-dimensional views. The methods of generation include synthetic stereo techniques to produce virtual three-dimensional images and hardcopy plotting methods developed to combine the vector and raster data at the output stage. DEM's are used to create synthetic stereo images for both the vector DLG and raster TM data. Combination of these stereo datasets at the output stage creates a three-dimensional scene of image information with vector overlays. Using electrostatic technology, a matrix of 4x4 or 8x8 plotter pixels on a medium resolution (400 dpi) device is used to represent red, green, and blue values for a single data pixel of a color image. Vector data are "threaded" through the 4x4 or 8x8 matrix to yield a true vector plot at the plotter resolution. Symbolized vectors use multiple pixels to achieve specific line widths. These techniques have been implemented with test sites in southwestern Wisconsin and the Grand Canyon area. For the Grand Canyon, a total coverage of three by eight degrees was plotted at 1:250,000 scale and yields an image map measuring five by nine feet. The large format, stereo image presents a view of the Grand Canyon which has significant applications for geomorphological analysis.

INTRODUCTION

The display of three-dimensional geographic datasets is commonly accomplished using a two-dimensional transformation of data to yield a cartographic product. These map products primarily use psychological depth cues such as relative size, linear perspective, height of objects above the line of sight, shadow, relative brightness, and atmospheric attenuation to convey three-dimensional information (Braunstein, 1976; Okoshi, 1976; Usery, 1991). Common examples of such products include contour maps, shaded relief images, and perspective block diagrams. These methods convey the volumetric aspect of the data but generally cannot display both effective three-dimensional visualization of the surface and provide the ability to make measurements of the vertical dimension. Conventionally, in the map-making process, measurements of three-dimensional data are performed using aerial photographs on stereoplotting instruments which construct a virtual three-dimensional model. A floating mark is moved in the vertical dimension of the model and heights are recorded (Wolf, 1983). Unfortunately, this process requires separation of the left and right images of a stereo pair of photographs and the direct translation to large-format hardcopy media is not possible. To generate an effective large-format three-dimensional display, anaglyphic techniques must be used.

Anaglyphic stereo display techniques separate the color spectrum into two parts, the short and long visible wavelengths yielding red and blue-green images. Display of the left image of a stereopair in red and the right image in blue or green yields a virtual three-dimensional panchromatic image. Color is not possible with anaglyphic displays since the entire spectrum is used to separate the left and right images. An alternative is image separation with polarized light which can be used for electronic displays and specific hardcopy polarized films (Beaton, et al, 1987; Moellering, 1989). True color three-dimensional images can be generated using polarized light; however, the requirement for specialized media prohibits this technique from being used in large-format hardcopy displays. Comprehensive reviews of these and other technologies for display of three-dimensional images are presented by Fornaro et al (1985), Hodges and McAllister (1985), and Kraak (1989).

Any geographic dataset containing quantitative values over an area can be displayed as a virtual three-dimensional image. Examples include elevation data, rainfall amounts, population, and percent urban. Datasets which are not volumetric, such as land use/land cover, can be combined with volumetric data to create a stereo image (Batson, 1976; Jensen, 1980; Usery, 1991). These synthetic stereo images provide an excellent method of visualizing natural resource and map data for analytical and interpretation purposes. It is the purpose of this paper to describe methods of combining TM, DEM, and DLG datasets to create synthetic stereo large-format hardcopy displays. The combination of these datasets requires vector-to-raster conversion which is accomplished in a manner to maintain the highest possible spatial resolution for each type of data.

OVERVIEW OF THE RESEARCH

The focus of this paper is on specific display techniques which are a part of a larger research project including input and output of geographic data in a variety of formats and media. A general purpose display program (PLOTIT) is under development including a mix of data types and processing options. The design of this software package includes format conversions, data structure conversions, raster/vector combination and overlay, and plotting (Figure 1). The following discussion will concentrate on the development of the combined vector/raster plotting on an electrostatic plotter using the Universal Host Computer-Based Software (UHCBS) from Calcomp. Other parts of the package which are already complete, such as the raster plotting on the Howtek printer and vector plotting with Hewlett-Packard's Graphic Language (HPGL) will not be addressed in this paper.

To combine raster and vector data for output generation is a different problem from combination at the database level or combination for analytical purposes. In

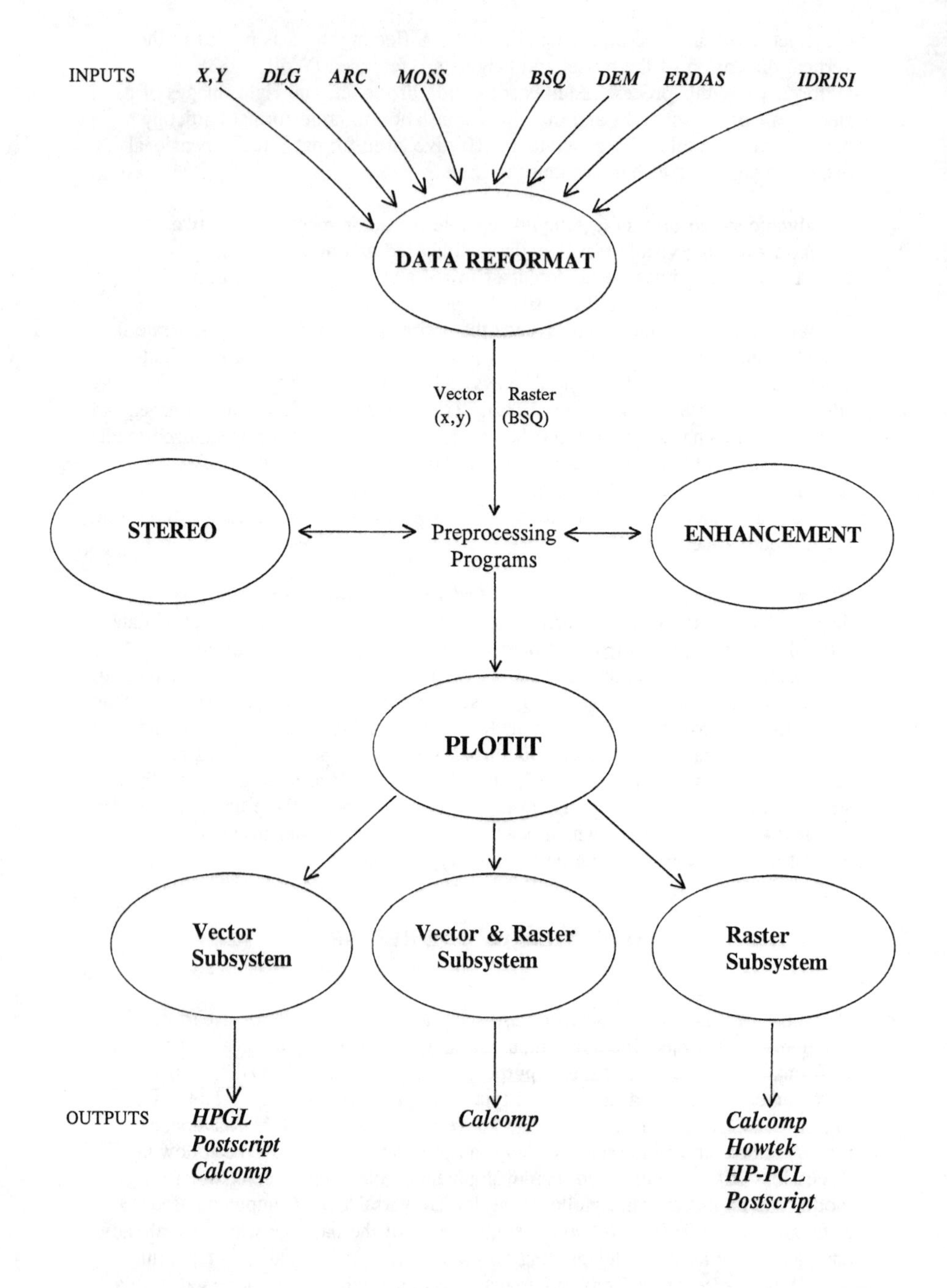

Figure 1. The design of the integrated vector/raster plotting system.

both of the latter cases, the combination is usually performed by vector-to-raster or raster-to-vector conversion to bring the datasets to a common resolution. For output generation, the conversion is also necessary but can be performed in a manner to preserve higher resolution of the original datasets. For example, raster data plotted on a hardcopy raster-based plotter often uses multiple pixels on the display device for a single data pixel. This allows more colors to be represented and yields a better image than representing each pixel in the data as a single pixel on the display device.

The use of process color with cyan, yellow, magenta, and black (CYMK) for hardcopy generation requires multiple pixels on the output for one raster data pixel. Vector data can be displayed with CYMK with single output device pixels representing single pixel width lines. True map symbology can be generated by using multiple pixels for the line width. The logical combination of these two representations can be achieved.

IMPLEMENTATION

The program design shown in Figure 1 is being implemented on an International Business Machines (IBM) RS/6000 Model 540 workstation in the Department of Geography at the University of Wisconsin-Madison. The code development is being accomplished in FORTRAN and C with the use of the appropriate libraries to support specific devices. For example, UHCBS is used to develop plot routines for a Calcomp Model 5700 electrostatic plotter. This library allows plotting of both raster and vector data on the 400 dpi color raster plotter.

Programs originally developed on microcomputers and on Prime minicomputers have been moved to the RS/6000 and form a part of the PLOTIT design. For example, a PC-based FORTRAN program called HOWPRT which generates page size color plots on a Howtek color printer from band sequential digital image files was moved to the RS/6000 and implemented as a part of the PLOTIT package. The raster plotting program for the Calcomp electrostatic was originally developed at the U.S. Geological Survey on a Gould minicomputer but was migrated to a Prime minicomputer where it was modified to be interactive. The interactive program was then moved to the RS/6000 as the raster plotting portion of PLOTIT for Calcomp electrostatic plotters.

While the PLOTIT software is a general purpose utility for hardcopy generation of vector and raster datasets, it contains no utility for image enhancement or synthetic stereo generation. Since these routines are necessary to generate a high quality three-dimensional visualization of the geographic datasets, separate programs have been implemented on the RS/6000 for these functions. The program STEREO accepts a digital image in band sequential format, a corresponding DEM to which the image has been registered, and user specifications for exaggeration factor and other parameters to control the stereo generation process, and creates a stereomate for the original digital image. Similarly, the program ENHANCE will perform a linear contrast stretch or histogram equalization on the specified digital image prior to plotting in PLOTIT.

DATABASE DEVELOPMENT FOR THE GRAND CANYON

The generation of a 1:250,000-scale synthetic stereo image of the Grand Canyon requires combination of many datasets. TM data from eight scenes were acquired from the National Aeronautics and Space Administration (NASA), the USGS, and the University of Colorado at Boulder; DEM's for the 12 1:250,000-scale quadrangles and transportation DLG's generated from 1:100,000-scale topographic maps covering the test site of 3x8 degrees were obtained from the USGS. To achieve a composite representation the TM scenes were rectified and resampled and horizontally joined. The DEM's were projected to the Universal Transverse Mercator coordinate system and both DEM's and DLG's were horizontally joined.

Once TM and DEM data were projected to UTM coordinates, the STEREO program could be used to generate a stereomate for any TM band. The process offsets each TM pixel in the stereomate by an amount proportional to the elevation of the pixel modified by the user-specified exaggeration factor (for details of the process, see Batson, 1976; Jensen, 1980; Usery, 1991).

The integration of the DLG data requires development of individual plot capabilities for the vector and raster data with sequential plotting to the same media. The raster TM data for each image of the synthetic stereopair is plotted using a 4x4 or 8x8 matrix of pixels on the 400 dpi Calcomp electrostatic plotter. The data are plotted directly by calls to raster filling subroutines (RASFIL) in UHCBS. The vector data are plotted as an overlay on each image at full plotter resolution. The vectors essentially thread through the 16 (for the 4x4 case) or 64 (for the 8x8 case) image data pixels. The vector DLG data are plotted by direct calls to the vector plotting routines (e.g., PLOT) in the UHCBS library. Registration of the vector and raster data is accomplished by transformation of the original UTM coordinates to the plotter coordinate system which is 400 dpi for the vectors and 100 or 50 dpi for the raster data. The transformation assures exact alignment of the vectors with corresponding image pixels.

Enhancement of the cartographic data including symbolized vectors for roads and hydrography, text for names, and specific point and area symbols can be implemented directly with the vector data. Titles, legends, and other marginal text such as grid references are added by a separate set of subroutines.

The data processing steps in this project have required significant resources and personnel time. Combination of 24 half-quad DEM files; rectification, registration, enhancement, and mosaicking of eight TM scenes; and horizontal integration of over 1000 DLG files were required to generate composite images prior to synthetic stereo generation and image plotting. Disk capacities for such file processing were constrained requiring TM data to be processed as half scenes. Initial processing of the DLG's on 80386- and 80486-based microcomputers required several weeks. A similar process was accomplished on the RS/6000 in a single day.

RESULTS

The results of the development have yielded a number of useful images in large format. Synthetic stereo using both shaded relief and TM data with the DEM's have been generated for test sites in Wisconsin, Georgia, and the Grand Canyon. The integration of the DLG vector data was first tested on small datasets of TM imagery over Madison, Wisconsin. The final stereo images for the Grand Canyon have been generated with several different TM bands. A band 3 (red) large-format hardcopy stereo image will be displayed at the meeting.

CONCLUSIONS

The ability to generate large format synthetic stereo displays of Landsat TM and other digital images has been developed. The generation requires the TM image to be registered to a DEM on a pixel-by-pixel basis. DLG data can be plotted as a vector overlay at plot time allowing both raster and vector data to be used without conversion of the databases. The raster and vector data are transformed so that each retains the highest possible spatial resolution. For example, on the Calcomp plotter with a hardware resolution of 400 dpi, the vector data are transformed to this resolution while each raster data pixel is represented by at least 16 plotter pixels. Because of the limited areal coverage of digital files such as DEM's and DLG's, horizontal integration of many datasets is required to produce synthetic stereo images of large areas. The problems of data processing, disk capacities, projection over large areas, and merger of the datasets are significant and consume much personnel time.

ACKNOWLEDGEMENTS

The authors thank Greg Allord of the USGS Water Resources Division (WRD) office in Madison, Wisconsin, for his support of this work including provision of the DLG and DEM data and access to the Calcomp electrostatic plotter. We acknowledge David Hooper of the USGS National Mapping Division office in Menlo Park, California, who developed the original digital image plotting program. TM data were supplied by the USGS, Geologic Division in Reston, Virginia; the USGS WRD in Las Vegas, Nevada; the Pilot Land Data System of NASA Goddard Spaceflight Center; and the University of Colorado in Boulder. The funding for the project was provided by the University of Wisconsin-Madison under a Hilldale undergraduate research fellowship.

REFERENCES

Batson, R.M., K. Edwards, and E.M. Eliason, 1976. "Synthetic Stereo and Landsat Pictures," *Photogrammetric Engineering and Remote Sensing*, 42/10, 1279-1284.

Beaton, R.J., R.J. DeHoff, N. Weiman, and P.W. Hildebrant, 1987. "An Evaluation of Input Devices for 3-D Computer Display Workstations," *True 3D Imaging Techniques and Display Technologies, Proceedings, Society of Photo-Optical Instrumentation Engineers*, Vol. 761, pp. 94-101.

Braunstein, M.L., 1976. *Depth Perception through Motion*, Academic Press, Inc., New York.

Fornaro, R.J., et al, 1985. "Interactive Geomorphic Processor," Final Technical Report, RADC-TR-85-142, Rome Air Development Center, New York.

Hodges, L.F. and D.F. McAllister, 1985. "Stereo and Alternating Pair Techniques for Display of Computer-Generated Images," *IEEE Computer Graphics and Applications*, 5/9, pp. 38-45.

Jensen, J.R., 1980. "Stereoscopic Statistical Maps," *The American Cartographer*, Vol. 7, No. 1, pp. 25-37.

Kraak, M., 1988. *Computer-Assisted Cartographical Three-Dimensional Imaging Techniques*, Delft University Press, Delft, The Netherlands.

Moellering, H., 1989. "A Practical and Efficient Approach to the Stereoscopic Display and Manipulation of Cartographic Objects," *Proceedings, Auto-Carto 9*, pp. 1-4.

Okoshi, T., 1976. *Three-Dimensional Imaging Techniques*, Academic Press, Inc., New York.

Usery, E.L., 1991. "Stereo Display Techniques for GIS," *Proceedings, American Society for Photogrammetry and Remote Sensing/American Congress on Surveying and Mapping Annual Convention*, Vol. 4, GIS, Baltimore, Md., pp. 227-236.

Wolf, P.R., 1983. *Elements of Photogrammetry, 2nd Edition*, McGraw-Hill Book Co., New York.

INTEGRATING VISITOR SURVEYS, ENVIRONMENTAL FACTORS, AND GIS INTO RECREATION PLANNING: SITE DEVELOPMENT AT GRAYROCKS RESERVOIR, WYOMING

William J. Gribb
University of Wyoming
Department of Geography and Recreation
Laramie, Wyoming 82071
(307) 766-3311

ABSTRACT

Grayrocks Reservoir, Wyoming is an undeveloped recreation site with over 35,000 visitors per year. State and local government officials and concerned citizens wanted to know what visitors desired at the site, what problems they were experiencing and where they should locate recommended facilities. To answer these questions a combination of techniques were utilized. A questionnaire was distributed to recreationists to determined what problems they encountered, what facilities they would like, what services they needed, and a map to delineate areas of activities by visitors. At the same time, documental information was gathered and mapped on eight environmental and jurisdictional factors to determine legal and environmental limitations and constraints to development at the reservoir. The integration of the questionnaire results, the activity maps, and the environmental and jurisdictional constraints into a GIS provided the needed information to make rational decisions for the development of Grayrocks Reservoir.

INTRODUCTION

Pressures for relieving the increased demands on a favorite "fishing hole" forced the Wyoming Game and Fish Department and the Wyoming Department of State Parks and Historic Sites to examine the problems at Grayrocks Reservoir (Map 1). The reservoir is normally 1740ha and contains approximately 105,000 acre-feet of water with no permanent recreation facilities. But, more than 30,000 fisherperson days have been recorded for the past several years (Snigg, 1990). The reservoir is used as a water reserve for the Missouri Basin coal-fired power plant. Problems were being encountered at its small boat ramp, lack of sanitation facilities, unimproved camping, and limited reservoir access. There are three major problems which have to be resolved at the site. First, what are the major problems experienced by the reservoir visitors. Second, what types of facilities and services do the visitors want at Grayrocks. And third, where should the facilities be located to serve the users.

METHODOLOGY

Four research techniques were utilized to obtain the needed information; questionnaires, maps and aerial photography, documents, and GIS analysis. This combination of techniques provided the data by which the site development recommendation could be developed.

Figure 1. Grayrocks Reservoir in Southeastern Wyoming.

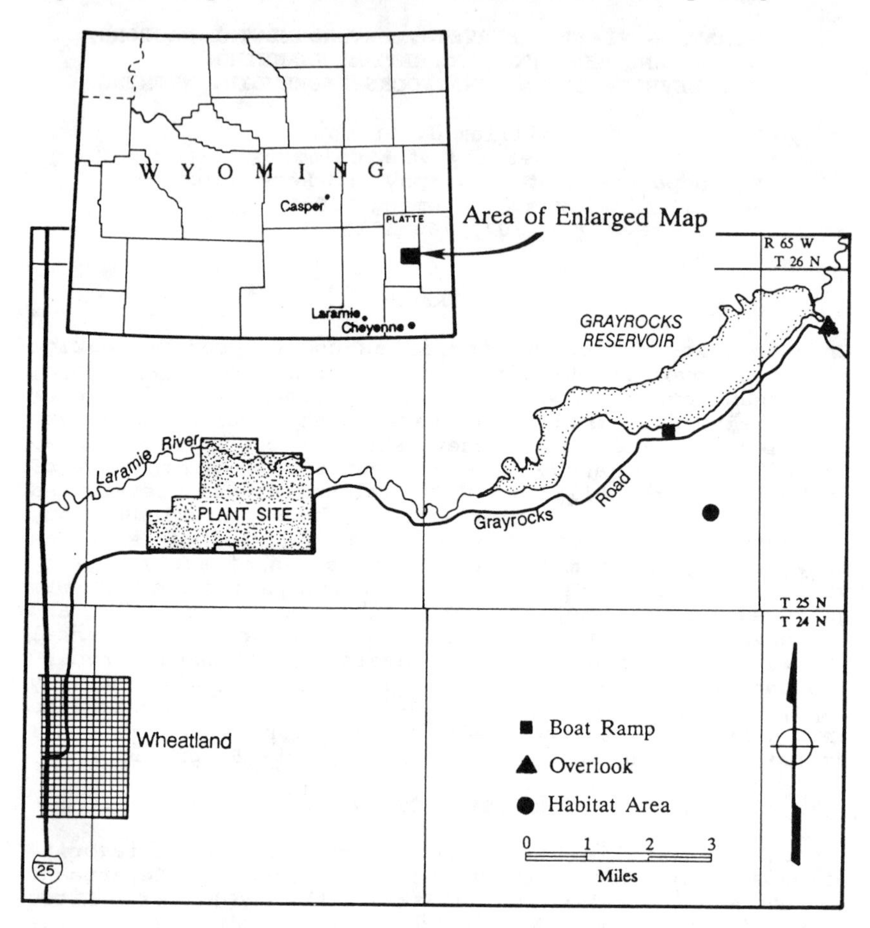

<u>Questionnaire</u>
To obtain visitor information, individually completed questionnaires were employed (TRRU, 1983). To obtain the data concerning the visitors of Grayrocks Reservoir, a four part questionnaire was developed and distributed to visitors. A random selection of seventeen days were utilized to collect user information. The four parts of the questionnaire are: basic visitor data; problems encountered; additional facilities needed; and, activity areas. A unique feature of the questionnaire was the employment of a map depicting the reservoir and surrounding areas so that visitors could literally delineate what areas they were using for sixteen identified activities.

<u>Maps and Aerial Photographs</u>
During the building of the Missouri Basin Power Plant a number of different state and federal permits had to be

obtained (WIS, 1973). To fulfill the permitting requirements, design and engineering maps had to be completed. The maps and information in the permit reports provided valuable information on topography, vegetation, wildlife habitats, soils, geology, archeology, and hydrology. These were basic considerations for the placement of any improvements suggested from the questionnaires. However, two major problems were experienced. First, some of the data did not extend into the full area impacted by the reservoir. And secondly, because the data was collected before the reservoir was completed, the reservoir "normal high-water" level was not depicted for hydrology or its impact on surrounding vegetation types. The use of a 1986 NAPP CIR photograph provided an excellent image of both.

Documentation

The documentation for the permitting procedures provided a substantial amount of information about the environmental and engineering characteristics of the Missouri Basin Power Plant. However, additional information was needed concerning what plans the Power Plant had for the reservoir, what the Wyoming Departments of Game and Fish and State Parks and Historic Sites were planning, and the jurisdictional restrictions on lands surrounding the reservoir were. Research was conducted in libraries, department offices, and county records to determine the limits and expectations for development of the reservoir.

GIS Analysis

Because of the range of data needed to determine the most appropriate areas for development or facility improvement, it was decided to enter the data into a GIS system. At the time, the system available for this research was IDRISI. Using a 4.05ha cell size, the Grayrocks Reservoir and surrounding area comprised 2560 grid cells or a total of 1036ha (10.36 sq. km.). All maps were converted to a computerized data base. Once the constraint variables were determined it was then possible to use Boolean logic and the overlay method to determine the most appropriate locations for development or facility improvement.

ANALYSIS

The data gathered by the different techniques provided the basis upon which to develop the appropriate recreational plan strategy. Five broad categories of information were utilized to select the sites and determine what facilities were needed. The five categories include the data from the questionnaire, the environmental material, the jurisdictional constraints, and the use demands from the projected number of users.

Questionnaire

During the seventeen days of sampling responses from visitors, 328 questionnaires were acquired. Because of incomplete forms, however, only 325 were employed in the analysis. The four main areas of the questionnaire revealed important descriptive information concerning the visitors. The visitors were generally couples (28%) or families (22%), going to Grayrocks as their primary destination (97%). The majority of the visitors camp using an RV or tent trailer (66%), with almost 50 percent bringing a boat. These are critical issues because it is imperative to know what type of camping is taking place because of the different amounts of space needed for the types of camping and vehicle parking. The boat issue is further compounded by the fact that a boat launch, mooring and storage is critical to their use. Grayrocks does have a boat ramp with an adjacent graded area for parking, but it is only .5 hectares in size, and there are no mooring capabilities.

There were several problems identified by the respondents. None, however, were extremely serious but create a multiplier effect on the enjoyment level. Figure 2 illustrates the leading problems experienced by visitors. On a scale of one (no problem) to five (excessive problem) overcrowding (2.9) and traffic congestion (2.9) were cited as the major problems. The overcrowding is experienced both at the most preferred camping areas and at the boat ramp. This is compounded by the fact that the area surrounding the boat ramp is the most preferred camping site. The second major problem, traffic congestion, also relates to the boat ramp. The flow of vehicles is disrupted during the launching and extracting the craft, waiting in-line to launch or trailer the craft, parking, and mooring. The highest density of camping and boat mooring occurs in an area of less than 20 hectares.

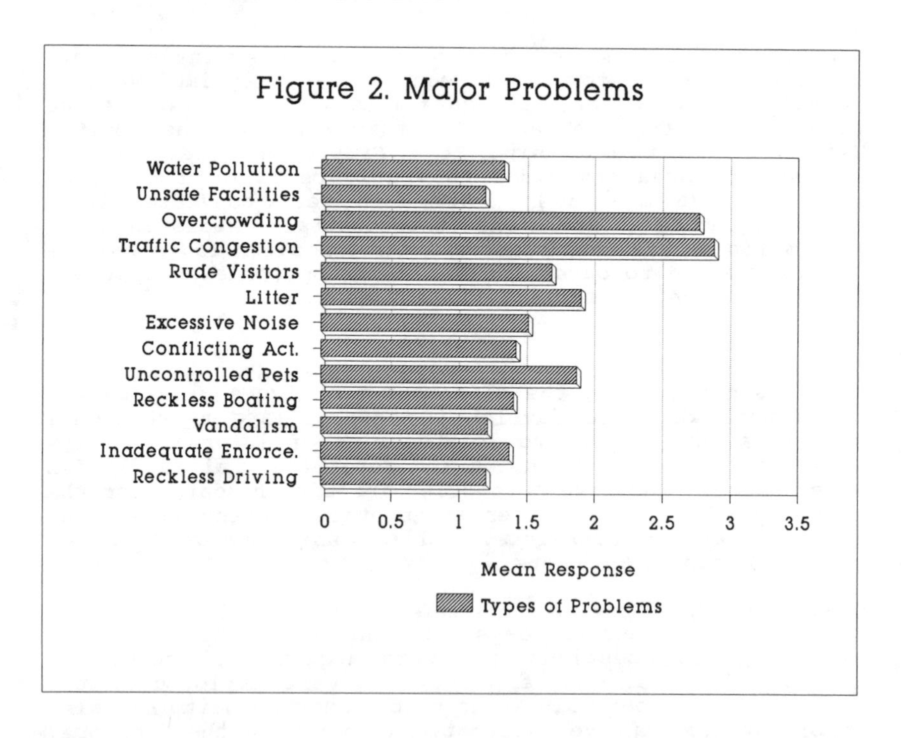

The visitors also registered several different types of

services and facilities they believed were needed. Over forty facility improvements, additional services, and expanded activities were identified by the visitors. However, three additional facilities were the highest priority; an additional boat ramp, designated camp grounds, and a picnicking area.

Environmental

The environmental data provided valuable information on the physical conditions in the project area and identified constraints to any development. According to Hirsch (1988), topography and soils are the two most important factors in recreational site development. However, hydrology of the reservoir is another critical factor. The topography of the area changes significantly but what is most critical is the slope adjacent to the reservoir. Flat-topped hills north of the reservoir are dissected by washes. While south of the reservoir the landscape rises over 200 meters and is dissected by a steep-walled canyon. Adjacent to the reservoir the shoreline is generally gently sloping except in the northeastern quarter which has almost vertical cliffs. Soils are divided into three major categories: the Dunday-Trelona sandy loam, rock outcrop, and the Dunday-Dwyer loamy fine sand (SCS, 1971). There is a high erosion rate experience by the Dunday soils if the vegetation is removed. Thus, in areas that are utilized extensively by campers and boaters erosion gullies are evident.

Though the Grayrocks Reservoir is designed to have a relatively small fluctuation, less than 7.6 meters, the areal change is significant (MBPP, 1979). The maximum high water mark is 1345.7m with a gross area of 1714.7ha. While the low water mark is 1338.7m and an area of 1156.2ha. All facilities have to be built above the high water mark, especially any sanitation services. This creates a problem because of the change in shoreline and distances to water.

Use Rights

The lands surrounding Grayrocks Reservoir are encumbered or restricted in various ways (Dombroski, 1990). Missouri Basin Power has either lease or extended rights to lands based on the elevation of the shoreline. The shoreline elevation also determines the extent of flooding easement. When the reservoir was built several adjacent ranchers sold their property to MBPP but retained their water access rights. Further, several parcels of land adjacent to the reservoir belong to either the US Bureau of Land Management, the Wyoming State Land Office, or the Wyoming Department of Game and Fish. Each of these different parcels have restrictions on them relating to access, use, or encroachment. All of the different use rights, ownership, and restrictions were integrated into the GIS. Thus, it was possible to determine what constraints each parcel of land had relative to site development.

Use Conflicts

To resolve the question of where to place the requested facilities, another concern was the interaction between different types of visitors and the activities in which they participated. Part of the questionnaire instructed the visitors to delineate on a map where they partook of 16 different activities. The purpose of this question was to determine not only the location of the different activities, but also where areas of potential conflict could occur. By mapping the location of each questionnaire and combining them through the GIS it was possible to identify the major conflict areas. Three types of confrontations were identified: between fisherpersons and water skiers, campers and shore fisherpersons, and boating and campers. Thus, regulations or locations could be developed which would eliminate the conflicts. Water skiers could be kept a minimum distance away from boat fisherpersons. Camping sites would be a minimum distance (61m) from the shoreline and from either the boat launching site (152m) or any locations for moorings. The restrictions were incorporated into the GIS so that potential sites could be determined.

Use Demands

Use demand was extrapolated from three different sources of data. During each of the seventeen days of surveying, an inventory of boats, vehicles, and campers was compiled. This information would give an indication of the demands on the reservoir and was used to estimate use for all recreation days. A second source of use demand came from electronic counters placed at the access points to the reservoir. The counters would record the number of large metal objects passing the gate. One hundred and sixty-four days of collection produced reasonably accurate counts on the number of objects passing through the different access points. The questionnaire survey revealed the percentage of the different types of vehicles, trailers, and the number of persons in each type of vehicle. This information was then utilized in association with the numbers from the counters to predict the total number of visitors, 38,716.

The third source of demand data came from the newly completed Wyoming State Comprehensive Outdoor Recreation Plan (WSCORP) (WDC, 1990). This document stated the standards to be employed for all the different types of outdoor recreation activities. Thus, based on the number of users, their activities, and demands, an estimate of how much space should be allocated for each use could be determined. Table 1 details the spatial demands for the major developments at Grayrocks Reservoir.

Table 1. Area Demands by Activity

Activity	No. of Users Ave./Max.	Average Use Area (ha)	Maximum Use Area (ha)
Camping	451/811	9.15	16.4
Picnicking	12.3/68	.2	1.2
Boat Ramp	79.5/187	1.25	2.4

Based on the number of visitors collected by the actual observational counts and the electronic counter, the frequency of visitors for holidays, weekends, and weekday use were tabulated. Maximum use was recorded on Memorial Day with 811 campers and 187 boats on the reservoir. The facilities should be designed not necessarily to meet maximum use levels, but to accommodate the 75th percentile use level. Further, the number of users had to be categorized by the type of use they demanded. For example, campers with tents have a different areal demand then campers with RV's. Thus, the number of users distinguished by the type of use had to be combined with the WSCROP standards for number of campsites per hectare. Thus, the camping area should be approximately 12.8ha, picnicking - .7ha, and the boat ramp approximately 3.6ha. But, because a single land boat ramp can only handle 8 boats/hr the 140 boats (75th percentile) would take 17.5 hrs. However, to accommodate the demand in less than 9 hrs., it is recommended to have two boat ramp areas, with parking, each approximately 1.8ha in size.

The size requirements for each of the activities were then included in the GIS. Along with the size requirements, other constraints were included for each of the different activities (Lancaster, 1987). Slope requirements for a boat ramp, both above and below the waterline, were a major consideration. Camping and picnic areas are best developed in areas with level ground, loamy to sandy soils, and vegetation cover. Also, to reduce development costs, closeness or ease of access to transportation routes and utility lines had to be considered.

<u>Summary</u>

By combining the four different types of analysis it is then possible to detail what types of facilities were needed, what constraints to development exist based on use rights and environmental factors, and the space requirements based on demand. The visitors wanted an expansion of the boat ramps and adjacent parking. This area caused the most traffic congestion and produced a negative experience for the visitors. To resolve the problem and to meet the demand, what is needed is an expansion of the existing parking area and development of a second boat ramp with adequate parking.

More than 84 percent of the visitors camp at Grayrocks, unfortunately there is not a designated site nor developed camp sites. A designated camping area should be developed with toilet facilities and appropriate utilities for recreational vehicles. Further, the camping should be located away from the boat ramp and the shoreline.

Currently, there are no picnic tables at Grayrocks. Day and overnight users have requested picnic grounds for the "short-term" user or organized group use. The picnic facilities should be away from the boat ramps and the camping area. But, should be located near the reservoir or in a scenic site.

Recommendations

The sites for the requested and needed improvements to Grayrocks were based on Boolean logic and design considerations. The general sites were distinguished by examination of the constraints imposed by the environmental and use rights. Further, the size demands and activity requirements were limitations that had to be analyzed. Finally, the restrictions to limit conflicts between users were included for camping, shore fishing, and the boat ramp. After all of this background material was collected, mapped, and analyzed the probable development sites were located (Figure 3).

The existing boat ramp would be expanded to accommodate increased parking facilities. The second boat ramp would be located approximately 2.8km to the east-north-east of the current facility. This area has the required slope for a boat ramp, yet is adjacent to a terrace for parking. Access by an existing road and reduction of utility easement problems will reduce development costs. The camping area should be placed between the two boat ramps. This area is above the normal high water level which can then be developed with low depth sanitation facilities. Further, the area can be developed in stages beginning at each of the boat ramps and proceeding toward each other. Each initial development would be approximately 8.2ha in size and accommodate 75 camping sites. Three picnic areas were identified which fulfilled most of the requirements. One of the picnic areas is approximately .5km to the west of the current boat ramp. This site is adjacent to the water and protected from site of the boat ramp. A small road from the entrance to the boat ramp area provides access without disrupting traffic at the boat ramp. The second picnic site is away from the reservoir but in an area with large trees adjacent to the Cottonwood Wash. This site is just off the road which leads to the Cottonwood Draw nature trail and waterfall. A third picnic area is approximately 1.1km northwest of the second boat ramp. This site is on a terrace which is approximately 10m above the shoreline offering an excellent panorama of the reservoir.

ACKNOWLEDGEMENTS

This project would not have been possible without the financial support of the Grayrocks Steering Committee and the Wyoming State Parks and Historic Sites Division. More importantly, the project would not have been completed without the field and library work of Jeff Hamerlinck, and the assistance of Doug Bryant, Katie Donovan, Art Foster, Ed Henke, Mary Kamby and Bill Westbrook.

REFERENCES

Dombroski, J., 1990, Personal communication, September, 1990.

Hirsh, J.L., 1988, Recreational Land Use Planning for Valley Center Parcel No. 1, San Diego, California, unpublished M.A. Thesis, San Diego State University.

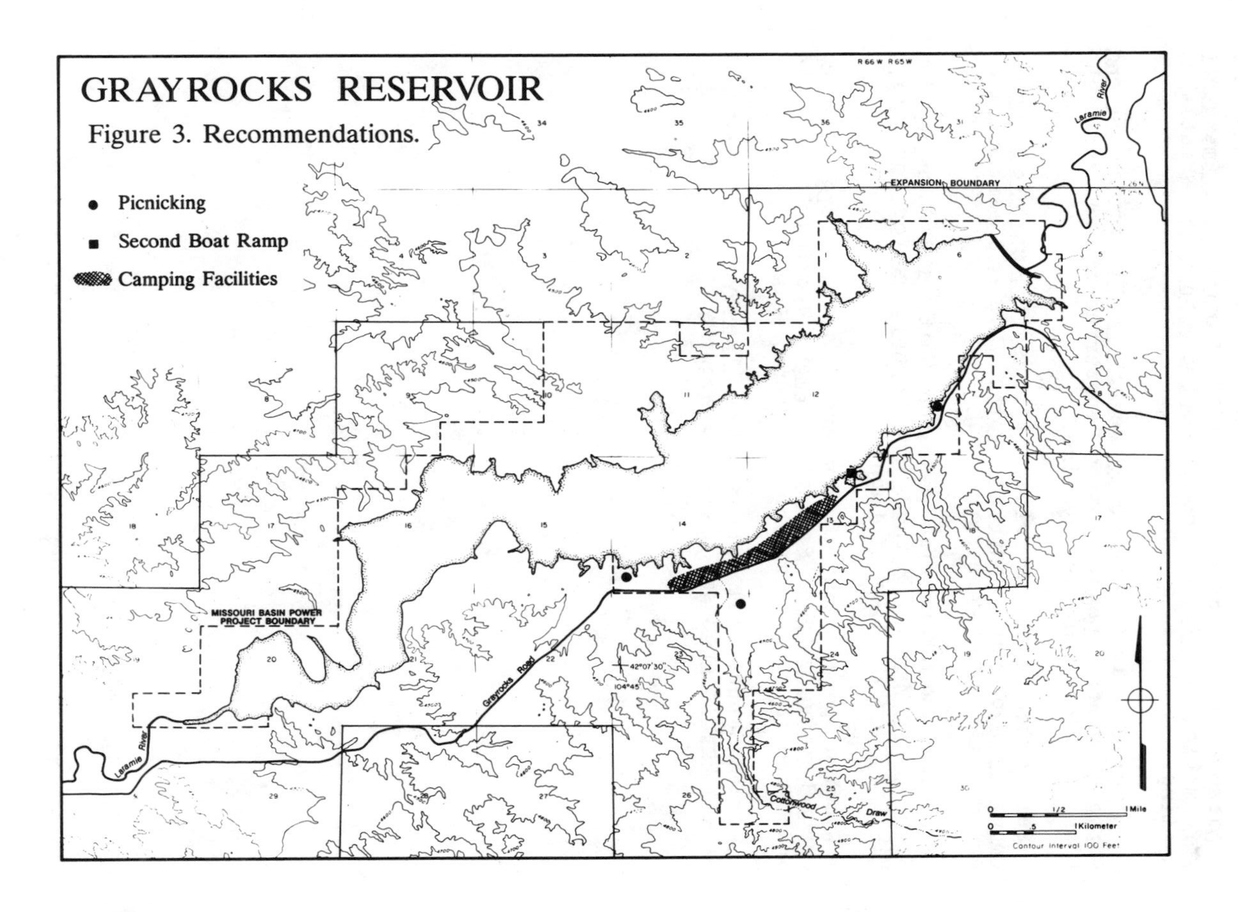

Figure 3. Recommendations.

Lancaster, R.A., 1987, ed.) Recreation, Park and Open Space Standards and Guidelines, National Recreation and Park Association, Alexandria, VA.

Missouri Basin Power Plant, 1979, Warning and Emergency Information Plan for Grayrocks Dam and Reservoir, Wheatland, Wyoming.

Snigg, M., 1990, Grayrocks Reservoir Fishermen Use and Harvest, draft report, Wyoming Department of Game and Fish-Fish Division, Laramie, Wyoming.

Soil Conservation Service, 1971, Soil Survey of Goshen County, Wyoming: Southern Part, U.S. Department of Agriculture, GPO, Washington, D.C..

Tourism and Recreation Research Unit, 1983, Recreation Site Survey Manual: Methods and Techniques for Conducting Visitor Surveys, Spon, London.

Wyoming Department of Commerce, 1990, Wyoming State Comprehensive Outdoor Recreation Plan, Wyoming State Parks and Historic Sites Division, Cheyenne, Wyoming.

Wyoming Industrial Siting, 1973, The Wyoming Industrial Siting Permit Process, Cheyenne, Wyoming.

MANAGEMENT AND ACQUISITION CONSIDERATIONS AFFECTING SELECTION OF PROPERTIES FOR NATURAL RESOURCE PROTECTION

Margit L. Crowell
Southwest Florida Water Management District
2379 Broad Street, Brooksville, Florida 34609-6899

ABSTRACT

The acquisition of lands for natural resource management and protection is guided by numerous concerns on the part of land management staff. These concerns reflect not only the resource-based characteristics of an area but also the man-induced trends and restrictions imposed on the landscape. The Water Management Lands Trust Fund (the Save Our Rivers program) outlines the objectives to be met when acquiring properties for resource protection. To assist District staff in making informed decisions regarding property acquisition for resource protection, the Southwest Florida Water Management District has developed and implemented a Geographic Information System (GIS) modeling application to evaluate numerous water resource- and land management-related factors.

The Management and Acquisition Considerations (MAC) subsystem of the model concentrates primarily on examining the effect of man-made pressures on lands within the District; these have been summarized into three related themes and include Land Ownership, Priority Watersheds, and Development Pressure. The MAC submodel and its development will be discussed here, as well as its significance within the larger framework of the Save Our Rivers GIS application.

INTRODUCTION

One of the more visible functions of the Southwest Florida Water Management District (SWFWMD), apart from its regulatory responsibilities, is the acquisition and management of natural lands for resource protection. Properties are evaluated on an ongoing basis in light of their value for water supply protection, flood protection/management, and wildlife habitat, as well as their feasibility for acquisition based on more traditional factors, such as parcel size and location, nomination by the property owner, and recommendation by District staff having specialized knowledge about a particular area.

While the objectives for acquisition are set forth and monies provided by the Water Management Lands Trust Fund (Section 373.59, Florida Statutes), prior to about 1990 it was in large part up to SWFWMD staff to determine which properties might be most suitable for acquisition using more traditional suitability assessments, such as those mentioned above. Original Save Our Rivers (SOR) legislation did provide some general guidelines for staff to follow, as well as recommend some particular projects for study, but there was a need for a more consistent and organized resource-based approach to determine which lands would most benefit from acquisition for resource protection.

SOR model development

In 1987 the District began the development of a computer-based model to be used as an aid in determining site acquisition suitability. This model considers numerous resource-based factors, and integrates several more common land acquisition principles as well. The model, developed and implemented on the District's Geographic Information System (GIS) and used for the first time in 1991 in the determination of site suitability, is known as the SOR Land Acquisition Site Identification Model.

Major model components

The SOR model is comprised of four major submodels, each of which considers an important aspect of water- or other natural-resource protection: Water Supply Protection, Flood Protection, Natural Systems Protection, and Management and Acquisition Considerations. Each of these submodels is in turn composed of several individual related themes. The development of the Management and Acquisition Considerations (MAC) submodel will be presented here, as well as its practical significance within the SOR model as a whole.

MANAGEMENT AND ACQUISITION CONSIDERATIONS SUBMODEL

The MAC submodel is comprised of three theme layers: Land Ownership, Priority Watersheds, and Development Pressure. Each of these layers is created according to a specific set of procedures, as described below. These layers are then overlayed in the GIS environment to produce the final composite for this submodel and show the entire District ranked for acquisition priority as determined by the combined effects of the man-induced trends and restrictions imposed on the landscape.

It is important to highlight the significance of these trends for the model since they, along with resource-based factors, determine the true feasibility of acquisition for any particular property. If parcels do not meet the minimum criteria for acquisition according to the Management and Acquisition Considerations submodel, the resource-related qualities alone will not make these lands any more appropriate for acquisition. At the same time, properties that do not exhibit promise as resource protection areas may not be practical for acquisition even if the management criteria are satisfied. Thus, there must be a practical balancing of themes in order for the entire SOR model to be effective in reaching its objectives.

The Florida Preservation 2000 Act of 1990 (Section 259.101, F.S.) provides additional motivation for emphasizing the significance of management and acquisition considerations. This program stipulates, among other conditions, that natural lands in imminent danger of development or subdivision may be purchased with funds set aside for this program; the MAC submodel highlights these areas through an analysis of the three land management-related themes.

MAC submodel themes

The execution of the MAC submodel produces a map layer that indicates the priority of lands for acquisition consideration based on three specific factors: (1) property acreage and proximity to public- or District-owned lands, (2) location of the property within a priority watershed area, and (3) location of the property within an area characterized by degree of development pressure. The map layer created by this submodel was combined using an overlay process with the layers from the other three SOR submodels to produce a summary map of acquisition priority over the entire District. All layer automation and processing was accomplished using the appropriate subsystems of Environmental Systems Research Institute's (ESRI) ARC/INFO software, unless otherwise indicated.

Land Ownership

The following sections describe the procedures for generating the Land Ownership theme data layer (see also Figure 1). This layer was created with two basic considerations in mind: (1) the size of properties in private ownership; and, (2) the proximity of these properties to public-

and District-owned lands. These factors are considered significant in determining acquisition suitability for two major reasons. Firstly, the larger the property, the more attractive it is for acquisition from a land-management perspective. It is more practical to acquire a large land holding than numerous small ones, since this practice minimizes the number of boundary difficulties and legal arrangements. Secondly, acquisition of properties located adjacent to public lands helps to encourage preservation of large contiguous natural areas important for maintaining wildlife habitat. The effects of property size and proximity are therefore considered below in the development of the Land Ownership theme. See Figure 2 for a map showing the results of ranking private properties according to these criteria (note that the public land boundary referenced in the map legend refers to public- and District-owned lands).

Create Private, Public & District Land Covers
↓
Calculate Preliminary Ranks Based on Acreage
↓
Generate 1-mile Buffers around Public and District Land Coverages
↓
Flag Private Properties in ARCEDIT
↓
Calculate Final Ranks based on AR•1•k•R
↓
Generate Map

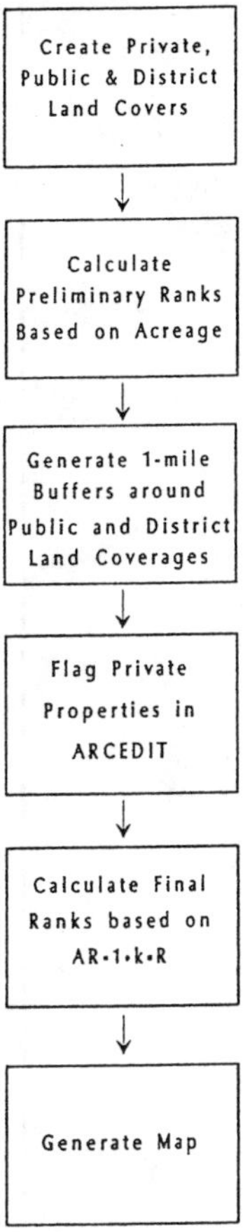

Figure 1. Theme 1 data flow

Privately-owned lands. This layer consists of digitized parcels in private ownership greater than 640 acres (one square mile) in size. All properties were tagged with a unique label ID, and outparcels were flagged with a "99999" so they would not be considered in any subsequent analysis of the properties. Private properties were first transferred from plat books for each county to 1:24,000 USGS quadrangles, from which they were then digitized. The USGS quads provided a common source with known tic-corner values from which to digitize the properties, and also eliminated the necessity to later transform numerous single-page coverages into one uniform District-wide Universal Transverse Mercator (UTM) coverage.

Publicly-owned lands. This layer contains digitized properties in public ownership. Properties were labeled according to the scheme described above for private lands, although the label values have no significance in the subsequent analyses. A buffer coverage, representing a one-mile (1609 m) zone around public properties, was generated for use in later determining proximity to private lands.

District-owned lands. These properties, owned by the Southwest Florida Water Management District, were entered into the database either by means of coordinate geometry (COGO) or digitizing. The labeling scheme for these properties does not affect subsequent model calculations, but all properties in this and the previous two layers must have polygon topology. As with the public property layer above, a one-mile buffer zone coverage around District lands was also required for determining proximity to private lands.

Calculations

Once the basic data layers above were automated, the following calculations were performed.

Preliminary rank values for private lands. Polygons (excluding outparcels) within the private lands coverage were reselected from the total coverage set by acreage, and then assigned a preliminary score, as

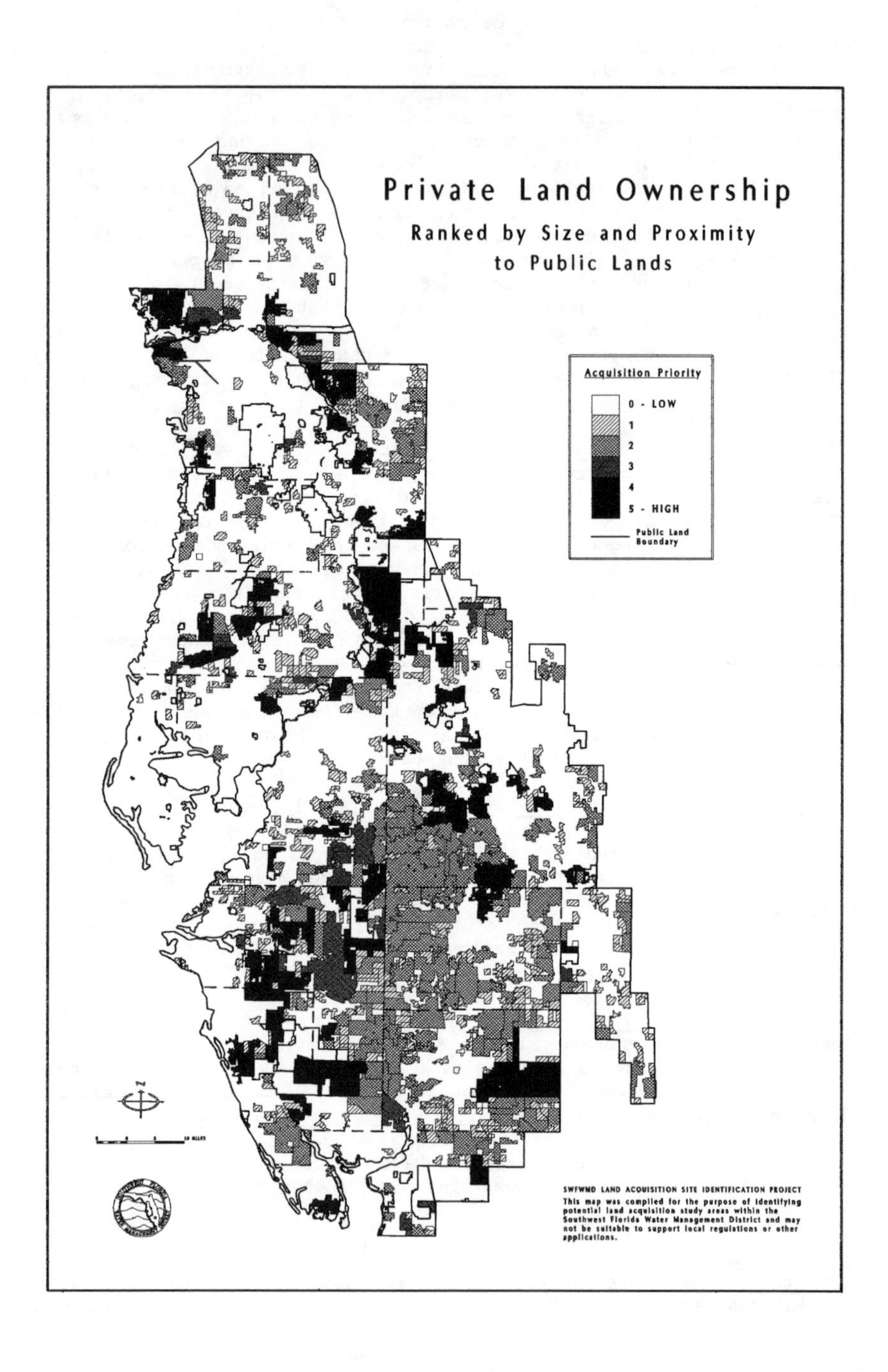
Private Land Ownership
Ranked by Size and Proximity
to Public Lands
Acquisition Priority
0 - LOW
1
2
3
4
5 - HIGH
Public Land Boundary
SWFWMD LAND ACQUISITION SITE IDENTIFICATION PROJECT
This map was compiled for the purpose of identifying potential land acquisition study areas within the Southwest Florida Water Management District and may not be suitable to support local regulations or other applications.

Figure 2. Private land ownership

indicated below. This procedure was automated using an Arc Macro Language (AML) program written for this purpose. The preliminary score was determined entirely by parcel acreage.

Acreage	Rank
< 640	0
640-1280	1
1280-1920	2
1920-2560	3
2560-3200	4
> 3200	5

Adjacency/proximity determination. In an interactive ARCEDIT session, flag values were assigned to valid (not outparcels) parcels within the private lands coverage. These flag values were assigned by determining the location of a private parcel relative to surrounding public and District lands and their one-mile buffer zones. Scores were then assigned to private parcels according to the following table.

Flag Value	Location
1	Adjacent to public property
3	Adjacent to District property
5	Within public buffer zone
7	Within District buffer zone

Adjusted rank values. These final scores are assigned to privately-owned properties and are a function of both the calculated preliminary ranks and the intermediately assigned flag values. These values reflect the priority for acquisition of these privately-owned properties based on these limited criteria. Final scores are calculated according to the following formula. (Eq. 1)

$$\mathbf{AR = 1 + k * R} \qquad (1)$$

where: **AR** = Adjusted Rank
k = Constant
R = Preliminary Rank

The constant **k** has the following values for the designated flag values and adjacency conditions:

Flag	k Value	Location
1	1.75	Adjacent to public land
3	2.00	Adjacent to District land
5	1.25	Within public buffer zone
7	1.50	Within District buffer zone

Priority rank values, once calculated according to the above formula, are normalized so that scores will be represented along a scale of 0 to 5. This will ensure scoring consistency throughout this and the other three subsystems of the SOR Land Acquisition Site Identification Model.

Priority Watersheds

The following sections describe the procedures for generating the Priority Watersheds theme data layer (see also Figure 3). This layer was created by manually attributing a coverage of all drainage basins within the District. The score assigned to each drainage basin was determined by the number of basins that could be determined to contribute to the watershed of a particular water body. For the purposes of the Site Identification Model, three types of water bodies/watershed areas were determined to be of interest for either resource management/preservation/restoration or aesthetic/recreational value. These priority areas are SWIM (Surface Improvement and Management) Priority Water Bodies, Outstanding Florida Waters (OFWs), and Aquatic Preserves. It is important that these and surrounding watershed areas be considered for acquisition priority due to man-made factors affecting both their water quality and ecological value. Figure 4 depicts the scored watersheds.

Attribution of drainage basin coverage. This is an entirely manual application; there is no programming involved in arriving at the scores for each basin in this coverage. The active coverage to be attributed is a copy of the drainage basins within the SWFWMD. Each basin that contributes to the watershed of a designated priority water body was flagged. A basin received one flag value for each water body falling within it or to which watershed it contributes. There is no maximum score which a basin may have.

The most direct method of accomplishing this scoring was to first create a set of 1:100,000 mylar plots of the drainage basin coverage. These plots were then used as overlays on 1:100,000 printed USGS quadrangles. The priority water body of interest on that quadrangle was located, and each of the basins contributing to that water body's basin was flagged. It was most logical to repeat the scoring procedure separately for each of the three types of water bodies using a different colored marker on the mylars for each category of water body. An item was added to the drainage basin feature attribute table to hold the total score value for each basin.

Once the basins were tagged for SWIM, OFW, and Aquatic Preserve water bodies on the mylars, the drainage basin coverage was brought up as an edit coverage in an interactive ARCEDIT session. The drainage basin polygons were then attributed with the total score determined on the mylars. Basins not contributing to the watershed of or not containing a priority water body received a score of 0 (default). Final score values, ranging from 0 to n, were then normalized to a scale of 0 to 5.

Create Copy of Drainage Basin Coverage

↓

Locate Priority Water Bodies on Drainage Basin Map

↓

Flag Basins Contributing to Watershed of each Water Body

↓

Total Scores for each Basin and Attribute Coverage with Scores

↓

Normalize Scores to Obtain Ranks from 0 to 5

↓

Dissolve on Rank Value

↓

Generate Map

Figure 3. Theme 2 data flow

Development pressure

The following sections describe the procedures for generating the Development Pressure layer for the Land Acquisition Site Identification Model (see also Figure 5). This layer depicts the distribution of development pressure around population centers. For acquisition purposes, it is important to consider those areas under highest development pressure that are still available for purchase and protection by the District, thereby preventing any increase in development pressure in these areas. This analysis was performed using the ERDAS image processing system.

A number of factors influence the development pressure exerted on an area. These include population size, density, and growth; major traffic

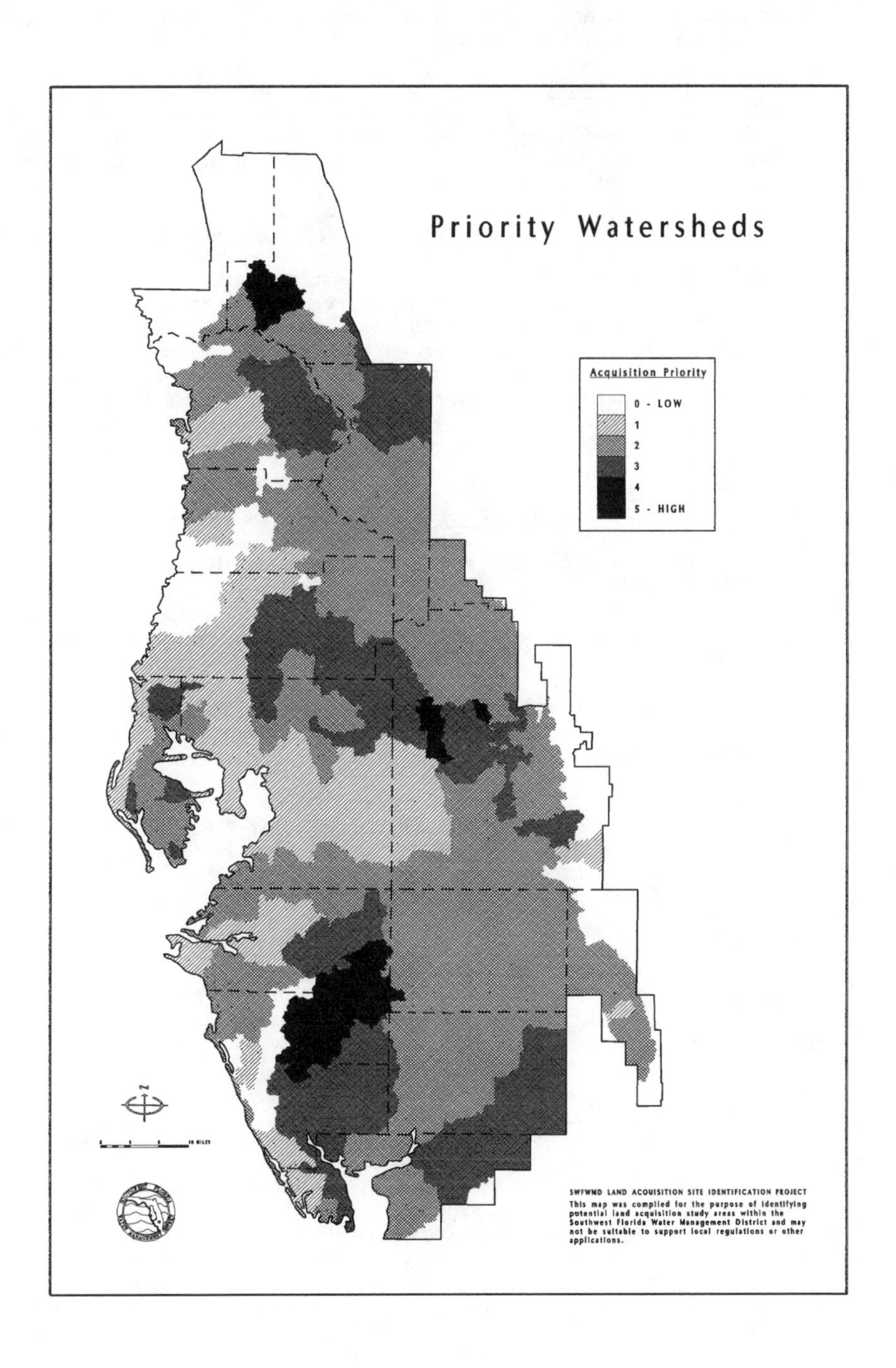
Priority Watersheds
Acquisition Priority
0 - LOW
1
2
3
4
5 - HIGH
SWFWND LAND ACQUISITION SITE IDENTIFICATION PROJECT
This map was compiled for the purpose of identifying potential land acquisition study areas within the Southwest Florida Water Management District and may not be suitable to support local regulations or other applications.

Figure 4. Priority watersheds

corridors; and natural and man-made attractions. In this model, only one factor, population size, was considered.

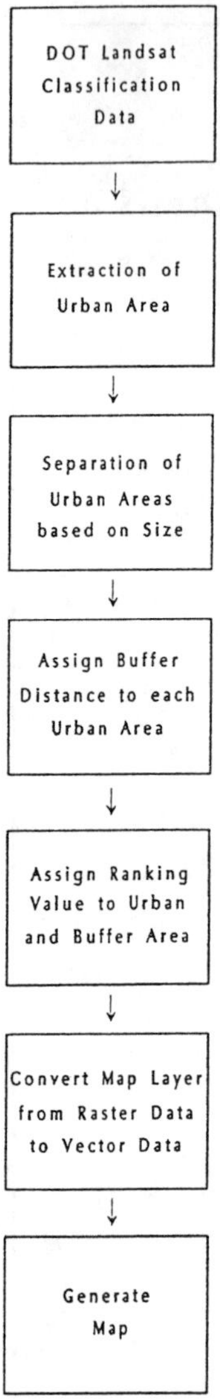

DOT Landsat Classification Data
↓
Extraction of Urban Area
↓
Separation of Urban Areas based on Size
↓
Assign Buffer Distance to each Urban Area
↓
Assign Ranking Value to Urban and Buffer Area
↓
Convert Map Layer from Raster Data to Vector Data
↓
Generate Map

Figure 5. Theme 3 data flow

Description of data. The raw data used in this section of the MAC submodel were classified raster-format Landsat Thematic Mapper data. From this data set, two land use categories were extracted: urban vegetated and urban non-vegetated.

Separation of urban centers. The size of the urban area was the sole factor governing the magnitude and extent of development pressure over the surrounding areas. The urban areas were separated into three classes according to their size: large, medium, and small. The classes were determined according to the following criteria: Large - urban area with diameter greater than 5 miles; Medium - urban area with diameter of 2.5 to 5 miles; Small - urban area with diameter less than 2.5 miles. A data layer was created for each size class.

Buffering of urban centers. Once the separate urban center layers were created, buffering distances were determined and scores assigned according to the table below. Two concentric buffer zones were generated around each urban center in each layer; the buffer distances were determined as a function of urban center size, in miles from the edge of the urban center. Areas outside of buffer zones received a score of 1. The assumption here is that development pressure on the area surrounding an urban center increases as the size of the urban center increases and vice-versa.

Urban Center Size	Urban Center Score	Inner Buffer Width	Inner Buffer Score	Outer Buffer Width	Outer Buffer Score
Large	4	8	3	16	2
Medium	4	4	3	8	2
Small	4	2	3	4	2

The layers were overlayed in the raster environment to produce a composite layer, with higher buffer zone scores superseding lower scores. Data were then converted from ERDAS raster to ARC/INFO vector format. The results are shown in Figure 6.

Final overlay

Once each of the three individual MAC layers had been generated, as described in the sections above, they were combined into one layer representing the Management and Acquisition Considerations submodel. Figure 7 shows the data flow for these procedures. This final layer was created using the ARC UNION overlay procedure, whereby one data layer (A) is joined with another data layer (B), resulting in a third coverage which has all of the features and attributes of both A and B. Once the Land Ownership (LO) layer and the Development Pressure (DP) layer were overlayed, the Priority Watersheds layer was then overlayed with the LO/DP combination, resulting in the final data layer for this submodel.

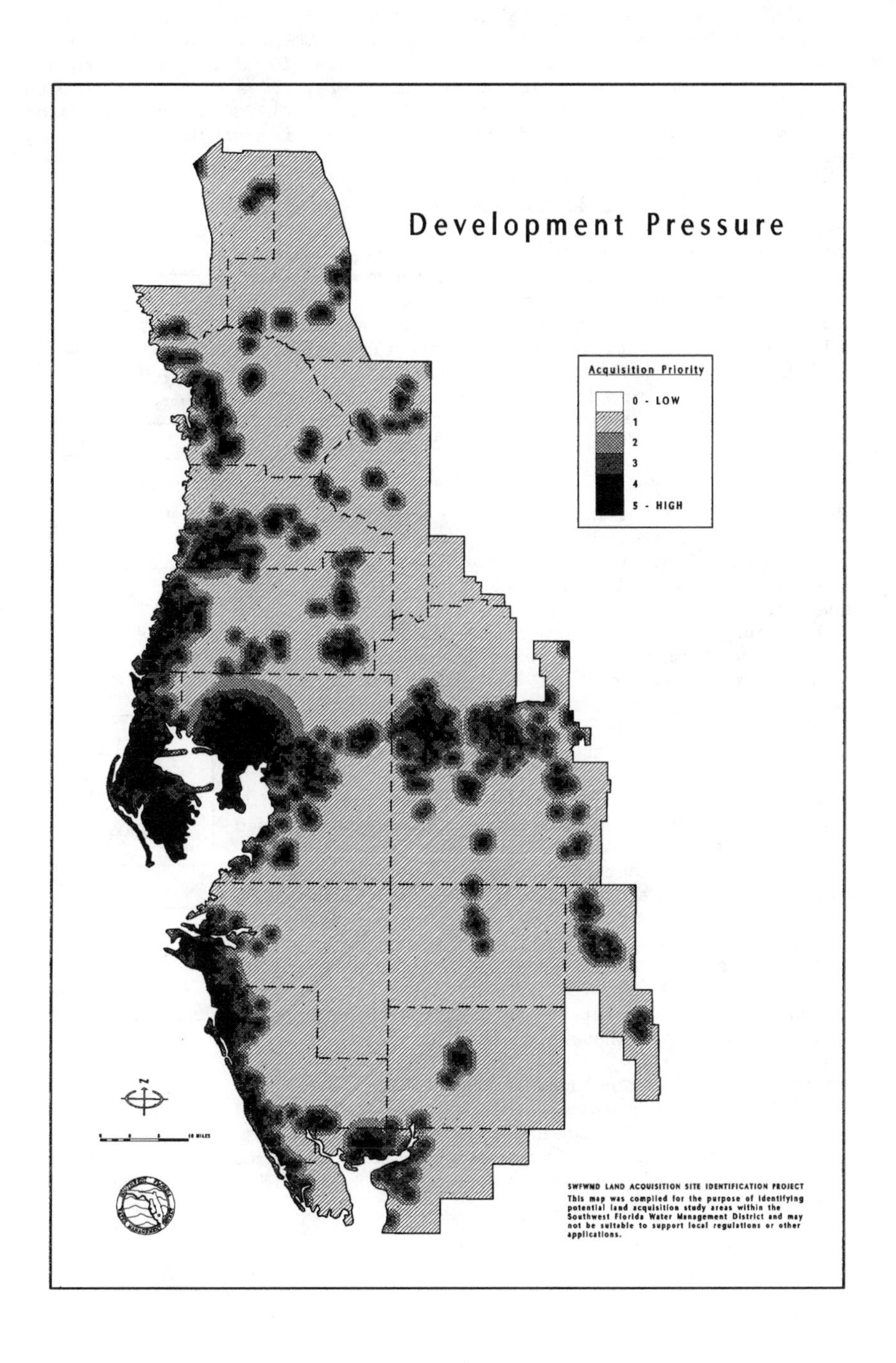
Development Pressure
Acquisition Priority
0 - LOW
1
2
3
4
5 - HIGH
N
10 MILES
SWFWMD LAND ACQUISITION SITE IDENTIFICATION PROJECT
This map was compiled for the purpose of identifying potential land acquisition study areas within the Southwest Florida Water Management District and may not be suitable to support local regulations or other applications.

Figure 6. Development pressure

In order to reduce data volume and the number of inconsequentially small polygons in the layer, the ARC procedure ELIMINATE was used at this point to remove polygons with an area of less than 10 acres (40470 square miles). The rank values, which after the overlay process ranged from 0 to 15, were normalized to 0 to 5. The ARC procedure DISSOLVE was then used to remove all polygon boundaries between areas having equal ranking values.

MAP PRODUCTION AND INTERPRETATION

The resulting map layer for the MAC submodel is shown in Figure 8. As can be seen on this map, there are several areas within the District that score high for acquisition priority based on the limited submodel criteria. This map provides a good starting point for land management staff who are interested in first locating potential properties, and then in learning more about the resource-related qualities of that area by combining this map with the other layers of the SOR model.

A major goal of the SOR model was to generate a series of GIS data layers which could subsequently be depicted in a series of map products. Maps were produced for each of the individually developed submodel themes, each submodel composite layer, and finally for the combination of all four submodels. The scoring on each map ranges from 0 to 5, with corresponding shades ranging from low to high in order of intensity, representing the priority with which depicted areas should be considered for acquisition based on the combination of all of the model components.

These maps provide an ideal medium for the presentation and interpretation of model results, and have been used in several ways to meet the aims of the Save Our Rivers program. One application includes using the final composite map with a graphic overlay of both recommended SOR study areas that are being considered for acquisition and also currently owned District property; this enables land resources staff to visualize areas with the highest acquisition priority and see how current and potential future acquisitions might compare most favorably with these high-scoring areas.

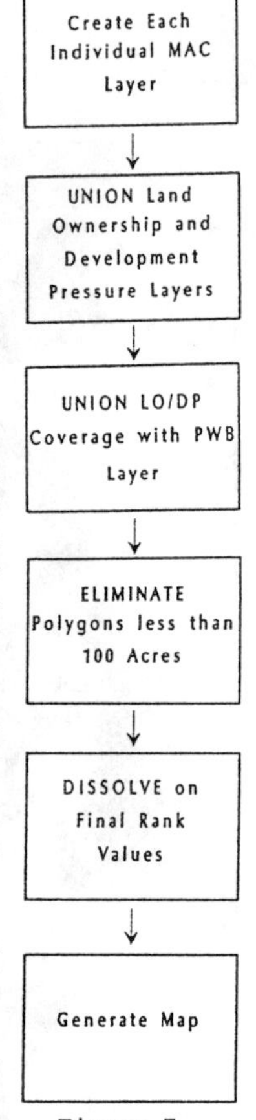

Figure 7. Final overlay data flow

Another application includes the determination of the top ten-percent, for example, of privately-owned lands based not only on acreage and proximity factors, but also on resource-based factors. This analysis allows staff to immediately target the highest-scoring properties that may be available and desirable for imminent acquisition.

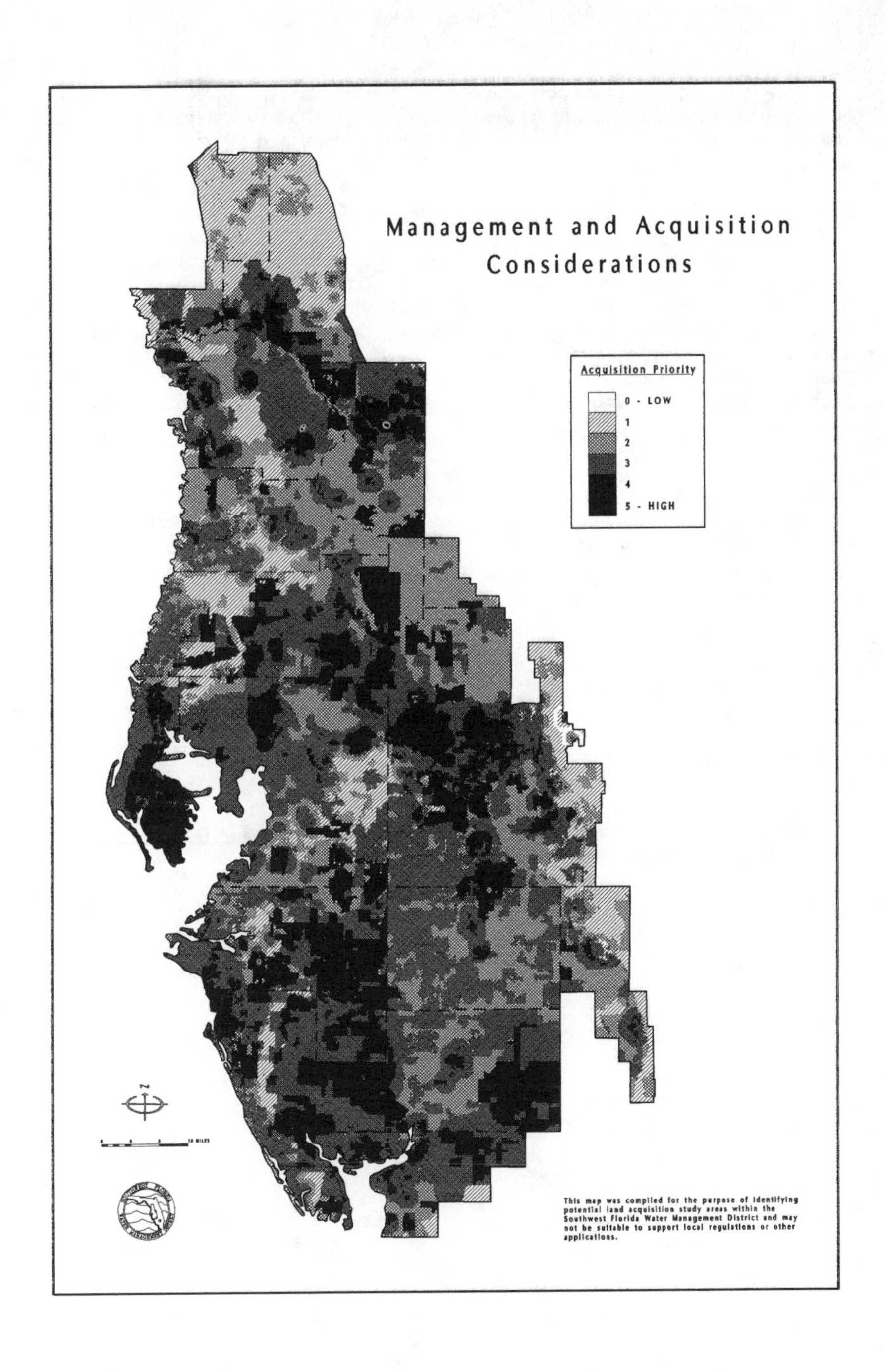
Management and Acquisition Considerations
Acquisition Priority
0 - LOW
1
2
3
4
5 - HIGH
10 MILES
This map was compiled for the purpose of identifying potential land acquisition study areas within the Southwest Florida Water Management District and may not be suitable to support local regulations or other applications.

Figure 8. Acquisition priorities for the Management and Acquisition Considerations submodel

CONCLUSION

Use of a GIS model in land acquisition studies has had several benefits for the District. District staff now have a tool with which they may assume a more proactive role toward targeting lands for acquisition. The SOR model results help to provide land management and resource protection staff with a District-wide view of the priority areas within the District's 16-county area, instead of only being able to compare smaller properties and projects with one another.

Also, special analyses may be designed for particular studies; the GIS data may be manipulated and mapped in many different ways depending on the application. As more current data for each of the SOR layers become available, the model may be updated using the programs and methodologies already developed. Implementation of the SOR Land Acquisition Site Identification Model is now an integral part of the District's Five-Year Plan for both the Save Our Rivers program and the Florida Preservation 2000 Act.

ACKNOWLEDGMENTS

The author wishes to thank Dr. Thomas H.C. Lo and Leah V. Polomchak for their valuable contributions to the development and automation of this application.

REFERENCES

Environmental Systems Research Institute, ARC/INFO Users Guide, ESRI, Inc., Redlands, CA, 1989.

Southwest Florida Water Management District, Land Acquisition Site Identification Model: A Geographic Information System Application, December 1990.

Southwest Florida Water Management District, Save Our Rivers/Preservation 2000 Five-Year Plan, January 1991.

A GIS APPROACH FOR THE ANALYSIS OF HUB-AND-SPOKE STRUCTURES OF AIRLINE NETWORKS

Shih-Lung Shaw
Department of Geography
Florida Atlantic University
Boca Raton, Florida 33431-0991
Bitnet: SHAWSL@FAUVAX

ABSTRACT

To extend the usefulness of geographic information systems (GIS) for transportation studies, it is critical to link GIS with transportation analysis. There are different strategies to aprroach this issue. This paper examines the feasibility of implementing analysis procedures within a general-purpose GIS for the structural analysis of transportation networks. Several basic implementation issues are discussed. This paper then presents an empirical study of analyzing the hub-and-spoke structure of airline passenger networks within a GIS environment.

INTRODUCTION

Geographic information systems (GIS) have been used in a variety of transportation applications. Many of the applications, however, have limited analysis features. The ability to create useful information for a better understanding of the spatial patterns of transportation systems and the spatial processes which cause these patterns is essential to many transportation studies. In order to achieve this goal, a geographic information system must be able to handle these models of spatial analysis. The strategies of integrating transportation analysis and GIS vary from the use of interfaces to link a GIS with existing analysis packages to a complete integration of analysis functions and models within a GIS (Lewis, 1990). A major advantage of interfacing a GIS with a transportation package is that we can avoid the difficult problem of designing a data model that will efficiently and effectively handle both GIS functions and transportation analysis. This approach, however, has limitations on the flexibility of using GIS functions during the transportation analysis process. Development of a full-feature transportation GIS can better integrate transportation analysis and GIS. The development time of such a specialized system could be too long to take care of immediate needs of the transportation community. Such a specialized system also could have limited general GIS functions available.

This study develops a series of transportation analysis and modeling procedures within a general-purpose GIS. One major advantage of this approach is that both transportation analysis and GIS analysis can be performed within the same environment. This has the potential of improving conventional transportation analysis procedures. For example, Shaw (1991) used network analysis and GIS functions available in a selected GIS package to facilitate a transit accessibility analysis. This study builds upon the previous study by examining the issues of implementing additional transportation analysis procedures in a general-purpose GIS.

For this study, the structural analysis of transportation networks was selected as a specific type of transportation analysis to be examined. Analysis of the hub-and-spoke structure of six U.S. passenger airlines' networks was used as an empirical study of implementing structural analysis procedures in a general-purpose GIS. In the next section the major characteristics of the "hub-and-spoke" structure of air passenger networks are reviewed. Section three discusses basic issues related to the implementation considerations. Procedures of implementing the empirical study are presented in section four. Section five gives concluding remarks and suggestions for future research.

HUB-AND-SPOKE NETWORKS

Since the airline deregulation of 1978, major airlines in the United States have restructured their flights toward hub-and-spoke networks. In such networks, hubs serve as central locations which collect and distribute passengers between a set of nodes connected to hubs. One major advantage of adopting a hub network is savings in the number of linkages necessary to connect all nodes on a network. For example, it requires n(n-1) linkages to provide direct connections between n airports in a network. The number of linkages required to connect these n airports can be reduced to n-1 if a hub is established in the network and direct flights are offered between the hub and individual airports.

Hub networks also result in economies of scale to airline operations due to increased passenger volumes at the hub locations. That is, airlines can enjoy higher load factors, use larger and more economical aircrafts, or schedule more departures (Kanafani and Hansen, 1985). On the other hand, a hub network tends to cause inconvenient transfers and lengthy routes between network nodes and create congestions at hub airports. Individual airlines choose their own network structures by making trade-offs among many variables. These structures often deviate from the idealized one-hub, two-hub, or multiple-hub networks (O'Kelly, 1986; Chou, 1990). It, therefore, is of interest to analyze the network structures of different airlines and to compare their similarities and differences.

GIS IMPLEMENTATION ISSUES

The structure of a network can be represented as a graph or as a matrix (Taaffe and Gauthier, 1973). The former approach is based on the graph theory, which represents a network as a set of nodes connected by a set of arcs. This graph-theoretic representation is an abstract of the real-world network and describes the topological properties of a network. The matrix approach, on the other hand, represents a network as a set of rows defined as origin nodes and a set of columns defined as destination nodes. The most basic matrix representation of network structures is a connectivity matrix, which records the presence or absence of direct connections between pairs of nodes. This basic connectivity matrix can then be used to derive other matrices, such as accessibility matrix and shortest-path matrix, for studying network structures.

The "topological" data model commonly employed in vector GIS to represent a network and its topological properties corresponds to the graph-theoretic approach. The matrix approach, on the other hand, is often missing in a general-purpose vector GIS. Since matrix data and matrix operations are fundamental to structural analysis of transportation networks, a GIS that can handle both the "graphic-theoretic" and the "matrix" representations of transportation networks and allow convenient transformations between the two representations is highly desirable. Some basic considerations of implementing network structural analysis procedures in a vector-based GIS for examining air passenger networks are discussed below.

Network Coverage Creation

Analysis is preceded by data input. In order to encourage transportation analysis within a GIS environment, one critical requirement is to offer easier and faster ways of entering data into a GIS for analysis. Airline flight data is usually available in a tabular form organized as origin cities and destination cities. Due to the nonplanar nature of air networks, the route maps of major airlines often are not adequate for digitization. A better input method is to have the GIS automatically create a network coverage from the flight data extracted from timetables published by individual airlines or from the Official Airline Guide. Preparing such a file for creating a network coverage requires data on direct and nonstop flight connections between city pairs only. All indirect flight connections can be derived by chaining these direct and nonstop flights together. In other words, creation of a network coverage based on the topological data model only requires the coordinate of nodes and a connectivity matrix that describes connections between nodal pairs. If a GIS does not accept data in a matrix form, it is necessary to expand the connectivity matrix into a flat file with one record representing each non-zero cell in the connectivity matrix.

This network coverage creation method fits particularly well with air networks because the exact locations of network linkages are not critical for either display or analysis purposes. The same data input method can also be applied to highway and other networks if the applications focus on analysis rather than displays of detailed network layouts. For example, many data files already created to work with available transportation analysis packages can be easily converted into a GIS coverage. Analysis can then be performed on the coverage.

Arc vs. Node Based Approaches

When a network is represented as a line graph in a vector GIS, it is a common practice in many available GIS packages to record attribute data with arcs rather than with nodes. For example, users would query "the traffic volume on arc xy" rather than "the traffic volume between node x and node y". This arc-based approach works fine with transportation applications that deal with fixed line segments. It however is not suitable for transportation analyses that reference directly to nodes in a network. For example, the average daily air passenger volume between city A and city B is X, and there is no direct connection between these two cities. This example presents a problem of recording the data in an arc-based system because there are multiple arcs, and possibly multiple paths, between the two given nodes. The same case can easily be handled if a GIS allows users to deal with data at both node and arc levels.

Analysis Algorithms

Structures of transport networks can be described by aggregate measures of network connectivity or disaggregate measures of nodal accessibility. Two commonly employed measures of network connectivity are the gamma and alpha indexes (Taaffe and Gauthier, 1973). The gamma index is defined as the ratio of the number of arcs in a network to the maximum number possible in that network. The alpha index is the ratio of the number of actual circuits to the maximum number possible in a given network. A circuit is a finite, closed path in which the beginning node of the path coincides with the end node. Therefore, in a minimally connected network, the number of circuits is zero. Each additional linkage added into a minimally connected network increases the number of circuits by one. Calculations of both gamma and alpha indexes are based on two variables only: the number of nodes and the number of arcs in a given network. A GIS that keeps track of these two numbers can easily implement procedures of computing these two indexes. One aspect that deserves attentions is the sensitivity of the two indexes to the presence of pseudo nodes in a network (Shaw, 1989). Pseudo nodes are often created due to partitions of an arc into smaller arcs for recording attributes at a finer level. This changes the numbers of nodes and arcs in a network which, in turn, generate different values of the gamma and alpha indexes.

With regard to the measures of nodal accessibility, conventional methods have been based on matrix operations. For example, a connectivity matrix (C matrix) can be powered up to the diameter of a network (n) to take into account indirect connections between nodes on a network. These connectivity matrices powered to different orders then can be summed to derive the accessibility matrix (T matrix). One major shortcoming of the T matrix is the inclusion of redundant connections caused by the powering of C matrix. For example, two nodes with a direct connection can reach each other through a one-linkage path or a three-linkage path. In the one-linkage case, travel is from one node to the other node. In the three-linkage case, the path starts at one node, travels to another node, then returns to the starting node. This connection is obviously a redundant connection because a better connection is available.

To remove the redundant connections, the shortest-path matrix (D matrix) is suggested. In deriving the D matrix, its matrix operations deviate from the standard matrix operations (see Taaffe and Gauthier, 1973). A GIS equipped with a "matrix" data model and standard matrix operation procedures would be capable of deriving the accessibility matrix, but not the shortest-path matrix. However, it is quite common to find a shortest-path-finding algorithm in many vector-based GIS. This algorithm can be used as the basic tool to generate a shortest-path matrix. If the topological distances are replaced by some real-world impedance measures in the shortest-path-finding algorithm, a matrix based on the concept of a valued graph is derived. This allows users to derive nodal accessibility measures based on travel distances, travel time, travel costs, or some other variables.

Visualization

Haggett et. al. (1977) suggested six stages in the analysis of nodal regional systems. These six stages are interactions, networks, nodes, hierarchies, surfaces, and diffusion. GIS, in general, can handle analysis of formal regions better than analysis of nodal or functional regions. This is mainly because many GIS functions are oriented towards polygons rather than interactions between nodes. In addition to analysis functions, GIS also can contribute to the visualization of the different stages in the analysis of nodal regions. For example, a GIS should be able to read data in an origin-destination (O-D) flow matrix and create a map showing the interaction intensities between different origins and destinations. Displays of networks and nodes are standard functions in most vector GIS today. Hierarchy of nodes in a network is usually determined from analysis results. This hierarchical pattern can be visualized by using graduated circles or different point symbols to show nodes at different levels in the hierarchy. If a display of variations over a continuous surface is desired, an isoline map or a 3-dimensional display can be generated. With regard to diffusion, a GIS needs to trace the spread of a phenomena on a surface over time. This involves temporal GIS issues for developing efficient and effective ways of showing the changing patterns over time, which is not covered in this paper.

AN EMPIRICAL STUDY

Six major U.S. airlines (i.e., American, Continental, Delta, Northwest, United, and USAir) were included in this study for the examination of the hub-and-spoke structure of air passenger networks. These networks covered the direct flight connections among the 100 most populous Metropolitan Statistical Areas (MSA) in the continental United States for the months of September and October of 1990[1]. The GIS used in the study was Arc/Info[2]. One major reason of choosing Arc/Info is because of its Arc Macro Language (AML), which provides useful tools for implementing analysis and modeling procedures.

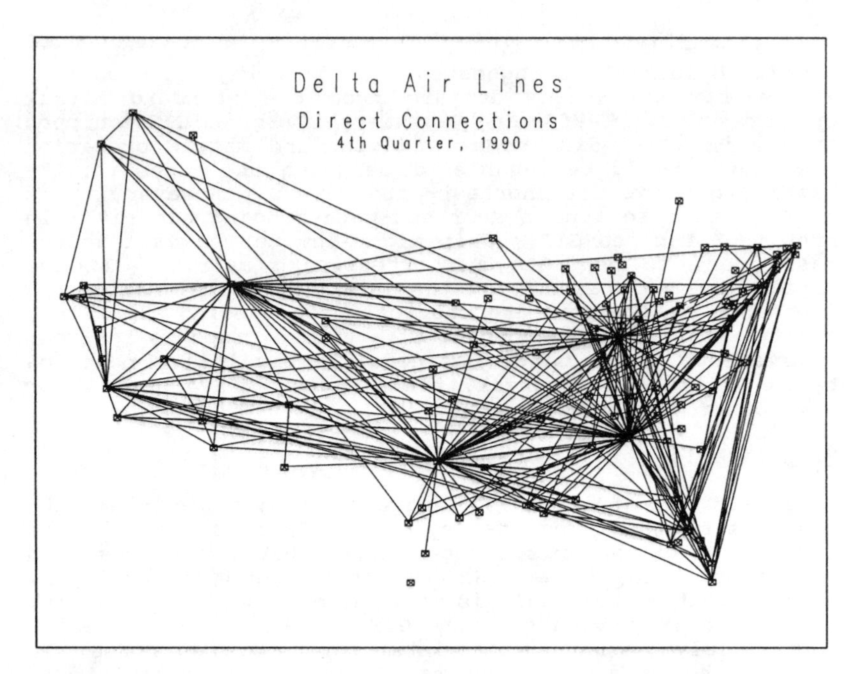

Figure 1: An example of air passenger network.

To create network coverages for the six airlines, a file consisting of UTM coordinates, projected from longitudes and latitudes, of the 100 MSA's was created. A second file that records direct flight connections between the MSA's was also created for each of the six airlines. The simplest form of recording the direct connection data is to use a connectivity matrix. However, since Arc/Info does not accept matrix input form, this file was created by recording the non-zero cells in the connectivity matrix as separate records in the format of "from" and "to" nodes. Other data items, such as number of daily flights, mileages, traffic volumes, etc., can also be included in this file. An AML program then was developed to read data from the two input files to automatically create a network coverage. An example of network coverage created from this process is shown in Figure 1.

In order to compare the structures of the six networks, the gamma and alpha indexes are used to evaluate the overall connectivity levels of the six networks. Since air networks are considered as nonplanar graphs, the topology of these network coverages was generated by allowing arcs to cross each other without creating new intersection nodes. An AML program was developed to retrieve the numbers of nodes and arcs in a network from the system and use them to compute the gamma and alpha indexes. According to the index values, USAir led the other airlines with significant margins in terms of overall network connectivity levels, with Delta Air Lines at a distant second place. This indicates that USAir provided more nonstop flights among the MSA's it served than other airlines.

In addition to the overall connectivity measures, it is also important to identify the relative positions of the MSA's in these networks for a better understanding of the hub-and-spoke structure. Measures of nodal accessibility are therefore needed. Implementation of this analysis in Arc/Info involves many more steps. This study first used topological distances to derive a shortest-path matrix. Measures of nodal accessibility were then calculated from the shortest-path matrix. This process was accomplished by first creating an Info table that resembles an O-D matrix. The Allocate program in the Network module of Arc/Info then was used to find out the shortest paths between an origin MSA and all destination MSA's on a network. The allocation results were written into the O-D matrix in the Info module. This procedure was repeated for all origin MSA's to derive a complete shortest-path matrix. Cell values in the shortest-path matrix then were summed to derive accessibility measures of individual MSA's.

One complication of this process is that the Allocate program uses the system-generated node ID's to figure out the shortest paths between nodes. The shortest-path matrix thus must be based on the system ID's to identify origins and destinations. These system ID's then need to be converted to user-specified ID's for interpreting the analysis results. Since users cannot access data at the node level in an Arc/Info network coverage, this conversion of system ID's into user ID's had to be accomplished at the arc level[3].

Based on the derived accessibility levels of individual MSA's in each airline's network, a hierarchy of the MSA's can be identified. Figure 2 uses point symbols to show the hierarchical pattern of the MSA's in one selected airline's network. In this figure, "cross" symbols represent the MSA's that are not served by the airline. The "star" symbol represents the MSA on the network that has the lowest nodal accessibility. The "square" symbols of different sizes indicate the accessibility levels of the MSA's that are 1-10%, 10-20%, 20-30%, 30-40%, and more than 40% above the least accessible MSA on the network, respectively.

To show the accessibility levels as a continuous surface, the TIN (Triangulated Irregular Network) module of Arc/Info was used. The TIN module is widely used to handle terrain data. Given the flexibility of data conversion and display capabilities for handling point, line and surface map features, the TIN module also is a good candidate to enhance both analysis and display aspects of transportation analysis. In this empirical study, the accessibility measures at individual MSA's were treated as point observations to create a TIN coverage. An isoline coverage was then generated from the TIN coverage (Figure 3). The TIN coverage can also be used to generate 3-dimensional displays of the accessibility surface viewed at different viewing angles and viewing directions. Figure 4 gives one example of 3-dimensional displays of accessibility variations over a surface. These different ways of visualizing the data and analysis results help us better comprehend the hub-and-spoke structures, hierarchy of nodes

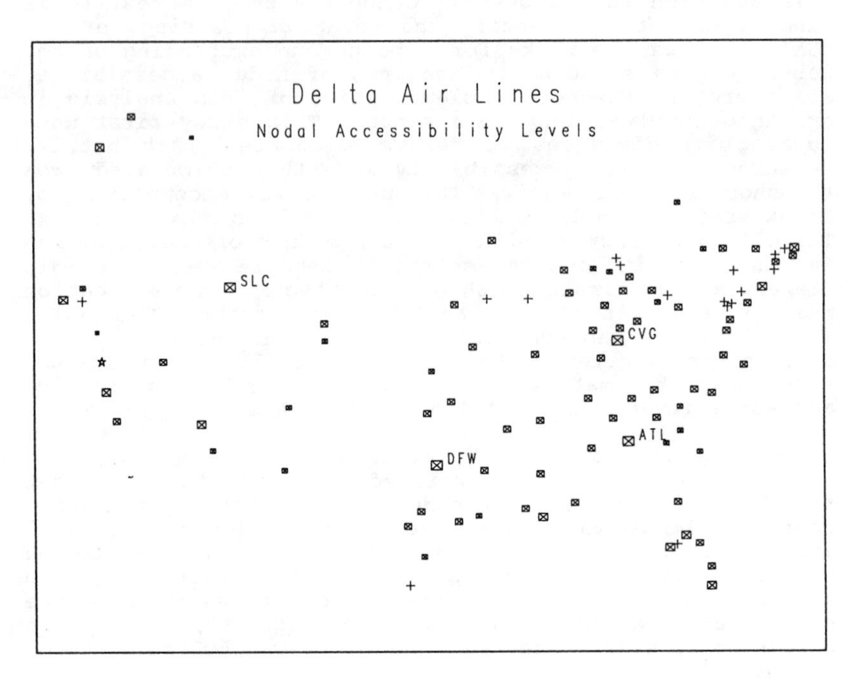

Figure 2: Hierarchy of nodal accessibility.

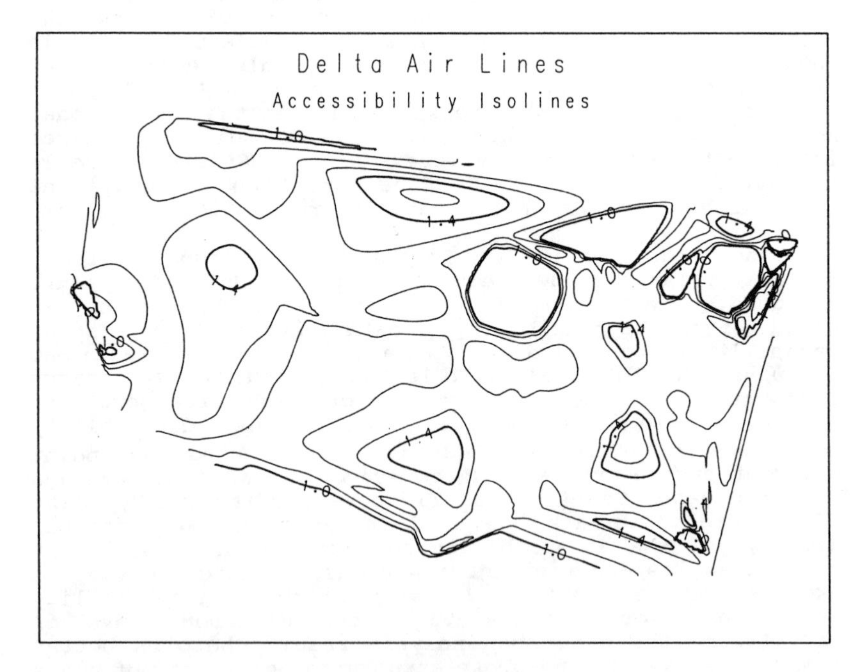

Figure 3: An isoline display of nodal accessibility.

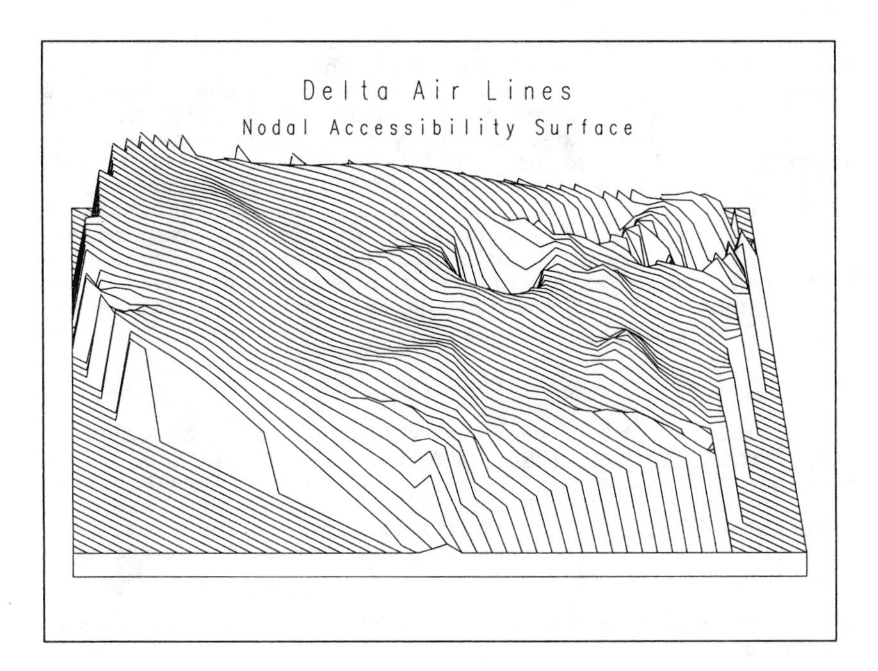

Figure 4: A 3-D display of nodal accessibility.

on a network, and the variations of accessibility between different regions.

CONCLUSION

Geographic information systems have been widely used in transportation applications. An important task faced by transportation and GIS researchers is to link transportation analysis and GIS together. There are different strategies to approach such a task. This study approached the task with the following two basic implementation considerations. First, the study took the strategy of implementing analysis procedures within a GIS environment rather than having them as separate programs interfaced with a GIS. Secondly, a GIS for transportation studies does not need to include various transportation analysis functions as readily available commands in the GIS. If the design of a GIS is flexible enough to accommodate the fundamental ways of representing transportation systems, a GIS equipped with some basic transportation analysis procedures and a powerful tool for users to create their own models will be sufficient for most transportation analysis.

This study evaluated the above approach through an analysis of the hub-and-spoke structure of air passenger networks. Arc/Info GIS, which includes basic network analysis functions and a macro language, was selected by the study to implement analysis procedures for deriving measures of network connectivity and nodal accessibility. With these derived measures and the graphic display capability of GIS, the similarities and differences of the hub-and-spoke

structures adopted by the different airlines were identified. The study suggests that the approach of implementing transportation analysis in a general-purpose GIS is feasible. However, the study also indicates the importance of considering analysis requirements at the initial design stage of a GIS. For example, what data model(s) would leave the most flexibility to transportation analysis? What would consist of a set of "basic" transportation analysis functions available in a general-purpose GIS? This study examined these issues for one type of transportation analysis. Similar studies are needed for other types of transportation analysis to fully evaluate the approach suggested in this study.

ENDNOTES

1. Population data is based on the Statistical Abstract of the United States, 1990, The National Data Book, U.S. Department of Commerce, Bureau of the Census, Washington, D.C., January, 1990. Flight data is based on the system timetables published by individual airlines for September-October, 1990. In the case of one MSA served by more than one airports, the data for those airports are combined.

2. Arc/Info release 5.0.1 for DEC VAX computers was used in this study.

3. Arc/Info release 6.0 introduces "node attribute tables" which allow users to associate tabular data with nodes.

REFERENCES

Chou, Y-H. 1990, The Hierarchical-Hub Model for Airline Networks, Transportation Planning and Technology, Vol. 14, pp. 243-258.

Haggett, P., Cliff, A.D., and Frey, A. 1977, Locational Analysis in Human Geography, Edward Arnold, London, UK.

Kanafani, A. and Hansen, M. 1985, Hubbing and Airline Costs, Institute of Transportation Studies, University of California, Berkeley, Research Report UCB-ITS-RR-85-12.

Lewis, S. 1990, Use of Geographical Information Systems in Transportation Modeling, ITE Journal, March 1990, pp. 34-38.

O'Kelly, M.E. 1986, Activity Levels at Hub Facilities in Interacting Networks, Geographical Analysis, Vol. 18, No. 4, pp. 343-356.

Shaw, S.L. 1989, Design Considerations for a GIS-Based Transportation Network Analysis System, GIS/LIS'89 Proceedings, Vol. 1, pp. 20-29.

Shaw, S.L. 1991, Urban Transit Accessibility Analysis Using a GIS: A Case Study of Florida's Tri-Rail System, Southeastern Geographer, Vol. 31, No. 1, pp. 15-30.

Taaffe, E.J. and Gauthier, H.L. 1973, Geography of Transportation, Prentice-Hall, Englewood Cliffs, N.J.

INTERFACING STAND ALONE TRANSPORT ANALYSIS SOFTWARE WITH GIS

Bruce A. Ralston
Xiaoming Zhu
Department of Geography
University of Tennessee
Knoxville, TN 37996-1420

ABSTRACT

GIS packages, while quite powerful, often lack the sophisticated modeling capabilities needed for transportation analysis. In this paper we report on our efforts in interfacing stand alone transportation analysis packages with commercial GIS systems. We address two broad classes of models: transportation simulation models, and network optimization models. Two stand alone systems, one of each type, are used as test cases. Key issues such as data compatibility, forward and backward linkages, and data editing and updating are explored.

INTRODUCTION

Using geographic information systems for spatial analysis is an important step in the evolution of the GIS field. Linking geographic information systems with analytic transportation models, particularly advanced models such as equilibrium models, optimization models, elements of the UTPS, and location and routing models, is likely to be a fruitful area of research and practice. There are several reasons for this optimism. Some GIS packages already contain simple transportation network analysis capabilities, such as shortest path analysis. The models often used in transportation network analysis are well developed and understood. At their core, network models and GIS use the same data primitives--points, lines, and polygons. Finally, many users of GIS already have developed databases of network characteristics.

There are, however, some obstacles to simply overlaying network models on a GIS platform. GIS systems were developed to serve several purposes, not just mathematical modeling of network operations. As a result, the data structures found in a GIS system may be inappropriate for use with network optimization code. The use of non-planar graphs in network models is another difference with many GIS databases. While GIS systems are based on the physical topology of a network, network models often use non-physical, logical entities to capture certain behaviors. Examples would include logical arcs and nodes to simulate intermodal transfers, delays at border crossings, and loading and unloading costs. These entities, which must exist within a network model's data inputs, are not likely to be digitized components of a GIS. They must be generated "on the fly." Finally, the objectives and constraints used in network modeling change from application to application. Formulating and solving network analysis problems can, in and of itself, be a difficult problem. Trying to integrate this task with a GIS system often complicates matters. As a result, network modelers have often developed their own stand alone systems, which may contain some GIS components, but seldom contain the full set of spatial operators, such as spatial joins, which distinguish GIS systems.

There are advantages and disadvantages to developing stand alone systems. The advantages are clear. The developer controls the database, so it can be built to use data structures which are conducive to network analysis. Since there is no need to go back and forth between the modeling software and a GIS system, database management and computational overhead are mininimized. However, there are drawbacks to stand alone systems. The added spatial operator capabilities of GIS systems are missed. Perhaps more importantly, the wealth of information to be found in existing GIS databases may be inaccessible to stand alone systems. Because of this, stand alone systems may languish on a shelf while agencies concentrate on using their GIS systems as fully as possible. For this reason alone, integration of stand alone systems with GIS packages is an important goal. It is our purpose here to describe some of the ways in which integration of stand alone systems with commercial GIS packages can be accomplished.

When we speak of integrating GIS systems with analytic transportation models, two types of approaches come to mind. In the first case, we could use GIS functions, such as buffering, spatial overlay, and polygon operations to perform some transportation modeling task. For example, trip generation estimates of agricultural products could be developed using productivity estimates over soil types in polygons which comprise supply areas. In the second approach, the GIS system could be used to manage network database information, and, possibly to display the results of network models. It is this type of integration we present here. We do this using two stand alone packages: the Food for Peace Inland Emergency Logistics and Distribution System (Ralston and Ray

1986) and the Bangladesh Transportation Modeling System (Liu and Ralston 1990).

FIELDS was developed in response to the African food emergency of 1984-85. It uses a multiperiod linear program to find the best routes, modes, and schedules for shipping food aid to feeding camps. Like all linear programming methods, using it means constructing an LP tableau, or matrix, based in part on the structure of the transportation network and the relationships between the nodes, arcs, supplies, and demands. The BTMS, which was developed for the World Bank, simulates the movement of commodities over a transportation network in order to assess the effects of proposed changes to the transportation system. Its algorithms are based on building tree structures between origins and destinations.

To integrate modeling software of these types with GIS packages, it is necessary to extract the required attributes from the GIS database, construct the appropriate problem statement, solve the problem, and update the GIS database to reflect the information generated by running the transportation model. In accomplishing these tasks, we must be able to distinguish between the logical entities generated for the models and those which relate to elements in the GIS database.

NETWORK ENTITIES

GIS databases contain attributes for each digitized element. The elements of concern in the GIS database are the nodes and links of a network. Information on these entities is stored in point and line attribute tables. These tables are related to each other by node IDs. That is, each link contains a from-node and to-node, and those nodes are in the node, or point, attribute table.

These files describe the physical network; that is, the actual nodes (towns, cities, junctions) and arcs (rail, road, and waterway links). We tend to "see" and store the network in this manner. Figure 1 illustrates this point. All the arcs entering and leaving nodes A and B are depicted.

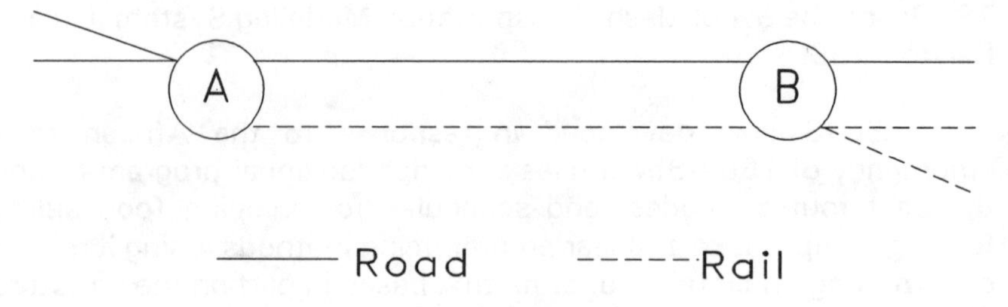

Figure 1 - A Sample Physical Network

However, most transportation models "see" a very different network, often called the logical network. The logical elements refer to nodes and links which may not actually exist, but which must be created in order to effectively model the characteristics of the transportation network. Thus Figure 1 is transformed into Figure 2. In this figure there are now separate modal networks, intermodal transfer arcs, loading and delivery arcs, and new nodes for each mode.

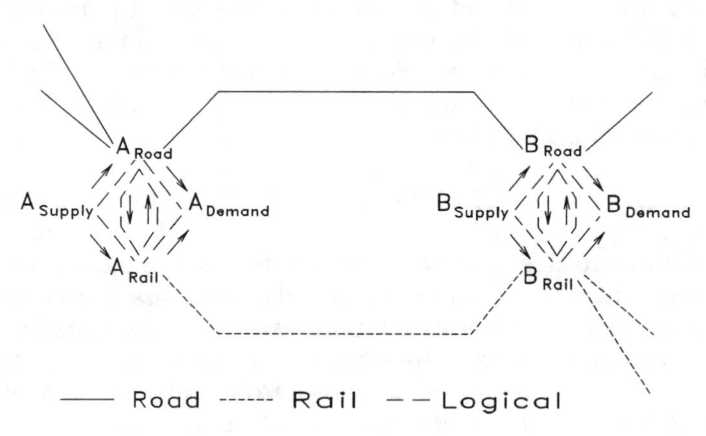

Figure 2 - A Sample Logical Network

A key step in integrating transportation models in a GIS is creating the logical network based on information in the GIS database. Once the logical entities are generated, the appropriate data structure can be built for the analytic technique to be used. We first consider the class of linear programming models.

LINEAR PROGRAMMING MODELS

A typical approach to matrix based problems is to write a matrix driver; that is, a program which reads the arc and node files and constructs the appropriate matrix. To accomplish this, a relational model must be developed which enables the geographic databases to be explicitly related to the LP formulations and solutions.

Specific records and data fields in the physical network files must be related directly to the elements of the problem matrix. In an LP tableau, each row and column is given an unique name. We exploit this by encoding each row and column name in the LP matrix so that the relevant information in the geographic databases can be associated with the corresponding constraints (columns) and variables (rows) of the tableau. In FIELDS, variable names and constraint names are constructed key data elements which logically point to specific files, records and data fields in the geographic database files they index (Stevens 1987). The use of variable names and constraint names as key data values for a relational database model allows both primal and dual information to be related to specific elements in the GIS database.

Consider, for example, intermodal transfers of food aid at Bamako, Mali in the third month of a multi-period planning horizon. At Bamako, there are three modes available: road, river, and rail transport. Assume that arc and node files exist with each period's data stored in blocks in the files. The LP driver, after scanning a period's node and arc lists (these lists might be derived from attributes based on modal coverages), would recognize more than one mode present at Bamako (Figure 3). Assuming no data flags are set which would rule out intermodal transfers at that node, intermodal transfer arcs would be generated, as would mode specific constraints. Names for these arcs and constraints would be generated, and each name would contain keys which "point" back to the GIS database. By encoding all variable names and constraint names in this manner, each column and row of the LP can be interpreted as an index to files, records and data fields in the geographic database.

Once the LP problem is generated it can be passed to an LP solver. The output of an LP solver contains, at a minimum, a listing of the variables and constraints and their primal and dual values at optimality. When interpreted, these results include information on which routes to use, the amount to ship on them, when to use

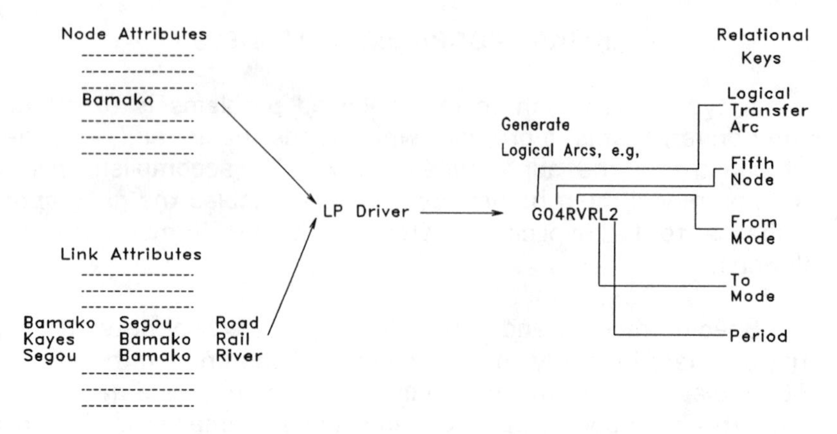

Figure 3 - Encoding Relational Keys in Constraint Names

them, and, if the LP is not degenerate, the value of changes in the right hand sides of constraints. Thus, we can indicate the value of having more trucks available at a given port.

The LP results can be related back to the proper locations and links in the correct time periods via an inverse transformation of that depicted in Figure 3. In other words, each value generated by the LP has a row or column address in the LP matrix. This address also acts as a pointer to specific files, records, and data fields in the database. Thus, a relational model exists between the geographic database and the LP results. This model can be used to generate report tables of the optimal solution, and look up tables for constructing flow maps. Relating the attribute and locational data in the geographic data files with the optimal primal and dual values from the LP allows for the construction of maps, via a relational map processing model (Aronson 1987) (Figure 4).

TREE BASED MODELS

A second class of transportation models consists of those based on constructing trees between origins and destinations. The classic shortest path problem can be solved on a tree. More complicated models, such as logit based modal split and assignment models, equilibrium models, incremental and iterative assignment models, and vehicle routing models, often have at their core a path problem solved over a tree.

In recent years it has become clear that the efficiency of such problems is dependent on the type of data structure used (Gallo and

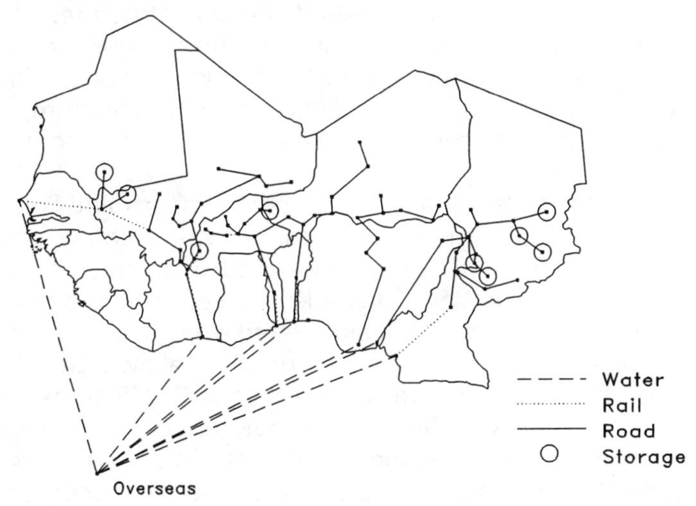

Figure 4 - Example Flow Map of LP Solution

Pallottino 1988, Van Vuren and Jansen 1988). The relational data structure found in many GIS attribute tables is not the best one for solving such problems. In the BTMS, we have found it advantageous to construct a forward star data structure which contains both physical and logical nodes and arcs.

The BTMS uses a program, called the logical network generator, to transform the digital representation of the physical network into a logical network. Building a logical network encompasses several steps. First, for each physical arc, its bi-directional twin arc is constructed. Secondly, to capture pickup and deliveries, each node that can be a supply and demand point (such nodes are called centroids) is transformed into two logical nodes, one for supply and one for demand. Next, for each mode that enters or leaves a node, a logical node for that type of mode must be constructed. Loading and offloading arcs from the supply and demand logical nodes to each logical node-mode combination are added to the network. Finally, intermodal transfer arcs are created. In the BTMS all nodes are logical entities. All physical links in the network have IDs which are found in the GIS attribute tables. Each logical node has, as part of its name, the name of its corresponding physical node. This acts as a key value for relating model activity at a logical node to the GIS database.

Once the logical structure at each node is constructed, the network is then rewritten in a forward star data structure. This type of data structure greatly increases execution speeds for path algorithms, which form the core of the analytic models. Thus, we keep two data structures. The GIS relational data structure, and the forward star data structure accessed by the analytic models.

The forward star data structure relates nodes and links in the following manner. A pointer is associated with each node to a link which leaves the node. The links are sorted in an array so that each link which leaves a particular node is in a contiguous block of the array. Each link, in turn, has associated with it a pointer to the node which it enters. This data structure allows tree building algorithms (the basic algorithms used in the BTMS transportation models) to quickly "walk through" the network data structure. The arc and node names imbedded in this data structure allow us to send model results back to the GIS maintained database (Figure 5).

TO CUSTOMIZE OR NOT TO CUSTOMIZE: THAT IS THE QUESTION

Commercial GIS packages contain utilities for displaying maps and updating spatial databases. PC Arc/Info, for example, can be used to digitize new arcs, and dBase programs, using either the dBase programming language or SQL, can be written to aid the user in updating link and node attributes. Since the transportation model databases contain information relating them to the GIS databases, model results can be displayed using the GIS graphics options. However, there may be instances when one wishes to use the graphics and data management options found in stand alone packages (Ding and Fotheringham 1991). For example, it is often desirable to generate maps of movements over logical entities at nodes. Trying to pop-up a map of logical entities is problematic in some GIS packages because the coordinates of the logical entities are at or near their physical location. Since topological relationships are based on proximity (e.g., node snapping) the use of logical maps is easier in stand alone systems.

The advantages of using a stand alone graphics package lie in the developer's ability to design graphics options and menus specifically related to the transportation modeling task. Similarly, data management tasks can often be customized to the specific problem at hand. Full integration would give users a choice of which functions to use, those found in the GIS or those which have been customized for the model being studied. The important task is to design utilities which insure that relevant information is passed between the modeling package and the GIS system.

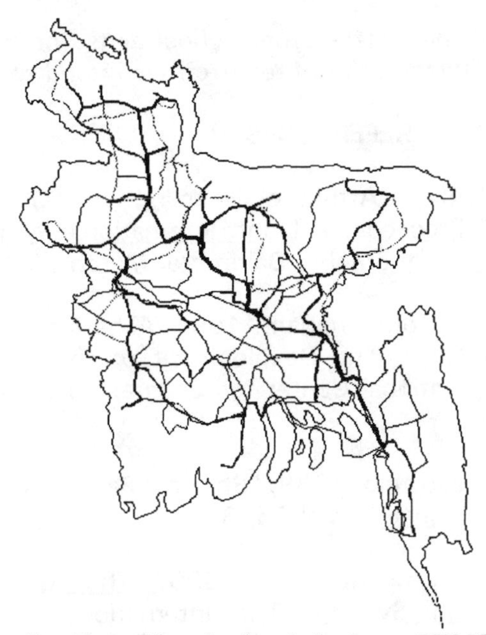

Figure 5 - Flow Map in Arc/Info from BTMS Models

This aspect of integration is a bit more cumbersome than the procedures described above because changes to the network often go beyond changes to the attributes of existing GIS entities. The result is that the digitized files and topological relationships found in the GIS must be accessed (in the case of graphics) or updated (in the case of data management). Many vendors treat the structures of these files as proprietary. Thus, users must often go through import and export routines whenever changes to these files are to be made outside the GIS system.

FUTURE RESEARCH

GIS packages will continue to have analytic engines built into them or grafted onto them. How difficult or easy this becomes depends in part on the ease of developing programs which interface mathematical models with GIS databases. A set of public domain object classes which could aid in such tasks is needed. Object oriented languages, such as C++, could be used to develop reusable classes which allow access to the GIS databases. Functions imbedded in these classes could contain the rules necessary for constructing data structures needed by modeling software, creating custom graphics, or updating attribute tables. The use of these classes would also make it easier to update the modeling software as new data types, constraints, or modeling objectives arise. It is this avenue of research we intend to pursue.

REFERENCES

Aronson, P. 1987, "Attribute Handling for Geographic Information Systems", Proceedings, Eighth International Symposium on Computer-Assisted Cartography, Baltimore, March 29-April 3.

Ding, Y. and A. S. Fotheringham. 1991, "The Integration of Spatial Analysis and GIS: The Development of the STATCAS Module for Arc/Info", National Center for Geographic Information and Analysis, Buffalo, NY.

Gallo, G. and S. Pallottino. 1988, "Shortest Path Algorithms," Annals of Operations Research, Vol 13, 3-79.

Liu, C. and B. A. Ralston. 1990, The Bangladesh Transportation Modeling System, The International Bank for Reconstruction and Development, Washington, DC.

Ralston, B. A. and J. Ray. 1986, The Food for Peace Inland Emergency Logistics and Distribution System, United States Agency for International Development, Washington, DC.

Stevens, A. 1987 C Database Development, MIS Press, Portland, OR.

Van Vuren, T. and G. R. M. Jansen. 1988, "Recent Developments in Path Finding Algorithms: A Review," Transportation Planning and Technology, Vol 12, 57-71.

A GIS-BASED MODEL FOR IDENTIFYING LANDS FOR ACQUISITION FOR WATER RESOURCE PROTECTION

Steven Dicks

Southwest Florida Water Management District
2379 Broad St.
Brooksville, Florida 34609-6899
(904) 796-7211

Robert Christianson

St. Johns River Water Management District
P. O. Box 1429
Palatka, Florida 32708-1429
(904) 329-4500

ABSTRACT

The protection and preservation of water-related resources can be accomplished using a number of means, including regulation, land and water use planning, capital improvement projects and the acquisition of critical lands. The last of these means, land acquisition, is becoming increasingly important in the protection of Florida's water resources. The Southwest Florida Water Management District currently purchases lands for flood protection purposes, surface and ground water supply protection and the preservation of natural systems as defined by the Water Management Lands Trust Fund (Section 373.59, Florida Statutes). To date, this program, commonly known as Save Our Rivers (SOR), has resulted in the purchase by the District of approximately 100 sq km.

The District has created a Geographic Information System (GIS)-based model that combines a number of natural and cultural map layers to assist in the identification of potential land acquisition sites. The primary goal of the model is to provide an objective, consistent and replicable means of ranking existing proposed acquisitions as well as to identify previously undefined areas that meet the criteria of the SOR Program.

Within the approximately 25,400 sq km area of the District, 24 potential acquisition sites covering more than 1425 sq km were identified using this model. Sites were selected based on the degree to which the model indicated that they were significant for water supply, flood protection, natural systems protection and various management and acquisition considerations. The model required the compilation and analysis of data sets representing floodplains, recharge, wildlife habitat and development pressure. This model will provide a tool to continue to evaluate and select acquisition strategies.

INTRODUCTION

The Southwest Florida Water Management District is one of five governmental agencies within Florida that are responsible for the protection of the state's water resources. This protection is accomplished through many means, including regulatory activities, structural flood control, water-related planning and the purchase of lands critical to the water resource. Described here is a Geographic Information System (GIS) model that was designed to assist in the latter of these means of protecting water resources, the purchase of lands to protect water resources.

The District acquires lands which meet the objectives of the Water Management Lands Trust Fund, commonly known as the Save Our Rivers (SOR) Program. These objectives are broadly outlined in statute as the acquisition of property for the purpose of water management, water supply and the conservation and protection of water resources (s. 373.59,F.S.). In 1990 the Preservation 2000 legislation expanded the SOR Program to include lands experiencing or threatened by imminent development pressure, lands which serve to protect or recharge ground water and protect other valuable natural resources or provide for natural, resource-based recreation, and lands containing a significant portion which serves as habitat for endangered or threatened species.

Historically, lands were identified for acquisition in one of three ways; either they were defined in legislation, proposed by District staff or by interested individuals or groups. This process was largely reactive and made long range planning of acquisitions difficult. The purpose of the project described here was to pro-actively identify those properties within the District which are most suitable for acquisition under the SOR and Preservation 2000 Programs. The underlying theory behind the project methodology is simple: the most important areas for acquisition are those that best meet the criteria for acquisition.

STUDY AREA

The District encompasses an area of approximately 25,400 sq km along the west-central coast of the Florida peninsula (Fig. 1). This area includes all or part of 16 counties and has a population of approximately 3.4 million. The primary urban areas are located around Tampa Bay and include the cities of Tampa and St. Petersburg. As with most of Florida the major population growth is found along the coast, though significant development is also occurring along the Interstate 4 corridor between Tampa and Orlando.

THE MODEL

Four major SOR land acquisition requirements are addressed by the model: Water Supply Protection, Flood Protection, Natural Systems Protection and Management and Acquisition Considerations. The model consists of four sub-models addressing these, with each of these sub-models containing two or more themes representing more detailed acquisition criteria (Table 1).

The combination of these themes and sub-models was accomplished using the Environmental Systems Research Institute's (ESRI) ARC/INFO (Version 5.0) software operating on a VAX 8550/6410 cluster and the ERDAS raster GIS software running on a personal computer. The following general processing steps were followed for the creation of the model:

1. Themes were input into the GIS (either as ERDAS or ARC/INFO files). Where necessary, raster-to-vector or vector-to-raster transformations were performed.

2. For each theme the raw data were assigned rankings between -5.0 and 5.0, with lower values representing those characteristics least valuable for a particular data set with regard to water resource protection and higher values being more important. For nominal data, the rankings were derived from reviews of the literature and discussions among District staff as to the appropriate values. For

ordinal or interval data, rankings were assigned after examination of the frequency distributions of the data.

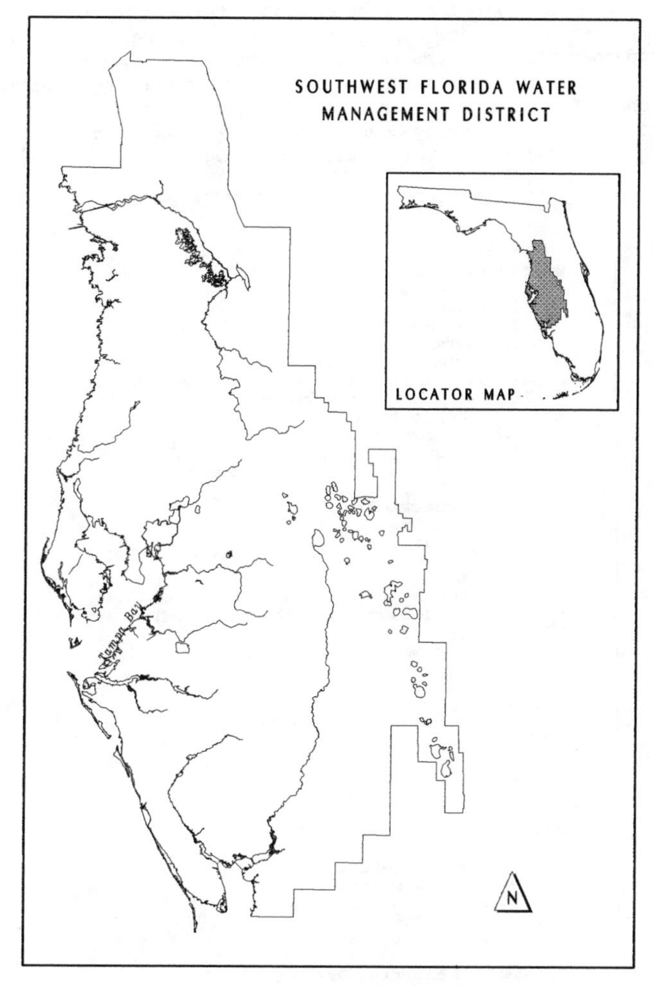

Figure 1: Study area.

3. The ranked individual themes were combined using ERDAS or ARC/INFO to create each of the four sub-models. The data values for each sub-model were then rescaled to values between 0 and 5.0 using a linear transformation. Weights on individual themes were applied at this step as deemed appropriate.

4. The four sub-models were combined in ARC/INFO to create the acquisition model with four attributes per polygon representing the scaled scores for each of the sub-models. The model was stored using the ARC/INFO Map Library structure, with approximately 200 7.5 minute map tiles containing over one-half million polygons. The final processing step was to eliminate all polygons less than four ha in size (approximately 10 acres). The resulting files contained approximately 130,000 polygons.

Once this overlay process was completed, the ARC/INFO software was used to weight each sub-model to reflect an increased emphasis on that particular consideration. For instance, if water supply protection and flood protection were considered the predominant interests for acquisition and natural systems protection was secondary to those, a map could be constructed to reflect that set of priorities. This map could be modified to reflect acquisition and management considerations to investigate the implications of these considerations. By having the capability to manipulate the data in varying combinations, decisions can be supported by evaluating their impact on the acquisition program.

Table 1. Model components.

1. Water Supply Protection
- 1.1 Ground-Water Supply Suitability
  - 1.1.1 Transmissivity
  - 1.1.2 Leakance
  - 1.1.3 Permitted Pumpage
  - 1.1.4 Potentiometric surface and
  - 1.1.5 Water quality (expressed as total dissolved solids
- 1.2 Protection Areas for Surface-Water Supply
  - 1.2.1 Drainage basins
  - 1.2.2 Stream network
  - 1.2.3 Soils infiltration rates
- 1.3 Protection Areas for Major Public Supply Wells
  - 1.3.1 Well diameter
  - 1.3.2 Transmissivity
  - 1.3.3 Potentiometric surface
  - 1.3.4 Porosity
  - 1.3.5 Aquifer thickness
  - 1.3.6 Pumpage.
- 1.4 Susceptibility to Ground-Water Contamination
  - 1.4.1 DRASTIC Floridan and Intermediate aquifers.
- 1.5 Recharge
  - 1.5.1 Recharge to Floridan aquifer.

2. Flood Protection
- 2.1 Wetlands Index
  - 2.1.1 Land cover and drainage basins.
- 2.2 Floodplains
  - 2.2.1 Land cover
  - 2.2.2 Drainage basins
  - 2.2.3 100 year flood plains.

3. Natural Systems
- 3.1 Wildlife Habitat
  - 3.1.1 Land use and land cover.
- 3.2 Water Quality Enhancement
  - 3.2.1 Land use and land cover.
- 3.3 Disturbed Lands
  - 3.3.1 Land use and land cover.

4. Management and Acquisition Considerations
- 4.1 Land Ownership
  - 4.1.1 Private ownerships greater than 2.59 sq km
  - 4.1.2 Public ownerships greater than 1.3 sq km.
- 4.2 Priority Watersheds
  - 4.2.1 Drainage basins
- 4.3 Development Pressure
  - 4.3.1 Land use

MODEL COMPONENTS

Following is a summary of the contents of each of the four sub-models and the 13 source models from which they are generated.

Water Supply Protection

The Water Supply Protection sub-model was designed to identify areas important for the protection of both existing and potential ground- and surface-water supplies. Evaluated in this sub-model was the potential of an area as a ground-water supply, the delineation of protection zones for surface water supplies, the protection of existing large public supply wells, the susceptibility of the aquifers to pollution from surface contaminants and the importance of areas for aquifer recharge.

Ground Water Supply Suitability. Ground-water supply suitability themes were prepared for the two primary aquifers within the District, the Floridan and the Intermediate. Based on an algorithm which combines the importance of water quality, transmissivity, leakance and wetlands location, these themes rank areas according to their suitability for ground-water resource development.

Protection Areas for Surface Water Supplies. Surface waters are used or have the potential to be used for public water supplies in portions of the District, particularly the coastal counties south of Tampa Bay. The water withdrawn from these locations is from rainfall flowing overland from throughout the watershed. Significant water quality degradation or enhancement can occur as this water flows, depending in large part on the nature of the uses of the land. This application is directed at identifying those portions of the watershed most critical for protecting the water quality at specified withdrawal points.

The accurate delineation of protection areas for surface water would typically require a detailed surface water flow model utilizing detailed topographic, soils and land cover data. Lacking detailed topographic and soils data, this theme combines drainage basin boundaries, stream network maps and soil infiltration rates to approximate protection areas. Using the ARC/INFO buffering and NETWORK software, zones adjacent to rivers and tributaries flowing into existing or potential surface-water supplies were delineated. Scores for areas of equal significance to water quality were based on proximity to a stream.

Protection Areas for Major Public Supply Wells. For each withdrawal location, the water reaching the well comes from what are referred to here as areas of contribution. These areas of contribution are defined by a combination of pumping rates and the movement of ground water in the regional ground-water flow system. A modeling effort of significant proportions would be required to accurately map the areas of contribution for all the public supply wells in the District. Efforts are being initiated at the District to undertake such a model in portions of the District, but are not complete as of yet.

As an interim effort to approximate these areas of contribution, a model developed for the Environmental Protection Agency was used (EPA, 1990). Several ground-water related parameters, including transmissivity, aquifer thickness, leakance and potentiometric surface were combined with permitted pumpage data to derive the extent of the areas of contribution. The resulting theme defines two zones of diminishing significance for areas successively farther from the well for each major public supply water use permit.

Susceptibility to Ground-Water Contamination. In many parts of the District the primary aquifers are either exposed at the surface or overlain by porous sands or limestones, making the potential contamination of ground-water supplies by surface pollutants a major consideration in protecting ground-water resources. To identify areas with a high susceptibility to contamination, a series of layers known as DRASTIC maps were used. These maps were developed for the Floridan and Intermediate aquifers by implementing an overlay technique developed for the U. S. Environmental Protection Agency (Aller et al., 1985). Seven data layers including **D**epth to water, net **R**echarge, **A**quifer media, **S**oils, **T**opography, **I**mpact of vadose zone and hydraulic **C**onductivity were combined to create a composite coverage that ranks areas according to their overall susceptibility to ground water contamination.

Recharge. In addition to the importance of recharge as a consideration in determining areas of susceptibility to ground-water contamination, high recharge areas play a major role in maintaining the water quantity characteristics of the ground-water system. A map of recharge to the Floridan aquifer was entered into the computer and scores assigned on the basis of relative contribution to recharge.

Layer Combination. The Water Supply Protection theme was created from the combination of the five maps described above. The hydrogeologic conditions within the District are such that the development of ground-water supplies are much more likely in the north half than in the south, with the result that scores for the Ground Water Supply Suitability layer are much higher in the north than in the south. To de-emphasize these regional differences, and to identify the "best of the best and the best of the worst," a weight of 1.5 was applied to the Ground Water Supply Suitability layer.

Other component themes to receive a weight of 1.5 were Recharge to the Floridan Aquifer and Susceptibility to Ground-Water Contamination. The additional 50 percent weight for these components reflects a long-term concern of the District for maintaining ground-water quality and protecting recharge areas.

Flood Protection

Flood protection is one of the more significant roles of the District in managing water resources. As our understanding of the nature of flooding and the effectiveness of protecting the functions of floodplains has increased, the strategy of non-structural flood control has taken a more central focus in our activities. Central to this strategy is balancing the regulation and acquisition of floodplain areas. Two themes were developed and combined to address flood protection. The first deals with establishing priorities for entire drainage basins based on the relative percentages of wetlands while the second highlights floodplain areas with a particular emphasis on uplands.

Wetlands Index. Drainage basins can be characterized by the percentage of land area which are wetlands. Wetlands represent areas that store or detain water and therefore are valuable for attenuating flooding. Areas with a high percentage of wetlands are important for these storage characteristics. Areas with relatively few wetlands may be critical for protection due to the lack of these natural flood control functions.

The percentage of area covered by wetlands was calculated for each drainage basin within the District. Based on the frequency distribution of all the basins, rankings were established which score those percentages that are highest and lowest as most significant for flood control. Basins with ratios closer to the mean value receive successively lower scores. Based on the ranking results, all wetlands within each drainage basin are assigned a score. Uplands are assigned a score of zero.

Floodplains. The primary areas important for acquisition for flood protection are floodplains. To balance regulation and acquisition as protection strategies, it is important to distinguish between uplands and wetlands within these floodplains. Wetlands receive special protection through various regulatory programs. Regulation of uplands is subject to significantly more resistance from property owners who demand compensation. It is the objective of this application to target those portions of the 100-year floodplain that have a relatively high proportion of uplands.

Using the ERDAS software, a wetland-upland data layer was created. A 15-by-15 cell moving window was used to calculate and assign to the center pixel the average percentage of uplands for that window. These values were converted to scores based on the distribution of values throughout the floodplain, with areas outside the floodplain assigned a score of zero. This technique reflects the observation that floodplains typically exhibit a mosaic of uplands and wetlands and that acquisitions should target regions with significant percentages of uplands and not specific locations of uplands within floodplains.

Layer Combination. The Flood Protection sub-model was based on an equally weighted combination of the Wetlands Index and the Floodplains themes. This combination represents the most significant areas for acquisition for flood protection at the regional scale.

Natural Systems

The preservation and/or the restoration of natural ecological systems is an important consideration within the SOR acquisition program. Themes were generated for wildlife habitat quality, water quality enhancement and the negative effect of disturbed lands.

Wildlife Habitat Quality. As a general principle, wildlife habitat quality is greatest in natural ecosystems. Two concepts, habitat quality by ecosystem type and the proximity of high quality habitats to other land covers, were combined in this analysis to yield a wildlife habitat quality theme. Using land cover as the base data, each category of land cover was assigned a score for wildlife habitat quality. These scores were derived from a previous District study evaluating various habitats found on existing District-owned lands (Christianson, 1986). The scores were adjusted to assign higher scores to large, contiguous ecosystems and to areas adjacent to higher-scored ecosystems. As a simple example, a high quality habitat type such as cypress forest received a high score if it was located adjacent to or in close proximity to large, undisturbed pine flatwoods than if it were surrounded by commercial development.

Water Quality Enhancement. This theme ranks land cover types according to a table of water quality enhancement scores derived from literature assessments. As an example, natural pine flatwoods receive higher water quality enhancement scores than do commercial or industrial land uses. The resulting theme identifies areas according to their water quality enhancement capability.

Disturbed Lands. Areas that have been disturbed by man's activities, e.g., cities, mining, agriculture, etc., have a diminished value to the District in the context of preserving natural systems. To devalue such areas within the model, land cover categories corresponding to disturbed lands are identified and assigned a negative score. Recognizing that impacts to natural systems adjacent to disturbed lands extend beyond the boundaries between the two, buffers are delineated around each disturbed area. When added to the other two themes in this natural systems sub-model, the priority for acquisition of these disturbed areas and their buffers is lowered.

Layer Combination. The Natural Systems sub-model was developed by combining the themes representing priority areas for Wildlife Habitat, Water Quality Enhancement and Disturbed Lands. Among these three components, the Wildlife Habitat map received a weight of 1.5 in the combined map.

Management and Acquisition Considerations

In addition to the three natural resource-based sub-models described above, an effective acquisition program must also consider various management and acquisition factors. Management of numerous small holdings presents significantly more boundary, security and resource protection obstacles than managing fewer large properties. Maintaining wildlife habitat quality requires large, contiguous tracts. One way to accomplish this is to acquire properties adjacent to existing large public ownerships with similar management objectives. From the standpoint of administering an acquisition program, maximizing project sizes for each transaction is attractive. Finally, it is important to recognize that some regions within the District are subject to more immediate development pressures with accompanying speculation of land values and diminishing opportunities for public acquisition.

This sub-model contains three themes: Land Ownership, Priority Watersheds and Development Pressure. The first consists of boundary maps of property ownership; the second maps watersheds of priority water bodies; and the third is derived from urban and suburban land areas to pinpoint areas of high development pressure.

Land Ownership. To concentrate on large ownerships, all privately-owned properties larger than 2.6 sq km (one square mile) within the District were entered into the computer. Each of these areas was scored on the basis of size in order to put a priority on large holdings. Linking together properties of similar, natural character is an important priority for creating wildlife refuges, for minimizing boundary difficulties and for maintaining large green areas in a rapidly developing state. This consideration is reflected in the theme by mapping publicly-owned lands. Large land holdings identified above which are contiguous to or within 1.6 km (one mile) of these publicly-owned lands received a higher score than those which are not.

Priority Watersheds. Several programs within the District and at the state and federal levels have resulted in specific water bodies receiving special status. These include the Surface Water Improvement and Management (SWIM) water bodies, Outstanding Florida Waters, and Aquatic Preserves. These designations are indicators of the significance of these water bodies. To reflect this, the watersheds of each priority water body were assigned a score in this theme.

Development Pressure. Certain regions within the District are experiencing significantly higher rates of growth than others. The basis for identifying these critical areas is that areas adjacent to or within close proximity to existing urban and suburban centers are subject to development pressures. Areas meeting these conditions were included in this application, resulting in a single map which approximates the pressures of development within the District.

Layer Combination. This final sub-model was generated by combining the Private Land Ownership, Priority Watersheds and Development Pressure maps. Each of these components was assigned an equal weight in the combination.

Final Model Creation

The last step in the creation of the model was to combine the four sub-models using ARC/INFO overlay routines to produce a single data set representing the final model results. In this data set, each polygon contained one score for each of the sub-models. Since these scores were contained in INFO database tables, it was straightforward to combine these to produce summations of the individual sub-model scores. The following two summations were produced and used for further analysis:

WSS + FP + NS = RES_SCORE

WSS + FP + NS + MAC = TOT_SCORE

where;

WSS, FP, NS and MAC are the scores for the Water Supply Suitability, Flood Protection, Natural Systems and Management and Acquisition Consideration sub-models, respectively,

RES_SCORE is the summation representing only resource-based considerations, and

TOT_SCORE is the summation representing resource-based and management considerations.

Though a pilot study had verified the effectiveness of the model in identifying appropriate lands, the results of the entire model were tested by examining the individual and summary scores for lands previously acquired under the SOR criteria. The assumption was made that if these lands scored well in the appropriate categories then the model would identify similar lands elsewhere in the District. Visual inspection of choropleth maps and polygon scores indicated that the model did indeed highlight areas with the intended characteristics. It was therefore determined that the model was functioning as desired.

MODEL APPLICATION

As stated previously, the goal of the model was to identify those areas for acquisition that best met the criteria of the SOR and Preservation 2000 program. In practical terms this meant that the model would be used in the development of a five year land acquisition plan. This was done in two general steps. First, area-weighted scores for RES_SCORE and TOT_SCORE were generated and used to rank 35 projects nominated by non-District individuals. Second, area-weighted scores for all private holdings greater than 2.6 sq km were calculated. Additional analytical tools included choropleth maps of RES_SCORE and TOT_SCORE and frequency distributions and descriptive statistics of area-weighted scores for all polygons created in the model.

It was quickly noted on the choropleth maps that overall scores were notably higher in the northern part of the District. An examination of the sub-models showed that this was primarily caused by differences in hydrogeologic and land cover characteristics. To ensure that regional differences were taken into account, the decision was made to split the District into north and south regions and that evaluation of the model would be done not only District-wide, but also according to this north and south split. Frequency histograms of north and south scores were then used to assist in the creation of choropleth maps reflecting these regional differences.

Next, the scores of the nominated projects and large private parcels were compared to the scores for all lands within the District. For the north and south regions, those nominations and parcels whose scores were equal to or greater than the ninetieth percentile of all polygon scores within their region were designated as areas for potential acquisition. These areas were delineated on 1:100,000 maps and initial inspections of these sights were made via helicopter. On the basis of this inspection two sites were deleted from further consideration. The remaining 24 sites totalling 1425 sq km were included in the District's SOR Five Year Plan (SWFWMD, 1991). It should be noted that prior to actual acquisition, detailed hydrologic, environmental and engineering studies will be completed to ensure that proposed sites meet all District acquisition criteria. Additional review of acquisition decisions is made through public hearings, review by appointed citizen boards and peer review by District and other state agency staff.

CONCLUSIONS

The model described here represents a two year data collection and analysis effort culminating in the development of a land acquisition plan for the District. The following benefits are derived from this process:

- Land acquisition has become a more pro-active process at the District.
- Potential acquisitions are now pre-screened to determine if they meet the basic criteria defined in the SOR and Preservation 2000 legislation. This makes the process more efficient and objective.
- Identification of potential acquisition sites was done District-wide. Previously, most lands purchased were located along certain riverine corridors in the northern half of the District. The model identified several viable projects in the southern District that had previously not been examined.
- A database has been built that improves the efficiency of detailed site evaluations.
- The model is designed to be refined as improved data become available or as land acquisition criteria change.

It is the opinion of the District that the land acquisition model represents a valuable and decisive step towards the use of GIS and other decision support tools for managing water resources in Florida. GIS databases and models are actively being developed at the District to support other activities such as regulation development, permit processing and management of the over 600 sq km of land currently owned by the District.

REFERENCES

Aller, L., Bennet, T., Lehr, J.H., Petty, R.J., 1984, DRASTIC: A Standardized System for Evaluating Ground Water Pollution Potential Using Hydrogeologic Settings, National Water Well Association, Worthington, Ohio.

Christianson, R. A., 1986, Uses of District-Owned Lands -- A Compatibility Analysis, Southwest Florida Water Management District, Brooksville, Florida

EPA, 1990. A Modular Semi- Analytical Model for the Delineation of Wellhead Protection Areas, Office of Ground-Water Protection, U. S. Environmental Protection Agency, Washington, D.C.

SWFWMD, 1990. Land Acquisition Site Identification Model: A Geographic Information System Application. Southwest Florida Water Management District, 48 pp.

SWFWMD, 1991. Water Management Lands Trust Fund Save Our Rivers/Preservation 2000 Five Year Plan, 1991. Southwest Florida Water Management District, 62 pp.

TRACKING TIGER:
THE USE, VERIFICATION, AND UPDATING OF TIGER DATA

Don Ferber
Wisconsin Department of Transportation
BS&DP, GIS section
4802 Sheboygan Ave.
Madison, Wi. 53707

ABSTRACT

The TIGER/Line files released by the U.S. Census Bureau for 1990 contain a wealth of geographic information that may be expected to be used for many purposes beyond the states' 1991 redistricting efforts. However, especially where data sharing or lineage are of concern, utilization of this data carries many risks as it becomes increasingly widely disseminated, updated and altered. Careful consideration should be given to developing structures and procedures to verify data accuracy and describe changes made to the data. The Wisconsin Department of Transportation in conjunction with the Wisconsin Legislative Redistricting Project developed and tested procedures to process, validate, and, where necessary, perform and document alterations made to the TIGER data. Data validation included checking both attribute consistency and gross spatial errors. Programs and data structures were created to address transactions and delineate significant information regarding alterations. This paper will discuss experiences of the Wisconsin DOT in verifying the accuracy of TIGER data and tracking the alterations and updates to the data.

INTRODUCTION

The Wisconsin Legislative Redistricting Project is a cooperative effort designed initially to utilize the U.S. Census Bureau TIGER data for legislative redistricting and ultimately to provide for a variety of data needs of cooperating agencies. Principal members of the team include the Wisconsin Legislature, the Wisconsin Department of Administration (DOA), the Wisconsin Department of Transportation (DOT), and the University of Wisconsin - Madison (UW). In addition to redistricting, benefits that are seen to accrue from the project include the dissemination of GIS as a planning and analytic tool, and an integrated approach to data management within Wisconsin. (Sullivan and Chow, 1990).

The role of the Wisconsin Department of Transportation in the project was to create the geographic data base from the Census Bureau TIGER/Line files. In order to develop a reliable set of procedures and to evaluate data accuracy and resource requirements, a prototype data base was created in 1990 from the Precensus TIGER/Line files. The experience gained from generating this prototype was used to create the 1990 Census data base during spring, 1991, which was the geographic base used initially for the redistricting application.

Redistricting in Wisconsin is initiated at the local level where municipalities create new wards and counties create new supervisory districts from these wards. This provides the building blocks upon which state assembly and senate districts are based. At the state level two Democratic and two Republican caucuses -- one each from the state assembly and senate -- independently develop statewide plans, and then attempt to negotiate the differences. In a politically "enriched" atmosphere where history has shown a propensity for these disputes to be reconciled under the auspices of the Wisconsin Supreme Court, the ability to provide a high-quality, defensible data base was given considerable priority. Additionally, the DOT's experience in developing a state GIS highway data base increased awareness of the need to maintain data quality by tracking alterations to the TIGER/Line file data. Therefore, procedures and programs were developed to ensure processing uniformity and a final product that could be held to a high standard. TIGER/Line file data was scrutinized during processing for accuracy and consistency related to the application at hand, and changes to TIGER data were carefully verified and documented.

DATA BASE CREATION

Creation of a data base involves several steps (Ries, 1991) -- conceptualization, definition of requirements, design development, generation of the data, and maintenance of the data base. When applied to the creation of the Wisconsin 1990 Census data base, these steps resulted in a variety of outputs and products as delineated in Table 1. While the data base is the primary product, documentation and programs are critical to the propagation and understandability of the data, the efficiency of data production, and the overall quality of the data base. The effort to develop the repeatable procedures and programs used to create each data base, and provide full descriptions of those processes was such that considerably more time was spent on procedures and documentation and tracking changes to the data than in creating the primary data base product.

Table 1

DATA BASE CREATION PROCESS

PRODUCTS	Concept	Structure	Process	Creation	Maintenance
Documents	General concepts	Data outline	Process tasks	Data base lineage	Transaction procedures
Programs			Process macros		Transaction macros
Data				Geographic data base	Transaction data

Use of the TIGER data was determined during the concept, structure and process steps and implemented during the actual creation phase. Verification and updating of the data base began with creation of the data base and continued during the maintenance phase of the data base.

TIGER IN WISCONSIN

The Redistricting Project prototype data base was created using ARC/INFO software on an at times painfully slow VAX 785. While adequate for certain functions, the VAX simply lacked the horsepower to create GIS data products that averaged 10 megabytes each for 72 counties. Hence the project turned to a network of IBM RS6000 workstations for production of the 1990 census data base and the redistricting application to follow.

A determination was made that several views of the data were required to serve redistricting needs. This entailed generation of the basic TIGER (TGR) spatial geography and associated arc feature attribute table (ACODE) by county, followed by construction of the selected data layers required for redistricting -- Census Blocks (CNS), Minor Civil Divisions (MCD), and Voting Districts (VTD). (See Figure 2.) A Census Tract/Block Numbering Area coverage (CTB) was also created for cartographic purposes only. Additionally, data that comprised the individual layers was reselected from TGR to form more complex views -- MCD and VTD arcs to form a MAV data set, and CNS, VTD, and MCD to create a CVM data layer. Associated TIGER arc characteristics were obtained from the TGR ACODE attribute table. VTD coding sheets sent by local governments and MCD data from the DOA were utilized to provide voting district and minor civil division polygonal attributes. INFO tables containing MCD and VTD names and codes were generated during the production process and used to label and verify polygon attribute codes and locations.

Figure 1

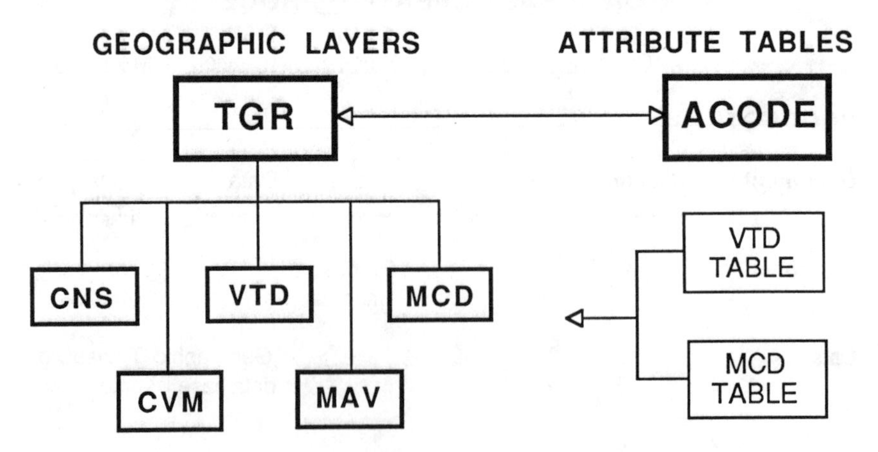

A key design decision was that the TGR data layer composed of all arcs in the TIGER/Line files would be treated as the master data layer from which all other layers would be derived directly. If any updates or corrections were necessary, all affected data layers and tables were deleted, changes were made to the master TGR data, and the derived layers and tables were recreated. This ensured that changes were made in a consistent manner and that they were accurately reflected in all subsequent data products.

VERIFYING TIGER DATA

Generation of a quality data product required verifying aspects of the TIGER data related to the redistricting application. This included the topology and accuracy of statistical and political boundaries and their associated attributes. Spatial tests were performed to check for topological accuracy and completeness using ARC verification procedures or data structures, and by comparing the processed data to external sources. Attributes were internally cross-checked for consistency and compared to external data sources to corroborate accuracy. A number of other attributes in the TIGER data were not utilized or checked since they were not applicable to the needs of the redistricting project.

The results of creating two data bases confirmed that Wisconsin TIGER data was of generally high topologic and attribute accuracy, but still warranted verification procedures. A noticeable reduction in data errors occurred, particularly in spatial data, from the Precensus to the final 1990 Census data. Corrections included eliminating a spatial mismatch between the GBF/DIME and DLG geography in one county, and correcting an intertwined road section in another. Given the primary concern of the Census Bureau to link population data to geographic areas, the focus on maintaining spatial topology can be expected. However, updates to spatial geometry were not given as high priority -- the GBF/DIME geometry in urban areas, which resembles the pattern a drunken driver might traverse, still persists.

Spatial errors from Precensus that remained in the 1990 TIGER data were usually the result of invalid intersections detected during the ARC 'BUILD' process. Included were arc shape or location inaccuracies and spurious or duplicate arcs. Occasionally this created polygonal topology problems that had to be resolved, but more frequently an arc deletion or shape change was simply required. However, numerous arcs were detected whose lengths could not be accounted for based on the scale of the USGS 1:100,000 data, the data precision, and the limitations of digitizing equipment -- less than 5 meters to no length in the arc attribute table. Similarly, polygons were detected nearly too small to stand in let alone be identified as a realistic piece of real estate.

The accuracy of the attribute data was more of a "hit or miss" proposition. Programming checks confirmed a high internal consistency of the arc attributes for each polygon. (The ACODE table contains both arc and polygonal feature attributes.) This was done by using an AML (ARC Macro Language program) to check that the CNS, CTBNA, VTD, and MCD attribute values matched within each polygon. With almost 200,000 blocks in Wisconsin, only a few questionable duplicate block codes were encountered and no Census Tract/Block Numbering Area coding problems occurred. In addition to verifiying consistency, a 'nesting' check identified no cases of individual blocks that were not fully contained within single voting district or minor civil division boundaries.

However, certain attribute errors were discovered at a sufficiently high frequency to question relying on those data characteristics without careful scrutiny. In Wisconsin, voting districts by definition do not cross minor civil division boundaries. Yet, numerous instances were detected of VTDs that did not nest within MCDs. A visual inspection of the spatial distribution of minor civil divisions showed many small areas that were incorrectly identified. They were often labeled as belonging to other towns where a check of geographic location alone indicated that these areas should have been suspect.

UPDATING TIGER DATA

Designing a Transactional Process

The errors and inconsistencies that were discovered (and also changes in desired views of the data) necessitated the implementation of an update process. As mentioned, this was given considerable attention due to the politcal nature of the project, the need for decisions to be defended, and DOT's understanding of the value of tracking data sources and quality. Unfortunately there was little in the way of role models or previous experience in stucturing and incorporating a full transactional process into data base design. Several articles were researched (especially Langran, 1988 & 1989), but there are both data base specific design requirements and software structural constraints that limit the ability to apply design theory to individual applications. It is difficult to design methods that allow flexibilty in processing and accommodate extensive data storage requirements and yet provide speed and ease of use.

The primary purpose of the transaction process was to differentiate TIGER/Line file data from DOT based data, document the rationale and processes used to institute the changes, and provide links from the current data base to the original data. For the Redistricting Project it was of more interest to verify data and processes than design a structure that allowed rapid movement between different temporal data views. The ARC/INFO data structure also restricted the speed and flexibility for many desired relationships when much of the transactional process required editing performed in the ARCEDIT module. The need to establish multiple relationships, sometimes for both shape and attribute data, in an environment that was not "relate friendly", had the ability to convince users that the VAX CPU had secretly replaced the IBM RISC processors.

The transactions were designed as a relational data base structure to track all TIGER data that was changed (deleted, updated, or added). The key was the retention of the original TIGER data as a historic record and the provision of a link between the new and old data. This link was provided by a transaction ID that indicated a particular process used on a feature or set of features. A transaction to the feature(s) was based upon a common rationale, source(s), and method or items changed.

Documentation was the other facet of the transaction design considered essential to track data base changes. Much of this data was stored in digital format in INFO tables. This permitted relates to be established to more easily track and recover changes. However, a paper form with a diagram and any additional explanation that might be helpful was used in conjunction with the tables.

Several aspects of structuring these transactions were considered critical to the data base product. First, suspected errors or inconsistencies in the data had to be verified. Often this entailed the use of outside reference sources -- DOT and DOA personnel knowledgable in minor civil divisions, maps from DOT, USGS, and the DeLorme 1:150,000 Atlas of Wisconsin, and county officials. Other problems were resolved simply by comparing the Precensus maps with the TIGER/Line file digital data. Secondly, the rationale for making a change had to be clearly defined and related to the data views required for the project. What spatial data and/or attributes were revised and the methodology used were also essential aspects of tracking updates to the data base.

Transactions Structure

The transactions structure is shown in Figure 2 below. The master data sets were comprised of the TGRCxx county layer ('C' for county and 'xx' was replaced by a two letter county abbreviation) and associated TIGER ACODE attribute table. All arcs that were deleted or altered in shape were first copied to the history geographic layer -- HISTCxx. All ACODE records were copied to the history attribute table -- the HISTCxx.ACODE before they were deleted or had items values altered. The two history data setscontained all data that was not considered accurate or pertinent to the present data base view.

The SHAPE and ATTRIBUTE change tables were the key tables used to track and document the transaction process. They contained items to link to all other types of tables in the transactions process. The item RECNUM (aka TLID), used by the U.S. Census Bureau as the unique statewide number assigned to each TIGER/Line file arc, was the link between the SHAPE or ATTRIBUTE table and the TGR geographic layer or ACODE table respectively.

The Census Bureau assigns VERSION to indicate a data base release number. Version was '1' for the TIGER/Line Files, 1990. The DOT increased this number by one each time a transaction was performed to track changes to an arc. A concatenation of VERSION and RECNUM provided the link to the history geography and attributes.

The TRX_ID transactions table was linked to the shape and attribute tables by a user assigned TRX_ID (a unique number for each county starting with '1'). This item linked features that were part of the same alteration process and also could track the number of transactions for each county. Since an arc could have shape and/or attribute changes, the item TRX_ID in the TRX_ID table provided the link. The TRX_ID table also contained information on the county to which the change was made, the data base creation process stage at which the problem was located, the person who performed the transaction, and the date that the data base change was done. (While not used for the Redistricting Project, this table is where real world change dates should also be located.)

Figure 2

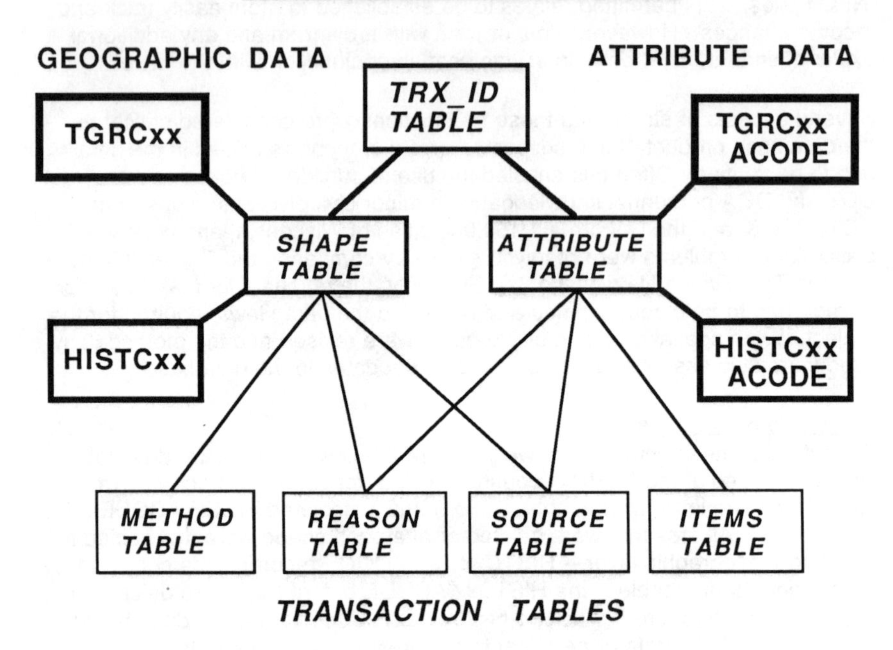

The REASON, SOURCE, METHOD, and ITEMS tables described the why, who, how and what of each transaction. They provided information on types of transactional events. These reference tables linked to the change tables by 'keys' (eg. SOURCE_KEY). Some codes were added during data processing as new situations were encountered. (Only shape changes had methods and attributes had items that were referenced.)

Transaction Use

Not all transactions to the data base were caused by errors in the TIGER data. Some were due to providing only selected views of hydrographic features. The state boundaries of Wisconsin include two Great Lakes -- Michigan and Superior. Therefore, boundaries of counties on these lakes were moved to the shoreline since not many (current) voters were expected to be found in these lakes. The decision was also made to add other water features that were linked to VTD or MCD boundaries. Much of this processing was done in an automated mode.

When situations were encountered that required alterations to the TIGER data, a copy of the TGR data layer was used. This necessitated returning to the process stage immediately after the initial creation of the TGR data sets from the TIGER/Line files. This processing usually entailed discerning the nature of the problem and the correction required, locating the area in the TGR layer, and performing the shape and/or attribute updates to the data. Errors were dealt with through editing procedures performed in ARCEDIT. During this phase much automated processing occurred at selected points to update data in various tables. Once the transaction was completed, any affected data sets were deleted and those processing sections were redone to recreate the updated view of the data.

A menu interface and many AMLs were developed to guide the user through the process and reduce the pain of the involved relates and data transfers that were required to update all tables and data sets. Even so, the processing, especially ;the automated portions, was often time consuming if the transaction involved a great number of arcs (eg. an entire miscoded minor civil division).

<u>Limitations and Suggestions</u>

One of the problems never fully resolved was the definition of a transaction. Since changes to Census Bureau data records werestored based on their unique IDs, the original concept was of feature tracking all individual Census features. While changes to indivdual arcs are tracked, the TRX_ID is process oriented -- not feature. A unique TRX_ID pertains to a type of change such as recoding an MCD or removing hydrographic features from the data view, regardless of the number of arcs involved. If there was more than one change to the same arc for different reasons, or if an error occurred or arcs were missed, there would be more than one transaction relating to that arc. When a change was made, the new process required a new transaction ID. Furthermore, a transaction may refer to a data error, or just to recoding done to provide a certain view of the data.

The features being tracked were based on the Census Bureau's definition of a feature -- for arcs this was basically a digital line segment with a unique identifier representing a real world linear object. Yet in many cases the real world feature is identified by a name and is not constrained by the limits of where Census places nodes. Furthermore, a TRX_ID was assigned based on a data base or editing event -- it may not match directly with the real world change. Tracking features by unique Census ID may be the most definitive means of tracking individual census features, but it may not be the most relevant method in some other situations.

While the intent was to provide a temporal record of data base changes, it was done only to a limited extent. While the date of data base changes was recorded, no real world time stamp was given (or in most cases available) to identify inception and, where appropriate, deletion dates of features. Hence, the view of the world on a temporal continuum is not available -- only discrete time stamps.

The transaction ID by itself is not very descriptive. It could refer to an attribute or shape change -- or both. Based on the rule of tracking any changes to the Census data, a transaction could be the result of a major spatial error or of misspelling an attribute. While each change table lists the type of action as 'A' (added), 'D' (deleted), or 'U' (updated), more definitive information on the type of change made is lacking in the tables.

Problems of insufficient information can be solved by adding further data to tables. However, as previously noted, there were times when the processing required was already overly slow and the addition of more data would not be help speed the process. Partially this due to the data structures that were being dealt with in ARC/INFO. Large amounts of data also had to be stored because INFO does not handle variable length records. However, the master data sets were already of a substantial size to begin with.

The relate structure also proved limiting for selecting and updating ACODE attribute records. Some arcs have alternate (record type 4) as well as primary records (record type 1). For reasons of simplicity and speed the decision was made to only update only primary records, but difficulties were encountered and it became necessary to place the alternate records into a separate file until processing was completed. The result was that primary records are updated but that alternate records may contain inconsistent information.

While the logical structure of the transactions procedures work well, their have been concerns over the physical structure that has been implmented. An alternative to placing history data out to separate data sets would be to retain the updated data with the history data and use selection procedures to obtain desired views of the data. Other options involve restructuring of tables and relationships, or performing minimal data changes in the editing environment and then finalizing the major processing in another environment. Some of these latter options are now being used to enhance the speed of the automated portions of the processing

SUMMARY

The TIGER/Line Files, 1990, contain information that will be widely applied in the public and private sectors. However, the usability of that information will depend in part on the particular application. The data seems to be more reliable for its topolgical data structures than for some of its attribute featue coding. The spatial geometry is much better suited to uses on the regional scale than to urban applications. However, the City of Milwaukee, for instance, has linked their own geometric data base to the TIGER topologic data to obtain a more usable product.

Regardless of the use, errors are likely to be discovered, or alterations of additions desired to the TIGER/Line file data that will require updates. A key element of a transactional structure is deciding on the parameters that are needed to track changes. Concepts of transaction design should be done within the context of the systems within which they will be applied. The experience of the DOT suggests that it is possible to design logical data structures to meet a select set of requirements, but that other issues such as hardware and software must be accounted for in the implementation of that design.

ACKOWLEDGEMENTS

The author would like to thank Rebecca Lee in particular (former DOT staff member now with Roy Weston & Associates in Seattle) who did much of the early research,design development and implementation of the transaction process, and Tom Ries, DOT TIGER Project leader, who has contributed his knowledge and insights into the transactions process design. The assistance of Andrew Hanson, Jerome Sullivan, (members of the Redistricting Development Team), and Tom Ries in reviewing this paper is also appreciated.

REFERENCES

Langran, G. 1988, Temporal GIS Design Tradeoffs: Proceeding of GIS/LIS 1988, Vol. 2, pp. 890-899.

Langran, G. 1989, A Review of Temporal Database Research and its Use in GIS Applications: International Journal of Geographic Information Systems, Vol. 3, No. 3, pp. 215-232

Ries, T. 1991, Data Quality in the Geographic Data Base Creation Process: Guidelines and Recommendations: Proceedings of Geographic Information Systems for Transportation Symposium, 1991

Sullivan, J. and Chow, A. 1990, The Wisconsin Legislative Redistricing Project: Design, Interface, Training, and Policy Issues: Proceedings of GIS/LIS 1990, Vol. 1, pp. 26-41

POSITIONING WITH THE GLOBAL POSITIONING SYSTEM SATELLITES

Clyde C. Goad
Department of Geodetic Science and Surveying
The Ohio State University
Columbus, Ohio 43210

ABSTRACT

Since the first commercial survey quality Global Positioning System satellite receivers were available in 1983, the surveying and geophysical communities have eagerly adopted this new technology mainly because of its precision, speed, and increased distances over which baselines can be measured. Now that the satellite constellation is approaching its designed limit of 21-24 satellites, other segments of society are beginning to notice its possibilities since twenty-four hour coverage is only the order of months away. In particular GIS projects are beginning to use these satellites as more moving applications are being implemented. One such project underway at The Ohio State University to develop the technology to map roadways will be highlighted. While initial results are very encouraging, one major conclusion formulated after the first eighteen months of study is that GPS receivers must be considered part of the technology package to be an economically viable system for infrastructure mapping.

INTRODUCTION

In 1983 the first commercially available Global Positioning System satellite receivers were delivered to the National Geodetic Survey. These receivers could not decipher the codes transmitted by the satellites and thus could only be used under very stringent conditions. The receivers had to be programmed to be able to measure the carrier phase states when the satellites were in view. No capability existed initially with these receivers to recover baseline information while the antennas were in motion. However for the survey community the limitations were offset by the added precision, mobility, and distance over which baselines could be determined. Precisions at the part per million level were commonly obtained at the onset of baseline observation campaigns. First-order precisions are an order of magnitude lower (one part in one hundred thousand).

By approximately 1985, survey quality receivers which could 'read' the information transmitted on the carrier were available but at fairly high cost (~$125,000 each). A minimum of two receivers is required for any precise baseline determinations. However these receivers exhibited some additional characteristics which allowed more experimentations for added capabilities (Kinematic Surveying). Essentially the code correlating receivers could maintain lock on the satellite signals while the antennas were moving. This led many researchers to exploit these receivers in many experiments which allowed development of requisite software and procedures. However since only a few satellites were in orbit (the order of 8-9), no real commercial applications other that surveying were developed. Today, there are 15-16 satellites in orbit and the designed constellation of 21-24 satellites is only the order of months away. In addition, improved GPS receivers have been offered by an increasing number of vendors. Thus today the navigation/ GIS markets are starting to take more notice of this exciting technology.

THE OHIO STATE HIGHWAY MAPPING SYSTEM

In response to this increasing awareness of the capability of GPS to provide precise positioning information to a large number of roadway infrastructure cataloging applications, The Ohio State University's Center for Mapping was contracted by the National Aeronautics and Space Administration (NASA) and the Ohio Department of Transportation acting as agent for itself and thirty-seven other state departments of transportation, the Province of Alberta, and the U.S. Department of Transportation to develop a highly efficient highway mapping system using the Global Positioning System as the core sensor.

In addition to the Global Positioning System receiver technology, the mapping van was equipped with magnetic sensing wheel counters plus an aircraft-quality three axis gyro providing angle changes in roll, pitch, and direction (yaw). The wheel counters sense front wheel rotation changes, both positive and negative for forward and backward motion respectively. These non-GPS measures can provide information about changes in van coordinates except for drifting in the gyros and wheel count scale factors.

The Global Positioning System receivers were used in differential mode. That is two receivers are used in tandem. One is with the highway mapping van and the other operates in stationary mode at a known location. Differencing the measurements collected at the same instant at each receiver provides the information necessary to determine the vector (or baseline) from the stationary receiver to the van. However one must realize that often the satellite signals are blocked which does not allow for any GPS baseline determination during these blockages. Trucks, mountains, buildings, road signs, and bridges are a few examples of such items which can block the satellite signals.

On the other hand the non-GPS instruments cannot provide absolute information about the van location. They only contain information about changes in postion. However these measures are very accurate. For instance the wheel counters have an integer count value of one centimeter. These instruments can exhibit drift however. For example, the wheel counter has no knowledge about slippage on wet roads, temperature variations of the air inside the tires, or mechanical friction in the bearings of the direction gyros.

As the reader has probably guessed, the strength of one system is the weakness of the other. So together the van system provides a very reliable mapping product. In essence the GPS determined locations provide dot-to-dot map of discrete locations of the van as a function of time. Using the wheel count and gyro information allows one to connect the dots similarly as in a dot-to-dot puzzle. The drift characteristics of the non-GPS instruments are found by forcing the paths derived from the wheel count and gyro measures to agree with the discrete GPS locations. Since many more measurements are available than required, a least-squares procedure is used to give a best-fit of the non-GPS measures to the GPS discrete values. Likewise when many satellites are in view simultaneously, another least-squares procedure is utilized to obtain a best-fit to the GPS measurements. It is quite common to collect data simultaneously from 5-7 satellites even though a minimum of four satellites is required. When the satellite geometry does not yield very reliable position determination, the least-squares procedure does not force the wheel count and gyro measures to fit the

GPS locations quite so strenuously. Thus some play is allowed to compensate for varying degrees of geometrical strength in the GPS location determinations.

Additionally both the front two undriven wheels are equipped with wheel count sensors. These sensors are the most likely to give problems due to their magnetization and closeness to the roadway. When both are working correctly, which is by far the usual situation, then they can be used to provide measures of distance traveled as well as turning information. Should one fail, the other can be used only to provide information about distance traveled. The gyro has yet to fail. It is a Sperry Tarsyn-333 aircraft-quality gyro.

EXAMPLES

Here two examples are given to show the ability of the combined highway mapping system to recover precise data files which can be used in the generation of map products of the van's path. The first example shows the results of data collected in San Diego, California in April, 1991. Fig. 1 shows the 'dots' recovered with the data collected by the two GPS receivers--one at a known location and the other with the van. This type of recovery is known as differential positioning. Its main characteristic is its rather insensitivity to common mode error sources such as tropospheric and ionospheric refraction and orbit errors. Notice the the dots sometimes form very smooth changing curves that are identifiable as roadways, culdesacs, and typical suburban streets. Also evident is the lack of detail at other parts of the plot where the van travelled. This has been found to be quite typical and hardly any data collection is void of such sections. However, even though there are 'holes' in the plots, if one had a continuous path of the van available such as is provided by the wheel count and gyro data, it is easy to see how connecting the dots is a very logical next step.

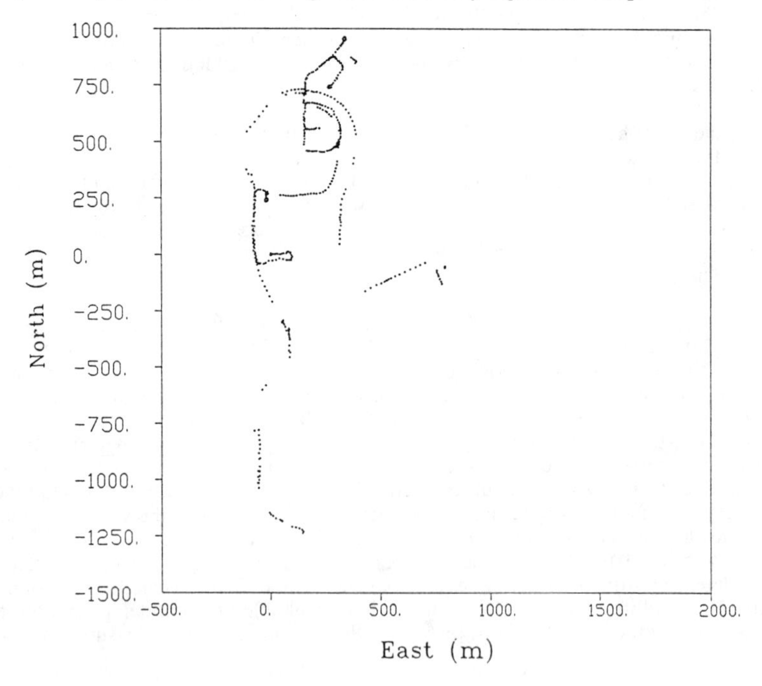

Fig. 1--Van Positions from GPS Data (Run 1)

Figure 2 contains yet another trace of the van postions recovered exclusively with GPS data. Some of the streets traversed in fig. 1 were also travelled in fig. 2. However it is also evident that the extent of the common coverage is difficult to discern due again to the outages.

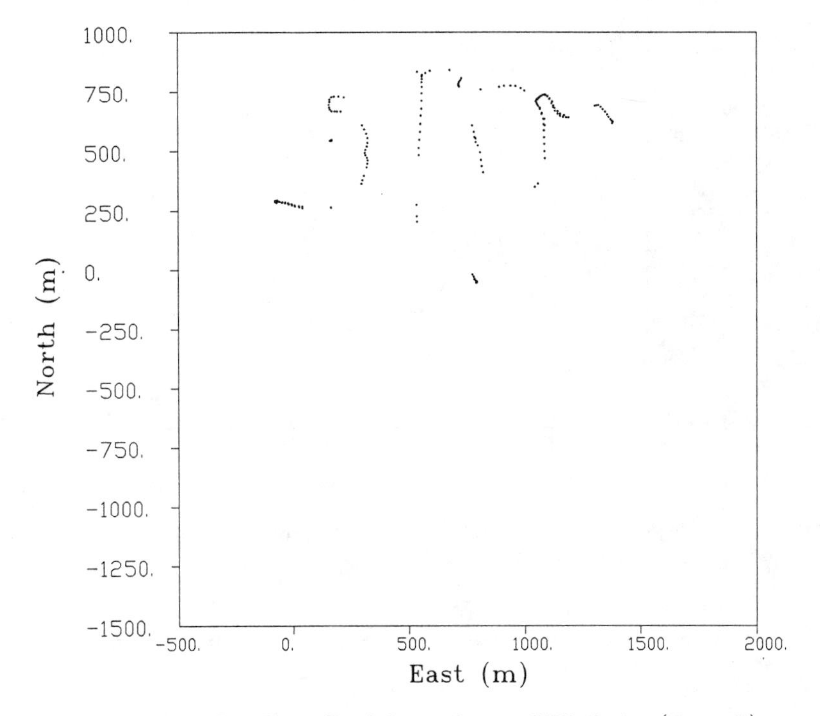

Fig. 2--Van Positions from GPS Data (Run 2)

Figures 3 and 4 are plots of the same streets travelled in figures 1 and 2 respectively using data from the wheel counters and gyro. Here one sees the continuous nature of the non-GPS measures. However these plots are skewed and stretched due to the drifts in the non-GPS instruments and variability of tire pressure and slip. However by solving for polynomial drift parameters, one can combine the first four figures into fig.5. Here the power of the total system is quite obvious. The larger the circles in fig. 5 the more uncertainty is assigned to the GPS position recovery. Thus sometimes the continuous path corrected with the better GPS postions is given greater weight than the uncertain GPS postions. Notice now how easy it is to recognize clearly the street layout, deadends, etc. Some closed paths and entire streets have no GPS data at all. But the calibrations work very well at these locations due to the slowly changing characteristics of the drifts.

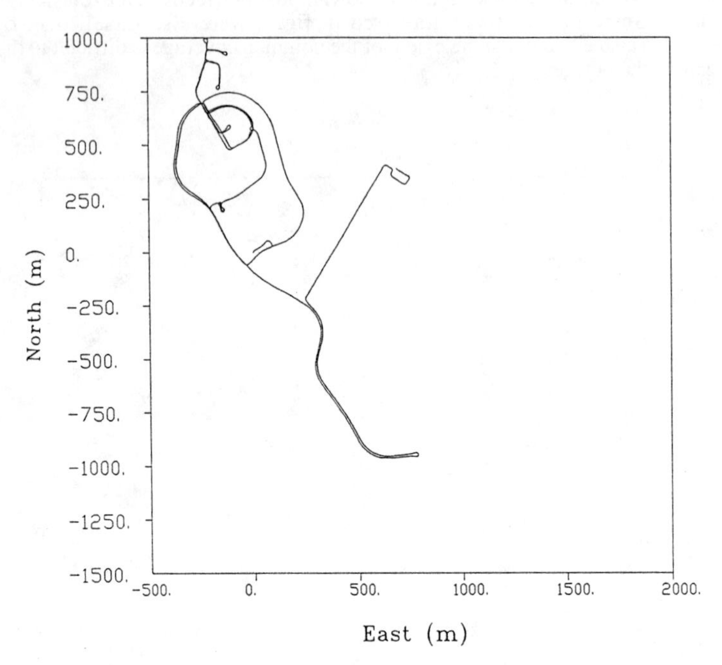

Fig. 3--Van Positions from Gyro Plus Wheel Counter Data (Run 1)

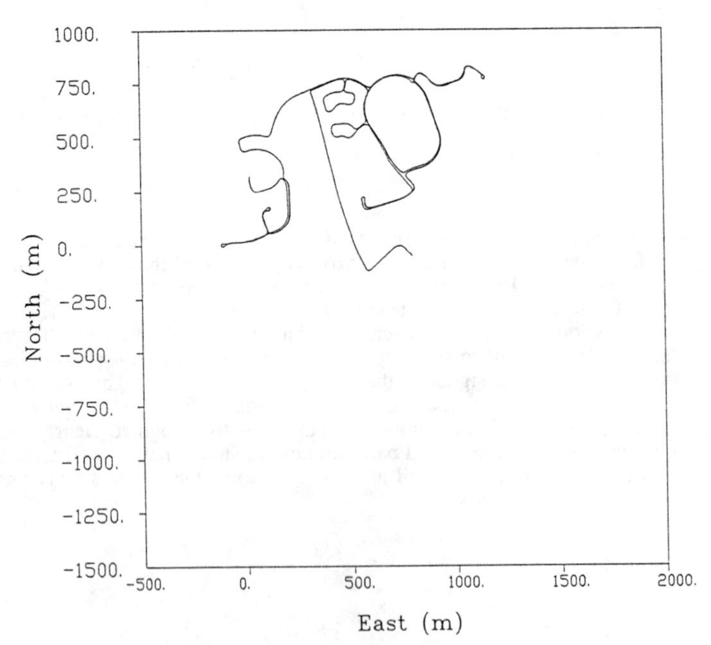

Fig. 4--Van Positions from Gyro Plus Wheel Counter Data (Run 2)

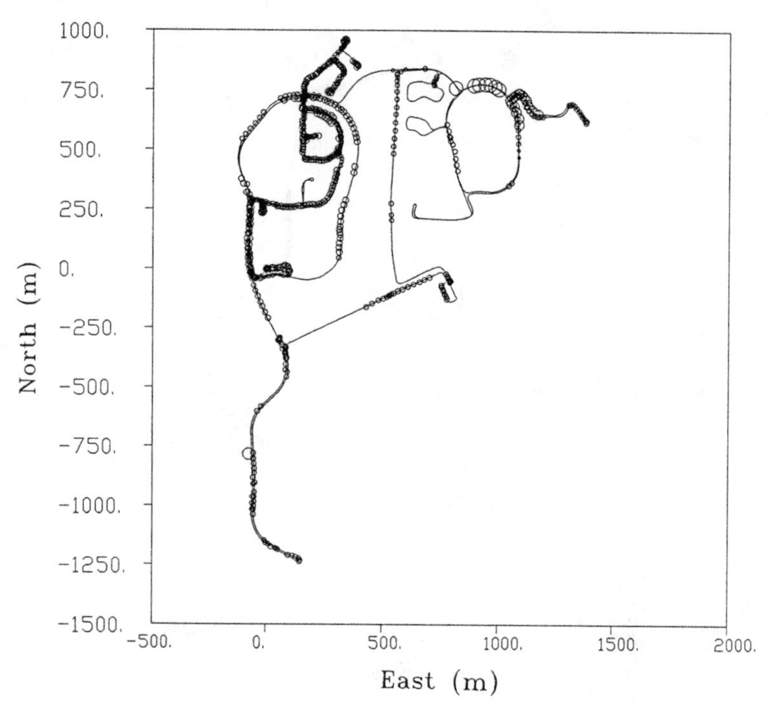

Fig. 5--Van Motion Positions with PDOP Circles Centered at GPS Point Positions

The scale of the first five figures was the order of 2.5 km on a side. This is of course a rather small area but respresents well how the system operates in the urban/suburban settings. For a better idea of more open road recoveries the van was driven around Columbus, Ohio on the eastern half of Interstate Highway 270 (fig. 6). Now the scale is the order of 35 km on a side. The fixed receiver collected data at the OSU Center for Mapping. Again it is clear that fairly long stretches of highway can be void of GPS recovered positions. Without the non-GPS instruments it would be impossible to map these areas. Also note that some GPS recoveries are also quite questionable denoted by the larger circles again representing large uncertainties of the GPS-determined locations.

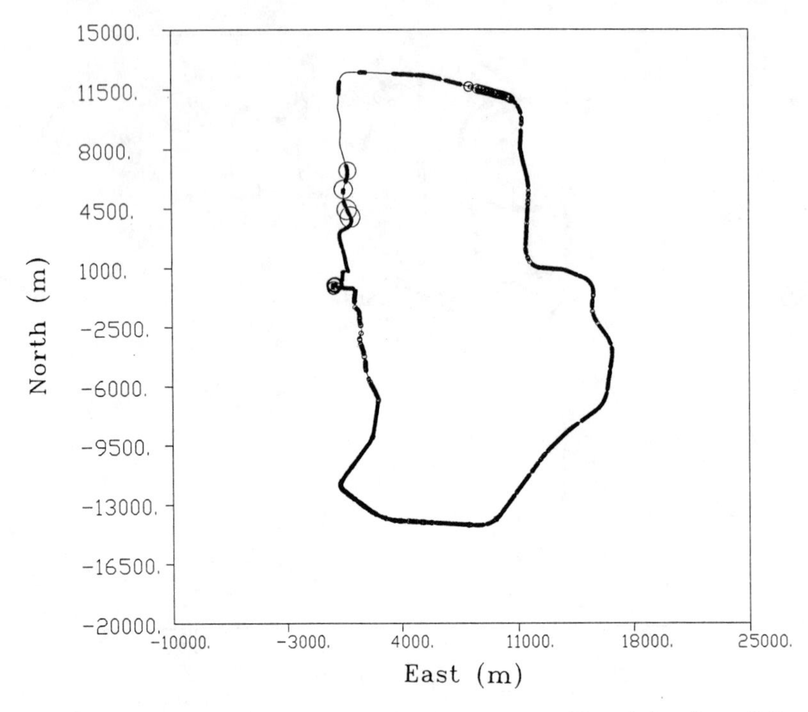

Fig. 6 – – Van Positions Around Eastern Half of Columbus, OH.

SUMMARY

This first study has allowed us to develop a reliable mapping system for all types of roadways so long as some GPS recovered postions are available. The system seems to be hearty, at least as far as the instruments are concerned. Of course we have learned that some instruments and associated equipment should be done differently next time. One is to collect data on solid state ram rather than on a spinning hard disk. Rough road conditions can hard on a spinning disk recording data. Ram is obviously better. Another element worth considering is the GPS receiver. Newer and more expensive receiver technology is now available which will allow us to position the antenna on the van to a centimeter relative to the nonmoving antenna. This then will allow one to perform precise surveying activities as well as meter-level positioning during the same run.

ACKNOWLEDGEMENTS

The support of the thirty-eight state departments of transportation, the Province of Alberta, the NASA Center for Commercial Development of Space, and the U.S. Department of Transportation are gratefully acknowledged. Project director Phillip Johnson and engineer Greg Orvets deserve special mention for the countless hours spent in putting the system together and testing it. Ming Yang, a master's degree student in the Department of Geodetic Science and Surveying wrote all software used in the postioning component; he also prepared the figures presented here.

GPS TO CONTROL COUNTIES AND MUNICIPALITIES

Jimmy D. Cain
Western Geophysical Company
A division of Western Atlas International, Inc.
3600 Briarpark Drive
Houston, Texas, 77048 USA

ABSTRACT

In recent years, a large increase in high-accuracy GPS surveys to generate geodetic control networks for various local and county governmental and quasi-governmental groups has been observed. The need for accurate control as a necessary starting point for successful GIS/LIS projects has been recognized at all government levels, including states, counties, and cities. In many cases, the need for control continuity has brought about cooperation between various government levels: state, county, & city. This presentation will examine various example projects of varying scope in some detail with a view toward assisting other local and regional governments in their own planning.

INTRODUCTION

With the recent explosion of GIS/LIS systems, the importance of providing an infrastructure of consistent control has been made clear. For a cadastral system to remain strong requires its basic control to be internally consistent at least to the accuracy needed to control the tightest scale mapping which will be needed in the foreseeable future. To establish a basic geodetic network only to the accuracy necessary to control a current 1:10,000 mapping project when future maps will have to be produced at 1' = 100 foot scale (1:1,200) is asking for problems. The advantages of setting adequate initial control are now obvious and accepted.

The recommendation is to provide an initial geodetic network with enough accuracy that it may be further densified as required for all reasonable future applications. In the past, such network integrity has not often been implemented due to economic considerations and the previous difficulties of creating networks of high order control in many areas and environments. With GPS surveys, these factors are no longer major roadblocks.

Many states are establishing very precise statewide GPS networks to accuracies better than one part per million (1 ppm) of interstation baseline lenth. Networks have now been completed in Tennessee, New Mexico, Florida, Wisconsin, Washington and Oregon and a number of others are either in preparation or under consideration. At least one state, Tennessee, has passed a law requiring all significant control surveys to be tied to the state's precise GPS network. Densifying county and municipal control has become very easy with GPS in states where such precise networks exist, particularly when GPS techniques are also used for the densification.

For local densification projects in other states, there is sometimes a lack of high order consistency in the pre-existing control on which the densification must be based. As all surveyors know, traverses run between specific "control" points do not always close at the expected levels of uncertainty. Often control was put in many years apart, in various different campaigns. Different adjustments over the area may compound discrepancies. Current GPS techniques offer an ideal and cost effective way of checking such control and this should be done wherever possible.

GPS OFFERS AN IDEAL SOLUTION

Interferometric techniques using carrier phase data from GPS were introduced for precision geodetic surveys in early 1982 (Counselman and Steinbrecher, 1982) and have been used on a production basis since mid 1983, with numerous improvements in technique and instrumentation along the way. GPS instrumentation and techniques have come a long way since the single-frequency MACROMETER® V-1000 system was approved by the United States Federal Geodetic Coordinating Committee (F.G.C.C.) in January 1983 for use as "a viable system that can be used successfully to establish geodetic control relative to the National Geodetic Reference System" (Hothem and Fronczek, 1983). Since that time, various manufacturers have introduced new systems and most have had their receivers and their associated software tested by the F.G.C.C.

Today, GPS control surveys with dual-frequency GPS receivers can consistently provide positional accuracies on a routine basis of up to 1 to 2 parts per million (ppm) of the baseline lengths surveyed, regardless of length, subject to a reasonable instrumental setup error. Such accuracies generally require a fairly long (3 to 5 hours) baseline observation and are often obtainable with broadcast ephemerides. Dual-band GPS accuracies can be readily improved to much better than 1 ppm by use of precise post-fit satellite ephemerides. Several surveys requiring contractual accuracies of 0.1 ppm or better have been performed successfully using GPS techniques (Cain *et. al.*, 1991).

Single-frequency receivers can provide comparable results for short baselines (less than 8 - 10 kilometers), but the accuracy of their results over longer baselines is often significantly reduced by the presence of traveling ionospheric disturbances.

The removal of interstation visibility requirements with GPS also allows for much better geometric configuration of networks than is often possible with terrestrial techniques, regardless of the associated control costs. GPS baselines can be observed across mountain ranges, swamps, and other natural or man-made obstacles which cannot be removed or conventionally traversed with ease.

GPS survey production is almost always much higher than that possible with conventional control surveys. Even with the partial GPS constellation currently in place, essentially all areas of the United States have good coverage have at least 6-8 usable hours per day and generally much more.

If a network is designed to yield 1:50,000 positional accuracies over a limited geographical area with good access, it may well be possible to do four or five observational setups per dual-frequency receiver per observation day yielding as many as twelve (12) to fifteen (15) non-trivial baselines per survey day if four receivers are deployed. This production does require good access between points and very few obstructions to satellite visibility. With somewhat limited access and occasional obstructions, three or four sessions per day are usually achievable. Even with First Order surveys several sessions per day are possible under most circumstances.

The above capabilities make such GPS techniques excellent for:

- Establishment of primary control networks;
- densification of secondary control from primary networks; and;
- checking existing control to be used on specific projects.

GPS surveys are not generally weather dependent. Strong inter-station differences in water vapor content should, however, be avoided if possible,

---

MACROMETER is a registered trademark of Western Geophysical Company, a division of Western Atlas International, Inc.

particularly for required accuracies better than 10 ppm (1:100,000). This situation might occur during high thunderstorm activity or during weather frontal situations. Nonetheless, it is generally possible to survey in rain, snow, fog, or heavy haze without significant degradation of survey accuracy.

All these items together make GPS surveying generally less expensive than terrestrial techniques for control surveys of 1:50,000 or higher. This is often the case for 1:10,000 or 1:20,000 surveys as well. More importantly, GPS surveys offer a better product for control surveys.

LIMITATIONS ON GPS

Although GPS is an excellent new tool, it should be viewed in that light and not considered as the answer to all survey problems. It is not! There are significant system limitations and someone knowledgeable with GPS should always assist in planning and execution of GPS projects. Like any good tool, GPS can be abused or misused. Lack of adequate system knowledge is the biggest single factor leading to GPS projects which won't meet specifications. Organizations planning GPS surveys are cautioned to make sure that those performing the surveys are qualified to do so. Make sure that the firm doing your work is experienced and knows what they are about. Don't hesitate to ask for example projects they have performed and names and contact information for responsible customers. These should then be checked out.

Some of the key system limitations and misunderstandings about GPS as a survey tool will be addressed below.

Satellite Visibility

The need for good visibility upward to the satellites is one stringent GPS survey requirement most misunderstood by new users. The importance of this is quite often underestimated --- many times to the degree that an acceptable GPS survey simply cannot be performed for selected points. Good visibility in the GPS sense doesn't mean being able to see some of the sky; it means being able to simultaneously see at least four or five satellites (or more) in different quadrants of the sky, some of which are sweeping across the sky at fairly low elevation angles. Remember, if the GPS units cannot receive signals from the same satellites at the same time, then they cannot produce data capable of yielding positions. The more obstructed the data set, the less accurate the results. As a result, GPS cannot be used under trees, very near tall buildings, or in many other situations with limited visibility upward. If a significant portion of the sky is blocked at a needed point, then an experienced GPS planner should review whether or not acceptable data can be obtained at that station.

Differential Ionospheric Delay

Using dual-frequency equipment will improve accuracy the of GPS surveys, particularly over longer baselines. With suitable processing software, the use of dual-frequency GPS equipment will allow the removal of most, if not all, ionospheric refraction effects. Ionospheric effects over baselines of intermediate length do not cancel out. Significant errors can remain after such differencing --- often as much as 10 - 12 ppm. Dual-frequency receivers are definitely recommended for all GPS surveys of First Order or higher which contain any reasonably long baselines (more than 15 - 20 kilometers). Improvements are also often seen on baselines as short as 10 kilometers.

Anti-Spoofing and Selective Availability Policies

Some users and manufacturers express great concern over the U.S. Government's Anti-Spoofing (A/S) policy to encrypt and deny the P-Code signal to civil users as well as the Selective Availability (SA) policy to degrade the C/A Code message to provide basic navigation of 100 meters, 2drms. Nonetheless, even when SA is activated, GPS interferometric surveys continue to provide high accuracy results on a regular basis.

The government's clearly stated policy on A/S will, when implemented, have a major impact on all pure P-code dual-frequency receivers, Those in the civil community using such receivers will simply not be able to receive the L2 signals. Use of a codeless L2 approach, as implemented by Western and others will eliminate this problem for the control surveyor.

GPS Network Planning

It is very important to remember that the GPS receivers are simply additional tools with which the surveyor can work Adequate checks on data quality and procedures are just as important with GPS as with optical instrumentation. The GPS survey itself should be designed to ensure that such quality control is incorporated. A thorough set of network design criteria and GPS observational checks are available (F.G.C.C., 1989). Adherence to the pertinent portions of these specifications is recommended for all GPS control densification surveys anticipated by municipal, county, and state authorities. For some example networks, see the discussion under Local Control Networks herein.

CONTROL NETWORKS ESTABLISHED WITH GPS SURVEYS

Precise Statewide Control Networks

The GPS experience of NGS by 1986 prompted an NGS official to state "GPS can establish geodetic control at significantly lower costs per station than classical survey methods, thereby reducing geodetic costs to a very small percentage of future engineering, land information systems, and mapping projects" (Zilkoski, 1986). Several other countries, including Australia and West Germany, were also implementing major GPS programs in the same time frame (Cain, 1987). Many others have done so since and the GPS program within NGS has continued to grow.

Several large-scale Order B statewide GPS surveys requiring project horizontal accuracies of one part per million (1 ppm) or better have been performed over the last several years. Three have been surveyed under contract by Western Geophysical[+] and three by the NGS. Statewide 1 ppm networks surveyed to date by Western include Tennessee in 1987 (Zeigler, 1988), final portions of New Mexico in 1989 (Henderson, 1988) and Wisconsin in 1990 (Hartzheim *et. al.*, 1991). Five of the Wisconsin stations were surveyed to Order A (0.10 ppm) accuracy. Comparisons are available between the above three state networks (Cain and Kyle, 1990). Statewide 1 ppm networks surveyed directly by NGS include Florida in 1989-1990 (Shrestha *et. al.*, 1990), Washington (Wegenast, 1989) and Oregon (completed in 1991), as well as early portions of the New Mexico network. Several others are underway or in preparation (Bodnar, 1990).

One common characteristic of each high-precision state GPS network surveyed was a target overall adjusted accuracy of one part per million of baseline length plus a small (8 mm) non-baseline-specific error component. This yields relative positions of adjacent statewide stations accurate at 1 - 3 centimeter level, (Strange and Love, 1990).

[+]Formerly operating as Aero Service Division, Western Atlas International, Inc.

All of these statewide network points are established with GPS usage in mind and hence are highly accessible by automobile and have good open visibility of the celestial dome. This makes them ideal for further densification by GPS for various county and city applications.

Local Control Networks

As GPS surveys have proven their cost effectiveness for control surveys (Antenucci, 1986), more and more municipalities, metropolitan areas and county governments are using this technology for control densification within their respective geodetic networks. A number of such programs are discussed below. Those discussed by no means comprise an exhaustive list, but are simply representative examples chosen to illustrate specific ideas or concepts.

GPS as a Check on Existing Control. The Arizona State Department of Transportation established a fairly extensive network of 85 First Order geodetic stations in the Phoenix area in 1986 using dual-frequency GPS receivers. These surveys were performed by Western$^+$ to F.G.C.C. standards for inclusion of the data in the National Geodetic Reference System (NGRS). A brief review of this project will emphasize one of the aforementioned advantages of GPS---the check of existing control. Figure 1 shows the network of GPS baselines surveyed. To adequately check this network to First Order standards by conventional terrestrial survey methods would have been a very expensive undertaking. With GPS however, it was possible not only to complete this project at a relatively low cost but also to run an excellent check on the pre-existing primary horizontal control. These check survey baselines, shown in Figure 2, were performed in one survey day. These directly measured GPS control baselines compared with the published First Order NAD-27 baseline results at the 1:150,000 to 1:360,000 accuracy level, except for one station, which quite obviously fell outside the expected tolerances. The discrepancy was discussed between the contractor and the NGS and, after review by NGS, it was determined that the new GPS measurement matched quite nicely with the provisional NAD-83 control coordinates, not published at that time. In this project, directly measured GPS baselines both pointed out a problem with one station's published NAD-27 coordinates and then confirmed the NAD-83 coordinates for the same station, relative to others in the area.

Cost Comparison with Non-GPS Approach. The previously referenced Phoenix project demonstrated the cost advantages of dual-frequency GPS surveys over single-frequency GPS and over conventional techniques. It only took ten days to complete the field observation program in Figures 1 and 2. The job would have taken about 6 months for a crew to complete using conventional surveying methods (Adams, 1986). In the same interview Adams indicated that the project was "costing about half as much as conventional surveying methods and producing better quality results". The same project would have taken between 15 and 20 days using single-frequency receivers, and would have resulted in a network of lower accuracy.

Another example: sixty-five GPS points were established for control densification in Louisville and Jefferson County, Kentucky in late 1985. The average proposal received witch included GPS was less than one-half of the price quoted for the average proposal received without GPS assistance (Antenucci, 1986). There were four proposers using GPS and seven who did not include GPS as a control tool. The price disparity for similar projects is much larger today due to increase in GPS survey productivity of the last five (5) years.

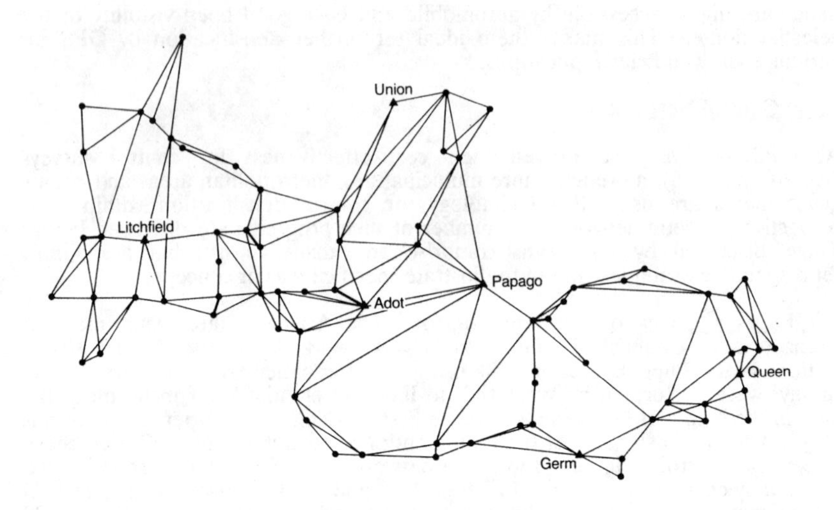

Figure 1. Complete GPS control network established in Phoenix (AZ) for the Arizona Department of Transportation.

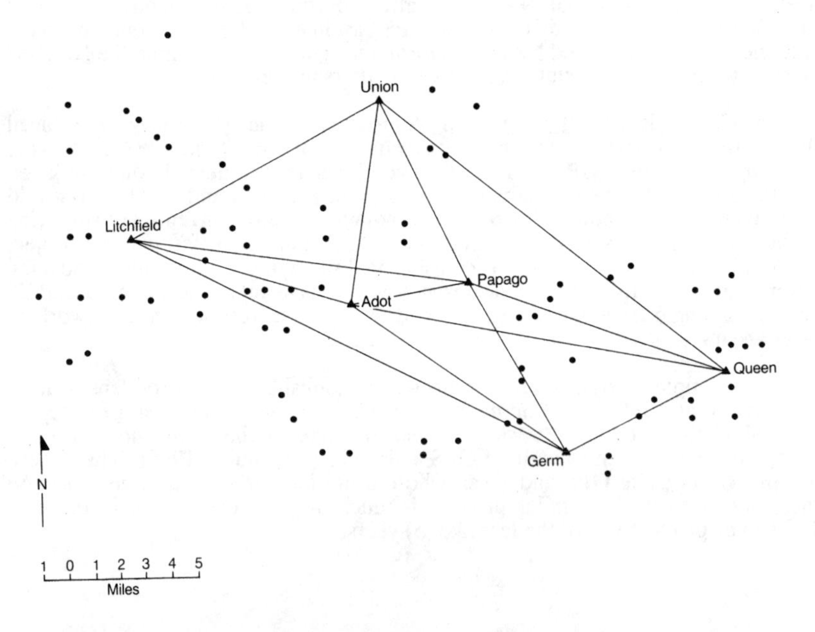

Figure 2. GPS test of geodetic control station consistency performed during survey of the network shown in Figure 1.

Cooperation Between Different Levels of Government. There have been several projects which demonstrate the benefits to be mutually gained by different entities working together to achieve a coherent control network for their common areas of interest.

A project which exemplifies the spirit and level of cooperation which can be obtained is found in the Metropolitan Kansas City Area GPS Control Network. Most of the various entities responsible for control in the Kansas City area (states of Kansas and Missouri, various County governments, and various city governments) jointly developed a multi-year plan for GPS densification of the entire metropolitan area in various stages. The first three years' programs of the State of Missouri portions were performed by Western+, with additional data provided in that same time frame by Liberty County (Missouri) and the Kansas counties of Johnson and Wyandotte. In the Missouri coordinated program, sites were selected in a cooperative venture between the State Department of Natural Resources and the local governing entities. Monuments were then set by volunteer surveyors within the local area. The GPS network was prepared by the contractor in consultation with the State Land Surveyor's Office to provide a network designed for further future densification, either by GPS or by conventional techniques. Figure 3 shows the way in which these programs coupled together for the first three years of this cooperative venture. Since 1989, required baselines have been added in certain parts of the network. All together, this project has comprised several hundred First Order survey points.

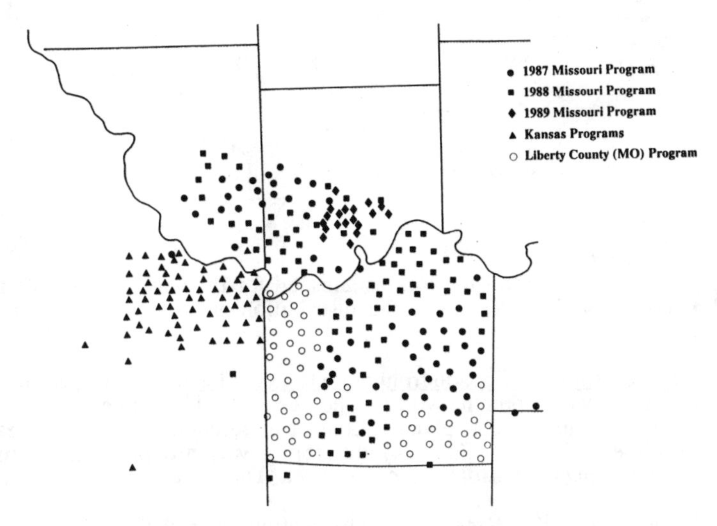

Figure 3. Metropolitan Kansas City Area GPS Control Network established over several years.

Another example of good cooperation between city and county control densification efforts came recently in the State of Washington. Pierce County (Washington) contracted with Western for a mixed First and Second Order network of over 100 stations covering the entire county except for Mount Rainier. Through the county, the cities of Tacoma and Puyallup also added additional points they wanted within their respective boundaries. Pierce County coordinated the project, involving both the contractor and the NGS state advisor to ensure that each sub-net would stand alone yet also couple together to form a large county-wide network further tied to a number of precise "supernet" stations of the Washington statewide Order B survey.

Good Distribution of Horizontal and Vertical Control. It is necessary to start with a good distribution of horizontal (and vertical if appropriate) control in order to get a good consistent network which ties well to the area's geodetic control. The area should be well bounded by horizontal control, in accordance with F.G.C.C. specifications. For reasonable vertical results with GPS, a good number of known vertical control stations should be occupied with the GPS receivers and these stations should be well distributed throughout the area. The network (Figure 4) surveyed by Western+ for end-use by the City of Nashville and Davidson County, Tennessee, clearly demonstrates such a distribution of both horizontal and vertical control.

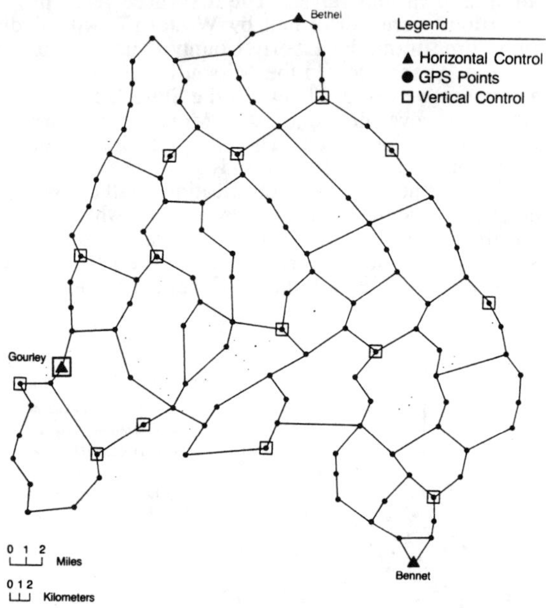

Figure 4. GPS control network established for the City of Nashville and Davidson County (TN). Note the distribution of horizontal and vertical control.

Consistent Control Densification from State to Local Levels. Many people ask, "Why do the states want 1 ppm networks - who needs that kind of accuracy?". An examination of the current and planned control hierarchy in Wisconsin clearly demonstrates the rationale. Four (4) different levels are planned in many instances, each serving as control for the next level. These are:

0.1 ppm	(1:10,000,000)	Five Station "Supernet"
1.0 ppm	(1:1,000,000)	Statewide Precise network
10.0 ppm	(1:100,000)	County GPS densification
20-100 ppm	(1:50,000-1:10,000)	Additional local densification (generally conventional)

The following project illustrates the "grand design" of such GPS densification projects. Following the establishment of the Wisconsin Order B survey in 1990, Western was awarded separate contracts in western Wisconsin by two individual counties (Pierce and Pepin) and by the River Country RC&DC for GPS densification within three additional counties (Dunn, Eau Claire, and Chippewa). These end- users all wanted to obtain First Order (10 ppm) surveys tied to the new statewide Order B (1 ppm) network. These counties all have some border commonality. Additional conventional survey densification at lower accuracies was then planned by the counties. The contractor suggested some revisions in the observation plan to offer one network covering the entire group of desired points. Through cooperation, the projects were all surveyed as one large project. Final results were obtained in one least-squares adjustment, providing a consistent densification throughout these five counties. This can be readily expanded in the future as the need arises without any fears of not tying well to adjacent county control. The resultant network was controlled by eight well distributed Wisconsin High Precision Geodetic Network (WHPGN) points adequately bounding the area, plus one (1) additional NGS station.

Another project illustrating a consistent tie to a statewide 1 ppm network (the Tennessee Geodetic Regional Network, in this case) is found in Knoxville, Tennessee. Figure 5 shows how this city network was tied to the state network.

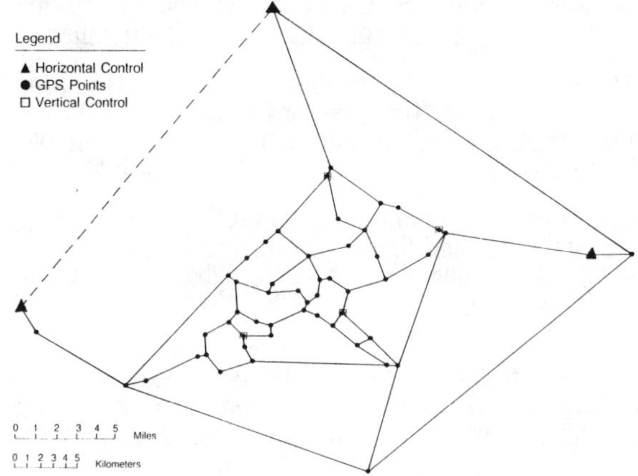

Figure 5. GPS control network established for City of Knoxville (TN). Horizontal control utilized was TGRN statewide 1 ppm network.

SUMMARY

Good geodetic control surveys are seen as key fundamental starting points for successful GIS/LIS implementation. These must have internal accuracy levels consistent with the applications and scales of the appropriate GIS/LIS which they will support in the future. For most applications, GPS surveys are recognized as the best way to achieve control surveys of the desired accuracy. Over the past several years, many cities and counties have established or densified control in their areas by GPS techniques, very often for the express purpose of providing a sound base upon which to build GIS/LIS programs.

In essentially every case examined, GPS surveys have yielded optimum control networks much less expensively than conventional surveys. John Antenucci (1986) said it best: "Organizations are being presented with the unusual choice of between lower cost/higher accuracy [GPS] surveys and high cost/lower accuracy [conventional] surveys". With the launching of additional satellites over the last five (5) years, GPS productivity has continued to increase.

REFERENCES

Adams, W., October 10, 1986, interview published in Arizona Republic Newspaper.

Antenucci, J.C., 1986, "Global Positioning: Foundation For Multipurpose Cadastre and Automated Mapping", Proceedings, 1986 ACSM-ASPRS Fall Convention, U.S.A., September 28 - October 3, 1986, Anchorage, Alaska.

Bodnar, A.N., 1990, National Geodetic Reference System Statewide Upgrade Policy, NOAA, National Geodetic Survey, Advance Copy, June 19, 1990.

Cain, J.D., 1987, "GPS surveying as an Effective Tool for Establishment of Control Networks", presented at First National Seminar on Global Positioning Systems, March 19, 1987, Bisbane, Queensland, Australia.

Cain, J.D. & D.W. Kyle, 1990, "Experiences with State-Wide High-Precision GPS Networks in the United States", Proceedings of Second International Symposium on Precise Positioning with the Global Positioning System, GPS'90, September 3-7, 1990, Ottawa, Canada, pp. 1025-1039.

Cain, J.D., W.C. Banks and D.W. Kyle, 1991, "Results of Orbital Improvement on Large Scale Networks within the United States and the Caribbean Islands", proceedings of Japanese Symposium on GPS (1991), Kyoto, Japan, pp 53-63.

Counselman III, C.C. and D.H. Steinbrecher, 1982, "The MACROMETER Compact Radio Interferometry Terminal for Geodesy with GPS", Proceedings of the Third International Geodetic Symposium on Satellite Doppler Positioning, February 8-12, 1982, Las Cruces, New Mexaco, Vol. 2, pp. 1165-1172.

Federal Geodetic Control Committee (F.G.C.C.), 1989, Proposed Geometric Geodetic Survey Standards and Specifications for Geodetic Surveys Using GPS Relative Positioning Techniques (Ver. 5.0 May 1988, reprinted with corrections, August 1, 1989), National Geodetic Information Branch, NOAA, Rockville, MD., 49 pp.

Hartzheim, P.J., L.D. Hothem & D.W. Kyle, 1991, "Results of Wisconsin High Precision Geodetic Network Survey", Presentation and Proceedings of ACSM - ASPRS 1991 Annual Convention, April 1991, Baltimore, Maryland, pp. 121 (Abstract).

Henderson, T., 1988, "The New Mexico GPS Reference Network", presented at the New Mexico Association of Surveyors and Mappers Annual Convention, February 1988, Las Cruces, NM.

Hothem, L.D. and C.J. Fronczek, 1983, Report on Test and Demonstration of MACROMETER Model V-1000 Interferometric Surveyor, Federal Geodetic Control Committee: FGCC-IS-83-2, Rockville, Maryland.

Shrestha, R., R. Taylor, S. Smith, and L. Stanislawski, 1991, "Preliminary Results for the Florida High-Precision GPS Network. Surveying and Land Information Systems, 1990, Vol. 50, No. 4, pp. 299-302.

Strange, W.E. and J.D. Love, 1990, "Statewide High Accuracy GPS Networks in the United States", presented at American Geophysical Union Spring Meeting 1990, Baltimore, Maryland.

Wegenast, D.A., 1989, "A GPS High Precision Network for the State of Washington." Proceedings of the 1989 ASPRS-ACSM Fall Convention, pp. 24-30.

Zeigler, J.H., 1988, "GPS Geodetic Reference System in Tennessee", Journal of Surveying Engineering, ASCE, Vol. 111(4), pp. 156-164.

Zilkoski, D.B., 1986, "Development of Alaska Geodetic Reference System Using the Global Positioning System (GPS)", Proceedings, 1986 ACSM-ASPRS Fall Convention, Anchorage, Alaska, U.S.A., September 28 - October 3, 1986.

Prediction of Disaster Occurrence Point on the Slope Land along the Railway Using Raster GIS

Tatsuo NOGUCHI, Tomoyasu SUGIYAMA
Railway Technical Research Institute
2-8-38, Hikari-cho, Kokubunji, Tokyo, Japan

Masahiro SETOJIMA, Masaru MORI
Kokusai Kogyo Co., Ltd.
3-6-1, Asahigaoka, Hino, Tokyo

Abstract

In Japan, many of the railways are located along the acute slope land, and therefore, a number of countermeasures have been provided against the slope land disaster from the past. Since the slope land disasters are caused by various factors such as the delicate geographic features including the inclination and shape of slope as well as the state of land cover of vegetation, complicated analysis is necessary for establishing the countermeasures for slope land disaster. In this paper, we report mainly on the study to predict the disaster occurrence points on slope land by digital image processing using raster GIS based on the information obtained from aerial photograph and topographic map as well as on the method for the collection of basic information and prediction of occurrence and the function of system developed to be put into practical use.

Background

The railways are the main transportation means in Japan, and its aggregate length amounts to huge distance. In Japan where the plain land is scarce, there are many acute slopes along the railway, and as a result, many slope land disasters are occurring such as landslides caused by heavy rainfall and earthquake. Aerial photographs have been used from the early stages in Japan for extracting and managing the disaster information of the slope land along the railway, and the approaches have been made mainly basing on the interpretation of aerial photographs. In recent years, on the other hand, remote sensing is going to be actively applied for the study of slope land disaster occurring along the railway thanks to rapid advancement and expansion of application fields of its technology. However, under the natural environment of Japan where geological structure is complicated and land surface is densely covered by vegetation, it is difficult to obtain the expected accuracy in most cases only through the exclusive approach of aerial photograph data or remote sensing data, and therefore, it is hard to say that such approaches are adequate enough. As a result, to obtain higher accuracy and manage the slope land disaster information more effectively, it is necessary to use the various kinds of information not only such as topographic and geological information which has close relation to the slope land disaster but also land cover information such as vegetation and land use, disaster history in the past, and location and kinds of civil engineering structures, and to collect and analyze the disaster related information from the comprehensive viewpoint. Each of these basic information is well collected separately, but in most cases, there is no uniformity in the accuracy and scale of them. Therefore, it is necessary to systematize the various kinds of information related to the slope land disaster along the railway so that they can be managed in the unified form. Also, the support should be provided through systematization to the prediction of future occurrence of disasters to be made from the huge volume of information and to the analytical works to clarify the mechanism of disaster occurrence in the past. This study discusses the application of raster type GIS for the information management of slope land disasters along the railways and extraction of possible points of disaster occurrence basing on these data.

About the slope land disaster information

The disaster information can be classified into (1) the basic primary information such the as actual state of damages and the terrain parameters which have caused such disaster, (2) secondary information which is obtained by analyzing and evaluating such primary information, and (3) tertiary information which would be used as supporting information for the establishment of disaster prevention plan.

The primary information is mainly composed of the information about the actual state of damages, etc. caused by the disaster and the disaster terrain parameter information based on the land conditions such as the topography, geology and vegetation of the land concerned. As a representative example of terrain parameter, there are the information on slope gradient, knick point and vegetation cover classification. Information on the state of disaster can be extracted from the aerial photograph and other information such as of remote sensing through the image processing and interpretation. On the other hand, it is difficult to extract the information of disaster occurrence factors only from the aerial photograph or remote sensing, and therefore, it is necessary to jointly use other geographic information such as DTM, etc.

The secondary information is the classified information of disaster occurrence risks in the future made by qualitative and quantitative evaluation standard basing on the primary information. Extraction of secondary information is also inadequate if it is made only by the image processing of remote sensing data such as aerial photograph and calculation processing of DTM, but is possible by the analysis of multi-dimension information using GIS. Fig. 1 shows the concept of the analysis of slope land disaster information.

Flow of study

In this study, we discuss the GIS established for the purpose to manage and evaluate the information of slope land along the railway in 3 steps:

Step 1: Discussion on the collection of disaster related information of slope land along the railway

From the aerial photograph and topographic map, terrain parameter such as of land shape and vegetation which are considered to contribute to the occurrence of disaster and the information of location of disaster and extent of damage, and discuss the methodology to manage such information as the raster data.

Step 2: Discussion on the method to predict the hazardous points where slope land disaster may occur

With respect to the slope land along the railway where landslide occurred actually in the past, we collect the terrain parameter before the occurrence of disaster and information of disaster occurrence points in the form of raster data, and discuss the spatial modeling to predict the point of disaster occurrence in the future basing on these data. Overlay analysis of raster type GIS is considered to be suitable

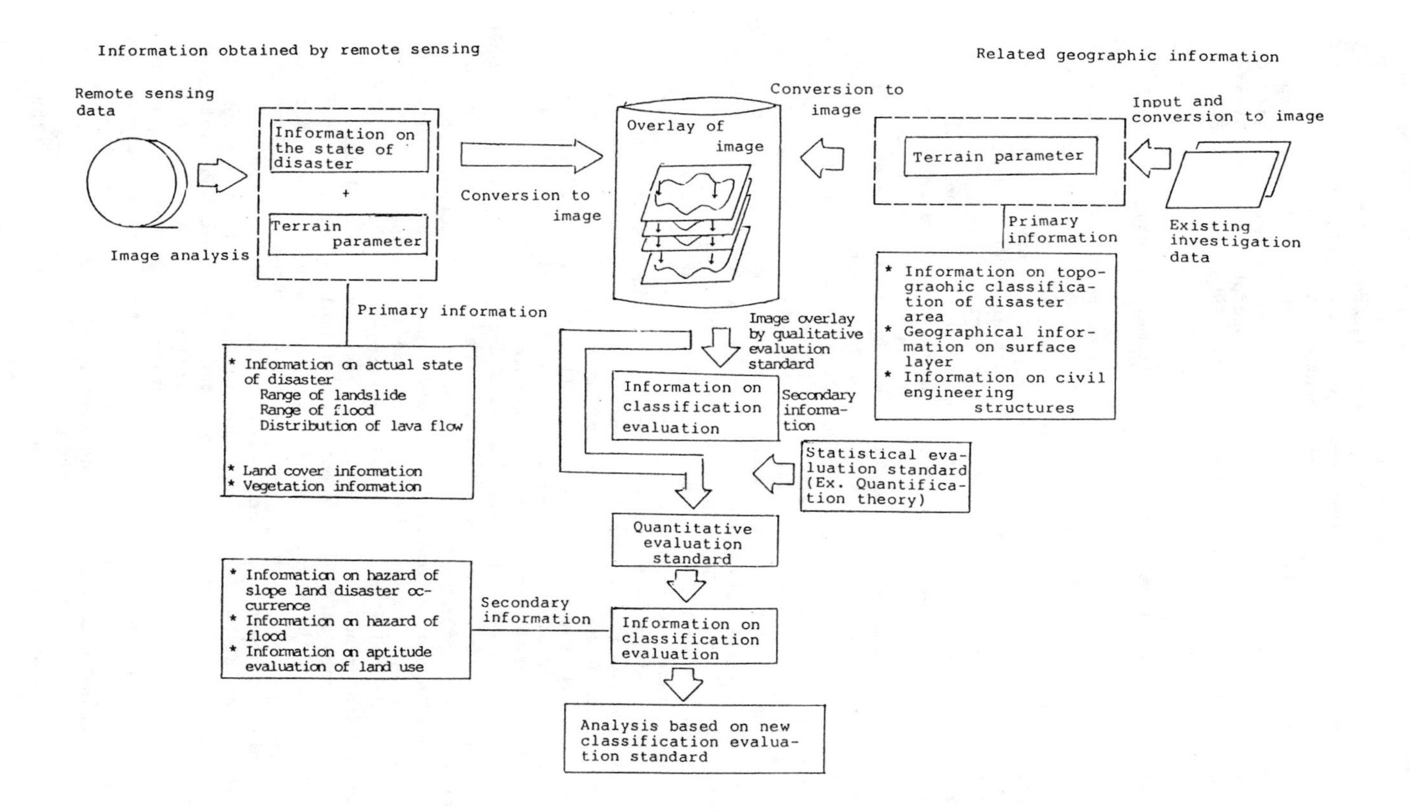

Fig. 1 Concept of slope land disaster information analysis by raster GIS

for this spatial modeling, and plural kinds of classification is attempted by using various kinds of terrain parameter and by adding empirical and statistic weight thereon, and then we discuss the applicability of them to the evaluation of hazardous degree.

Step 3: Discussion on the management/evaluation system of slope land disaster information using personal computer

For the practical use of raster GIS based on personal computer which has the function to collect/extract the slope land disaster related information and to evaluate the susceptibility of disaster occurrence, we study and summarize the information processing/analyzing function necessary for this purpose.

Method to collect slope land disaster information using raster GIS

We discuss the methods to collect disaster related information obtained from the information source of aerial photograph and topographic map classifying these kinds of information into (1) macro information including background areas, (2) very limited local information and (3) information on the disaster occurrence points.

(1) Collection method of macro information including background areas

In managing the disaster information of slope land along the railway, it is necessary to grasp the information on macro ranged area not only of the slope land along the railway but also of the areas including the background slope lands which are considered to exert influence as a factor to the slope land concerned. It is necessary, for this purpose, to extract the slope land disaster information from the vertical aerial photograph, especially the color aerial photograph. Fig. 2 shows the flow of analysis. The vertical aerial photograph is converted into digital image by image scanner, and then converted by orthographic projection so as to coincide it with map information and other geographic information. This conversion is a rather laborious work, but it becomes possible to overlay the aerial photograph image with the map image and DTM by conducting the geometric correction with strict accuracy. Since it is possible to adequately grasp the information which would be overlooked by the interpretation of aerial photograph alone by displaying it with other information, it would provide every effective information for the management of disaster information of slope land along the railway and its background area from the macro viewpoint. After consolidating the coordinates system of information sources such as topographic map and aerial photograph, etc. which form the basic information and converting such information into digital form, the factor information is extracted through vegetation cover and land cover classification by image processing, calculation of topographic shape by DTM, and addition of attribute values by image interpretation.

(2) Collection method of information on very limited local area

Since the scale of subject slope land is small in many cases, it is necessary to use the approach from more micro viewpoint based on the oblique photographs to manage the disaster information of slope land immediately adjacent to the railway. In managing the disaster information of slope land along the railway based on the oblique photographs, it is important to grasp the topographic and geological information which is likely to contribute to the occurrence of disaster at single slope land, to arrange such information as the uniformed data, and at the same time, to jointly manage the information on the history of debris/avalanche occurrence as well as the location and kinds of civil engineering structures.

The slope land disaster information based on oblique aerial photograph is grasped by the similar analytical method as in the case

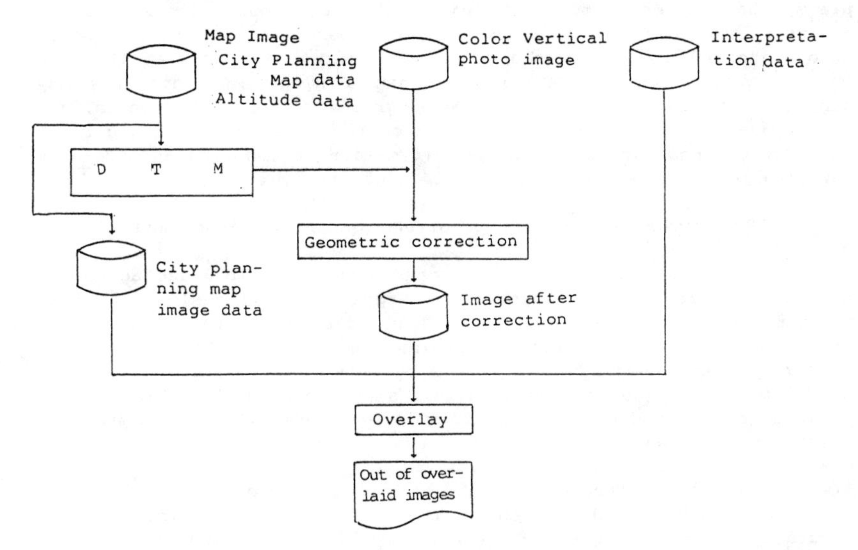

(Extraction of information on slope and background area)

Fig. 2 Flow chart of disaster information arrangement using vertical color aerial photograph

of vertical aerial photograph. Namely, the slope land disaster information is extracted and arranged by the collection of image information and interpretation of aerial photograph.

Comparing with the vertical color aerial photograph, the oblique photograph is more suitable grasping the continuous and linear topographic features such as the knick point. Furthermore, it is quite effective for the information management because it enables to extract very accurate information about the location, kinds and extension of civil engineering structures.

Information is extracted from the oblique photograph by conducting geometric correction with strict accuracy, and then correcting it and adding it to the terrain parameter data made mainly from the vertical aerial photograph through the interpretation of digital image.

(3) Collection method of information on the disaster occurrence points

In general, when a slope land disaster has occurred, the aerial photographs are taken to grasp the actual state of disaster. In most cases, these photographs are taken from different platforms (air plane, helicopter, etc.) at different locations. It should be necessary to convert these photographs taken from different angles into uniformed scale and into uniformed optical axis so that they can be easily and visually available. When a slope land disaster may happen in the future, the oblique photographs will be taken again, and time sequential analysis will be made on the same point using current and previous photographs, and the information of location and scale of new disaster will be added to existing information.

Prediction of disaster occurrence point of slope land

Slope land disaster takes place caused by the compounded action of various factors such as topography, vegetation and geology. Spatial modeling to predict the occurrence of disaster is performed by adding proper weight to the attribute information extracted from the aerial photograph and topographic map, and by obtaining comprehensive value therefrom. The method proposed in this study to predict the occurrence point of slope land disaster is to make such prediction by 2 steps of (1) addition of weight to the terrain parameters and (2) comprehensive evaluation of terrain parameters.

Step 1: Method to add weight to the terrain parameters

In our study, we discussed 1) the method to add empirical value according to the judgment of expert and 2) the method to use the statistic analysis. The method 1) is to give relative ranking to the attribute value of disaster factor information basing on knowledges obtained from the investigation results of slope land disaster in general (for example, the landslide tends to occur more frequent at the acute slope land; therefore, higher weight is added where the inclination angle is more acute).

In case of method 2), the combination of factors at the points where the disaster took place in the past is determined by the statistic calculation and multiple variable analysis to provide the objectivity to the prediction, and the weight is decided according to its result.

Step 2: Comprehensive evaluation of terrain parameters

Comprehensive evaluation of factor information is performed by first adding the weigh to the attribute values of each factor information, and then by overlay operation. It is important to select operation method which suits the mechanism of collapse occurrence of the subject slope land. For example, if one factor is at the hazardous level, the landslide will happen; in this case, therefore, it is suitable to add the weighted attribute value of each factor. On the contrary, in case where the disaster may happen when all of the various factors are present, multiplication calculation will be more proper.

To verify the propriety of the above-mentioned model, we evaluation results of hazard made by the site investigation and hazardous points determined by calculation by applying each kind of methods to the slope lands along the railway where land collapse had occurred actually in the past.

As the weighting method among the methods applied, we employed both the method to add weight according to the empirical judgment and the method to add weight according to statistic analysis. The following two calculations were used in the statistic method.

1) The method to use the factor occurrence rate at the disaster occurrence point in the past as the weight

Mask processing was made to each disaster factor data using the data of past disaster occurrence point, and the occurrence rate of each factor was obtained from the histogram of the factor attribute of the disaster occurrence points. These occurrence rates were used as the weight. Also, each occurrence probability was normalized by the probability of total slope land, and used as the weight.

2) The method to classify the subject slope land into types by multiple variable analysis

Calculation was made by performing multiple variable analysis (Quantification III) to the factor information of subject slope land as a group, and combination of factors were unified into small number of groups. And the relative weight was added to each of these groups.

Also, 3 methods of addition, multiplication and logical addition were applied as the overlay operation for comprehensive evaluation. The calculation was attempted by properly changing the combination of weighting methods and overlay methods.

The factors used for the analysis were (1) distance from the knick point, (2) distance from the water system, (3) existence or otherwise of water depositing land shape, (4) slope gradient, (5) geological feature, (6) vertical and horizontal cross section of slope land, (7) vegetation cover classification and (8) distance from the vegetation border which were considered to have close relation with the landslide.

We selected 3 slope lands which had different land shape and geological features and attempted the analysis on them.

The results of analysis can be summarized as follows:

(1) The prediction with weighting by empirical judgment was easy because the calculation such as statistical processing was unnecessary, and coincided with the site investigation results in main points. But there were some points which did not conform with the actual state in some subject area. Therefore, it can be said that this method is suitable for the stage of rough investigation over the wide range of subject area.

(2) The calculation with weighting by statistic method produced fairly different results for 3 areas subject to the analysis, and is likely to have extracted the disaster characteristics of each slope land relatively faithfully. The method using the occurrence rate of factors and the method using the slope classification has produced different results with different weighting. The former method produced relatively proper results for all of the three test areas, and can be said to be a general method, but there were some points where the result obtained by the method using slope classification was considered to be more proper. As a result, it can be said that both methods have merits and demerits. Therefore, it should be desirable to apply both methods, and to use their results for the local prediction

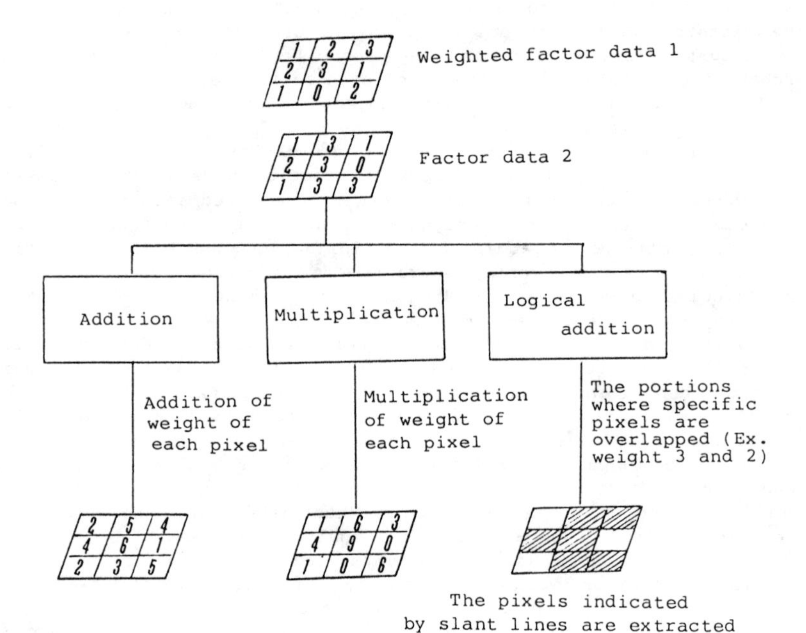

Fig. 3 Concept of factor information evaluation by overlay operation

after comparing and evaluating the both results.
(3) In the comparison of comprehensive evaluation methods by overlay, there was a tendency that the results obtained by addition were in conformity with the site investigation results in general as far as the three study areas which we analyzed this time were concerned. In case of the results obtained by multiplication, the minute points which had extremely hazardous factor were over exaggerated and remaining majority grid cells were scored at "0". On the contrary, the most grid cells were scored highly in case of the evaluation by logical addition, and therefore, it was impossible to perform detailed investigation of slope land.

Since the discussion made in the above is based on the results obtained from only 3 study areas, we will continue the similar analysis in the future to to obtain more general conclusion.

As it is necessary to conduct various kinds of analysis to predict the occurrence of slope land disaster, it should be necessary to develop and disseminate raster GIS so that the information can be analyzed, prepared and updated by the unit of each competent authority of slope lands.

For building up the information management system of slope land along the railway basing on the personal computer

The information concerning the slope land disaster along the railway includes wide range of information such as topographic and geological features, vegetation on the land surface and land use in addition to distribution of artificial structures, climate distributions such as of rainfall, snow fall and temperature, and in recent years, it also includes the environmental information. Furthermore, it is also necessary to have secondary information which is created by classifying and evaluating the these various kinds of information. It is important to manage these multi-dimensional pieces of information concerning the slope land disaster collectively, and to process them in the forms to be easily used, and it is strongly desired to develop a system for this purpose. In Japan, there is a plan for expert system and its trial production has been attempted at present, but it seems to necessitate a substantial length of time until the practical use. At present, it is considered to be most proper to develop a personal computer system which is easy to use and has high operability and which can adequately reflect the knowledges of experts.

Now therefore, when we assume as the precondition to construct a system based on personal computer for the purpose to analyze and manage the slope land disaster information which is mainly composed of raster data, we will need the following basic requirements.

(1) Various formats of slope land information should be converted into uniform format so that they can be processed and analyzed under the same standard. Therefore, the system must be equipped with input devices such as image input devise and digitalizer and external memory device with large capacity which all comply with this purpose.
(2) The system must be equipped with the index database of raster data so that the points subject to the analysis may be easily retrieved among the huge length of areas along the railway.
(3) The precondition should be to construct the system configuration with low cost and high operability. It is desirable that this system should be able to support everything from the input of data to analysis and output of analysis result. The functions to perform geometric correction of image information, classification

and statistical analysis should be necessary as the analysis function of the system.

(4) The system must be able to accept the information and judgment based on the knowledges and experiences of analyzer. In concrete, the system will be equipped with image interpretation function of interactive type and be operated by mouse, etc.

At present, the personal computers are becoming low in cost, and are equipped with increasingly high functions; and therefore, they can adequately satisfy the above requirements. If the persons in charge of slope land management can use this kind of system, not only the analysis and management of slope land disaster information but also updating of information thereafter will become far more effective.

Conclusion

In this study, we discussed the utilization of raster GIS for the purpose to manage and evaluate the information of slope land disasters including the landslide along the railway in Japan. As a result, we obtained the following findings:

(1) Aerial photograph interpretation is important as the information source for the management of slope land along the railway in Japan. Raster type GIS is considered to be suitable to coordinate this information with other map information, and to manage them effectively.

(2) It has been proved that effective extraction of information is possible by providing the GIS with image processing function and interpretation function of vertical and oblique aerial photographs in arranging the disaster information.

(3) With respect to the prediction of occurrence of slope land disaster, it is possible to obtain more objective calculation than the conventional spatial modeling based on past experiences by jointly employing the statistic method. It will be possible to provide more effective prediction by selectively employing conventional method and statistic method according to the application of investigation.

(4) From the viewpoint to manage the disasters occurring along the railway, it is desirable to construct simple raster GIS based on the personal computer, and to disseminate it to the actual sites of slope land management.

Bibliography

1) T. Noguchi, M. Setojima, K. Okada, H. Muraishi: Extraction of Topographic and Geological Information on Slope Land along Railways using Aerial Photo and Geographic Information, IGASS '89, Vol. 3, pp. 1669 - 1673, 1989

2) Masahiro Setijima, Tatsuo Noguchi (1991); Trial of Disaster Information Management on Slope along Railways Using Remote Sensing Technology, JSRS Vol. II, No. 2

3) Masahiro Setojima, Yukio Akamatsu, Kenichi Shibata (1985); Trial of Collapse Hazard Classification Using Overlay Processing, Proceeding of 10th Symposium on Engineering Information Processing System, pp. 187 - 192, Committee on Engineering Information Processing System, Japan Society of Civil Engineers

4) Masahiro Setojima, Yukio Akamatsu, Yoichi Oyama (1986), Trial of Landslide Hazard Classification Using Image database, Proceeding of 11th Symposium on Engineering Information Processing System, Committee on Engineering Information Processing System, Japan Society of Civil Engineers

OBJECT-ORIENTED GIS FOR LOCAL GOVERNMENT

Masako Yamashita
Fuji Xerox Co., Ltd.
KSP / R&D Business Park Bldg. 100-1, Sakado, Takatsu-ku
Kawasaki-shi, Kanagawa-ken, 213 Japan

Takaaki Kuboki
Fuji Xerox Co., Ltd.
KSP / R&D Business Park Bldg. 100-1, Sakado, Takatsu-ku
Kawasaki-shi, Kanagawa-ken, 213 Japan

ABSTRACT

In these days, several types of GISs have been developed and used as practical application in Japan. In these application areas, geographic data are used in various way. As one of these examples, in a local government, they also use maps and other geographical data in many sections.

This paper describes a system application named Urban Planner that focuses on the urban planning in local governments, and some of its unique characteristics. This system is developed in ParcPlace System's Smalltalk-80 running on a Sun Microsystem's Sparc Station 1. We selected Smalltalk-80 because of some reasons as follows:

1 To realize WYSIWYG by icon operation
2 To realize easy operation by person who is not familiar with the computer
3 Easy (for developper) to customize the system.

Here we will introduce the system and the advantage of using object-oriented environment for system development.

INTRODUCTION

Urban Planning Job in Local Governments

Generally speaking the job at an urban planning section in Japanese local government will be categorized into two groupes. One is the daily work at the office counter, the other one is the real planning with maps. The former is a kind of reference job. For example,

1. Legal Information Reference
   - To refer the whole legal information of the selected area.

2. Individual Attribute Reference

- To refer the individual attribute of buildings and land

The latter is a kind of caliculation and analysis job. For example,

1. Data Inspection
   - To inspect the data and display in a desirable form.
   - To make subjective map (which used for planning and reporting)

2. Basic Analysis
   - To analyze the data concerns the population. To check the change over a period.
   - To analyze the circumstances of land use.
   - To analyze the circumstances of purpose which the buildings are used for.

3. Advanced Analysis
   - To check if there is a building which charactaristic is unacceptable accoding to the law.

4. Simulation
   - To change the legal bondary and forcast if something wrong will happen.
   - To estimate the growth of building and population and analyze the impact and so on.

<u>Our Concept in Development</u>

In Japan three years ago when this project had begun, there were few geographic information systems for governments. Even though there were, the function was too simple or too complicated.

The simple system is only used for reference. User asks the system about the legal information of selected area submitting to the menu guidance system offers. The system is useful but they can do nothing more than reference.

On the other hand, very high level and complicated system also exists. It is really useful and powerful if you know well about the function and operation, but actually the system demands such a difficult operation that user can't use even half of its treasure.

Regarding these factors, we decided to develop the new type system (that is Urban Planner) with these concepts;

1 To integrate each function properly for the real practical work at urban planning section. According to the job analysis, we were going to make system not with too much function nor too little for urban planning section.

2 To present the system with excellent user interface which is easy to learn, easy to operate so that a

user who isn't familiar with a computer can make use of it without mistakes.

URBAN PLANNER

Characteristic in Function

1.All data is polygon. All of the geographical data is produced as polygon data. The polygon data tends to have many advantages. First of all, you can recognize the area easily because the polygon makes area by itself. Therefore it is easy to classify by color. In Urban Planner not only the geographical data but also the attribute is made into polygon data.

2.Layering concept. General map has many kinds of information on one sheet, boundaries, addresses, names of building and area, and so on. We separate each kind of information each other and make them independent. We call each separate information a layer. One layer has only one information. For example, one layer has only building polygon data, the other road polygon data, the other a kind of legal restriction data.

Each layer can be overlapped as many as you want so that you make your own map with enough information freely. To see the overlapped result in the system, we prepare a tool named AtlasBrowser. AtlasBrowser controls all the geographical data in the system. Imagine an over head projector and transparancy sheet. Layer is a transparency sheet and AtlasBrowser is an advanced OHP. If you overlap the redevelopping area layer on the building polygon layer, you can find the current condition which buildings are in the redevelopping area.

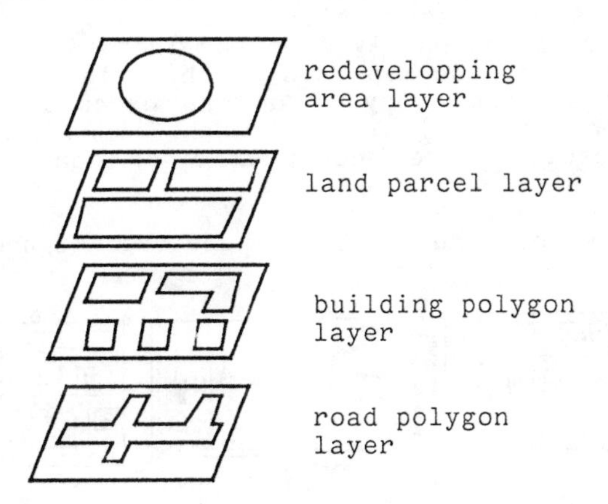

Figure 1.Layering Concept

3.polygon caliculation. With more than two polygon data layers, you can get precise data by caliculation using geometrical manupulation. In the case of redevelopment, the total size of the area can be found by geometrical caliculation, then you can make a budget how much it costs to redevelop this area.

Characteristic in User Interface

We classify the basic function needed in urban planning into twelve and represent each function by icon. As a result to adapt icon modeling, following feature has come.

1.Easy to be adapted to unregulary ordered job by icon operation. In icon operation, system has no rule about the operation order; which icon shoud be used after the other icon is used. User is allowed to use icon freely. when inappropriate operation occurs, system draws user's attention to that error and shows the proper way to operate with messages.

We surveyed cases in local governments of real job at urban planning. According to that research, we find that they have to do many try-and-errors in planning from various points of view, so if the system fixes the operation flow using job menu, it will make no use.

2. Uniformity in Operation. In Urban Planner operation trigger is only one thing: to copy an icon on the other.

Whenever you want to do something, for example, to make charts, to make subjective maps, to print, all you have to do is to copy the icon on the other that represents the function which you need. You can identify the function easily from the icon piucture. To print the map, copy the map icon on the printer icon, to statistically analyze, copy it on the chart icon. The same operation (to copy) with different icons causes the diffenrent result. User don't have to concern about procedure.

Here we show you some of icon images (see Figure 2).

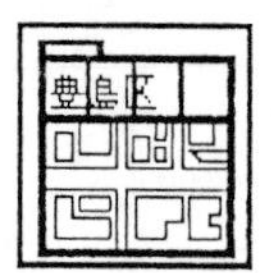

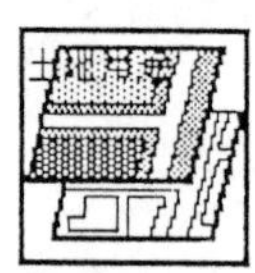
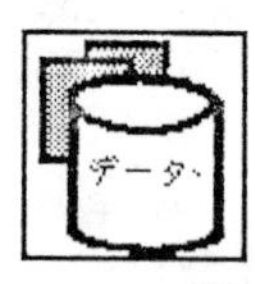

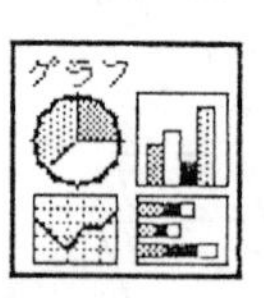

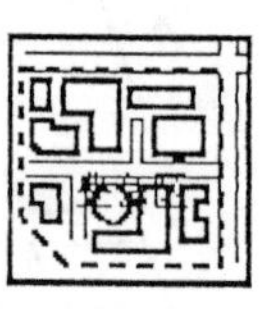

Figure 2. Icon image examples
(AtlasBrowser/Layer/Database/ChartCreator/SpawnMap)

Overview of the system

1.Job List of Urban Planner. Here is a list of jobs which you can do with Urban Planner. Of course adding other data, you can apply other kinds of jobs with the same icons initially prepared. If you want additional function, we will be pleased to integrated in original Urban Planner.

1. Legal Information Reference
   - To refer the legal information

2. Individual Attribute Reference
   - To refer the individual attribute of buildings and land.
   - To check the administrative guidance for development in special area.
   - To list and check the guidance records of special buildings.
   - To investigate the current circumstances of arrangement of urban planned buildings.

3. Data Inspection
   - To survey fundamental data for urban planning.
   - To make subjective map used for planning and simulation materials.

4. Analysis
   - To analyze the circumstances from various points of view (land use, purpose which the buildings are used for, building materials, and so on).
   - To check if there is a building which charactaristic is unacceptable accoding to the law.
   - To check if there is a area which charactaristic is unacceptable accoding to the law.
   - To plan the most effective usage of land.

5. Report
   - To printout the subjective maps in full color.
   - To printout the result of legal information reference in full color.
   - To printout tables.

6. Filing
   - To file the result of legal information reference.
   - To file the result of individual attribution of buildings and land.

2.Icons in Urban Planner. Urban Planner offers many functions in icons (see Table 1.).

ICON NAME	FUNCTION
Catalogue	Catalogue of the icons prepared by the system. First of all you have to copy the icon from this catalogue icon to prepare your working environment.
Atlas Browser	To control all the geographical information. Used to get a subjective map, a legal information, a SpawnMap and so on.
Layer	Geographical data and attributes. Each layer has only one kind of information.
Database	Database access (using SQL). In Database individual data is stored. General legal information is hold as a layer, not in Database.
Table Creator	To create tables.
Chart Creator	To create charts.
EC10	Color printer. EC10 is Fuji Xerox digital color copier duplicater.
Folder	Utility. A filing folder.
FiliIn/ FileOut	To store the icon on the disk. To retreive the icon from the disk.
SpawnMap	Map spawned from the AtlasBrowser. This has geographical and attribute of selected area. It also stands for the result of legal information reference. This icon has legal information of selected area and building.
Reference Result	Result of inquiry to database. This icon has data used for creating charts and tables. This icon has an individual data (not legal information).
Analysis Chart	Spacial tool used in planning. It uses triangle chart.

Table 1. Icon list prepared in Urban Planner

<u>Job Examples by Urban Planner</u>

Let me introduce two examples with Urban Planner.

<u>1.Legal Information Reference.</u> When you open the AtlasBrowser icon, the map displayed in AtrasBrowser.

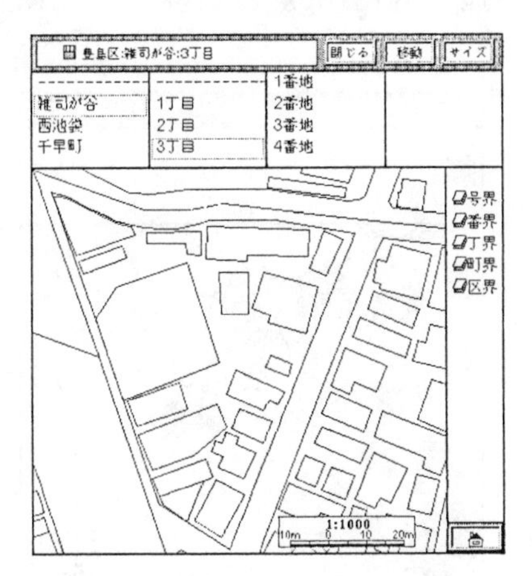

Figure 3. AtlasBrowser

To refer the whole legal information of the selected area, point the area from the map directly.

As the result, you will get the SpawnMap. In this icon, there is a map selected and all legal information conserns with that area.

Figure 4. SpawnMap
(showing the result of legal information reference)

2.Making subjective map. To check if there is a building which characteristic is unacceptable according to the law, officer uses AtlasBrowser and two kinds of layers; the current status of building material layer and legal restriction layer of building material. Overlapping two layers on AtlasBrowser you can select only unacceptable buildings filled in color (see Figure 5).

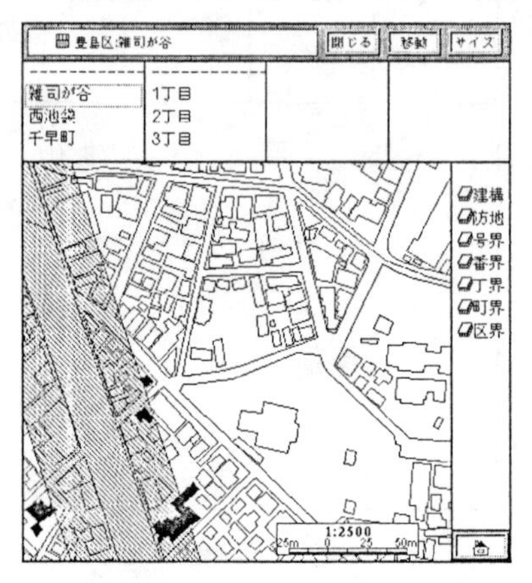

Figure 5. A subjective map

Expandabilities of Urban Planner

Urban Planner has much possibility in expansion of system.

1.Fundamental function is useful for other system. Urban Planner is an application for urban planning but its basic function can be applied for different kind of system that deals with polygon data and additional data. For example, the system used to decide where we should open our new shop. For this system, area map layer and current market condition layer are needed. Using these layers, we can know our rival's territory and choose the place where we should open our new shop.

2.Advantage in Object Oriented System. They say that making system is modeling the real world in the software. We use Smalltalk-80, one of the most famous object-oriented programming languages because it is easy to make model.

An object-oriented programming language is based on a single universal data structure (the object), a uniform control structure(message sending), and a uniform description structure(the class hierarchy). An object is a way to represent, in one place, properties about a data structure and the operations permitted on the structure. Programs obtain information from an object, or request an object to do something, by sending a message to the object.

Description of a system in terms of objects, and of objects in terms of possible behavior (as defined by messages), permits modular design. The power of the Smalltalk-80 system derives from the modularity in object-oriented description.(What is Smalltalk? ParcPlace Systems)

In Urban Planner, the object means icon and message means copy operation. If you want to add new function in the system, you may design the new icon and implemented into the system. Each icon is independent in the system, therefore it won't cause any evil influence by adding. It is easier in Smalltalk-80 than in other language to add new function into the existing system.

CONCLUSION

We are now building the model system with Urban Planner for one local government in Japan. At the same time, we have begun to survey another field of mapping which Urban Planner can be applied. It is our pleasure to make miscellaneous systems based on Urban Planner.

REFERENCES

Adele Goldberg 1984 Smalltalk-80 The interactive programming environment, Addison-Wesley Publishing Company

Adele Goldberg and David Robson 1983 Smalltalk-80 The Lnaguage and its implementation, Addison-Wesley Publishing Company

ParcPlace Systems, What is Smalltalk?…some questions and answers from ParcPlace Systems.

Roger F. Tomlinson 1985, 地理:地理的情報システム(Geographic Infromation System- A New Frontire), Vol30, pp.14~24 古今書院

1986, 地理, Vol31 古今書院

建設省国土地理院 財)日本建設情報総合センター 1991, 地理情報システム技術動向調査作業報告書

建設省 1987, 都市情報データベース,ケイブン出版

社)日本都市計画学会 1985, 都市計画マニュアル,(株)ぎょうせい

NIPPON COMPUTER GRAPHICS ASSOCIATION 1988, コンピュータマッピング入門, 日本経済新聞社

MAP RECOGNITION WITH INTERACTIVE EDITING FOR GIS

Michihiro Inagaki, Masakazu Nagura
and Syuzo Yamakawa

NTT Human Interface Laboratories
1-2356 Take Yokosuka-shi
238-03 Japan

ABSTRACT

Map recognition is one of the fundamental technologies needed to create a truly Geographical Information System (GIS). Existing computer-based map recognition methods offer relatively low recognition accuracies and the results cannot be corrected easily.

This paper introduces an effective map recognition method and uses it to create a system that offers interactive editing. We use a compound likelihood measure to ensure more flexibility in recognizing map entities. The system almost equals the recognition ability of humans. A totally automatic system is made by combing scanning, automatic vectorization, full-auto contour and house recognition, and semi-automatic figure editing. Evaluations show that the system recognition rate can be as high as 90%. A high performance system can be constructed around a low cost personal computer, because the algorithms that decide classification use fuzzy measure evaluations and are quite sophisticated.

INTRODUCTION

The information within maps serves as the basis of the mapping systems created to provide various social services. The usage of map-based information is becoming more important in conjunction with personal or commercial data. Maps, however, demand special management techniques if their information is to be accurately and efficiently utilized. Moreover, there are many other demands placed on a GIS to manage the customer and network information based on spatial data. These factors promote the automatic data capturing of map entities.

Many mapping systems have been the target of Japanese research and development activities these past few years. NTT has already developed several automated mapping (AM) and facility management (FM) systems. Most current map data capturing systems seem to consider only hand digitizing for its advantages of precision and user attribute data binding. In other words, systems employing automation were considered to be unsatisfactory, even though automated recognition was known to be inherently more labor efficient. However, our opinion was that automated recognition could be enhanced enough to replace manual digitizing when the map contains many instances of common entities. Input time and cost reduction are the primary barriers to revolutionizing mapping systems so that an effective automatic approach is needed now.

Middle scale type maps (scaled from 1 : 500 to 1 : 5000) are most suited for automatic recognition because the maps include many isolated vectors and common entities. Automatic recognition is more efficient if such vectors and common entities occupy the major part of a map. This paper discusses automatic recognition on this basis. (Nagura, 1990)

MAP RECOGNITION CONCEPT

Japanese city planning drawing maps, drawn on a scale of 1 : 2500 are used most frequently in Japan, and provide a useful example of the recognition condition. There are more than 50 categories of entities in such maps. Map entities are recognized through the following operations. (Table 1)

Table 1. Map Entity Classification with Technologies

Data Class	Automatic Recognition	Semi-Automatic Recognition	Interactive Editing
1.House and Building	Houses / Buildings	Houses / Buildings under construction, Houses / Buildings without walls	(Correction, Cosmetics)
2.Contour Line	Curves	Sub-curves	(Correction, Cosmetics)
3.Road	N.A.	Street roads, Sidewalks, Footpaths, Paved alleys, Unpaved alleys, etc.	(Correction, Cosmetics)
4.Line	N.A.	Railroad lines, Blockfaces, River lines, Seashores, Longitude and latitude lines, etc.	(Correction, Cosmetics)
5.Area Boundary	N.A.	Block zone boundaries, Administrative boundaries, etc.	(Correction, Cosmetics)
6.Name Text	N.A.	N.A.	Building names, Street names with ID, etc. (by CAD input)
7.Map Symbol	N.A.	N.A.	Planting zone symbols, Building symbols ID, etc. (by CAD input)
8.Map Polygon	N.A.	N.A.	River side slopes, Face of slopes, etc. (by CAD input)

- Automatic recognition extracts houses, buildings, and contour lines. In Japanese city planning drawing maps, these elements usually account over 80 percent of all entities.
- Semi-automatic recognition extracts road lines and boundary lines. Extraction is started after manual indication of the starting segment.
- Interactive editing processes all unrecognized figures in a manner similar to that used in CAD systems.

For automatic and semi-automatic recognition, we introduce a compound likelihood measure, which is like a fuzzy function, to enhance evaluation and classification. This approach has the advantages that the system produces statistically weighted recognition results and that recognition parameters can be easily adjusted by employing a sort of feedback process. All the figures in a map can be converted into data using these three operations. This approach maximizes total system efficiency.

MAP RECOGNITION ALGORITHMS

Data Generation for Recognition

Our recognition algorithms use both structured polyline vector characteristics and group figure data characteristics to evaluate each figure unit in a manner similar that used by humans. The recognition process proceeds globally as follows.

a) generate a structured polyline vector, the so called arc vector, in a batch process. Arc vectors are constructed with only terminal points, points terminated at crossings, or closed figures. (Figure 1)

b) dynamically generate a group figure data set which is a set of arc vectors representing a meaningful figure during the recognition process.

c) measure compound likelihood values for both arc vectors and group figure data, and then classify entities with the fuzzy measure evaluation.

The automatic recognition process uses these steps automatically without any intermediate human operation for the categories of houses/buildings and contours. The semi-automatic recognition process is interactive and relies on human decisions. All entities remaining unknown after these recognition operations are processed with interactive editing.

Evaluation and Classification Algorithms

As some kinds of the likelihood value, measurement parameters are measured commonly for each arc generated from the simple vectors created by the scanner as follows. These measurement parameters are calculated as probabilistic values, not as single on-off values.

- the existence of turning points within an arc.
- the existence of turning points at the crossing end of an arc.
- the length and curvature of an arc, including corners.
- the straightness of an arc, including corners.

In the next step, a group figure data set is evaluated by the measurement parameters; gradual curvature rate, area value, perimeter length, polygon shape degree, and so on.

We introduce a compound likelihood measure to perform fuzzy measure evaluations. Likelihood p for measurement parameter m is presented as a probabilistic function f ($p=f(m)$). For examples, p is defined as a function as

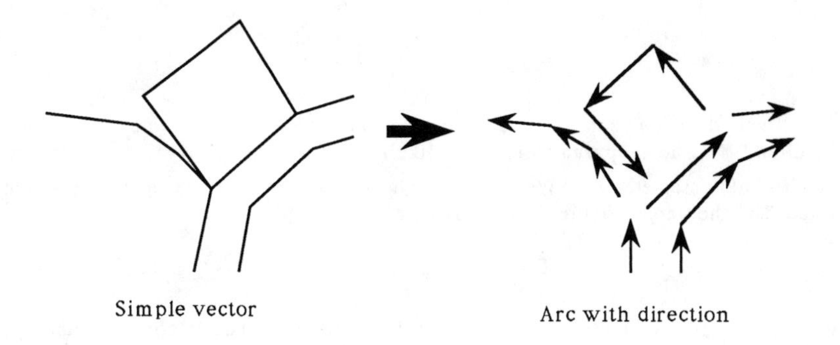

Figure 1. Arc Vector Generation

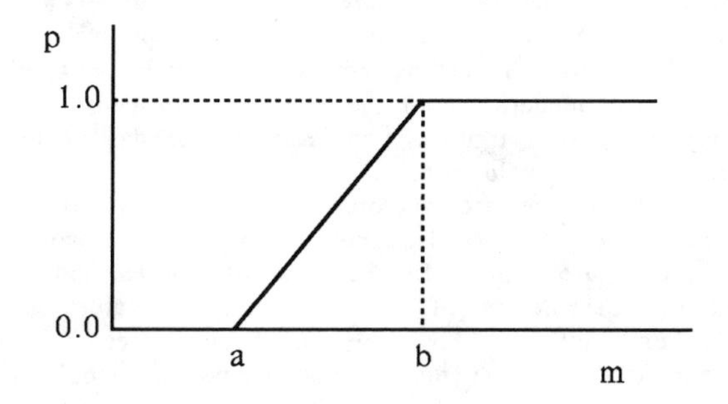

Figure 2. Likelihood Measure

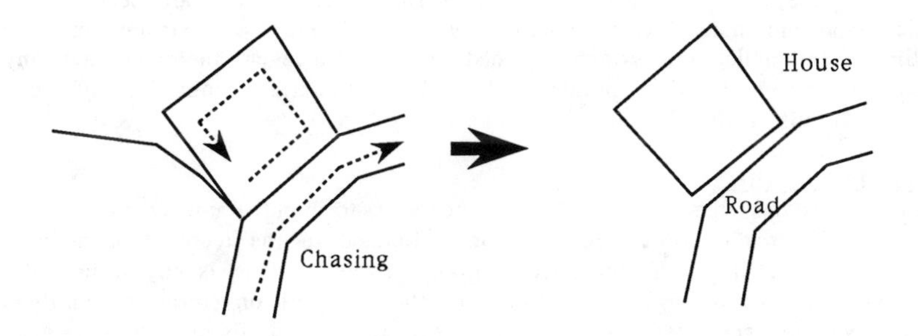

Figure 3. Recognition by Chasing

follows when a,b $(0<a<b)$ are threshold values. (Figure 2)

$$
\begin{aligned}
p &= 1.0 & &: b \leqq m \\
&= (m-a)/(b-a) & &: a<m<b \qquad (1) \\
&= 0.0 & &: m \leqq a
\end{aligned}
$$

The final decision compares a threshold value Th with a total evaluation function F whose parameters are likelihood values p_i (i=1,..n, n is the number of parameters). When the evaluated value of F exceeds the given value Th, the target figure is classified. For examples,

$$F(p_1, p_2, \ldots p_i, \ldots p_n) > Th \qquad (2)$$

F and Th are uniquely defined in each category. The algorithm employs only information of geometric parameters, not heuristic knowledge about map entities. This enables the algorithm to be applied to drawings of different scale or type by simply adjusting p, F and Th.

Automatic and Semi-automatic Recognition

The automatic recognition algorithm was created to recognize houses /buildings and contours. These two categories satisfy the condition of automatic recognition in that they occupy the major part of a map and are readily distinguished. It proceeds in the following steps.

Step1: Automatically choose a starting arc which includes a turning point.

Step2: Chase lines or boundaries until the arc closes or ends.
(skip already recognized figures. chase dotted or dashed lines)

Step3: Measurement and evaluation.

These steps can handle the arc vectors shared by two different types of entities such as houses and roads. (Figure 3) The processing order is independent of the figure type. In the case of houses and buildings, the algorithm uses the parameters of i) no gradual curvature rate, ii) area value, iii) perimeter length, and iv) polygon shape degree. In the case of contours, the parameters are i) no polygon shape, ii) single polyline, iii) curved line length, and iv) gradual curvature. Step 1 to 3 are automatically repeated until all vectors have been examined.

The semi-automatic recognition algorithm employs the first two steps in an interactive manner. (Suzuki, 1988) If the system encounters an ambiguity when chasing a road line, the operator can decide to continue or not by answering the request message from the system. Road lines are seen as the most important data for a mapping system. There is a strong need to optimize automatic processing for road lines. However, the fact that any road has many crossing points makes the automatic recognition of road lines difficult with current algorithms.

Interactive Editing

In CAD systems, the regular editing process is to input Kanji characters and geometric figures or to correct mistakes. Related to the recognition process, there is another type of interactive editing process. This is to cosmetically process recognized figures. Because the recognition process employs simple vectors generated by a scanner, the recognized figures usually need to be modified to remove unnatural curvature of simple vectors at crossing points. These interactive editing functions support automatic recognition and make the system more useful.

TOTAL PROCESSING SYSTEM

System

The system (AI-CHASER) is composed of three subsystems: a scanning subsystem for map input, a personal computer performing recognition and editing, and an output plotter. After reading the map, simple vectors created in the scanning subsystem are sent to the personal computer. The personal computer performs all recognition and editing functions. The recognition data is stored as a file and can be sent to the plotter for output. (Figure 4)

Preprocess

Before the recognition process can commence, a preprocessing operation is needed to prepare vector data from the scanner output. The recognition system, which runs on a 32-bit personal computer, uses the run-length-vectors and simple vectors produced by an A0 type intelligent scanner. The precision of such scanners is usually 400 dots per inch which is satisfactory for Japanese city planning drawing maps. Because scanned map data is often skewed by scanning, it must be transformed using reference points given by the operator. Also, the operator can input the division number which matches the map mesh control of the database management system. Arc vector generation and measurement of geometric characteristics are included in this stage.

Recognition and Editing Processes

The output of the preprocessing stage undergoes automatic recognition in a batch process. Recognized entities are isolated within the vector set. After automatic recognition terminates, the operator can select either semi-automatic recognition or interactive editing as needed. These processes are selected through a menu using the keyboard of the computer.

Postprocess

A complete file of the recognition data is made by combining the data sets captured by the above processes. This data file is transformed into a standard format data file for database processing or data transfer between systems. In the database system, the captured data set can be converted into a layered structured map database which can be accessed by various mapping applications. (Yamakawa, 1991) Each entity, such as a house, can be utilized as a management target by assigning data attributes to it in the system.

EVALUATION

The evaluation used a city planning drawing map drawn on the scale of 1 : 2500, A0 size. The result is summarized in Table 2. Automatic recognition achieves recognition rates higher than 95 percent for houses/buildings, and higher than 90 percent for contours. (Figure 5) The time spent on automatic processing is negligibly small compared to semi-automatic recognition or interactive editing; less than 2 percent of the total processing time.

Regarding semi-automatic recognition or interactive editing, the processing time strongly depends on the number of vectors and/or entities. Locating and identifying points to be processed also takes considerable time.

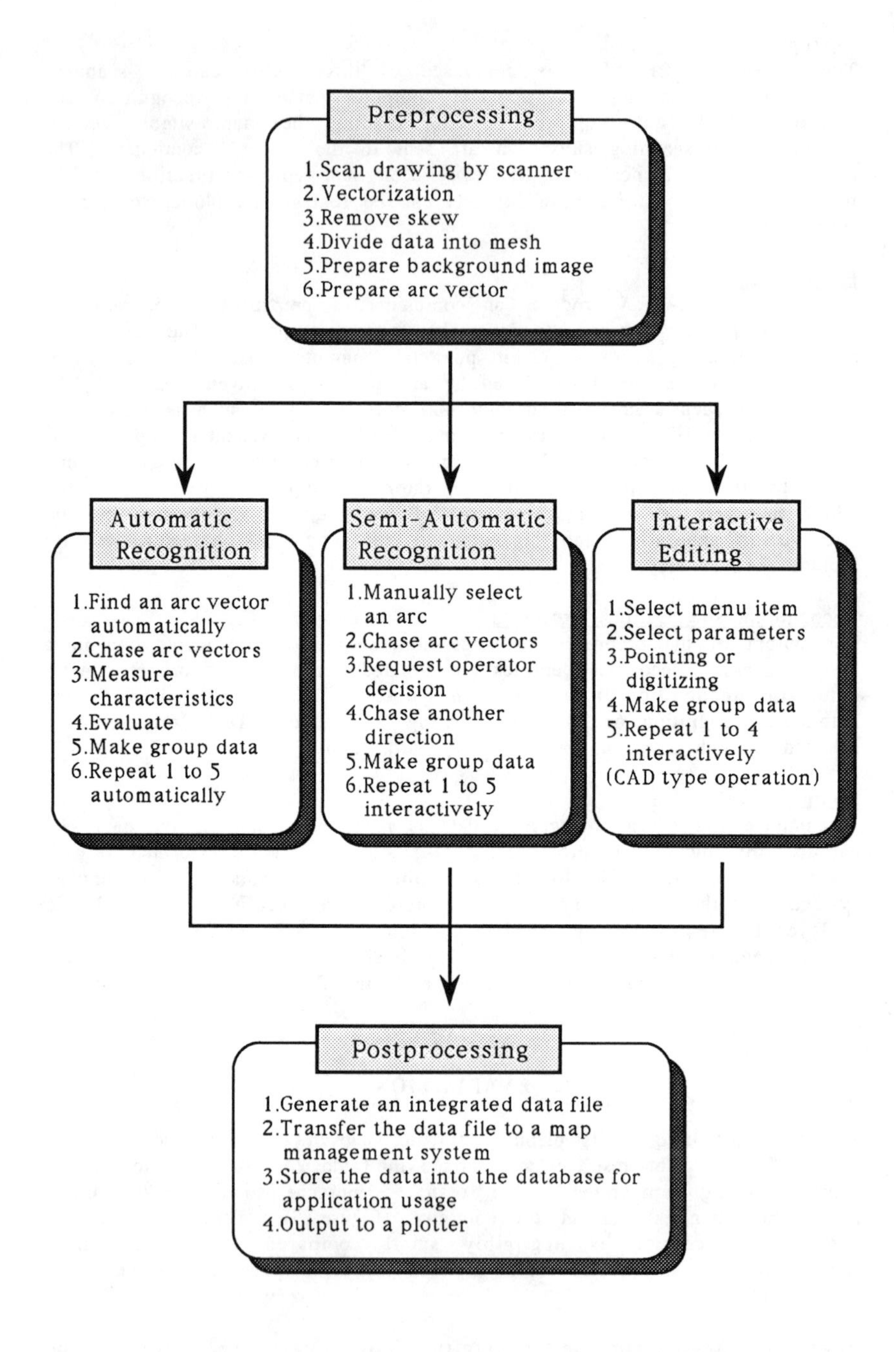

Figure 4. Processing Flow of Recognition System

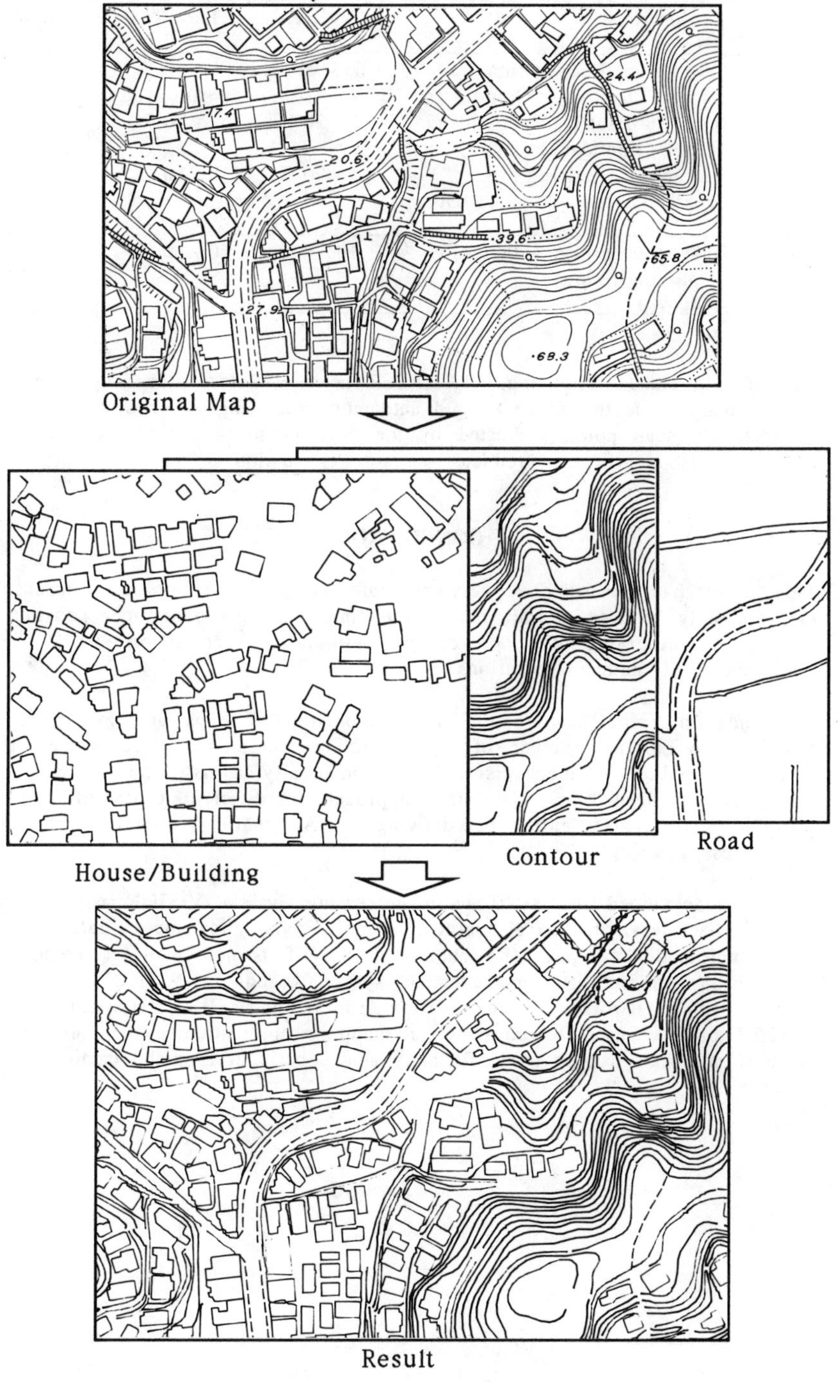

Figure 5. Example of Map Recognition

Table 2. AI-CHASER Evaluation

Level	Technologies	Cover Rate	Time Reduction Rate
1	Automatic Recognition (House/Building, Contour)	60 - 90 %	90 - 95 % (=Accuracy)
2	Semi-Automatic Recognition (Line, Boundary)	5 - 20 %	50 - 70 %
3	Interactive Editing (Text, Symbol, Polygon, Others)	5 - 20 %	10 - 80 %
	Total (Comparison with CAD Type digitizing)	100 %	60 - 80 %

Compared to manual digitizing, AI-CHASER takes 60% less processing time. This is mainly due to the very rapid automatic processing. The system can automatically snap points indicated by the operator to the nearest vector or entity. This significantly reduces operator exhaustion.

CONCLUSION

This paper has described a very accurate automatic recognition approach that not only reduces overall recognition time but also operator exhaustion. We have integrated the power of automatic recognition with the intelligence of human operations.

The remaining problems are to handle sophisticated maps and to increase efficiency when the number of figures to be automatically recognized is relatively small. A mechanism that automatically adjusts the recognition parameters is needed to adapt this approach to different types of maps, different map scales and type drawings. Automatic recognition should tackle these areas.

NTT has developed the multi-media processing system VISION which can handle characters, numerical values, graphic data and image data in a single multimedia database. (Kawano, 1990) NTT is now using a telephone line design and facilities management system based on VISION that can manage line networks, communication routes and GIS statistical data. VISION can be used to construct a mapping application system and the captured map data can be used efficiently as the basis of a spatial database.

REFERENCES

Kawano, S., Oda, Y. and Yamakawa, S., 1990, "VISION", A Multi-media information Processing System, NTT REVIEW, Vol. 2 No. 1, pp. 79-87.

Nagura, M., Inagaki, M. and Yamakawa, S., 1990, Map Recognition and Editing System, 6th NICOGRAPH, pp. 359-369 (in Japanese).

Suzuki, S. and Yamada, T., 1988, "MARIS", Map Recognizing Input System, Proc. IAPR Workshop on Computer Vision, 1988 (10), pp. 421-426.

Yamakawa, S., Kawano, S., Oda, Y. and Inamori, K., 1991, Facilities Management System for Distributed Database Environment, AM/FM Conference XIV, pp. 34-41.

CONSIDERING THE EFFECT OF SPATIAL DATA VARIABILITY ON THE OUTCOMES OF FOREST MANAGEMENT DECISIONS

by

Dr. James L. Smith
Department of Forestry
Virginia Tech
Blacksburg, VA 24061-0324

Dr. Stephen P. Prisley
Westvaco Timberlands
Box 458 Highway 121
Wickliffe, KY 42087

Dr. Robert C. Weih
ITD
Space Remote Sensing Center
Stennis Space Center, MS 39529

ABSTRACT

The role of spatial data in decision making is increasing at a remarkable pace. In forestry, spatial data plays a critical role in nearly every major decision. The question then becomes how inherent variability in spatial data affects these decisions. Fundamental concepts of this dilemma will be discussed, and two actual case studies will be described. Further, different algorithms used in GIS's may result in different decisions even if the same data are used. We will illustrate this effect by describing a decision-making situation that utilizes DEM information as input.

INTRODUCTION

It is generally accepted that the major use of GIS thus far has been as a data display tool, or as a simple data base manager. However, the time has come in many organizations to make the next developmental step, i.e., to use the GIS as an analytical, planning or operational tool. In natural resource management, the most important uses of GIS will be to tell us where to perform what operation, how much will it cost, what will be produced, how will it impact surrounding areas, how will surrounding areas affect the decision, etc. GIS will be the information integrator, as well as the medium for producing final reports and graphic displays.

Incumbent in taking this next step is to ascertain how "good" are the resource management decisions that are being made in the GIS environment. Clearly, "good" is an ambiguous term, and will be defined differently by different people, or in different situations. While GIS is not the only factor affecting the quality of a decision, it is clear that GIS in a broad sense has a role to play in this "goodness." To use the old joke, there is good news and bad news. The good news is that GIS may be used to determine some measure of "goodness" for a decision where none existed before. We will illustrate two examples of incorporating measures of spatial data variability into standard forest management decision-making situations.

The bad news is that the algorithms used within a GIS may themselves create some "goodness" problems, and we will describe a decision-making situation which illustrates this concept. Specifically, we will be describing the effect on a resource management decision of choosing a different algorithm for calculating cell slope values.

We ask that all readers note three important facts. First, in all cases, real decision criteria and real values were used wherever possible. Second, for policy reasons regarding proprietary information, the names of any data sources or contributors have been purposefully deleted. Finally, we ask all readers to note our emphasis on the impact of GIS on decisions. This, we feel, is the ultimate test of a spatial data base and a set of GIS tools. Practitioners need to know how previously decided-upon factors affect final on-the-ground decisions. This paper is not an academic exercise, but a practical demonstration of that effect.

EFFECT OF DATA UNCERTAINTY ON DECISIONS

Resource managers have long accepted the inevitability of uncertainty in inventory data. They recognize that the sampling and measuring processes create this inevitability, and so they counter with measures of statistical certainty such as variance. However, we believe that it is clear that map information also contains inevitable uncertainty imparted by the processes used to produce it. Spatial data uncertainty is created by the process used to collect the mapped information (ie., photointerpretation, digital image processing, etc.), the drafting of the map, and if a GIS is used, uncertainty in the spatial data is created by the analog-to-digital conversion process. Yet, spatial data uncertainty is assumed not to exist by most practitioners. One of the reasons this assumption is often made are the difficulties encountered when trying to assess map accuracy in a practical way. Statements such as the national map accuracy standard do not help users understand how much the values they derive from maps actually vary under that standard. Resource managers are more interested in area, distance and direction than position, yet most accuracy standards are stated in terms of position. However, GIS provides an environment for determining and incorporating spatial data accuracy measures. GIS provides unprecedented opportunities for harnessing, rather than ignoring natural spatial data uncertainty. What follows are two actual examples of forest management decisions which can be "improved" by the use of a GIS, and the use of spatial data accuracy measures.

Chemical Application

Natural resource managers frequently express much of what they do in terms of areas treated. Acreage estimates are used in planning, budgeting and carrying out operations. Without knowledge of the uncertainty inherent in these area estimates, decisions are made as if all acreage figures are known exactly. If a manager could know the magnitude of area variability, decisions could be modified to minimize the risk associated with this uncertainty.

For example, forest managers are increasingly using chemicals to promote the efficient growth of desirable vegetation. Chemical application (fertilizers or pesticides) is often performed by contractors who are paid on a per-acre basis using *a priori* estimates of acreage. An underestimate of acreage may result in too little chemical being applied to the tract because the contractor expected less acres. If the contractor thinks there are 40 acres, but there are really 44 acres, the same amount of chemical is used, but it is spread more thinly over the entire area. Conversely, if the _a priori_ acreage figure is too high, the contractor uses too much chemical on the tract because it is being applied to fewer acres than expected. Therefore, there are risks associated with both under- and over-estimation of acreage.

Area variability would be an integral part of a risk assessment. Suppose that we could state with 90% certainty that the area of a tract is between X acres and Y acres. Further suppose that we have models which predict the response of the crop to the level of chemical application that would occur if there were X acres and Y acres. The result would be a range of responses to the chemical application. For example, the spatial uncertainty may be enough to create an _a priori_ area overestimate which would produce an overapplication of chemical that would kill the trees. Conversely, variability in area may create a serious _a priori_ underestimate of area which would create an under-application of chemical which would not have the desired effect upon the competing vegetation. By incorporating the inherent variability in area, the manager has more information at his/her disposal, and may decide to intentionally overapply or underapply depending on the risks involved.

A second example of the impact of spatial variability can be found in Prisley and Smith (1991). Timber values varied by 5 to 15% depending on the levels of spatial data variability assumed. For more information on this example, please refer to the cited publication.

Inventory Point Location

Area is not the only variable computed in a GIS that contains uncertainty. If the basic spatial locations are known imprecisely, then estimated distances between features also contain uncertainty. Distances are used, among other things, in a point-in-polygon analysis, which will become more common in forestry as organizations digitize the location of field inventory locations. Many have outlined the advantages of storing plot locations and associated inventory information. Recently, Stumpf (1991) expounded on the usefulness of GIS for inventory analysis.

Let's examine a realistic situation where spatial data uncertainty affects the management of the forest resource. Suppose that a forest products company uses digitized inventory plot locations to specify the use of timber growth models. Silvicultural activities which impact timber growth are mapped and overlaid with the plot locations to select appropriate models for projecting timber volume. Often, only a portion of a stand is thinned or fertilized, and the treated area is delineated on maps, and digitized. Subsequently, plots in the thinned area are "grown" using a post-thinning model, and plots in the unthinned area are grown using an unthinned model. In order to determine which model to apply to which inventory plots, the "silvicultural treatment" and "inventory plot location" themes are overlaid and a point-in-polygon analysis is performed. Clearly, this process contains significant uncertainty. Boundaries between treated and untreated areas can vary significantly depending on how they were determined. Inventory plot locations are not known exactly, even if great effort is expended in locating them. What is the effect of distance uncertainty on point-in-polygon analysis? It is intuitive to believe that the further an inventory point is from an enclosing boundary, the more confident we are that it is contained within that boundary. Thus, if some measure of certainty of distance from point to boundary is known, there is also a measure of certainty in polygon membership. Using a probability model developed by Prisley (1989), it is possible to make a probability statement of the type described above.

An example was developed from real inventory and mapped information. Inventory plots were located along a 4 by 5 chain grid within an actual 218 acre tract that contained several different forest stands. The techniques of Prisley (1989) were used to determine the p-value which describes the confidence with which a plot may be located within a particular stand. Under a fairly generous set of assumptions regarding point variability, only slightly more than half of the 107 plots could be located within a given stand at the 80% probability level. Under more liberal error assumptions, only 38% of the 107 plots were contained within a stand at the 80% probability level. Clearly, we do not know which plots are within which stands as well as we may have thought.

EFFECTS OF ALGORITHM DIFFERENCES ON DECISIONS

We expanded the original scope of the paper from dealing exclusively with spatial or locational uncertainty to include that uncertainty created by the algorithms used in a GIS to compute values needed by users. There are almost always two or more ways to perform an operation, with each method having its associated advantages and disadvantages. You can usually find proponents of each method. Importantly, different calculation methods may produce different final answers, and different final answers may produce different decisions even if the same basic data are used.

In this portion of the paper, we will describe the effect of changing the cell slope calculation method on categorizations of land suitability. More details of this study can be found in Weih (1991). A USGS DEM for a mountainous Appalachian area was acquired and imported into ARC/Info. Eleven slope calculation methods were then applied individually to each cell. The results of using six of these methods will be reported here. Method 1 was taken from Fleming and Hoffer (1979), Methods 3, 4 and 5 from Struve (1977), Method 6 from Horn (1981) and Method 11 from Weih (1991).

The cell-by-cell comparisons for five of the methods are presented in Table 1. Method 11 could not be compared here because it does not calculate individual cell slope values. It should be clear to all that using a different cell slope calculation algorithm results in different final cell slopes. Even the means of over 77,000 cells are often quite different. Individual cell slopes occasionally varied by more than 100 percentage points. These differences are entirely explainable and predictable. For instance, Method 1 uses simple average of the eight surrounding cell slopes, while Method 3 uses the maximum of neighboring cell slopes. It would be surprising if they did not produce quite different answers. The question is, which algorithm is the best one to use? We might also say that these cell-by-cell differences do not really tell the user how much the decisions they reach using these slope values will vary due to algorithm differences. That is the next step in the discussion.

In forest management, the slope of the land can limit its potential use by making it too difficult or costly to harvest. Stands that lie

Table 1. Mean (minimum, maximum) differences in calculated slope on a cell-by-cell basis for all cells (>77,000) in the study area using six slope calculation methods.

SLOPE Method	SLOPE Method 3	SLOPE Method 4	SLOPE Method 5	SLOPE Method 6
1	-15.5 (-136, 9.82)	-7.24 (-177, 0.00)	-5.79 (-78.1, 0.00)	0.72 (-37.2, 34.1)
3	---	8.29 (-65.3, 86.7)	9.74 (-22.8, 14.7)	16.2 (- 7.07, 181)
4	---	---	1.44 (-74.5, 107	7.95 (-33.7, 127)
5	---	---	---	6.51 (0.00, 51.4)

on steep land may require special harvesting equipment. Thus, some measure of gradient can be a critical factor in a forest management plan. Suppose that we had the following criteria:

Suitable: Stand slope < 35% and site index > 50 feet
Stand slope > 35% and site index > 60 feet

Unsuitable: Otherwise

and further, if suitable, the following harvest mechanisms are recommended:

Ground based skidding: 0-35% stand slope
Advanced ground skidding: 35-50% stand slope
Cable yarding: 50% or more stand slope.

This set of decision criteria was applied to each of 240 polygons covering the same area as the DEM described above. The boundaries of the polygons were used to partition the DEM into the 240 pieces representing individual polygons. Individual cell slope values for each of the 240 areas were averaged to form polygon or "stand slopes." Table 2 contains the results of this decision model. Unsuitable acreage varies by a factor of 10, from 175 to 1735 acres. Total suitable acres varies by 1560 acres, which is 25% of the total acreage of the 240 stands. Some stands changed harvest equipment recommendations three times, and some even changed from suitable to unsuitable categories or vice versa. Remember, each method used exactly the same basic elevation values.

Table 2. Number of polygons in each land suitability class for six cell slope calculation methods.[1/]

Decision Category	SLOPE METHOD					
	1	3	4	5	6	11
Suitable: Grnd Skid	183 (4438ac)	49 (921ac)	120 (2650ac)	135 (3277ac)	188 (4579ac)	212 (5338ac)
Suitable: Adv. Grnd Skid	32 (818ac)	91 (2116ac)	70 (1676ac)	57 (1415ac)	29 (682ac)	13 (278ac)
Suitable: Cable Yd	5 (260ac)	41 (1132ac)	12 (379ac)	14 (448ac)	5 (260ac)	3 (113ac)
Suitable: Total	220 (5516ac)	181 (4169ac)	202 (4705ac)	206 (5140ac)	222 (5521ac)	228 (5729ac)
Unsuit-able	20 (389ac)	59 (1735ac)	38 (1200ac)	34 (765ac)	18 (383ac)	12 (175ac)

The evidence is indisputable that the land suitability/harvest method decision is affected strongly by the cell slope calculation method used. (How many practitioners know which calculation method their software is using?) Each of the algorithms tested above is cited in the literature, and may be used in any number of software packages. We believe that in cases such as this, the algorithm selected to perform a particular task may have more impact on the decision than common levels of positional inaccuracy. Each practitioner must become familiar with the algorithms used by their software, and become familiar with the characteristics of those algorithms. Otherwise, they really do not know how their decisions are reached, and what affects them.

SUMMARY AND CONCLUSIONS

Forest management decisions are affected by spatial data uncertainty and differences in calculation method. Notice that we did not say that they are affected by GIS, because that is the case only in the strictest sense. GIS is the vehicle, not the driver. This variability in decisions has always existed because spatial data uncertainty and different algorithms have always existed. GIS actually created little of the uncertainty demonstrated above. In fact, GIS allows us to determine our variability and to utilize that knowledge to our advantage. It is up to each practitioner to do so, and the pressure to do so will increase as budgets tighten, regulations expand, and as outside groups take more interest in forest management activities.

1/ Individual cell values within a polygon were averaged to produce "polygon" values.

LITERATURE CITED

Fleming, M. D. and R. M. Hoffer. 1979. Machine processing of Landsat MSS data and DMA topographic data for forest cover type mapping. LARS Technical Report 062879. Purdue University, West Lafayette, Indiana.

Horn, B. K. P. 1981. Hill shading and reflectance map. Proceedings of IEEE 69(1):14-47.

Prisley, S. P. 1989. Statistical characterization of area and distance in an arc-node geographic information system. Ph.D. dissertation, Department of Forestry, Virginia Tech, Blacksburg, VA. 136 pp.

Prisley, S. P. and J. L. Smith. 1991. The effect of spatial data variability on decisions reached in a GIS environment. In: Proceedings of GIS '91 "Applications in a Changing World." Vancouver, B.C., CANADA. pp. 53-57.

Struve, H. 1977. An automated procedure for slope map construction. Technical Report M-77-3, S. Army Engineer Waterways Experiment Station. 98 pp.

Stumpf, K. A. 1991. The integration of inventory field data and thematic spatial databases. In: Proceedings of GIS '91 "Applications in a Changing World." Vancouver, B.C., CANADA. pp. 59-66.

Weih, R. C. 1991. Evaluating methods for characterizing slope conditions within polygons. Ph.D. dissertation, Department of Forestry, Virginia Tech, Blacksburg, VA. 230 pp.

A PROPOSED APPROACH FOR NATIONAL TO GLOBAL SCALE ERROR ASSESSMENTS

Lynn K. Fenstermaker
Desert Research Institute
Las Vegas, NV 89119

ABSTRACT

The error assessment of large area classified imagery presents unique problems to the remote sensing scientist. The primary problem is usually a lack of resources to collect sufficient ground information. This paper proposes a field verification and accuracy assessment approach that uses field, aerial, and existing data from a representative area. The proposed approach has three phases. The first phase partitions the large area into relatively homogenous strata. For example the United States may be partitioned into ecoregion types. Then within those strata, representative areas that contain most of the classes of interest are selected. The third phase is a stratified random sample by class, ensuring at least 50 points per class or 100 percent representation of small area resources. At each verification site, an area at least 3 by 3 pixels in size will be examined. This minimizes positional accuracy issues since the verification area will be larger than sensor resolution, and the center point location is recorded with GPS. Existing data will be used for verification where available if it meets project requirements. Areas not covered by existing data will be verified by an appropriate scale aerial photography. At least 20 percent of the verification sites will also be examined on the ground. The aerial and existing verification data will be compared with ground data to assess their accuracy. Final accuracy assessment results will be reported in confusion matrices with errors of commission and omission, overall accuracy, and the Kappa coefficient.

INTRODUCTION

It has become common practice over the past decade to assess the accuracy of remote sensing imagery classifications. However, until recently, most research using remotely sensed data has been on a local or regional scale. Now with interest in global change research and the availability of low resolution imagery such as the Advanced Very High Resolution Radiometer (AVHRR) data, the development of national and global scale datasets has increased. Some applications will require an error assessment to effectively and reliably use these large datasets. Several issues face remote sensing scientists before initiating such an assessment. This paper will discuss those issues and propose an approach to thematic accuracy assessment. Equally important but not discussed in detail in this paper is locational accuracy assessment.

NOTICE

Although the research described in this article has been supported by the United States Environmental Protection Agency through assistance agreement CR816826 to the Desert Research Institute, it has not been subjected to Agency review and therefore does not necessarily reflect the views of the Agency and no official endorsement should be inferred. Mention of trade names or commercial products does not constitute endorsement or recommendation for use.

Global/National Error Assessment Issues

There are numerous issues that need to be addressed to perform an appropriate and statistically sound error assessment. The assessment of national or global scale data accuracy has some unique concerns. One of the most problematic issues related to large datasets is scale or resolution. For example, how does one assess the classification accuracy of an AVHRR dataset given the coarse resolution of one kilometer when an area larger than sensor resolution, i.e., more than one pixel, is recommended (Curran and Williamson, 1986). Not only is there the question of funding to verify enough sites, but how to label an area larger than one square kilometer as one distinct class. It becomes rapidly apparent that accuracy objectives must take into account the sensor resolution, the availability of time and funding for field verification, and guidelines must be established for the entire accuracy assessment process.

If the objectives of the study including accuracy are defined at the beginning, then an appropriate field verification and accuracy assessment will be easier to design and implement. For example, let's say that a land cover classification of the United States will be used in a model to estimate national groundwater recharge. The model requires an Anderson level II classification with a minimum map unit of ten hectares, a thematic accuracy of 80 percent with a 90 percent confidence interval, and a spatial accuracy of plus or minus one hectare. Not only does the desired accuracy and confidence aid in defining the verification sample scheme, it also dictates data resolution and analysis requirements. Therefore, prior to project initiation, the time and resources necessary to successfully meet the objectives may be defined. All too often the issue of data quality is tagged on at the end of a project, and the verification effort is limited to the remaining resources.

The final issue that will be discussed is the use of existing data and aerial photography for verification. Due to the aerial extent of national or global datasets, it is cost prohibitive and often inefficient to acquire 100 percent of the verification data on the ground. Medium-scale aerial photography or video is a cost effective means of acquiring sufficient verification data if the project goals permit. Biging and Congalton (1991) examined the accuracy of photointerpretations for three types of forest classifications, namely, species composition, size class, and crown density. Using 100 percent ground verification data and two photointerpreters, their results were: 68 to 88 percent overall accuracies for species composition; 68 to 83 percent accuracies for size class; and 34 to 46 percent accuracy for crown closure. This clearly indicates that at least for somewhat detailed classifications, photointerpretation may not be suitable as a verification data source. Therefore, it is important that a subset of the verification sites be examined on the ground as a quality control check on the photointerpretation. If there are significant discrepancies between the ground and photointerpreted data, then photo data should not be used.

There are some instances where ground checks of aerially acquired verification data may not be desirable, for example, an AVHRR accuracy assessment. More bias may be introduced into an AVHRR accuracy assessment by ground observations that provide too much detail than using high-altitude photography. (The old saying 'that you can't see the forest because of the trees' applies here.) The high-altitude photography would provide a synoptic view of the sample sites at a finer resolution than the AVHRR and make it easier to define one class name for the area.

The reality of most national to global scale error assessments is that existing data or data acquired by other agencies/groups will be used for error assessment. It is probable that supplemental data will have to be acquired to cover areas or classes that are not represented by the existing data. Existing data should meet the criteria established at the beginning of the project for data accuracy assessment. For example, what were the objectives for collecting the data? What is the aerial extent of the sample site? What are the sample size and scheme? What are the dates of data acquisition? Are there potential biases in the data set, and did the agency collecting the data perform any type of quality assurance/quality control (QA/QC). The answers to these questions should be within the range of what may be considered "truth" for the accuracy assessment.

A PROPOSED ERROR ASSESSMENT APPROACH

When faced with the task of error assessing a national to global scale classification, the ideal is to develop an approach that is both statistically sound and inexpensive. One method of accomplishing this is by a multistage sample. Multistage sampling entails definition of hierarchal strata into which a large population may be divided. Williams (1978) describes multistage sampling and demonstrates how it may be used to reduce sampling variance by stratification. Cochran (1977) also states that stratification may provide more precise estimates of population characteristics. However, it is important that the strata be homogenous and soundly based on the target population traits. By stratifying a large population into homogenous primary, secondary, etc., strata, it is possible to describe the entire population with a smaller number of samples.

The definition of strata requires knowledge of the population that will be assessed. Therefore classification of the remotely sensed data must be performed before field verification. Obviously, resources may undergo change between the time of data acquisition and verification. Care must be taken in the collection of field data whenever possible to account for any temporal changes. Once the classification has been satisfactorily completed, primary strata may be defined. Examples of primary strata include morphologic provinces, ecoregions, and geologic formations. The definition and rules for division into each stratum must be clearly defined.

It is probable that further subdivision within the primary strata will be desirable. Since typically the purpose of a thematic error assessment is to describe individual class accuracy as well as overall accuracy, the selection of representative areas within the primary strata should be considered. Secondary strata may be designated by either a systematic or simple random sample. Numerous authors have stated pro's and con's for both sample schemes; this is briefly discussed in the following paragraph. The actual field verification samples would comprise the tertiary strata, and may also be selected by a systematic or simple random sample. The number of samples to be selected for the final strata again is dependent on the goals of the project. If only a right-wrong assessment is needed then the binomial distribution may be used to calculate the sample size (van Genderen and Lock, 1977; Rosenfield and Fitzpatrick-Lins, 1982). Hay (1979) recommends that sample size per class should always be at least 50 to test the accuracy of determinations. However, if the project objective is to test not only right versus wrong but look at the multiple classes of wrong then a multinomial distribution should be used to calculate sample size (Congalton, 1991; Rosenfield, 1982). Both the binomial and multinomial distributions require that an expected or acceptable accuracy and a maximum error be defined.

Sample design

The choice of a systematic or simple random sample should be based on the statistical requirements of the project and characteristics of the data. If there is a substantial amount of spatial autocorrelation in the data then a simple random or stratified simple random sample will be the best choice. Congalton (1988) found that for forest, range and agricultural populations simple random sampling always performed adequately. Both Congalton (1988) and Fitzpatrick-Lins (1981) stated that a stratified simple random sample would be preferred if representation of small but important resources was required. Congalton further stated that a stratified systematic unaligned sample and a systematic sample could be used depending on the complexity of the data. If spatial autocorrelation analysis indicates periodicity within the data, then use of systematic sampling schemes may result in poor estimates of classification accuracy. Hay (1977) and van Genderen (1978) also recommend the use of a stratified simple random sample for testing thematic map accuracy. Another consideration is whether the Kappa statistic will be calculated. The Kappa statistic assumes a simple random sample and it is not known how other sampling schemes will impact Kappa (Bishop et al., 1975).

Accuracy assessment

After all field verification data has been acquired, the construction of an error (or confusion) matrix is typical for performing an error assessment. The use of an error matrix for this purpose has been well documented in the literature. One of the best explanations of this approach is contained in Story and Congalton, 1986. Use of this approach not only provides an estimate for each class, but also estimates errors of omission and commission or using other terminology, producer's and user's accuracies. Additional accuracy information may be acquired by calculating the Kappa statistic or other multivariate analyses. The Kappa statistic is a coefficient of agreement that is a measure of actual minus random agreement. Kappa increases to one as chance agreement decreases and becomes negative as less than chance agreement occurs. A Kappa of zero occurs when the agreement between classified data and verification data equals chance. Other multivariate analyses may be used to test factors for potential impact on classification accuracy (Bishop et al., 1975).

A TEST CASE

Now that the ideal multistage sampling scenario has been presented, let's take a look at reality. The following test case is from the U.S. Environmental Protection Agency's Environmental Monitoring and Assessment Program (EMAP). A component of that program is landscape characterization using remotely sensed data. The EPA and its contractors are in the process of developing a Thematic Mapper (TM) classification for the Chesapeake Bay Watershed; a total of 16 TM scenes are being mosaicked. During the classification process, the TM mosaic will be partitioned into primary strata either by ecoregions or landscape pattern type. (Pattern types are defined as areas of similar land use and cover which form a unique homogeneous pattern, e.g., an urban area.) After the classification is completed, secondary and tertiary strata will be defined for accuracy assessment. However, the Landscape Characterization Group will not be able to devise their own primary sampling scheme.

EMAP is divided into seven ecosystem groups based on resource areas, namely, Agroecosystems, Arid, Forests, Great Lakes, Near Coastal, Surface Waters, and Wetlands. Some of these groups will be collecting field data regarding the health of their individual resource within the Chesapeake Watershed. Rather than funding several data collection activities within the same area, the seven ecosystem groups will acquire Landscape Characterization verification data simultaneously with their field activities. EMAP, as an overall program, has developed a systematic sampling grid consisting of uniformly spaced hexagons. The hexagons have a 40 square kilometer area. Each resource group is required to field sample within the hexagons. The Landscape Characterization Group will be able to randomly select hexagons within ecoregions. However, within those hexagons, they will be limited to the other resource groups' sample sites. Figure 1 displays the Chesapeake Bay Watershed, the ecoregions contained within the watershed, and the location of each hexagon. Only a subset of the hexagons represented in that figure will be selected for verification. The final number of verification samples will at a minimum be 50 samples per class as recommended by Hay (1979) or preferably a size calculated by use of the multinomial distribution. This will permit assessment of where the classification errors are distributed.

A field verification form has been designed to acquire all the descriptive information necessary to perform both an accuracy assessment and an evaluation of misclassification errors. The important components of this form include the items listed in Table 1. Field crews will be trained in the use of Global Positioning Satellite (GPS) systems and how to objectively examine the landscape for land cover and use. Since the minimum map unit defined for Landscape Characterization is one hectare, the field crews will be instructed to examine a circular area with a radius of 50 meters (0.78 hectares).

Table 1. A List of Information Acquired during Field Verification.

* Sample site identification	* Date and field crew names
* GPS measurement of location	* GPS filename and weather conditions
* Descriptive information on:	* 35 mm slides of the site
surface conditions	* land cover/use classification code
features under 6 meters (shrubs)	
features over 6 meters (trees & buildings)	
water regime	

Presently, it is anticipated that only the states of Virginia and Maryland will be sampled by the other EMAP resource groups within the next year. For the remainder of the watershed, a survey of existing data sources will be conducted. Known sources of existing data include the U.S. Forest Service's Forest Inventory and Analysis (FIA) plots, the National Wetlands Inventory (NWI), U.S. Census Bureau data, and the Soil Conservation Service's soil surveys. For each of these data sources certain parameters will be examined to determine whether the data is suitable for verification purposes. Table 2 lists the parameters that will be investigated. If the existing data meets the criteria established for each parameter, then it will be used in the verification process. However, as a quality control check, a subset of the sites will be ground verified by the Landscape Characterization Group.

Table 2. Evaluation Parameters for Existing Data Sources.

* Date of acquisition	* Data source: field, photos, etc.
* Size of sample sites	* Location: level of detail and quality
* Attribute information	* QA/QC procedures implemented
* Methodology	* Data format: analog, digital
* Availability	* Cost

It is anticipated that the Landscape Characterization Group will have to independently collect verification data for some portions of the watershed to meet the required number of samples. For those areas, a stratified simple random sample will be employed. Since the classification level of detail is approximately an Anderson level 1.5, aerial photography should provide accurate verification information. Aerial photography at a scale of 1:12,000 will be acquired. As a quality control check, at least 20 percent of the sites will also be examined on the ground according to the procedures mentioned in the previous section. After the verification data has all been acquired, quality control checked, and compiled, the accuracy assessment will be performed.

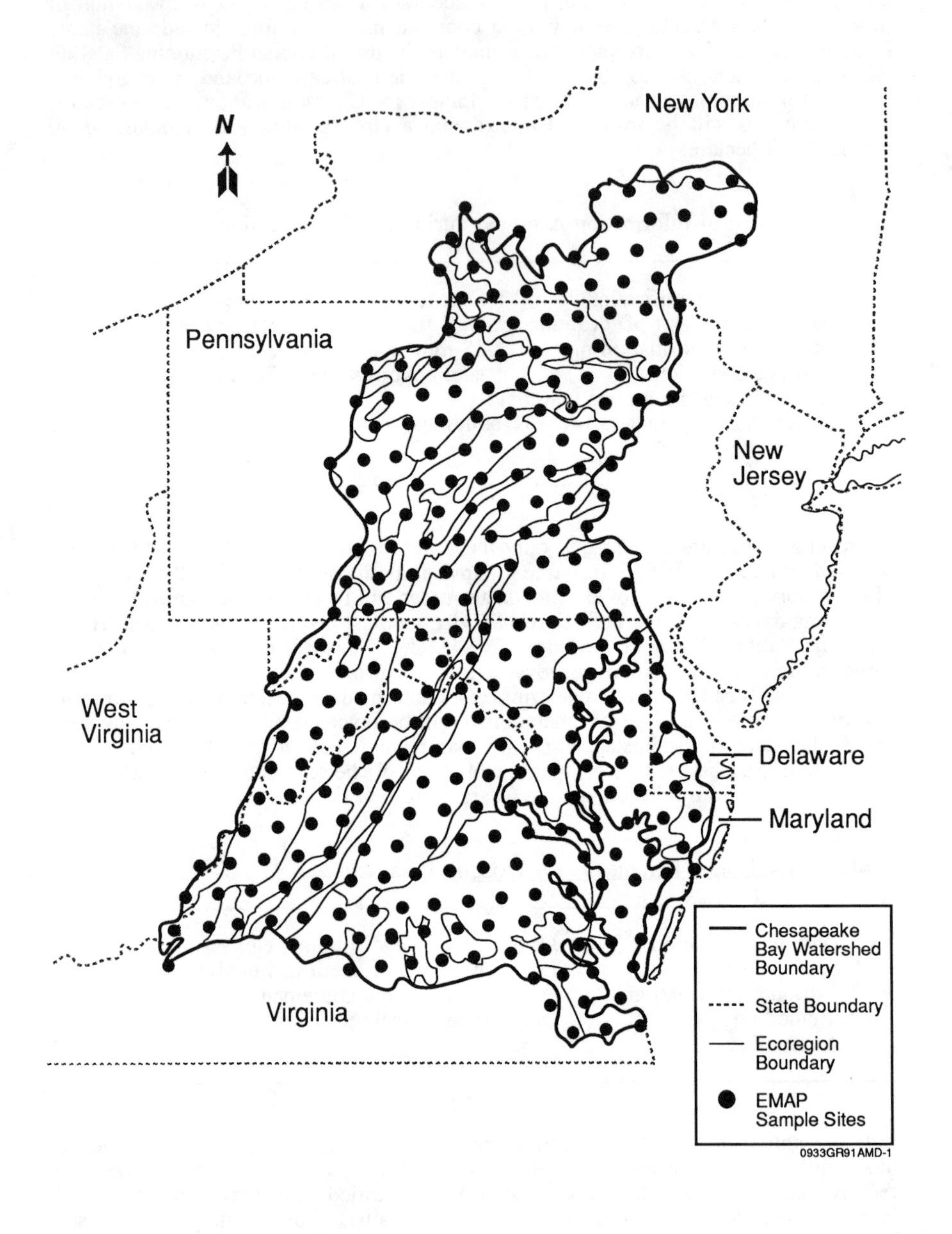

Figure 1. Boundaries for the Chesapeake Bay Watershed, Ecoregions within the Watershed, and EMAP Hexagons.

The accuracy assessment will be performed in two stages. First, the existing and aerial data will be compared to actual ground information. If that comparison results in significant differences, then either new verification data will be acquired or the quality objectives will be re-scoped to permit use of the existing data. The second phase would develop error matrices, errors of commission and omission, and the Kappa statistic to assess classification accuracy. Additional multivariate analyses may be performed on the data if the assessment results warrant more information. During the assessment process, procedures will be carefully monitored to ensure that statistical assumptions are not invalidated by the mixed sampling scheme and data sources. Also, each step will be documented to prevent or limit error in the assessment.

CONCLUSIONS

While this paper did not address all the aspects for global and national scale error assessments, it did attempt to list the most important considerations. These considerations are summarized in the following paragraphs.

* Clearly define all project objectives, especially data quality objectives, before project initiation. At a minimum the following should be defined: classification level of detail; required data resolution; temporal restrictions; minimum map unit; and required spatial and thematic accuracy, i.e., percent accuracy at a particular confidence interval.

* Choose an appropriate sampling scheme, site size, and number of samples to meet project objectives. In most cases a multistage stratified simple random sample will be the best sample scheme. Size of the sample sites should be larger than sensor resolution, e.g., 3 by 3 pixels. If only a correct versus incorrect assessment is desired a binomial distribution may be used to calculate sample size. However, if an assessment of which class a pixel was erroneously classified as is important, then a multinomial distribution should be used for sample number. At a minimum, at least 50 samples per class should be collected.

* It is inevitable that existing data will have to be used. Therefore, it is critical that acceptance criteria be developed to assess whether an existing dataset may be used as a verification dataset. One component of the acceptance criteria should be a quality control check. At least 20 percent of the sites in the existing data should be ground checked to ensure accuracy.

* Document and quality control check the entire accuracy assessment process to help prevent or limit error.

REFERENCES

Biging, G.S. and R.G. Congalton, 1991, A Comparison of Photointerpretation and Ground Measurements of Forest Structure, Proceedings, 1991 ACSM-ASPRS Annual Convention, Baltimore, MD, Vol. 3, pp. 6-15.

Bishop, Y.M., S.E. Fienbert, and P.W. Holland, 1975, Discrete Multivariate Analysis: Theory and Practice. MIT Press, Cambridge, MA, 575 p.

Cochran, W.G., 1977, Sampling Techniques. John Wiley & Sons, New York, 428p.

Congalton, R.G., 1991, A Review on Assessing the Accuracy of Classifications of Remotely Sensed Data. Remote Sensing of Environment, In Press.

Congalton, R.G., 1988, A Comparison of Sampling Schemes Used in Generating Error Matrices for Assessing the Accuracy of Maps Generated from Remotely Sensed Data. Photogrammetric Engineering and Remote Sensing, Vol. 54, No. 5, pp. 593-600.

Curran, P.J., and H.D. Williamson, 1986, Sample Size for Ground and Remotely Sensed Data. Remote Sensing of Environment, Vol. 20, pp. 31-41.

Fitzpatrick-Lins, K., 1981, Comparison of Sampling Procedures and Data Analysis for a Land-Use and Land-Cover Map. Photogrammetric Engineering and Remote Sensing, Vol. 47, No. 3, pp. 343-351.

Hay, A.M., 1979, Sampling Designs to Test Land-Use Map Accuracy. Photogrammetric Engineering and Remote Sensing, Vol. 45, No. 4, pp. 529-533.

Rosenfield, G.H., 1982, Sample Design for Estimating Change in Land Use and Land Cover. Photogrammetric Engineering and Remote Sensing, Vol. 48, No. 5, pp. 793-801.

Rosenfield, G.H. and K. Fitzpatrick-Lins, 1982, Sampling for Thematic Map Accuracy Testing. Photogrammetric Engineering and Remote Sensing, Vol. 48, No. 1, pp. 131-137.

Story, M. and R.G. Congalton, 1986, Accuracy Assessment: A User's Perspective. Photogrammetric Engineering and Remote Sensing, Vol. 52, No. 3, pp. 397-399.

van Genderen, J.L., B.F. Lock, P.A. Vass, 1978, Remote Sensing: Statistical Testing of Thematic Map Accuracy. Remote Sensing of Environment, Vol. 7, pp. 3-14.

van Genderen, J.L. and B.F. Lock, 1977, Testing Land-Use Map Accuracy. Photogrammetric Engineering and Remote Sensing, Vol. 43, No. 9, pp. 1135-1137.

Williams, B. 1978, A Sampler on Sampling; Wiley Series in Probability and Mathematical Statistics. John Wiley & Sons, New York, 254 p.

INTEGRATION OF SPATIAL AND RELATIONAL INFORMATION
RESIDING ON MULTIPLE PLATFORMS
IN A DISTRIBUTED ENVIRONMENT

Frank W Jacquez
McDonnell Douglas
Infrastructure Solutions Division
Mail Code 6800610
13736 Riverport Drive
Maryland Heights, MO 63043

Abstract

The City of Winnipeg, Manitoba, conducted a pilot to demonstrate the integration of spatial and relational land records information across multiple vendor platforms in a distributed data processing environment. The pilot project, completed January 1991, demonstrated the integration of spatial and textual databases for the City's mainframe based DB2 Land Records System with the City's GIS workstation based system and the Planning Department's downtown property PC database. The pilot project was a major step toward achieving one of the main objectives of the City's GIS project, which is "to facilitate the linkage of all land based data in the City so that civic departments could integrate and share data."

This paper presents the rationale behind striving for an integrated solution, review of the current technology, the specific objectives of the pilot project, and a review of the technology employed in the pilot project. A review of the applications used in the project, a review of the results of the project, and future applications and activities resulting from the lessons learned by this project.

Rationale

The City completed a Land Based Information System Feasibility and User Needs Study in June 1990. The study indicated approximately 70% of the City's 1990 Operating Budget is spent in "land-based" Departments.

The common requirement for all the "land-based" business functions is the use of maps and data which is geographically referenced. Map data was found to be the most significant data category for the Land Based Information System. However, data which is geographically referenced was found to be equally important as it provides additional information about items represented on the map. An example of geographically referenced data would be a water main identifier which indicates map location. Information such as year of construction, diameter, last inspection, etc., would also be available to manage the water facility. Parcel identifiers associated with property address, assessment data, permit tracking, etc, is another example of geographically referenced data.

Even though land-based functions are the most prominent in the City of Winnipeg, the degree of automation to facilitate the management of land-based resources and functions is minimal. There were 343 databases identified, and only 136 (40%) were listed as automated. This has severely reduced the City's and individual Department's abilities to share land-based information, perform long range planing and budget effectively.

Analysis of the City's departmental business functions indicated 100 out of 150 (67%) required inter-departmental data exchanges to conduct the departmental business function. The most severe, and most often identified problem, was that the maps and land related information were "not being up to date". This problem was interfering with the City's ability to perform its functions in a correct, timely and efficient manner. The following is a brief summary of some of the duplication of effort and redundancy which currently occurs in the City of Winnipeg when dealing with land based information:

1 Duplication of effort to collect and verify information,

2 Redundant data elements recorded in various Departments,

3 Lack of data standardization between systems,

4 Poorly maintained and poorly duplicated maps,

5 Map information exists in a variety of forms,

6 Data in departments is isolated and difficult to share,

7 Data is not manageable in its current form, and

8 Duplication of automation effort, between and within Departments.

After cost/benefit analysis the City has determined the preferred course of action is to develop a Land Based Information System. This system would integrate existing systems and better facilitate inter-departmental sharing of information to reduce the current levels of duplication of effort and redundancy.

In order to protect current investment, the strategic direction chosen for the Land Based Information System (LBIS) is not to utilize a single hardware and software configuration. LBIS is to be a collection of different systems and computer hardware and software all integrated such that data can be exchanged between systems. Therefore, one of the key configuration issues is communications between similar and different hardware platforms.

The study concluded that the GIS database will consist of two basic types of data, graphic and non-graphic. The GIS graphic data is digital representations of the graphic map features, including lines, symbols, and annotation. The GIS non-graphic data include attribute data which describe the characteristics of the map features, geographically referenced data containing a geographic location key, and geographic indices. The non-graphic data used with the GIS may be managed on either the GIS itself, or on the other computer systems operated by the participating City departments that will be linked to the GIS. The requirements for the GIS database may therefore include a number of attributes and identifiers which are resident on the GIS, as well as links that tie the map features to other files of attribute or geographically referenced data not residing on the GIS.

The pilot project described in this paper was a demonstration of feasibility for the above mentioned LBIS conceptual configuration.

Current Technology

The City currently supports a number of different platforms containing land based information. The largest platform is an Amdahl 5890 Model 300E (IBM S/370 technology) containing land based databases on information such as property assessment, building permits, and customer billing. The databases utilized on the mainframe include DB2 (relational), IMS (hierarchical), and VSAM (sequential).

A large number of PC's (mostly IBM compatible) supporting small land based data files currently exist throughout City departments. Some PC networks are being installed throughout the City. The current standard is Novell running on a Token Ring network. Databases utilized on the PC's include Dbase, Foxbase, Paradox, and Lotus 123.

The Geographic Information System utilized for property base maps is on a Digital VMS platform (VAX 3100 workstations) using the McDonnell Douglas Graphic Design System (GDS). The databases utilized on the workstations include Oracle (relational), and Rdb (relational).

Pilot Objectives

The pilot project was initiated to investigate the feasibility of distributed communication between the City's mainframe DB2 database and other distributed systems residing on different platforms. The project's three primary objectives were:

1. Investigate distributed data architecture technology between different hardware and software platforms such as:
   - a. Mainframe DB2 database to VAX system using McDonnell Douglas' GDS product as the application interface program.
   - b. Mainframe DB2 database to PC database running on a LAN.
2. Demonstrate distributed data architecture technology as an implementation tool for the City's Land Based Information Systems (LBIS) project using existing land based data files residing on different platforms.
3. The distributed processing pilot was for demonstration purposes only and will not become production systems. However, three important secondary objectives would be accomplished by the pilot:
   - a. Establish a permanent communications line between the VAX system used for base mapping and the City's mainframe computer to facilitate data transfers between the two systems; and
   - b. Establish PC database standards for departments wishing to distribute data between PC's and the City's mainframe computer.
   - c. Develop an understanding of the PC hardware and software requirements in order to evaluate future purchases.

The GIS graphic component of the pilot involved the purchase of all the necessary hardware and software and is installed in Land Surveys and Real Estate Department to support their base mapping program. There is to be only one copy of the mapping information (either stored distributed or centrally) which is shared by departments using high speed telecommunications. The PC components of the pilot involved the loan of hardware and software for evaluation purposes.

The pilot project was mandated to focus on database products which supported the SQL language. This involved DB2 on the mainframe, Rdb on the workstation, and a new SQL Server on the PC. A brief discussion of the technology employed in the pilot follows.

Pilot PC - Mainframe Architecture

The PC used was a Compaq 386-33L Deskpro running Micro Decisionware's Database Gateway for DB2. Database Gateway allows PC-based applications using Microsoft's SQL Server Application Programming Interface (API) to access DB2 databases through the same interface those applications use to access SQL Server databases.

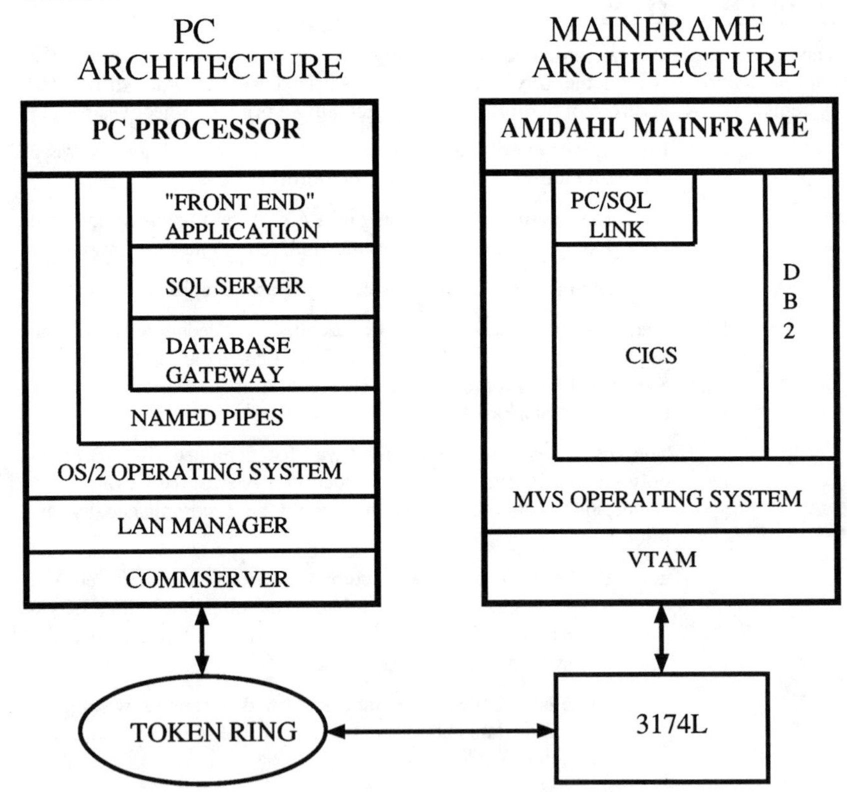

Pilot Workstation - Mainframe Architecture

The following hardware and software description defines an interface between the workstations using GDS and the mainframe DB2 relational database using VIDA2.

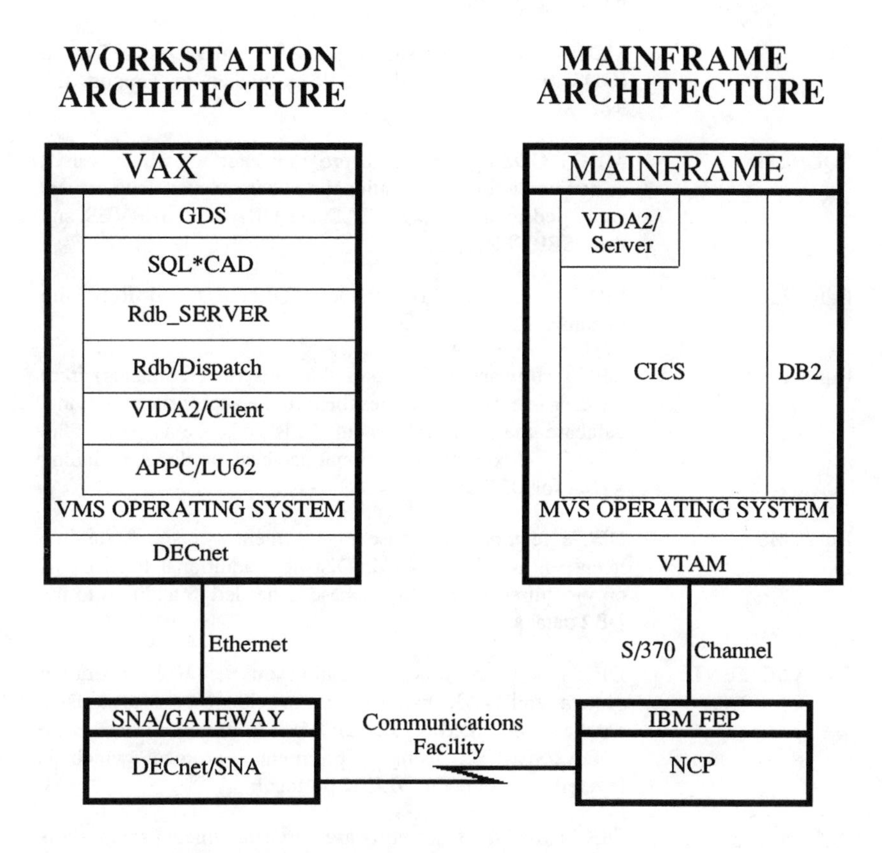

Figure 2 - Workstation/Mainframe Architectural Schematic

The major hardware and software components of the PC and Workstation solutions presented in the above diagrams are described in the next section.

MAJOR ARCHITECTURAL COMPONENTS

WORKSTATION ARCHITECTURE

VAX:	DEC's CPU hardware and associated peripherals.
VMS:	DEC's Operating System Software for VAX hardware, including DECnet.
GDS:	MDC's core product software, the Graphics Design System (GDS) must be Version 5.0 or higher to support the Rdb_SERVER.
SQL*CAD:	MDC's GDS dynamic SQL program interface which is used in conjunction with a relational database server. Two servers supported today by SQL*CAD are ORACLE_SERVER and Rdb_SERVER.
Rdb_SERVER:	MDC's software to interface SQL*CAD and Rdb with dynamic SQL.
Rdb/Dispatch:	DEC's software which takes the calls (or commands) from the user interface and routes them to the appropriate relational database management system. This software is part of all DEC DSRI-compliant relational database products, including VIDA for DB2 and Rdb/VMS.
Rdb/VMS:	DEC's relational database management system. A runtime license is required for Rdb/Dispatch, additional licensing is only required if an Rdb database is needed in addition to the DB2 database.
VIDA2/CLIENT:	DEC's software product running on the VMS operating system and VAX hardware to provide access to a DB2 database on IBM. VIDA2/CLIENT communicates with VIDA2/SERVER using a proprietary protocol which is layered upon the APPC/LU62 protocol.
APPC/LU62:	DEC's SNA/DECnet software product which provides the facility for Application to Application Program Communication (APPC) from a DEC Application to an IBM Application via IBM's LU 6.2 protocol. The APPC/LU62 software requires the SNA/GATEWAY hardware and software.
SNA/GATEWAY:	DEC's hardware and software which provides the DECnet/SNA protocol interface between the IBM and DEC network architectures. From the IBM/SNA perspective the SNA/Gateway is a PU2 type device with the user specified mix of LU's. From the DEC/DECnet perspective the

SNA/Gateway is a DECnet node. Two models of the SNA/GATEWAY are available, one supports an ethernet to IBM S/370 channel interface, and the second supports an ethernet to synchronous (RS232 or V.35) interface.

MAINFRAME ARCHITECTURE

FEP: IBM's Front End Processor supports remote SNA communications. A FEP is not required if a channel attached SNA/GATEWAY is used. The hardware model numbers may be 3745, 3725, (37x5), and all require Network Control Program (NCP) software.

ACF: IBM's Advanced Communications Facility to support SNA networks. Primarily comprised of NCP and VTAM, and is assumed whenever discussing either NCP or VTAM.

ACF/NCP: IBM's Network Control Program (NCP) software to support IBM's System Network Architecture (SNA) on the FEP hardware by providing the physical management of the remote elements of an SNA network.

ACF/VTAM: IBM's Virtual Telecommunications Access Method (VTAM) software to support the SNA network on the mainframe processor by providing the interface between VTAM application programs and other resources in the SNA network.

MAINFRAME: IBM's big processor and associated peripherals. The actual processor may be an AMDAHL, but still runs IBM software.

MVS/XA: IBM's mainframe operating system software required for the VIDA2/Server software. May be a VM guest operating system.

CICS: IBM's Customer Information Control System (CICS) performs task and resource scheduling, and controls sessions between application programs and devices associated with the application program. CICS is defined as a VTAM Application and is used to interface the CICS VIDA2/Server with the SNA network resources. CICS may be spelled out or pronounced "kicks".

VIDA2/SERVER: DEC's Server for DB2 is software running on the MVS/XA operating system on the IBM mainframe. VIDA/Server runs as a CICS application program which accesses DB2 in read-only mode, and communicates with VIDA/Client on the DEC with a proprietary protocol layered upon CICS/SNA APPC/LU62 communication protocols.

DB2: IBM's relational Database 2 (DB2) product for the MVS operating system. User applications can connect to DB2 through several environments. However, VIDA/Server requires CICS connection facilities for access to DB2.

PC/SQL: Micro Decisionware's host server software for DB2. Runs as a CICS application program which accesses DB2 (implemented in read-only) and communicates with DATABASE GATEWAY on the PC with a protocol layered upon CICS/SNA APPC/LU62 communication protocols.

PC ARCHITECTURE

COMMSERVER: DCA/Microsoft's 3270 communication software for PC to IBM SNA communication environment running on Token Ring.

SQL SERVER: Microsoft's relational database product for OS/2 based PC's.

Db GATEWAY: Micro Decisionware's advanced LAN-based software link that gives SQL Server applications transparent access to DB2 under MVS/CICS using APPC/LU62 protocol.

LAN MANAGER: Microsoft's LAN management facility which supports Named Pipes.

APPLICATION OVERVIEW: WORKSTATION TO MAINFRAME

DESCRIPTION:

The maps currently being developed by the Land Surveys and Real Estate Department were tied to the Assessment Department's DB2 database. The application allows the GDS workstation mapping software to access any field in Assessment's mainframe DB2 database through either a pre-defined or dynamically created "ad-hoc" SQL query. Two areas of major functionality illustrating the integration of the graphical and relational databases were demonstrated.

The first application prompted the user for a map parcel which is stored graphically with PCN structure intelligence in the GDS database. Once the parcel was graphically selected by the user, the program then determines the parcel's unique identifier from GDS and requests and displays the appropriate information from the DB2 database at the parcels centroid. Response time was consistently less than 3 seconds.

The information returned from DB2 and displayed at the parcel centroid by the first application was the Parcel Roll Number, Zoning, Adjusted Assessment Rate, Oldest Sale Year, and Front Foot Sales Rate (sales price/frontage).

The second application prompts the user for parcel selection, and then automatically builds the appropriate SQL command to retrieve the user requested (menu assisted) information from the DB2 database for the selected parcel. The DB2 data is then displayed in the GDS dialogue window, all in less than 3 seconds.

FUTURES:

This capability will eventually allow the Assessment Department to query and display information spatially to support property assessments or assessment appeals. Information such as land assessment per square foot, assessed value or sales price per parcel front foot, or any other combination of data on the Assessment database could be analyzed.

Currently some attempts are made by the Assessment Department to manually map this information, however the dynamic nature of the data makes it difficult to maintain using manual mapping techniques. In the future, information such as property address, and roll number may be acquired in a "bulk annotation mode" providing considerable time savings along with improved accuracy.

Based upon the successful pilot results, the McDonnell Douglas workstation component was purchased and is now installed. Additional departments are planned in 1991.

ADDITIONAL FUTURE BENEFITS:

The integrated land based information also provided the immediate benefit of assisting the Land Surveys Department in converting their assessment maps. Longer term benefits are the ability for the Assessment Department to geographically analyze their data. This technology will eventually benefit other departments which have large databases and mapping requirements such as the Winnipeg Hydro and Waterworks, Finance (tax data), or the Planning Department (building permits).

APPLICATION OVERVIEW: PC TO MAINFRAME

DESCRIPTION:

The PC application performed two major functions. The first function was to check for data integrity between data fields duplicated in the Planning Department's PC database and the Assessment Department's Mainframe database. The duplicate data fields used for comparison were Plan Area, Zoning, Gross Upper Area, Gross Floor Area, Number of Units, Number of Stories, and Year of Construction ("Age of Building" on mainframe).

The "Address Validation" SQL query was developed to print records from the Planning database where the property address on the Planning database does not match the address in the Assessment database. The building name from the Planning database was also displayed to give a better indication of the premise in question.

The "Data Validation" SQL query was developed to print records which do not match the corresponding value on the assessment database.

The second major function was the downloading of information from the Assessment database to the PC in order to "expand" the type of information utilized by the

Planning Department. The new fields added to the Planning Departments PC database were Assessed Land Value, Date of Last Sale, Adjusted Sale Value, Building Use Code, Building Condition, and Building Description.

Due to inconsistencies between the PC and DB2 database designs, the above functions were only able to be performed on records which met the following PC database criteria:

1. Records which do not have more than one building per roll number
2. Records which do not have more than one roll number; and
3. Records for commercial properties only.

An SQL query, "Add New Information to Downtown Database", was developed to use Assessment data to enhance the Planning Department's current capabilities by producing a report on land value and sales activity in the Downtown area. Combined with unique data from the Planning file such as Heritage Classification for historic buildings or special zoning bylaw designations.

FUTURES:

The PC pilot required significantly more time and effort to complete than the integrated workstation solution. The city is continuing evaluation of additional products to fulfill this requirement.

APPLICATIONS: FUTURE

A Property Addressing System is planned which could take advantage of both the workstation and PC components of the pilot project. The Property Addressing System is a computerized inventory of addresses for property, namely land parcels, buildings or units. These three types of property will also be monitored over time giving a perspective on properties which previously existed (past), currently exist (present), and which may exist someday (proposed).

The City's property based systems are an important component in the LBIS as they represent the largest level of automated data currently existing in the City. There are approximately 20 systems on the mainframe and 40 PC systems which use property address as a "geographic" identifier. An analysis of all manual and automated City databases showed 55% (187 of 343) databases use property address to access the data.

ACKNOWLEDGMENTS - The following individual contributed to the development of this report through either valuable comments and review or through extracts provided from other reports commissioned by the City.

Peter Bennett City of Winnipeg

Bob Thiessen City of Winnipeg

Mike Houston City of Winnipeg

THE DESIGN OF A GRAPHICAL USER INTERFACE FOR KNOWLEDGE ACQUISITION IN CARTOGRAPHIC GENERALIZATION

Robert B. McMaster
Department of Geography
University of Minnesota
Minneapolis, Minnesota 55455

David M. Mark
Department of Geography and NCGIA
SUNY at Buffalo
Amherst, N.Y. 14261

ABSTRACT

This paper presents a research project designed to elicit the procedural knowledge used in generalizing maps. The project is based on the development of a specific graphical user interface (GUI) for map generalization. Using a Sun SPARCstation, a user interface (designed under X-windows) is being developed. The basis for the user interface is a series of pull-down windows, each with a series of generalization operators, specific algorithms, and parameters. For instance, a user selecting the GENERALIZATION OPERATOR, SIMPLIFY, would be given a series of specific algorithms to select. After selection of a specific algorithm, the user would, after being given another menu, be able to select a tolerance value. The same structure is provided for each of the twelve operators. A component of the user interface--generalization tools--also allows a user to select individual objects or entire areas on the map to be generalized. Development of the structure and code for both the SIMPLIFY and SMOOTH operators is now complete. Eventually, the user interface will be used with the generalization test data set, developed by the NCGIA, to gain procedural knowledge on generalization from trained professional cartographers. As a cartographer works with the image via the interface, a generalization "log" will be maintained. Such a log will record, for each feature, the application and sequencing of operators, along with specific tolerance values. Such knowledge will be used to determine, in a holistic manner, how maps are generalized by evaluating the relationship between operators, parameters, and features.

INTRODUCTION

Although the process of digital cartographic generalization, a significant aspect of visualization, developed quickly during the period 1965 - 1980, little progress has been made during the 1980s. Most of the initial progress resulted from work in the development of algorithms (such as the well known Douglas and Peucker line simplification routine, a variety of techniques for smoothing data, and algorithms for displacement, such as those by Nickerson), and attempts to analyze both the geometric and perceptual quality of those algorithms. Recent attempts at developing a more comprehensive approach to the digital generalization of map features--such as the application of simplification, smoothing, and enhancement routines either iteratively or simultaneously--have not been, for the most part, successful. This stems in part from our lack of procedural information--or knowledge--on generalization. Such procedural knowledge includes decisions on which techniques are applied to actually generalize map information, the sequence in which these techniques are applied, and what tolerance values, or parameters, are used. Until researchers working in the spatial sciences have access to such procedural knowledge, a comprehensive approach to cartographic generalization will not be possible.

CURRENT RESEARCH

Although a recent book (Buttenfield and McMaster, 1991) addresses the issues of knowledge representation in generalization, much work lies ahead in the actual acquisition of this knowledge. The book focuses on the development of the rule base and contains a series of papers on (1) rule base organization, (2) data modelling issues, (3) formulation of rules and (4) the computational and representational issues. Several existing rule bases for map compilation and generalization are also provided. Mackaness and Beard (1990) discuss some of the fundamental considerations in the design of a user interface for user interaction in rule based map generalization. Armstrong and Bennett (1990) develop a knowledge-based object-oriented approach for generalization. In this paper they also emphasize the need for a categorization of cartographic knowledge to "facilitate its representation, management, and use" (Armstrong and Bennett, 1990, 50).

KNOWLEDGE CLASSIFICATION

Armstrong (1991) has proposed that three kinds of knowledge are necessary for effectively implementing rule-based systems designed to perform generalization: geometrical knowledge, structural knowledge, and procedural knowledge.

Geometrical knowledge, as defined by Armstrong, involves the feature descriptions on location and density. For instance, not only do linear features consist of strings of x-y coordinate pairs, but density measures can be developed that provide a sense of congestion amongst the features.

Structural knowledge, which involves the intrinsic expertise of the geographer, the hydrographer, the demographer, or the soils scientist, for example, is a difficult, if not nearly impossible, form of knowledge to formally encode. As Armstrong notes, "Structural knowledge arises from the generating process of an object, and is used to provide guidance in automated generalization because the generating process determines, in part, the way the feature is depicted (Armstrong, 1991, 92). To a large degree, the structural knowledge is influenced by the purpose of the map. For instance, a hydrographer will generalize a stream network using a different set of decisions than a geographer constructing a thematic map, since both individuals' structural knowledge is much different. Likewise, a set of contour lines would be generalized differently by a soils scientist, who would focus on the relationship between soils and topography, and an engineer planning a road cut. Thus the structural knowledge is both user and context specific.

Procedural knowledge allows control of the individual generalization operators and algorithms. It has been well-established in the cartographic literature that distinct "generalization" operators exist, although there remains disagreement on the exact set (McMaster, 1991). The procedural knowledge must: (1) identify the fundamental operators (such as simplification, smooothing, displacement, or enhancement) that are necessary, (2) select a specific algorithm in order to invoke the operator, and (3) determine the required algorithmic parameters. For certain operators, such as simplification, selection of an algorithm is more difficult than with operators such as displacement, where only one or two approaches have at this time been proposed. There is also, of course, intense geometric interaction amongst the operators which requires a knowledge of the sequencing and interrelationship. Although this problem has been acknowledged for some time, there is a paucity of research to assist those individuals select operators, algorithms, and parameters. Most of the effort thus far in the formalization of knowledge lies in the development of government agency rule bases.

RULE BASES

One set of rules has been established by the Canada Centre for Mapping (CMM), as reported by Nickerson, 1991:

Rule R1: If a feature would be less than the designated minimum size for this feature at the target map scale, then it can be eliminated.

Rule R1-E1: If there is not much drainage in the surrounding area, keep a lake feature which is smaller than the minimum size.

Rule R2: Exaggerate (by up to 1/3) prominent features which would be visible from the air

Rule R3: Maintain a minimum line separation at the published scale

Other rules, established by the United States Geological Survey, look similar:

Roads and trails: Private roads, access roads, and driveways less than 500 feet (152.4 m) in length will not be shown unless of landmark value in areas of sparse culture.

Railroads: Within the [railroad] yard, main-line through tracks are shown correctly placed, but other tracks are symbolized, preserving as much as possible the distinctive pattern presented by the yard.

Urban areas: The [urban] area must be no less than 0.75 of a square mile (1.9 x 1.9 km). Isolated islands adjacent to large urban areas may carry the urban area tint when they are as small as 0.25 of a square mile (0.6 x 0.6 km). (Mark, 1991, pp 108-114).

While such rule bases are useful for developing descriptive and geometric rules, they do not provide cartographers with the necessary "procedural" knowledge or rules. It is necessary that research continues in acquiring general rules, but at the same time methods must be designed for the acquisition of the procedural knowledge.

GENERALIZATION OPERATORS

In a forthcoming work, McMaster and Shea define the process of generalization as, "the process of deriving, from a data source, a symbolically or digitally-encoded cartographic data set through the application of spatial and attribute transformations. Objectives of this derivation process are: to reduce in scope the amount, type, and cartographic portrayal of the mapped or encoded data consistent with the chosen map purpose and intended audience; and to maintain clarity of preservation at the target scale. In this same work, ten spatial transformations are identified which control this modification:

[1] simplification
[2] smoothing
[3] aggregation
[4] amalgamation
[5] merging
[6] collapse
[6] collapse
[7] refinement
[8] exaggeration
[9] enhancement
[10] displacement

Other sets of operators have also been suggested, including those by Mackaness and Beard (1990). Current work by a series of researchers will undoubtedly refine and enhance these listings in an attempt to identify a set with which the user is most comfortable.

BUILDING EXPERT SYSTEMS: KNOWLEDGE ACQUISITION

In their chapter in the book, Building Expert Systems, Waterman and Hayes-Roth (1983, p. 170) discuss some applications of expert systems: "The types of tasks addressed [by expert systems] include interpretation, diagnosis, prediction, instruction, monitoring, planning, and design." The term 'expert' may be misleading, as it connotes elite of exceptional individuals. But, as Herb Schorr, an IBM Vice-President, pointed out in his keynote address at the Eighth International Workshop on Expert Systems and Their Applications, encoding the expertise of low-level employees is the most important area of application for expert systems technology in business computing (Schorr, 1988). Gains from the application of an expert systems at relatively low levels in the institutional hierarchy are also likely to be important in cartography (Mark, 1991).

Perhaps then the term 'specialist' would have been more appropriate that 'expert' in the term 'expert systems': "Specialists are distinguished from laymen and general practitioners in a technical domain by their vast task-specific knowledge, acquired from their training, their subsequent readings, and especially their experience of many hundreds of cases in the course of their practice" (Barr and Feigenbaum, 1982, p. 80-81). In the research project described in this paper, most of our subjects will be 'specialists' in map making, although only some will be 'experts' in cartography.

Knowledge Acquisition

The acquisition of domain knowledge, and the conversion of this knowledge into rules, is a key problem in the development of expert systems. Buchanan et al. (1983, p. 129) have claimed: "Knowledge acquisition is a bottleneck in the construction of expert systems," and this claim is echoed by Feigenbaum, as paraphrased in the Handbook of Artificial Intelligence: "Feigenbaum (1977) suggests that the painful process of knowledge engineering, which involves domain experts and computer scientists working together to design and construct the domain knowledge base, is the principle bottleneck in the development of expert systems. Efficient interfaces for acquiring this domain-specific knowledge ... need to be developed before significantly larger expert systems can be constructed." (Barr and Feigenbaum, 1982, p. 84).

Sources of Knowledge for Cartographic Expert Systems

Tanimoto (1987, p. 286) states that "The question of where a system's knowledge comes from is an appropriate one to begin with before examining any processes of knowledge acquisition." The major sources that Tanimoto goes on to describe are: experts; books or other texts; or direct experience. (A fourth source, not mentioned by Tanimoto, would be a study of things produced by experts; in cartography, rules for map generalization and selection might be inferred from a careful study of maps at two or more scales, an approach being used currently by Barbara Buttenfield and her colleagues.)

Acquiring knowledge directly from interviews with experts is often difficult, since much of their knowledge is procedural, and can be best taught through demonstration, or by trials by the apprentice followed by critique from the expert. However, many cartographers have attempted to build components of cartographic expert systems using this method, with themselves as the 'experts'.

Books are usually written by experts, and thus presumably contain similar knowledge, but the information has already been selected and organized. A programmer who is not a domain expert may be able to acquire enough knowledge from text books to build an expert system. As an example, most programs for automated cartographic label (text) placement have included rules based on the cartographic literature, including previous systems for map annotation. Derivation of map-making rules from manuals and product specifications for map production at government agencies, as discussed elsewhere in this paper, would fall under this type of knowledge acquisition.

By "direct experience", Tanimoto was referring to system development in which either the system itself, or the programmers developing it, interact with real problems, and knowledge is acquired by generalization from particular cases and responses to general rules. This research uses a 'direct experience' approach, with subjects making maps using the computer program, and the program monitoring their actions. We will not, however, build learning modules for automated rule generation into our system. If the reader has not studied artificial intelligence or expert systems, he or she may think that artificial intelligence is mainly about computer programs that learn from experience. In fact, machine learning does not play a major role in knowledge acquisition for expert systems. Instead, "the primary objective of most research in machine learning has been to gain a better understanding of learning itself" (Tanimoto, 1987, p. 283); expert systems are usually built by directly coding rules, based on knowledge acquired in the various methods discussed in this section.

A USER INTERFACE FOR MAP GENERALIZATION

In order to acquire procedural knowledge on map generalization, it is necessary to first build a comprehensive graphical user interface (GUI), and then monitor the activities of trained cartographers as each generalizes a map. It is expected that such a GUI would be a first step towards the ultimate goal of creating automated generalization software. This preliminary user interface, which is designed for obtaining knowledge, should: (1) provide users with a full array of generalization tools and operators, (2) focus on the individual, as well as the sequential and interactive, effects of generalization, and, (3) perhaps most importantly, keep a detailed "trace" or history of the activities involved in creating the generalized image. Weibel (1991), in discussing the concept of amplified intelligence, identifies the need for a powerful user interface "based on the design paradigm of direct manipulation and visual feedback" (Weibel, 1991, 178). Specific objectives for such an interface include:

[1] Provide the user a comprehensive set of generalization operators that will allow the user to manipulate the map image.

[2] Provide a comprehensive set of tools which will assist in identifying map features

[3] Provide the user with assistance in selecting tolerances and parameters for individual operators and algorithms

[4] Provide the user with warnings when an illogical selection is made

[5] Provide a "trace" or accounting of feature generalization

[6] Provide the user with supporting documentation and diagrams

[7] Provide the user, when possible, with a "measure of success"

[8] Provide the user with features on the map in need of generalization, or what may be called generalization "hot spots". Such regions where the density of features is high may be located by the "conditions of generalization" identified by Shea and McMaster, 1989. Such conditions are necessarily evaluated through the application of measures which act as descriptors of the geometry of individual features and assess the spatial relationships between the combined features.

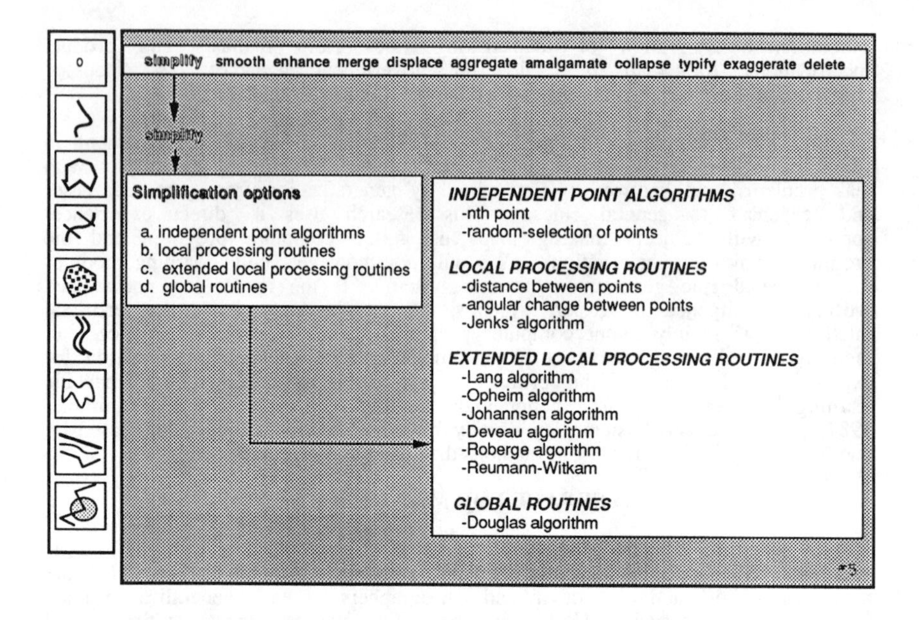

Figure 1. A user interface designed specifically for generalization with simplification selected.

An idealized user interface is depicted in Figure 1. Along the top of the interface are the operators and along the side individual "tools" for manipulating features are provided. For instance, the first tool allows the identification of a single point feature, while the second tool allows a line feature to be identified. Other tools allow for the selection of entire features, line segments, aggregations of point features, selection of parallel linear features, as well as tools for establishing a hierarchy for displacement, and the removal of spatial conflicts. In Figure 1 the selection of the simplification operator is illustrated. The user is given a menu of simplification categories and individual algorithms, such as Lang or Douglas.

A critical component of such a user interface is the inclusion of an on-line help system. As an example, Figure 2 depicts the help window for simplification. If, at any point in the map generalization process, the user has difficulty in understanding the selection of an operator, algorithm, or parameter, the help function may be invoked. Optimally, such help windows should be developed in a hypermedia environment.

An entire session with the user interface is depicted in Figure 3. The basic window includes: (1) the map image, (2) the generalization operators, and (3) the selection tools. In this example a single line segment has been initially selected. The user has selected the SIMPLIFY operator from the menu and has been provided with a set of algorithm choices. After selecting (1) extended local processing routines, and (2) the Lang algorithm from this set, the user is given an option to CANCEL or CONTINUE. If the user continues, additional information, involving a specific tolerance selection, must be provided. For instance, in the case of the Lang algorithm the user must provide both the number of points to look ahead as well as the distance tolerance, since Lang is a multiparameter algorithm. In terms of user interaction, a slider bar is

currently being implemented to assist with tolerance selection.
Simplification:

The algorithms provided under the simplify option enable the weeding of uneccessary coordinate information. A digitized representation of a map feature should be accurate in its representation of the feature, yet also efficient in terms of retaining the least number of data points necessary to represent the character. Simplification operators will select the characteristic, or shape-describing, points to retain, or will reject the redundant points considered to be necessary to display the line's character.

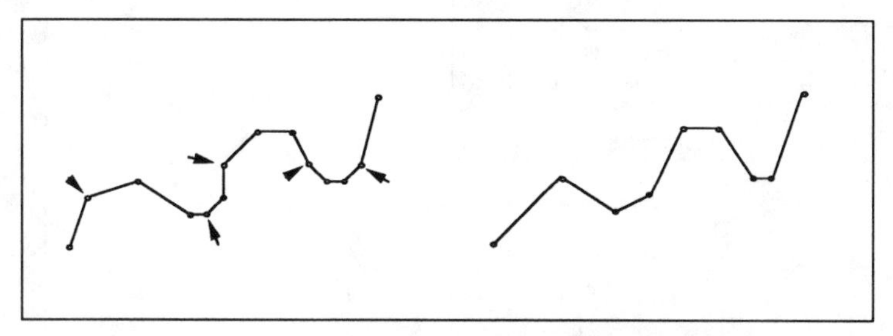

Figure 2. An example of a help window for the simplification operation.

Figure 4 illustrates the menu for the merge operator/tool, where the user is able to select several alternative merge techniques, including merge to center and merge to edge. As the user interface is designed and tested, and is modified through observation of the generalization users, both tools and operators may be added or dropped from the initial set.

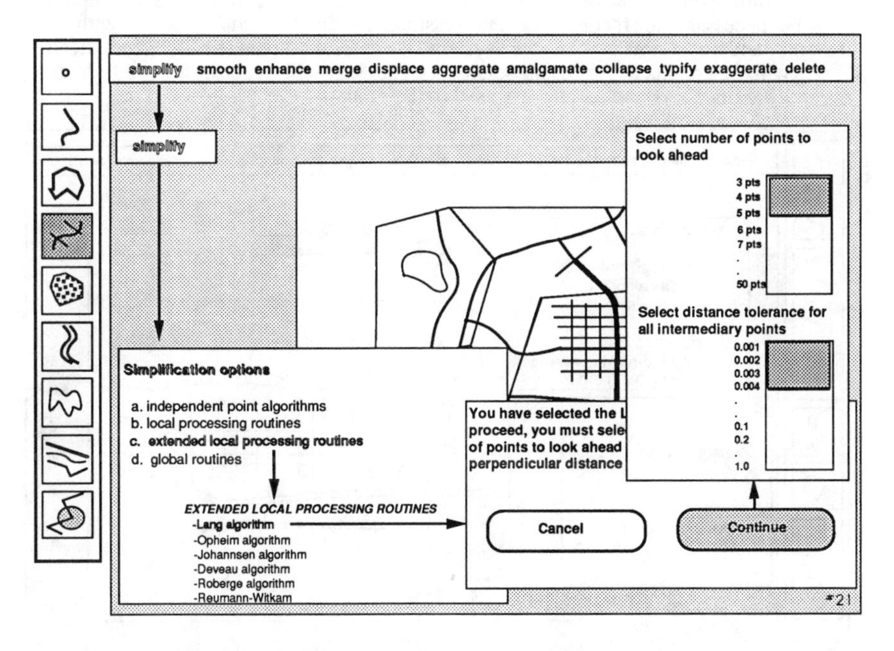

Figure 3. A complete session with the user interface including tool selection (line feature), simplification operator (Lang), and parameter selection.

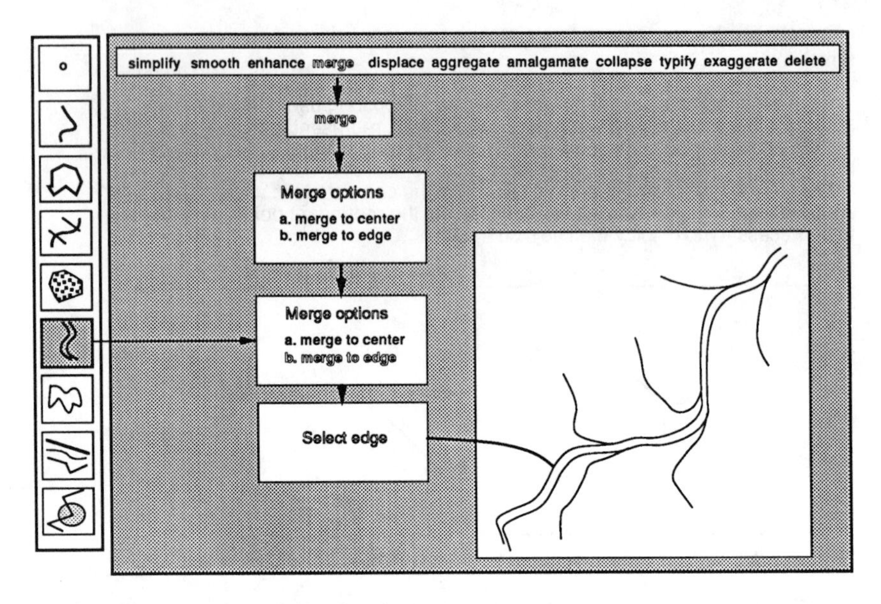

Figure 4. The merge operator and tool.

The final objective for such a system is the acquisition of procedural knowledge for generalization. Thus the design and development of the user interface, while a critical consideration, is secondary to the actual acquisition of the knowledge. In order to gain insight into the "cartographers' mind", it will be necessary to record working sessions as the individual works with map features and operators. As the system is currently planned, a generalization "log" will be retained for each session. The log will record, for each feature, the specific operators and parameters used in the process.

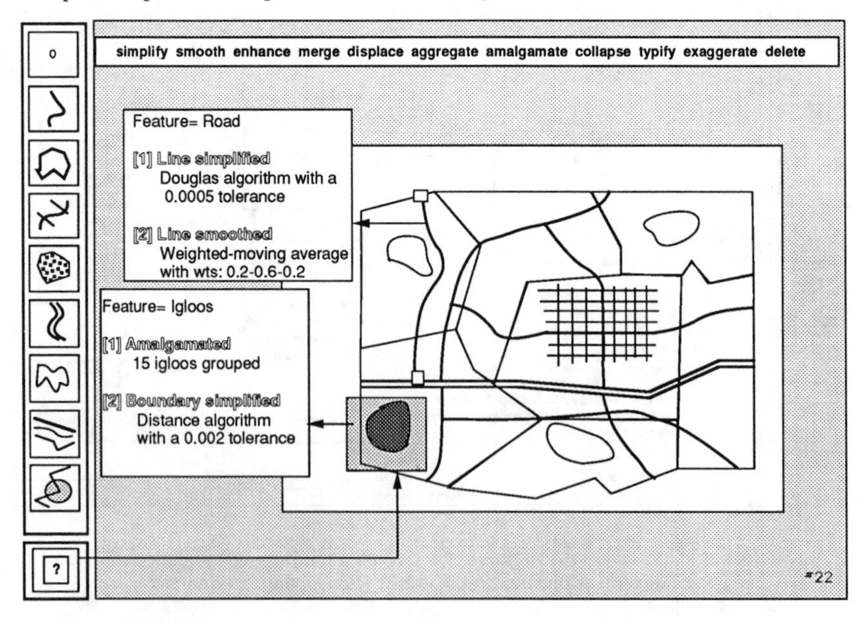

Figure 5. Illustration depicting creation of a generalization log for the acquisition of procedural knowledge.

For instance, a sample log for two features is created on Figure 5. Here, using the hypothetical map and cartographic features, the road has first been simplified using the Douglas algorithm with a 0.0005 tolerance and subsequently smoothed using a weighted-moving average technique. In the second example, a group of point symbols (igloos) are first AMALGAMATED into a polygon. The polygonal boundary is then SIMPLIFIED using a simple distance algorithm. These logs, then, provide both the types and sequence of procedures individuals use in generalizing maps.

SUMMARY

Although the research design for the project is nearly complete, as the actual system is built adjustments will be made. The most important aspect of the design is knowledge acquisition. In the final design, we have attempted to follow the structure provided by Buchanan et al. (1983, p. 140). They identify the three major stages in knowledge acquisition as:

1. Identification Stage.
    - 1.1 Participant Identification and roles
    - 1.2 Problem identification
    - 1.3 Resource Identification
    - 1.4 Goal Identification
2. Conceptualization Stage
3. Formalization Stage
4. Implementation Stage
5. Testing Stage
6. Prototype revision

The user interface now has functionality for both simplification and smoothing, thus we are completing stage 4, the implementation.

ACKNOWLEDGEMENTS

David Mark's contribution to this paper represents part of research initiatives 8 and 13 of the National Center for Geographic Information and Analysis, supported by grant SES-88-10917 from the National Science Foundation.

Robert B. McMaster wishes to thank The College of Liberal Arts at the University of Minnesota for travel support and to Kuo-Chen Chang for his assistance with developing the prototype interface.

REFERENCES

Armstrong, Marc P., 1991. Knowledge classification and organization. In Buttenfield, B. P., and McMaster, R. B., editors, Map Generalization: Making Decisions for Knowledge Representation. London: Longmans Publishers, in press.

Armstrong, Marc P. and Bennett, David A., 1990. A Knowledge Based Object-Oriented Approach to Cartographic Generalization. Proceedings, GIS/LIS'90, pp. 48-57.

Barr, A., and Feigenbaum, E. A., editors, 1982. The Handbook of Artificial Intelligence, Volume 2. Los Altos, California: William Kaufmann, Inc.

Beard, M. Kate., 1991. Constraints on rule formation. In Buttenfield, B. P., and McMaster, R. B., editors, Map Generalization: Making Decisions for Knowledge Representation. London: Longmans Publishers, in press.

Buchanan, B. G., Barstow, D., Bechtal, R., Bennett, J., Clancey, W., Kulikowski, C., Mitchell, T., and Waterman, D.A., 1983. Constructing an expert system. In Hayes-Roth, F., Waterman, D. A., and Lenat, D. B., editors, Building Expert Systems. Massachusetts, Addison-Wesley, pp. 127-168.

Feigenbaum, E.A., 1977. The Art of Artificial Intelligence: 1. Themes and Case Studies of Knowledge Engineering. Proceedings, 5th International Joint Conference on Artificial Intelligence, pp. 1014-29.

Buttenfield, B. P., and McMaster, R. B., editors, 1991. Map Generalization: Making Decisions for Knowledge Representation. London: Longmans Publishers, in press.

Mackaness, William and Beard, M. Kate, 1990. Development of an Interface for User Interaction in Rule Based Map Generalization. Proceedings, GIS/LIS'90, pp. 107-116.

Mark, D. M., 1991. Object modelling and phenomenon-based generalization. In Buttenfield, B. P., and McMaster, R. B., editors, Map Generalization: Making Decisions for Knowledge Representation. London: Longmans Publishers, in press.

McMaster, Robert B., 1991. Conceptual frameworks for geographical knowledge. In Buttenfield, B. P., and McMaster, R. B., editors, Map Generalization: Making Decisions for Knowledge Representation. London: Longmans Publishers, in press.

McMaster, Robert B. and Shea, K. Stuart, 1991. Generalization in Digital Cartography, forthcoming.

Nickerson, Bradford G., 1991. Knowledge engineering for generalization. In Buttenfield, B. P., and McMaster, R. B., editors, Map Generalization: Making Decisions for Knowledge Representation. London: Longmans Publishers, in press.

Schorr, H., 1988. Expert systems: An IBM perspective. Proceedings, Eighth International Workshop on Expert Systems and Their Applications, Avignon, France, May 30 - June 3, 1988, vol. 1, p. 39-40 [summary only].

Shea, K. Stuart and McMaster, R.B., 1989. Cartographic Generalization in a Digital Environment: When and How to Generalize. Proceedings, Auto-Carto-9, pp. 56-67.

Tanimoto, Steven L., 1987. The Elements of Artificial Intelligence: An Introduction Using LISP. Rockville, Maryland: Computer Science Press, Inc.

Waterman, D. A., and Hayes-Roth, F., 1983. An investigation of tools for building expert systems. In Hayes-Roth, F., Waterman, D. A., and Lenat, D. B., editors, Building Expert Systems. Reading, Massachusetts, Addison-Wesley, pp. 169-215.

Weibel, Robert, 1991. Amplified intelligence and rule-based systems. In Buttenfield, B. P., and McMaster, R. B., editors, Map Generalization: Making Decisions for Knowledge Representation. London: Longmans Publishers, in press.

CARTOGRAPHIC VISUALIZATION AND USER INTERFACES IN SPATIAL DECISION SUPPORT SYSTEMS

Marc P. Armstrong
Departments of Geography and Computer Science
316 Jessup Hall
The University of Iowa
Iowa City, IA 52242

Paul J. Densham
National Center for Geographic Information and Analysis and
Department of Geography
State University of New York at Buffalo
Buffalo, NY 14261

Panagiotis Lolonis
Department of Geography
The University of Iowa

ABSTRACT

Spatial decision support systems enable decision-makers to analyze complex locational problems. When such systems are used, problems often undergo reconceptualization as the effects of altering criteria and model specifications become known. Model results, however, must be placed in a geographical context for decision-makers to understand key spatial relationships among components of their problem. Several distinct tasks typically must be accomplished during locational decision-making, and the interface presented to the user must adapt to different display requirements. Decision-makers also must be allowed to interactively visualize the effects of making adjustments to parameters in the solution space. We present a model of an interface that relies on the establishment of linkages between database entities and displays to support visualization requirements.

INTRODUCTION

A spatial decision support system (SDSS) helps decision-makers explore the nature of semi-structured locational problems by enabling them to iteratively change model parameters and to examine the effects of these changes on the resulting solutions (Armstrong *et al.*, 1991a). Semi-structured problems contain aspects that are not amenable to a formal specification and therefore cannot be included directly in a computer-based solution (Langendorf, 1985; Harris, 1988). As a result, they are difficult to solve using conventional GIS techniques. Though SDSS have undergone a series of conceptual improvements in the past several years and are now being used to analyze semi-structured problems in a variety of application areas (e.g. Guariso and Werthner, 1989; Honey *et al.*, 1991; Peterson, 1990), the degree to which such systems can be used effectively by those other than expert users often is limited. In one SDSS (Armstrong *et al.*, 1991a), for example, a system developer acted as an "intermediary" through whom requests for analyses were filtered. While there is considerable room for debate about the ways in which powerful geographical analysis tools can be abused at the hands of those without adequate training in the principles of geographical data analysis, decision-makers are more likely to use and adopt GIS and SDSS technology when they directly interact with the solution process and see the results of their modeling endeavors in map form. Our goal, therefore, is to construct a user interface that will provide graphical support to decision-makers as they engage in the process of solving semi-structured problems. The focus of this paper is placed on districting problems in which the service area of facilities must be delineated. This requires an interface that is designed to accommodate the set of tasks that are needed in this application domain. We construe the interface broadly to focus on the types of maps and other graphics that should be provided to decision-makers and also describe the ways through which they interact with these graphics to arrive at decisions.

REPRESENTATION OF DISTRICTING PROBLEMS

Typical districting problems require that decision-makers consider the demand for, and supply of, services when developing plans. Within the SDSS context, demand and supply characteristics are stored in a geographical database and are used to perform analyses. Specifically, demand characteristics are often represented either as nodes or polygons, and are associated with a set of attributes. Facilities and candidate locations are represented similarly. These demand and supply characteristics are usually organized in different layers, which permits users to manipulate various entities, obtain data to perform optimization analyses, and create maps and reports for different districting scenarios.

Current SDSS have two major limitations, however. First, demand and supply information represented by loosely coupled layers fails to capture explicitly relations that are inherent in facility location problems. For example, when facilities (nodes) and service areas (polygons) are organized in two or more independent layers, and changes are made in the polygon layer (boundary shifts), the attributes in the facility layer are not updated automatically. Instead, analysts may have to perform several steps to preserve consistency in the database. Clearly, such low level technical operations offer nothing positive to the development of location plans. On the contrary, they slow the planning process, reduce the number of alternatives that are developed, and distract the attention of decision-makers. An effective SDSS, therefore, must exhibit a tighter coupling between such entities to improve the effectiveness of the system.

The second, and more important, limitation is the lack of multiple views that represent the state of the service system for a given set of parameter values during the planning process. Some information about the service system (spatial relationships between demand and supply) is presented effectively using maps, while other information may be presented better using graphs or tables. Decision-makers must access needed information, make changes in the state of the system, and visualize the consequences of those changes in real-time. Such needs require a careful choice of display tools, good organization in screen layout, availability of functions that allow decision-makers to manipulate objects in the service system, and sufficient computing speed to perform the tasks rapidly (Lambert, 1984; Laurel, 1990).

INTERACTIVE USER INTERFACE FOR SDSS

Several principles can be followed when designing the user interface in order to achieve the goals described above (Turk, 1990; Schneiderman, 1986; Apple Computer, 1985). For example, graphical objects that resemble real world objects (Mark, 1989) can be used to reduce learning costs and increase user acceptance, provided that appropriate metaphors are used (Kuhn, 1991). Users unfamiliar with the interface can use intuition to infer the function of such interface objects. In addition, graphical objects must be organized so that they are ergonomically laid out, and those tools that are used frequently must be accessible. SDSS interfaces also must provide flexibility in manipulating objects and allow users to avoid or recover from unintended actions.

An additional critical component of a SDSS interface, besides its appearance and organization, is its behavior. A fundamental principle guiding the design of interface responses is that users must be provided with constant and consistent feedback (Tognazzini, 1990). Specifically, when users manipulate an object on the screen then, depending on the kind of manipulation, there should be a change in the appearance of the object to indicate to users that their actions have had an effect. When computing tasks require more than several seconds to be performed, then graphical feedback can be used to indicate the progress of the task (Myers, 1985). Such behavior by the interface allows users to concentrate on the real problem they are attempting to solve rather than on the mechanics of the interface.

In addition to these general design guidelines, the SDSS interface has special characteristics that must be considered during its development. Specifically, the needs of facility planning and service area delineation require the simultaneous availability of at least three different kinds of displays to show information about the state of each scenario: map, graph, and table displays. Maps are necessary for presenting site and situation characteristics of entities in a study area, graphs are effective for displaying attribute information, and tables are suitable for containing attribute information in an accurate, compact form.

SDSS TASKS AND THE USER INTERFACE

The interface under development has been designed after evaluating the application of SDSS to several districting problems. Although each is unique, these problems exhibit common characteristics and tasks that must be accomplished. These commonalities enable the specification of generic decision support functions and the development of effective user interfaces (Rasmussen, 1986; Turk, 1990).

The different ways of viewing data must present the state of a plan as it is modeled in the SDSS database. Actions performed by users on the screen are translated to changes in the database and those changes are reflected in screen objects that provide users with needed feedback. Table 1 shows how the tasks that decision-makers wish to perform are linked to

Table 1. Tasks, actions and displays in the SDSS interface.

User Task	User Action	Map Window	Graph Window	Table Window
Focus on a facility	Select option (*Mouse; Palette*)	Display demand and supply	Display facility attributes	Display facility attributes
	Select facility (*Mouse; Map, Graph, Table*)	Highlight selected facility	Highlight bar of selected facility	Highlight attributes of selected facility
Close facility	Select option (*Mouse; Palette*)			
	Select facility (*Mouse; Map, Graph, Table*)	Highlight selected facility	Highlight bar of selected facility	Highlight attributes of selected facility
	Confirm choice (*Mouse; Dialog*)	Display new service areas (*Chorochr., Delta*)	Delete bar Adjust length of bars	Delete attribute row Display new facility attributes
Open facility	Select option (*Mouse; Palette*)	Display candidate locations		
	Select candidate location (*Mouse; Map*)	Display & highlight new facility	Create a new bar	Create a new attribute row
	Confirm choice (*Mouse; Dialog*)	Display service areas (*Chorochr., Delta*)	Update attributes	Update attributes
			Highlight bar of new facility	Highlight attributes of new facility
Move facility	Select option (*Mouse; Palette*)			
	Select facility (*Mouse; Map, Graph, Table*)	Highlight selected facility	Hilghlight bar of selected facility	Highlight attributes of selected facility
	Drag facility symbol (*Mouse; Map*)	Move facility symbol Highlight candidate location Display new service areas (*Chorochr., Delta*)	Replace bar annotations Update bar lenghts	Inititialize attributes of new facility Update attributes
	Confirm new location (*Mouse; Dialog*)	Display service areas Highlight new facility (*Chorochr., Delta*)	Display bar lengths Highlight bar of new facility	Update attributes Highlight attributes of new facility

actions on the screen when a map, graph, and table window are open. When the plan under consideration involves closing a facility, for example, users activate the "close facility option" from a palette, select the facility they want to close from the map, graph, or table, and confirm it by responding to the prompt in a dialog window. When a facility is selected, its corresponding symbols in the open windows are highlighted to inform users that they are working with that object. When users close the facility and the SDSS interface is in automatic update mode, the demand served by the closed facility is allocated to other facilities and the new state of the service system is presented in the different windows. For example, the map window may show a chorochromatic map (Monkhouse and Wilkinson, 1971:38) of the new service areas or one showing changes in service areas, depending on which map option is active. Similarly, the graph and table windows will show the new values of facility attributes. The previous example shows how the separate views which are linked through a common data space can be altered in each window, with the resulting changes depicted in the other windows. This approach is similar to that used by the current generation of spreadsheet software.

Map Display and Interaction

Several map types can be used to support decision-making in districting applications (Armstrong *et al.*, 1991b):

- **Demand and Supply Maps** enable decision-makers to visualize shifts in demand over time and to plan for service system changes. Figure 1, for example, shows the projected distribution of elementary school students in the SE part of the Iowa City, Iowa school district in 1995. The demand data used to produce these maps also can be aggregated to determine whether the students in an existing attendance zone will exceed a school capacity constraint. Supply maps are similar in many respects to demand maps, but show locations of service centers, service area boundaries, and capacities of facilities.
- **Spider Maps** show the simultaneous relationship between supply of services and the pattern of demand that is assigned to supply centers, in this case schools (Figure 2). Such maps may show the existing service system or the effect of adopting different assumptions and criteria on patterns of service provision. They also can be used to identify areas that are underserved by visually assessing differences in the size and spread of allocations to different facilities.
- **Delta Maps** highlight salient aspects of differences between alternative plans. Figure 3, for example, shows the difference between a solution in which students are assigned to schools using their existing school attendance areas and one in which students are assigned to their proximal school.

These maps are used to support several tasks that users perform during location decision-making. One task is to select locations as candidates for facilities (e.g. schools). Candidates may be selected one at a time, or as a group. They may also be selected using spatial query and display capabilities of the type commonly provided by database management systems (e.g. lot size > X acres and frontage on a road). After a location selection model has been run, the user may wish to return to a map display to evaluate marginal changes to the solution. Once users have created such a map, additional, often subjective, criteria can be used to evaluate the suitability of facility locations and service regions. Furthermore, users, after identifying certain regions as "candidate-poor", may decide to search for additional locations that do not fully meet the stated, objective selection criteria but otherwise appear promising.

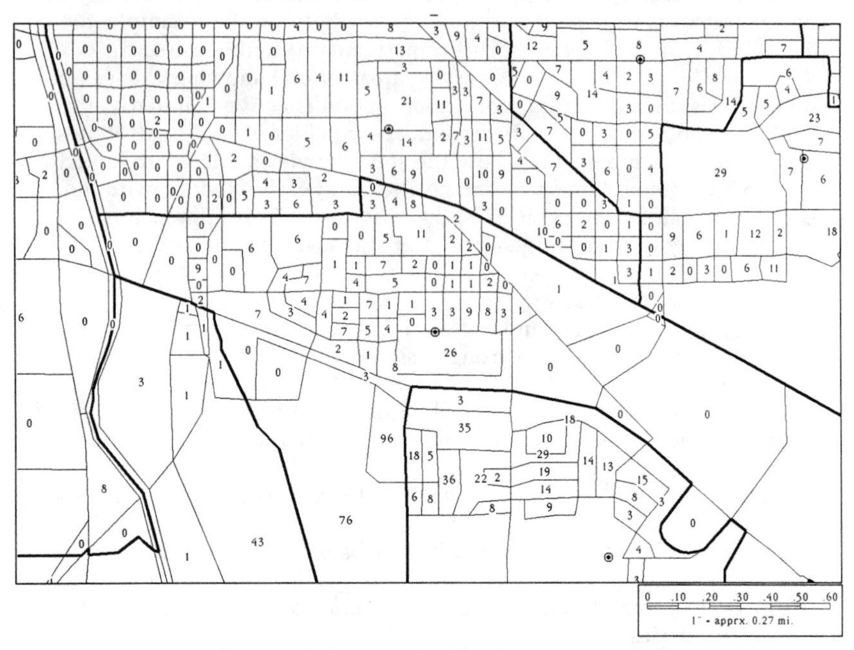

Figure 1. Projected K-6 enrollments, by block in SE Iowa City, 1995-96.

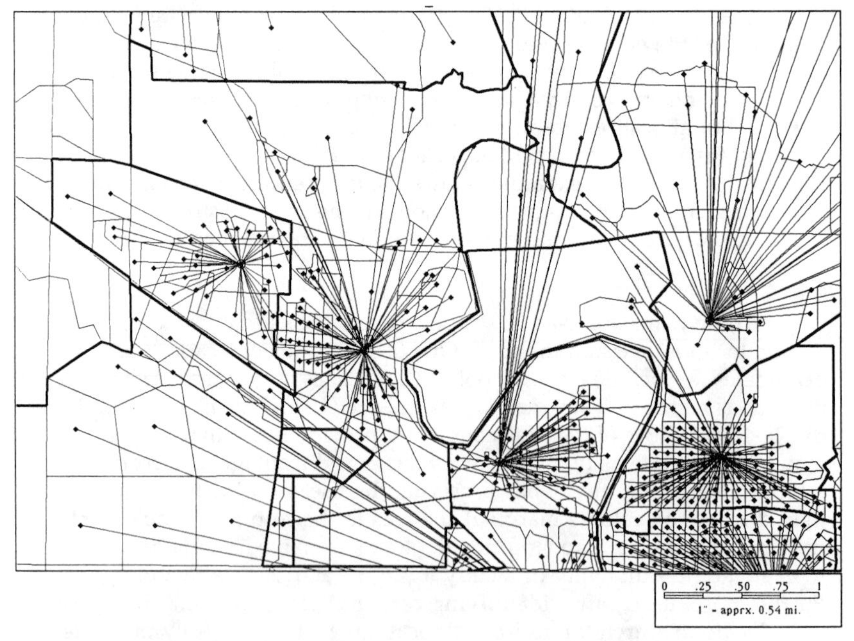

Figure 2. Spider map for NE Iowa City, 1990-91 enrollments.

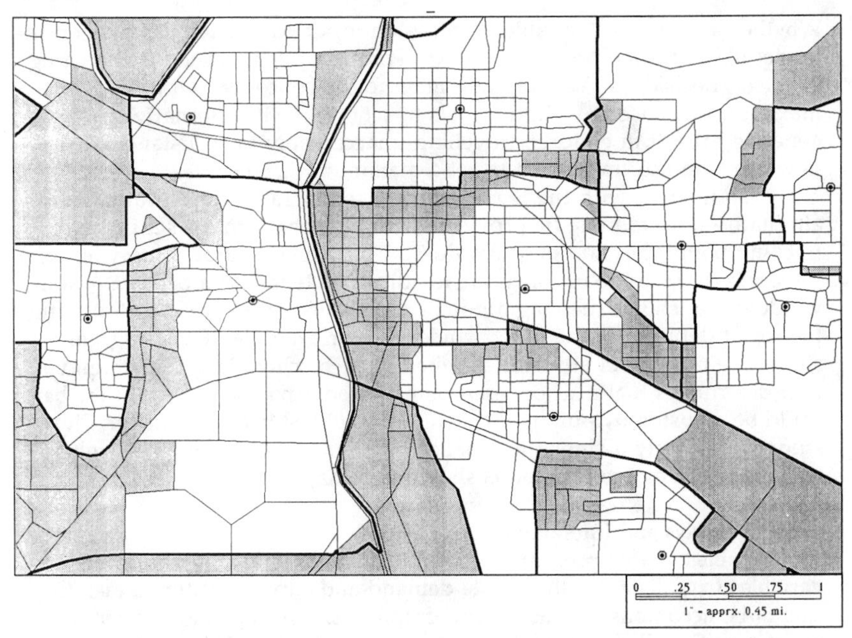

Figure 3. Delta map showing differences between the current and proximal assignment of students to schools.

When decision-makers adjust an existing solution, they normally will wish to either add, drop, or move a facility (Table 1). A user, for example, may wish to drag a school to a new candidate site and the system will determine which households are affected by the move, what reallocations of children to schools are necessary, and what effects the move will have on indices such as enrollments and grade profiles. When they wish to move a facility, decision-makers may choose to locate it at any candidate site, or restrict it to a specific candidate site or set of sites. Alternatively, they may wish to interactively adjust boundaries between service areas.

Graph Display and Interaction

Graphs are used to present different views of demand and supply attributes stored in the database. For example, one view may present several attributes of a demand location measured on the same scale (e.g. a bar graph of the number of students in each grade). A second view may present the values of an attribute (e.g. school enrollments) for several locations. With such a graph, decision-makers can obtain an idea about the values of attributes in a compact form. Note that more than one attribute may be displayed simultaneously with this type of graph (e.g. demand served and capacities of facilities).

Graphs also can be used to specify the values of input parameters to optimization models. Input parameters often define model constraints and represent characteristics of the problem in the abstract model space.

Whether a particular variable becomes an input parameter depends on the model used and the problem context. For example, in a school districting problem, decision-makers initially may not be concerned with the maximum distance any child travels to school. If the alternatives generated result in children travelling unacceptably long distances, however, the user may wish to set a maximum distance.

To interact with the SDSS during the generation and evaluation of alternatives, users must be provided with a mechanism for setting and tracking both input and output values. Consequently, our prototype system is designed to enable users to adjust the length of a bar in a graph window that shows the magnitude of a variable and to "readjust" it with the result depicted in the other windows. For example, if a histogram shows the number of students assigned to each school in a district, and a user determines that a school exceeds its stated capacity, the top of the bar could be adjusted by pulling it down to a level just below capacity. Thus, students formerly assigned to that school are now forced to go elsewhere, and the resulting reallocation is shown in a map and table.

Table Displays and Interaction

Tables enable presentation of information about the values of variables (e.g. distance) that relate demand and supply entities, a task that cannot be accomplished easily with graphs. The system allows users to select specific cells in the tabular data space and make changes that are reflected in the map and graph windows. For example, when the table contains information about facilities and candidates, the capacity of a school could be adjusted to examine the resulting districting scenario. Graduated symbols used to depict the capacities of schools (supply map) would reflect such changes made to the table. The operations on the table are done on a temporary (scratch) copy of the data, and a continuous log of each transaction is written to disk to maintain a record of the changes that have been examined by the user.

Interaction and Performance

An important assumption underlying this discussion is that computing technology provides sufficient performance to enable the described tasks to be accomplished in real time. Otherwise the benefits resulting from the development of an interactive system are reduced substantially (Lambert, 1984). Currently, real-time interaction has proven to be difficult since the locational models used by SDSS are computationally intensive. However, because these models contain components that can be logically disconnected and processed concurrently, we are developing the user interface and optimization algorithms in a parallel microcomputing environment using Transputers as MIMD processors (e.g. Healy and Desa, 1989) to speed task performance. Spreading the computational load simultaneously over several processors reduces solution times and increases user interactivity.

CONCLUSIONS

By using decision support environments that link modeling and visualization tools, decision-makers are better equipped to handle complex locational modeling problems. Our initial efforts to describe the tasks typically performed in districting problems have enabled the specification of this first attempt at SDSS interface design. The interface has been developed to support those tasks that are commonly performed in generating and evaluating alternative districting scenarios. Further work is needed to more fully specify these tasks and to uncover common structures shared by different application domains.

When decision-makers are able to visualize and evaluate the interplay between parameters, they gain new insight into the nature of trade-offs inherent in semi-structured problems. At present, such comparisons are essentially done in batch mode using hard copy maps, graphs and tables. The interface ultimately will allow users to interactively manipulate and see the effects of varying parameters in much the same way that a driver of an automobile with a manual transmission oversees the smooth interplay between clutch and accelerator. At present, however, we are limited in our ability to perform real-time analyses of combinatorially complex locational problems. Further research is needed to define the organization of data to support such actions, and the ways that application domains can be decomposed to enable the effective application of parallel processing to geographical problems.

ACKNOWLEDGEMENTS

We wish to acknowledge support provided by a grant from the National Science Foundation (SES-9024278). MPA wishes to acknowledge a broadening of insight obtained from those who attended the NCGIA Specialist Meeting on User Interfaces for GIS, especially the "Rasmussenians" in that group. Thanks to Claire E. Pavlik for comments on a draft of this paper.

REFERENCES

Apple Computer, Inc. 1985. *Inside Macintosh Vol I.* Reading, MA: Addison-Wesley. pp. 27-70.

Armstrong, M.P., Rushton, G., Honey, R., Dalziel, B.T., Lolonis, P., De, S., and Densham, P.J. 1991a. Decision support for regionalization: A spatial decision support system for regionalizing service delivery systems. *Computers, Environment and Urban Systems,* **15** (1): 37-53.

Armstrong, M.P., Densham, P.J., Rushton, G., and Lolonis, P. 1991b. Cartographic displays to support locational decision-making. Manuscript submitted for publication.

Densham, P.J. and Rushton, G. 1991. *Designing and implementing strategies for solving large location-allocation problems with heuristic methods.* NCGIA Technical Report 91-10, Santa Barbara, CA.

Guariso, G., and Werthner, H. 1989. *Environmental Decision Support Systems.* New York, NY: John Wiley.

Harris, B. 1988. Prospects for computing in environmental and urban affairs. *Computers, Environment and Urban Systems,* **12** (1): 3-12.

Healy, R.G. and Desa, G.B. 1989. Transputer based parallel processing for GIS analysis: Problems and potentialities. In *Proceedings of Auto-Carto 10*, pp. 90-99.

Honey, R., Rushton, G., Armstrong, M.P., Lolonis, P., Dalziel, B.T., De, S. and Densham, P.J. 1991. Stages in the adoption of a spatial decision support system for reorganizing service delivery regions. *Environment and Planning C: Government and Policy,* **9** (1): 51-63.

Kuhn, W. 1991. Are displays maps or views? In *Proceedings of Auto-Carto 10*, pp. 261-274.

Lambert, G.N. 1984. A comparative study of system response time on program developer productivity. *IBM Systems Journal*, 23 (1): 36-43.

Langendorf, R. 1985. Computers and decision making. *American Planning Association Journal*, Autumn: 422-433.

Laurel, B. (ed.) 1990. *The Art of Human-Computer Interface Design*. Reading, MA: Addison-Wesley.

Mark, D.M. 1989. Cognitive image-schemata for geographic information: relations to user views and GIS interfaces. *Proceedings of GIS/LIS '89*, pp. 551-560.

Monkhouse, F.J. and Wilkinson, H.R. 1971. *Maps and Diagrams: Their Compilation and Construction*. London, UK: Methuen.

Myers, B. 1985. The importance of percent-done progress indicators for computer-human interfaces. *Proceedings of the Conference on Human Factors in Computing Systems*, pp. 11-17.

Peterson, K. 1990. Toward the specification of a spatial decision support system for real estate investment analysis. In *Proceedings, GIS/LIS '90*, Volume 2. Bethesda, MD: American Congress on Surveying and Mapping, pp. 467-475.

Rasmussen, J. 1986. *Information Processing and Human Machine Interaction: An Approach to Cognitive Engineering*. New York, NY: North-Holland.

Schneiderman, B. 1986. *Designing the User Interface: Strategies for Effective Human-Computer Interaction*. Reading, MA: Addison-Wesley.

Tognazzini, B. 1990. Consistency. In B. Laurel (ed.) *The Art of Human-Computer Interface Design*. Reading, MA: Addison-Wesley, pp. 75-78.

Turk, A.G. 1990. Towards an understanding of human-computer interaction aspects of geographic information systems. *Cartography,* **19** (1): 31-60.

DEVELOPING GIS DATA LAYERS FOR ESTUARINE RESOURCE MANAGEMENT

William H. Jefferson, William K. Michener, David A. Karinshak
Belle W. Baruch Institute
for Marine Biology and Coastal Research
University of South Carolina
Columbia, SC 29208 USA
(803) 777-3926

William Anderson and Dwayne E. Porter
South Carolina Marine Resources Center
P.O. Box 12559
Charleston, SC 29412 USA
(803) 795-6350

ABSTRACT

Past efforts to manage many fish and shellfish resources have been based primarily on catch and effort statistics. Resource management tools which incorporate the spatial distribution and abundance of a resource can significantly enhance management decisions. Specific applications for the management of estuary-wide and regional shellfish populations, harvested both commercially and recreationally, are needed. Geographic Information System (GIS) techniques have only recently been adopted for coastal resource management efforts. As part of a larger study designed to examine and mitigate the impacts of urbanization on southeastern US estuaries and salt marshes, GIS data layers were developed to characterize the distribution and size of oyster reefs present within the Murrells Inlet Estuary, SC. In addition, population recruitment patterns of oysters throughout the estuary were examined in 1990/1991. Global Positioning System (GPS) technology, spatial search and overlay operations, and a broad-scale sampling program were utilized to analyze oyster reef distributions and document spatial recruitment patterns within the estuary. The distribution of established oyster populations and oyster reef recruitment were related to hydrography and adjacent land use patterns. The integration of GIS, GPS and innovative sampling strategies can provide an invaluable decision support system for the future management of coastal resources.

INTRODUCTION

Southeastern estuaries are valuable resources for the recreational and commercial fish and shellfish industries. Increased tourism, greater utilization of the fish and shellfish resources, and impacts associated with coastal development have generated the need for management tools that support management and enhancement of these valuable resources at a relevant spatial scale. Efforts to manage the shellfish industry in South Carolina, have relied primarily on catch and effort statistics (Burrell, 1982; Maggioni et al., 1982; Low et al., 1987; Low, 1989). The construction of marinas and associated dredging directly affect shellfish populations by physically removing suitable growing areas (Wendt et al., 1990). Point and non-point source pollution from increased land use and development have affected the hydrographic properties and water quality of estuarine environments stressing nearby shellfish areas (Swearingen and Marcus, 1983; Newell, 1990; Wendt et al., 1990).

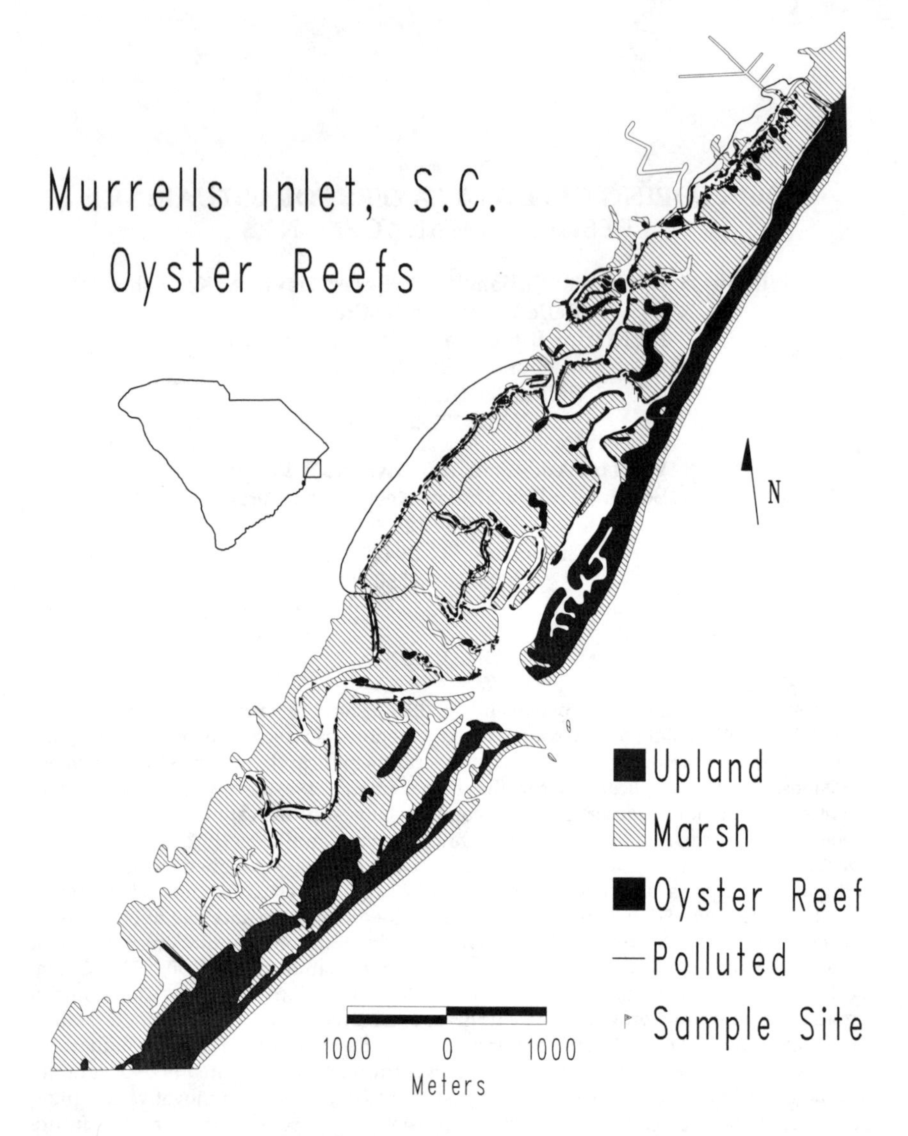

Figure 1. Map showing location of Murrells Inlet Estuary and its position along the South Carolina coast. All oyster reefs (surveyed) and polluted areas are depicted.

Geographical Information System (GIS) technology has recently been adopted as a tool for use in coastal and estuarine resource management (Michener et al., 1989). A GIS provides an automated mechanism to incorporate vast amounts of land use, development, water quality and biological data into one streamlined decision support system. The development of accurate GIS data layers can assist the long-term management of commercially and recreationally valuable resources. The mapping of oyster reefs, marinas and potential pollution sources, utilizing Global Positioning System (GPS) techniques in a GIS environment, provides resource managers with a means to assimilate and analyze estuarine resource data in relation to projected or potential environmental impacts. The incorporation of ancillary data (e.g., docks, bulkheads, land use, etc.) provides the means to monitor change and develop management strategies to promote healthy, productive resources among the pressures of increased land use and development.

This paper discusses the development of GIS data layers, using GPS ground truthing, that were produced as part of a study to examine the effects of urbanization on southeastern estuaries. Data layers included the location and characterization of all oyster reefs, land use patterns, marinas and sites of point and non-point source pollution. Oyster recruitment and growth were examined throughout the Murrells Inlet Estuary. Patterns in recruitment and growth were related to reef quality, proximity to existing marinas and other environmental factors. Other potential applications are described.

STUDY AREA

Murrells Inlet is a shallow, ebb tidal dominated coastal inlet located on the northern South Carolina coast (33°32'N, 79°2'W) (Figure 1). There is no riverine input and the estuary is characterized by vertically homogenous high salinity water (31.4 ppt, 20°C; Blood, unpubl. data). Precipitation for 1989 averaged 130 cm, with an annual average temperature of 18°C (Blood and Vernberg, in press). Murrells Inlet Estuary is surrounded on all sides by development, except the southeastern edge, which is adjacent to Huntington Beach State Park. Six marinas are located throughout the estuary. Periodic dredging has taken place to maintain navigability of the creek channels. Murrells Inlet is heavily utilized by both commercial and recreational fishermen.

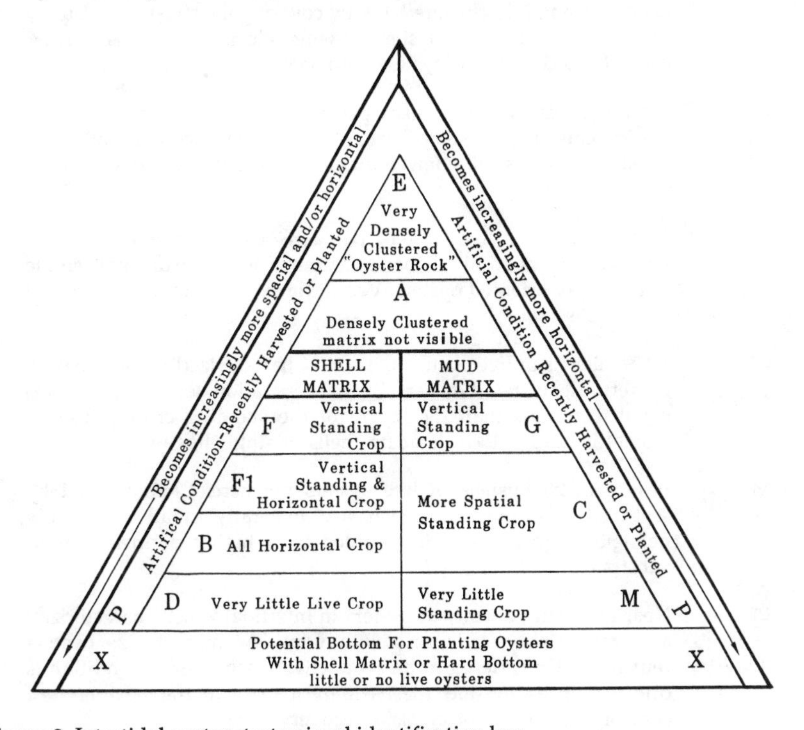

Figure 2. Intertidal oyster strata visual identification key.

STRATA	DESCRIPTION
"A"	12,093 bushels of live oysters per acre. Greatest yield per acre of densely clustered live oysters. Exhibits little exposed dead shell or mud and the shell matrix is not visible.
"B"	2,608 bushels of live oysters per acre. Characterized by having no vertical clusters in the standing crop. Found mostly in the lower intertidal zone, oysters are frequently single. Located on heavily shelled grounds with thin shell matrices.
"C"	1,957 bushels of live oysters per acre. Characterized by vertical clusters with spatial separation. Substrate is usually mud with little or no surrounding shell. Spatial separation between clusters ranges from a distance equal to the height of an individual cluster to approximately one meter.
"D"	295 bushels of live oysters per acre. Characterized by scattered live oysters usually integrated with large quantities of "washed" or dead shell. Found in the lower intertidal zone on hard substrate. Hard clams are found sympatrically in this area.
"E"	7,277 bushels of live oysters per acre. Characterized by overgrowth. Oysters are tightly clustered, totally covering the substrate. Usually found at the highest oyster growing elevation and is further characterized by small oysters with sharp, thin shells.
"F"	4,021 bushels of live oysters per acre. Characterized by mostly vertical clusters of oysters. Similar to "C" strata, except the substrate consists of shells with few horizontal live oysters and very little mud.
"F1"	1,926 bushels of live oysters per acre. Characterized by small, vertical clusters evenly dispersed within a substrate of small, single horizontally oriented oysters. Very little exposed mud is associated with "F1" strata.
"G"	5,199 bushels of live oysters per acre. Characterized predominantly by vertical, clustered oysters. Spatial separation between clusters is equal to or less than the height of the standing crop. Substrate habitat is mud with little or no shells or single live oysters.
"M"	Less than 20 bushels of live oysters per acre. Characterized by scattered live oysters, which are generally small and show negligible aggregation. Surrounded by a highly permeable mud substrate.
"P"	Near minimum density of oysters in intertidal zone. Characterized as recently harvested areas with very few market size oysters remaining. Considered to be productive since present condition is due to harvesting and they will propagate to the next highest category by natural or artificial recovery.

Table 1. Intertidal oyster strata descriptions.

GIS DATA LAYER DEVELOPMENT AND ANALYSIS

Primary GIS Data Layers

A National Wetlands Inventory (NWI) map of the Murrells Inlet Estuary was used as the base layer. The NWI map, in ARC/INFO digital format, was provided by the South Carolina Land Resources Commission. The study area was a subset of the larger map. Habitat classes were aggregated to water bodies, lowland and upland areas for purposes of illustration.

The South Carolina Wildlife and Marine Resources Department conducted an Intertidal Oyster Survey for the Murrells Inlet Estuary during 1982. Each reef was located on a 7.5' topo quad, and the oyster population associated with each reef was classified according to strata characteristics (Table 1), using the intertidal oyster strata visual identification key (Figure 2). The strata differ with respect to reef structure, substratum characteristics, presence or absence of clams, depth of shell matrix, density of live oysters and proportion of shell to live oysters (Table 1, Figure 3). All oyster reefs examined during this survey are included in Figure 1. Ancillary data (bottom type, size of oysters, shell matrix depth, elevation, strata, water depth and reef area) for all reefs in the intertidal zone were collected during the survey. The reefs were then given unique identification codes, which were used to identify the reef polygons in the ARC/INFO data layers.

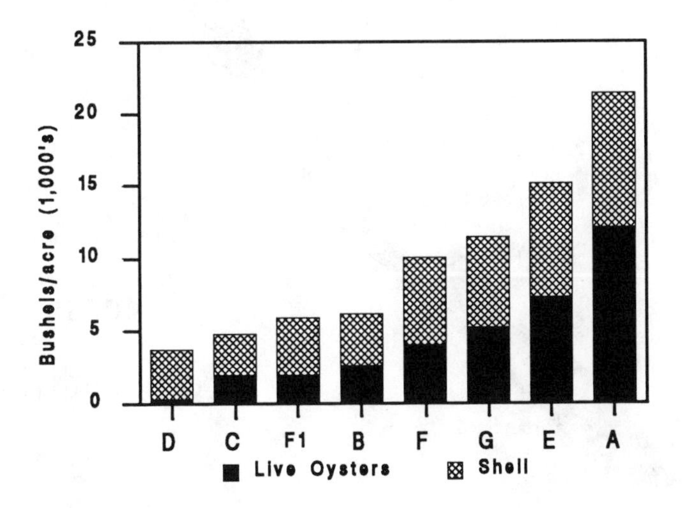

Figure 3. Bar chart showing the amount of live oysters and shell present in the different strata.

Oyster Recruitment and Spat Size

Plastic tubes with lengthwise mottled grooves were used as artificial, vertical settling substrata during this study. The tubes were obtained from IEC Collaborative Marine Research and Development Limited (Broadley, 1989) in 2 m sections and cut into 33 cm sections. All tubes were maintained in aquaria with flowing, filtered seawater for at least one month prior to deployment in the field. Tubes were only used once during this study. The tubes were hammered into the mid-intertidal zone of oyster reefs, with approximately 15 cm remaining exposed above the sediment to insure that the tubes

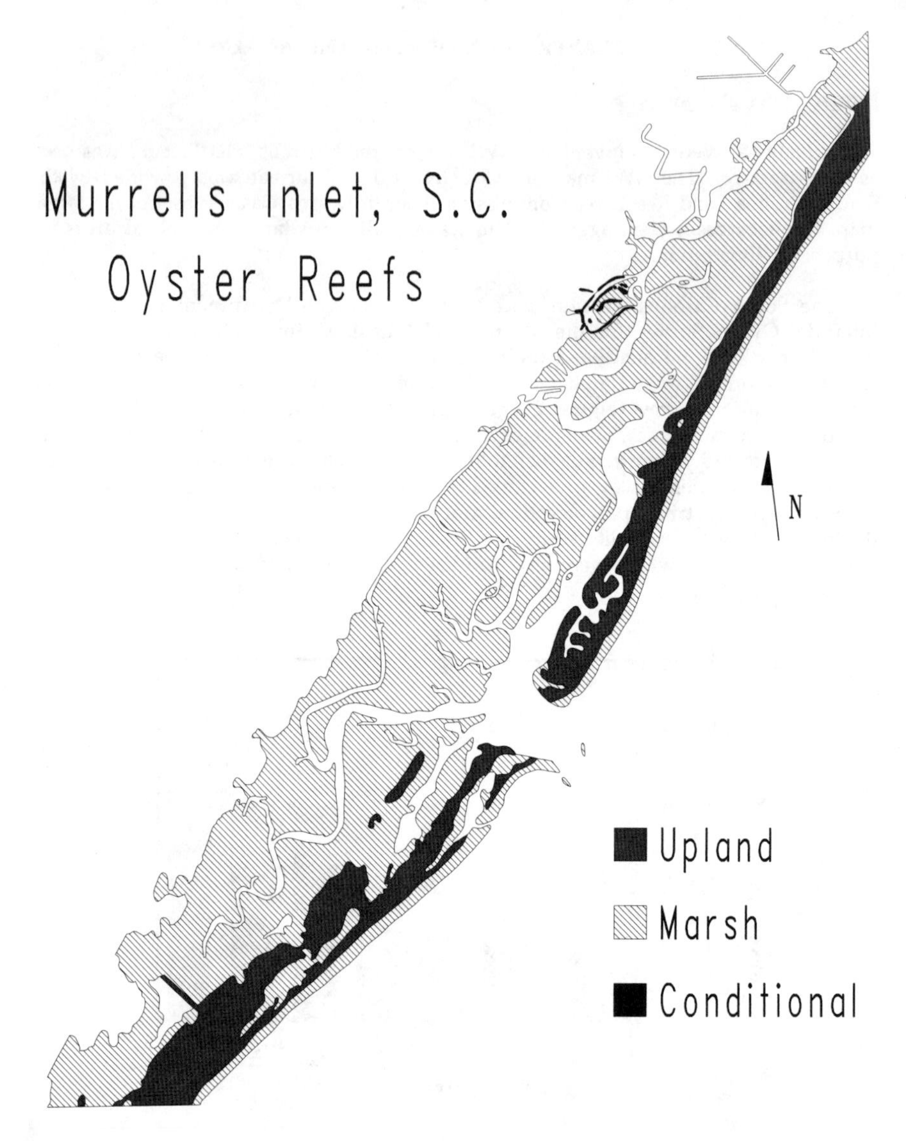

Figure 4. Map of Murrells Inlet Estuary showing oyster reefs located within conditionally polluted waters.

were embedded firmly in the oyster reefs. Tubes were placed approximately every 50 m along the ocean side of every creek in the Murrells Inlet Estuary during March 1990.

Tubes were collected the following year (March 1991), transported to the lab, and gently rinsed with water to remove sediment. Oyster recruitment and spat size were monitored by enumerating and measuring all oyster spat that occurred on each settlement tube. Observations were made with a dissecting microscope (125X) on exposed portions of tubes. Each oyster's length, width and height above sediment were measured with vernier calipers.

Each settlement tube was located using a remote GPS Pathfinder unit, along with a base GPS unit which was maintained on a known benchmark. GPS receivers record and decode ephemeris and time information transmitted by NAVSTAR satellites. Two GPS units were necessary to factor out selective availability (SA, a source of error introduced by the Department of Defense). Differential corrections were applied to remove the error introduced by SA, and average position accuracies of 2.2 m were obtained. The tube positions, in latitude and longitude, were converted to seconds and then projected in ARC/INFO to form a data layer in Universal Transverse Mercator (UTM) coordinates. Generally, tubes were assigned codes which corresponded to the oyster reef in which they were placed. In some cases, when tube positions did not correspond to a surveyed reef, they were not assigned codes. Oyster size and recruitment data were incorporated into the GIS database. Subsequent spatial search and overlay procedures were performed in ARC/INFO.

RESULTS AND DISCUSSION

Incorporation of field and survey data into the GIS allowed us to readily identify and characterize reefs located in polluted, conditionally polluted or non-polluted waters. Oyster reefs in polluted waters (Figure 1) were located near marinas, areas of extensive boat traffic and urban runoff from service industries and high density housing (e.g., large trailer park located along the northern border of Murrells Inlet). Reefs in conditionally polluted waters (Figure 4) were located near a small marina and a point source of intermittent storm runoff from an adjacent urban watershed. Since most of the viable oyster populations were located in non-polluted or conditionally polluted waters, the remainder of the discussion is restricted to considering reefs in these two zones. Reefs characterized as belonging to the "B" stratum accounted for almost 33% of the total area occupied by reefs and 45% of the oysters present in non- or conditionally polluted waters (Table 2). Conversely, "D" and "M" strata oyster reefs made up 39% of the total number of reefs and occupied approximately 30% of the total reef area, but contained less than 4% of the total live oysters.

Strata	# of Reefs	Mean Area	% Total (ft^2)	% Total area	# of Bushels
"A"	none	—	—	—	—
"B"	34 (13.4%)	13,620	32.6	44.6	27,725
"C"	27 (10.6%)	4,480	8.5	8.7	5,435
"D"	82 (32.3%)	4,268	24.6	3.8	2,370
"E"	none	—	—	—	—
"F"	28 (11.0%)	4,460	8.8	18.6	11,530
"F1"	34 (13.4%)	3,659	8.7	8.9	5,501
"G"	12 (4.7%)	6,666	5.6	15.4	9,547
"M"	17 (6.7%)	4,104	4.9	<0.1	<32
"P"	20 (7.9%)	4,459	6.3	—	—
Totals	254	—	100.0	100.0	62,140

Table 2. Statistical characteristics of oyster reef strata located in nonpolluted or conditionally polluted waters in Murrells Inlet Estuary, SC.

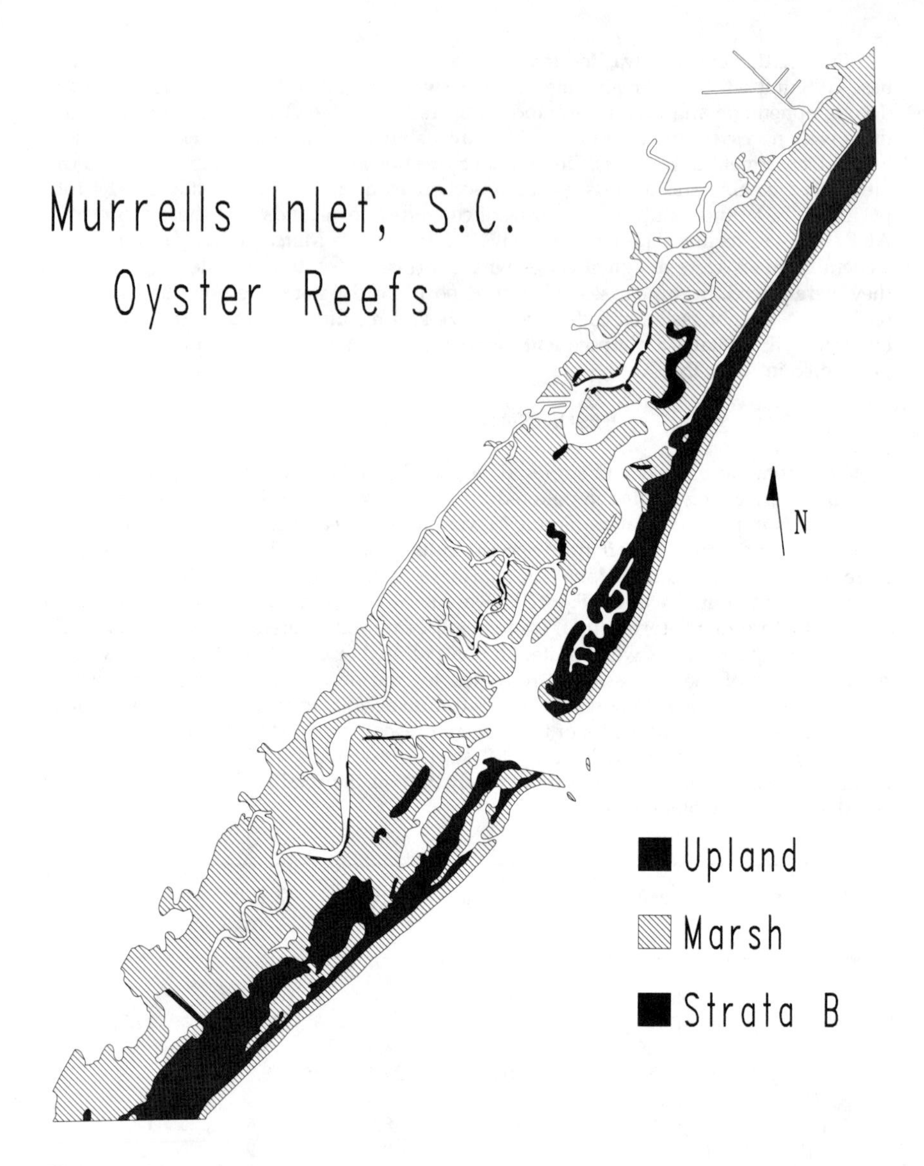

Figure 5. Map of Murrells Inlet Estuary showing "B" strata oyster reefs (areas of projected high population recruitment).

Oyster recruitment was highest on reefs characterized as belonging to the "B" stratum (Table 3; Figure 5). These reefs were relatively large and were not located in areas of high boat traffic. Furthermore, they were generally not located in the immediate vicinity of marinas or other more highly polluted waters. Many were located on commercially leased ground.

Low recruitment and small oyster size after one year were associated with reefs belonging to the "P" and "M" strata (Table 3, Figure 6). Generally, these were small reefs

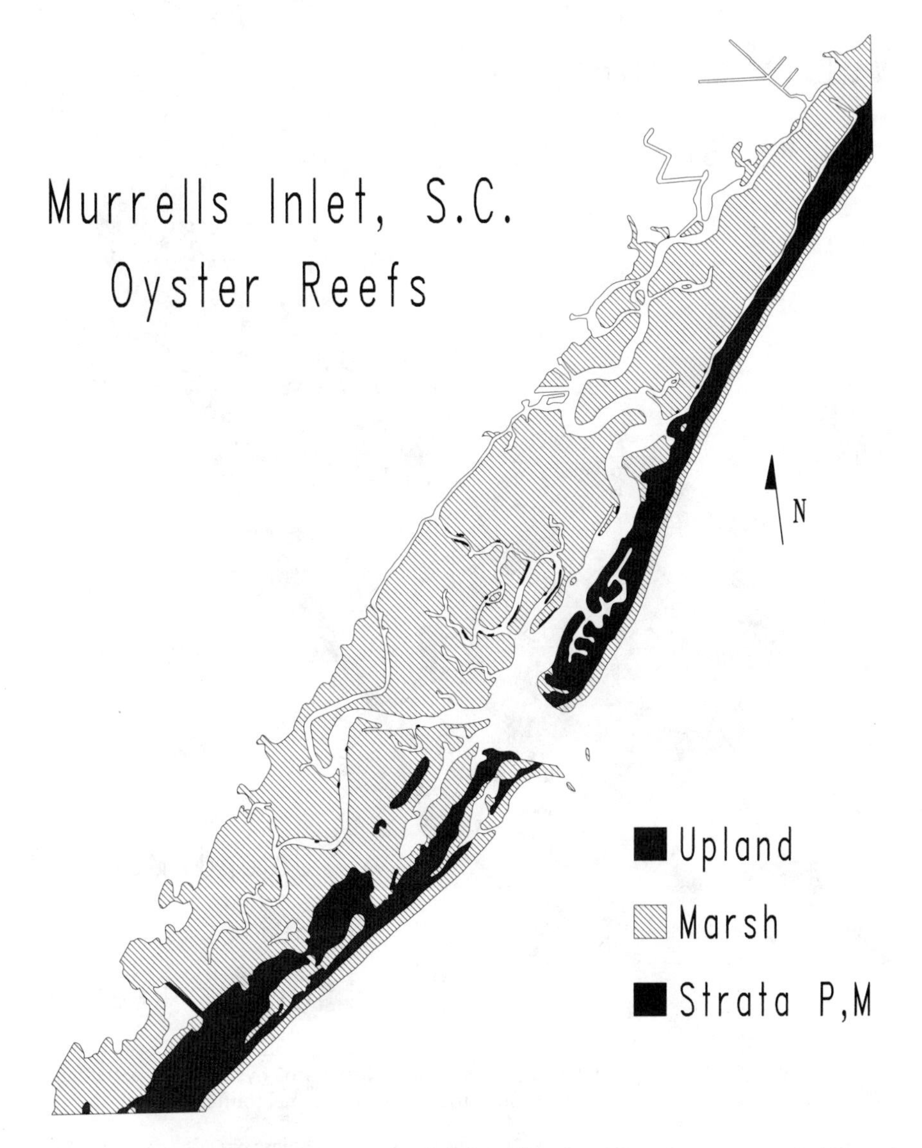

Figure 6. Map of Murrells Inlet Estuary showing "P" and "M" strata oyster reefs (areas of projected low population recruitment and low growth).

which were located in shallow portions of creeks. These areas were characterized by muddy substratum and minimum densities of live oysters (Table 1).

The data layers developed as a result of the field and survey efforts can provide resource managers with the information necessary to address many questions. For example, reefs located in areas of high recruitment (e.g., "B" stratum) could be used to provide a seed source for other reefs which have been overharvested or damaged by dredging or other activities.

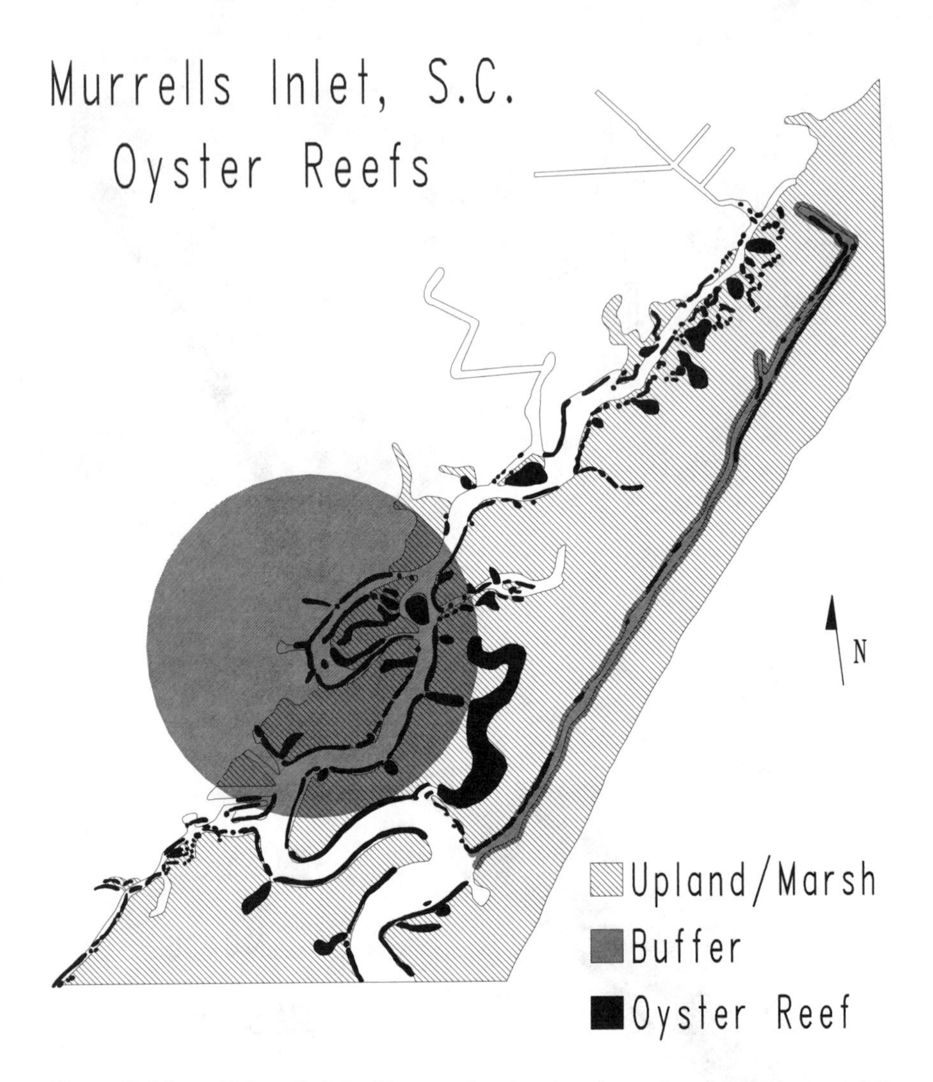

Figure 7. Map of Murrells Inlet Estuary showing locations of oyster reefs potentially affected by dredging a small marina basin (circle) or boat channel (buffer strip).

Spatial search and overlay procedures available in the GIS allow us to address potential impacts of different development scenarios. For illustrative purposes, a buffer strip was created along a shallow boat channel and a 750 m buffer zone was drawn around a site (small marina) where maintenance dredging may be required (Figure 7). Under these scenarios, analyses indicated that 52 oyster reefs comprising 154,000 m^2 could potentially be affected by dredging the small marina site. Similarly, 22 oyster reefs comprising 43,000 m^2 could be affected by dredging the boat channel. These analytical procedures support numerous potential applications. For example, similar techniques could be used to site new docks or bulkheads in areas where there would be minimal impact on healthy, productive fish or shellfish resources.

Strata	Number of Tubes/Strata	Average Recruitment/Tube	Average Oyster Size (length, mm)
"B"	9	21.1	23.0
"C"	3	19.5	31.3
"D"	14	20.5	26.7
"F"	6	14.8	24.2
"F1"	1	18.0	19.0
"M"	2	16.0	17.6
"P"	4	11.8	12.5

Table 3. Density and oyster size on tubes from different strata.

The GIS developed for this project provided an efficient, organized way to manage large amounts of spatial data. Such a system is particularly suited to estuarine resource management. Data layers can continuously be developed and added to a GIS providing a means to monitor long-term development, land use and the potential effects on the overall health of valuable resources. Spatial scale is becoming increasingly important in estuarine resource management and research. A GIS provides a standard technique for data storage, interpretation and representation. The incorporation of GPS technology enables research sites, resource locations, developed areas, pollution sources and other geographical features to be accurately located and incorporated into GIS data layers. This provides resource managers with a means to make educated management decisions to protect and enhance commercially and recreationally valuable resources amidst increasing development and urbanization pressures.

ACKNOWLEDGEMENTS

Funding for this project was provided by NSF grant BSR-8514326 and NOAA grant NA90AA-D-SG672. The South Carolina Coastal Council provided funding for the intertidal oyster survey. Digitization and editing of the basemap and data layers was performed by Sharon J. Lawrie. Jim Monck, Mike Yianopoulos and Ray Haggerty completed the intertidal oyster resource assessment in the study area. F. Danny Spoon assisted in the field and laboratory components of this study. GPS Pathfinder is a trademark of Trimble Navigation, Limited. ARC/INFO is a trademark of Environmental Systems Research Institute, Inc. ERDAS is a trademark of ERDAS, Inc. This paper is Contribution Number 889 of the Belle W. Baruch Institute for Marine Biology and Coastal Research.

REFERENCES

Blood, E.R. and F.J. Vernberg., in press. Characterization of the physical, chemical, and biological conditions and trends in Winyah Bay and North Inlet Estuaries: 1970-1985, In, Characterization of the Physical, Chemical, and Biological Conditions and Trends in Three South Carolina Estuaries, SC Sea Grant Consortium, NOAA.

Broadley, T.A., 1989. The Remote Setting of Oyster Larvae: Lecture/Laboratory Handbook, IEC Collaborative Marine Research and Development Limited, Victoria, British Columbia, 61 p.

Burrell, V.G., 1982. Overview of the South Atlantic oyster industry, Proceedings of the North American Oyster Workshop, World Mariculture Society, Special Publication 1:125-131.

Low, R.A., 1989. South Carolina Marine Fisheries, 1987-1989, South Carolina Marine Resources Dept. Data Report 6, 50 p.

Low, R.A., D. Theiling and E.B. Joseph, 1987. South Carolina Marine Fisheries, 1977-1986, South Carolina Marine Resources Center Technical Report Number 67, 78 p.

Maggioni, G.J. and V.G. Burrell, 1982. South Carolina Oyster Industry, Proceedings of the North American Oyster Workshop, World Mariculture Society, Special Publication 1:132-133.

Michener, W.K., D.J. Cowen and W.L. Shirley, 1989. Geographic Information Systems for Coastal Research, Proceedings of Sixth Symposium on Coastal and Ocean Management/ASCE, 4791-4805.

Newell, C.L., 1990. Area IV Sanitary Survey, Shellfish Sanitation Program, South Carolina Department of Health and Environmental Control, 69 p.

Swearingen, G.R. and J.M. Marcus, 1983. A Water Quality Assessment of Marina Activities at Murrells Inlet, Georgetown County, South Carolina, South Carolina Department of Health and Environmental Control Technical Report No: 027-83, 50 p.

Wendt, P.H., R.F. Van Dolah, M.Y. Bobo and J.J. Manzi, 1990. Effects of Marina Proximity on Certain Aspects of the Biology of Oysters and Other Benthic Macrofauna in a South Carolina Estuary, South Carolina Marine Resources Center Technical Report Number 74, 50 p.

GIS ASSESSMENT OF LARGE-SCALE ECOLOGICAL DISTURBANCES (HURRICANE HUGO, 1989)

William K. Michener, Elizabeth R. Blood, L. Robert Gardner, Björn Kjerfve, Mary E. Cablk, Catherine H. Coleman, William H. Jefferson, David A. Karinshak, F. Danny Spoon

Belle W. Baruch Institute
for Marine Biology and Coastal Research
University of South Carolina
Columbia, SC 29208 USA
(803) 777-3926

ABSTRACT

Hurricane Hugo struck the SC coast near Charleston on 22 September 1989. Landsat Thematic Mapper (TM) imagery, aerial reconnaissance missions (low altitude color infrared photography) and ground truth surveys along 10 forest transects were used to assess hurricane related impacts to the barrier island system, salt marsh and a large forested region bordering the North Inlet Long-Term Ecological Research site (near Georgetown, SC). Analyses indicate that the impacts on undeveloped barrier islands and back-barrier marshes were mild. Adjacent low-lying maritime forest, however, experienced extensive damage due to wind stress and salt water intrusion. The utility of the data sources for assessing hurricane-related impacts varied and generally could be related to the spatial and spectral resolution of the sensors as well as the size of the area affected and type of damage. Whereas satellite imagery and aerial photography revealed similar extent and magnitude of the forest damage, geomorphic modification of the barrier island system was best discriminated from aerial surveys. Ground observations were necessary for ascertaining the damage to forest understory species and interpreting causal mechanisms of the acute and chronic impacts of this large-scale ecological disturbance.

INTRODUCTION

Hurricane Hugo made landfall at Charleston, SC at 0001 hours EDT on 22 September 1989 (Figure 1). The hurricane, which started as a depression off the west coast of Africa on 9 September, was the most devastating storm to hit South Carolina in modern times. Hurricane Hugo reached its maximum strength several hundred kilometers east of the Leeward Islands on 15 September, with a central pressure of 918 mb and observed surface wind speed of 72 m s^{-1} (National Hurricane Center, 1989). Hugo caused extensive destruction on the island of Guadeloupe on 17 September and Puerto Rico the following day. The hurricane lost some strength as it moved away from Puerto Rico toward the southeastern United States, but gathered renewed energy as it crossed the Gulf Stream. Hugo advanced toward the South Carolina coastline at a rapid 12 m s^{-1}, exhibited a central pressure of 935 mb and had measured sustained winds of 39 m s^{-1} with gusts up to 48 m s^{-1} in Charleston upon landfall (National Hurricane Center, 1989).

Fifty hurricane-related deaths were reported in the Carolinas and 129,687 families had their homes either destroyed or damaged (Brennan, 1991). In South Carolina,

Hurricane Hugo caused $6 billion damage ($500 million, flooding; $1.2 billion, forestry and agriculture; and $4 billion in property damage; Sparks, 1991). Of South Carolina's 46 counties, 23 experienced timber damage (7, extensive; 3, moderate; 13, light). The 1.8 million ha of forest damaged in South Carolina during the hurricane exceeds the area affected by the Mount St. Helens eruption (60,750 ha) and the 400,000 ha burned during the Yellowstone fires of 1988 (Hook et al., 1991).

Hurricane Hugo resulted in extensive forest damage (both coastal and inland) as well as modification of coastal barrier islands. In this paper, we report results of a multi-disciplinary effort to examine hurricane-induced modification of a large forested tract (Hobcaw Barony), an undeveloped barrier island (North Island) and the North Inlet salt marsh system, all located northeast of the point of landfall and within a zone of widespread damage caused by the wind and storm surge. We further examine some of the mechanisms which account for the acute and chronic forest damage, and characterize the applicability of different data sources (satellite imagery, aerial photography and ground observations) for discriminating hurricane-related impacts.

MATERIALS AND METHODS

STUDY AREA

North Inlet, located 90 km northeast of Charleston (Figure 1), is one of 18 sites in the Long-Term Ecological Research (LTER) network sponsored by the National Science Foundation (Callahan, 1984; Franklin et al., 1990). The site covers about 80 km^2 and consists of barrier islands, intertidal salt marsh and low-lying coastal forest (Hobcaw Barony). The site has been the focus of studies on coastal ecology, forestry and geology for over 20 years (Miller et al., 1989). An extensive long-term database (>10 years) containing hundreds of biological, chemical and physical parameters has been developed (Michener et al., 1990).

The primary research areas are a 2,630 ha high salinity *Spartina alterniflora* marsh and 715 ha of tidal creeks and intertidal flats which are separated from the Atlantic Ocean by sandy barrier islands. The estuary is bordered on the west by loblolly and long leaf pine forests. Hydrographic characteristics of the North Inlet estuary include an annual average seasonal salinity range of 30 to 34 ppt (monthly mean salinities average 19 to 36 ppt), average channel depth of 3 m and a seasonal water temperature range of 3° to 33° C. Wetland habitats include exposed and sheltered sandy beaches; marsh; intertidal flats and oyster beds; submerged macroalgal mats; sand, shell, and mud benthic habitats; shell middens; and bird rookery islands. At mean tide, *Spartina alterniflora* marsh comprises 73.0 percent, tidal creeks 20.6 percent, oyster reefs 1.0 percent and exposed mud flats 5.4 percent of the marsh-estuarine zone (Dame et al., 1986). More than 1,200 ha of brackish and freshwater marshes border the Winyah Bay side of Hobcaw Barony.

The forest ecosystem at North Inlet has developed on sandy Pleistocene beach ridges and swales, which trend northeast-southwest (Gardner and Bohn, 1980). The highest ridges lie about 6 m above mean sea level (MSL). The forest system primarily consists of variable-age pine forest growing on the ridges. The swales between ridges typically contain intermittent blackwater streams, support stands of cypress and gum, and have fringes of mixed bottomland hardwoods.

REMOTE SENSING

The Hobcaw Barony area was subset from the first cloud-free Landsat Thematic Mapper (TM) scene (path 16, row 37) available after the passage of Hurricane Hugo (3

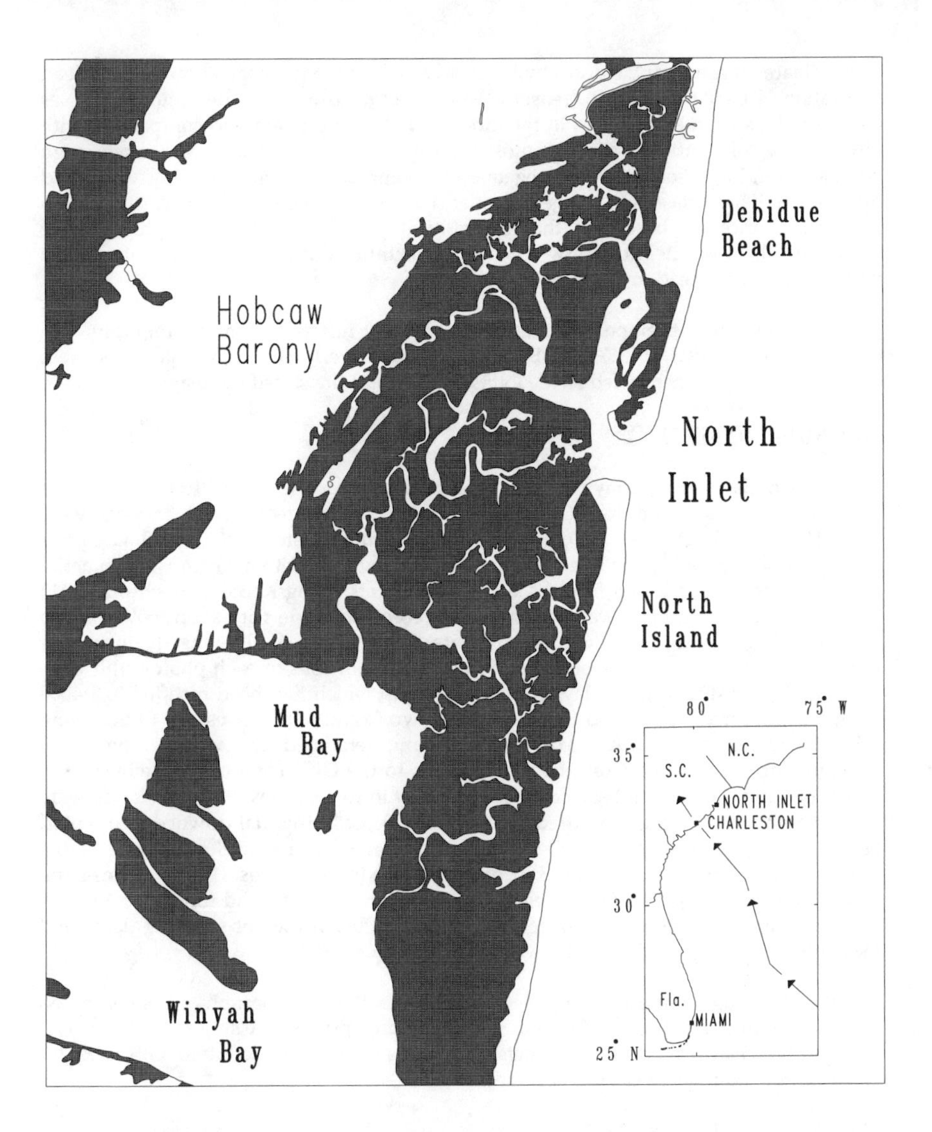

Figure 1. Map showing path of Hurricane Hugo and geographic features of the North Inlet salt marsh-estuarine system.

October 1989). The area was rectified to a Universal Transverse Mercator (UTM) coordinate system using ground control points taken from a Geographic Information System (GIS) layer of the study site. Atmospheric scattering due to haze was corrected for by subtracting the bias based on histograms from the data in the visible bands, Landsat TM 1,2,3 (Jensen, 1986). After rectification, the study area was extracted from the image for further analysis. Dredge spoil areas on the west side of the property and the DeBordieu golf course were excluded.

Bands 4,3,2 of the remaining data were clustered using STATCLAS, a statistical clustering algorithm in ERDAS, and classified using a maximum likelihood classifier. This combination of bands were chosen since the near-infrared band was able to discriminate (based on chlorophyll absorbance) healthy vegetation from dead vegetation or background soil (Jensen, 1986). The image was grouped into ten classes based on the spectral signatures of the forest and adjacent marsh and comparison with low altitude color infrared aerial photos: healthy, low damage, medium damage, high damage, standing dead, sparsely vegetated, barren/road, marsh, standing water and open water. These classes were then recoded into seven classes which correspond to a similar classification based on the low altitude color infrared aerial photography (described below): healthy, low damage, medium damage, high damage, salt damage/standing dead, marsh and barren/road.

The data layer developed from low altitude color infrared aerial photography was converted from ARC/INFO into ERDAS for map overlaying. Class statistics were derived for both the converted ARC/INFO overlay and the classified image.

AERIAL PHOTOGRAPHY

Three aerial photography surveys were conducted to assess the effects of Hugo on the geomorphology and forest resources of the study area. Aerial surveys were conducted in early October 1989, early February 1990 and late October 1990, just prior to leaf fall on the deciduous trees. Photographs were taken with a 23 cm x 23 cm aerial camera suspended in a pod from a Cessna 150 aircraft, using Kodak Aerochrome 2443 infrared film. Photographs were taken in a stereoscopic mode with about 60% overlap and at an altitude that gave an approximate scale of 1:6000, allowing distinction of individual uprooted and broken trees. Scale correct portions of each photograph were obtained by constructing overlays from a base map which had been rectified to 1:4800 scale cadastral maps produced for the Georgetown County Tax Assessor's Office using ARC/INFO GIS software. A map of discolored and defoliated trees was then produced by digitizing information from the photographs to the GIS. These maps delineate the extent of wind damage and salt-induced mortality in tree species. Healthy stands were devoid of fallen trees and appeared red on the photos. Individual downed trees could be distinguished on the photos. Salt-stressed trees appeared yellow-brown on the photos. Three damage classes could be discriminated in the photos: (1) Low - consisting of stands with up to 25% of the trees blown down or standing dead (broken crowns or salt damage); (2) Medium - characterized by 25-75% blowdown or standing dead; and (3) High - consisting of 75-100% blowdown or standing dead.

For mapping geomorphic changes, the October 1989 photographs were compared with true color aerial photos of similar scale taken in April 1987. Using a GIS base map, geomorphic features on the two sets of photos were digitized into GIS files for rectification and adjustment to a common scale.

GROUND OBSERVATIONS

To study the effects of soil salinization on forest vegetation, ten sampling transects were established around the perimeter of Hobcaw Barony (Figure 2). Each transect starts at the boundary between forest and high marsh (0 m inland) and extends inland up to the approximate limit of surge penetration (3.0 m contour). Most transects have one or more stations in the first 30 m (0, 3, 10, and 25 m). Thereafter stations are typically located at 25, 50, or 100 m intervals. The elevations along each transect were surveyed with a transit and compass and leveled with respect to known bench marks. Sampling along transects consisted of measurements of water table depths, chemical analysis of ground water samples and descriptions of the type and condition of the vegetation at each station. Water table depths were obtained by augering holes at each station and allowing water to rise to a static level. Samples of ground water from each hole (or standing surface water in some cases) were taken in acid-cleaned scintillation

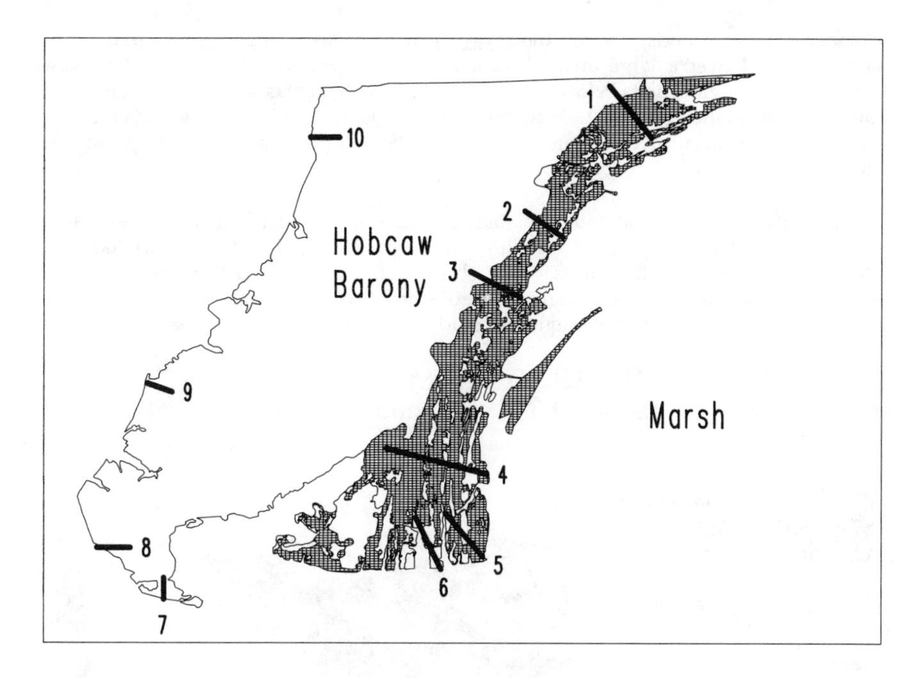

Figure 2. Map showing extent of salt damaged vegetation on Hobcaw Barony and locations of study transects. Shaded area represents that portion of the forest located approximately within the 3 m contour on the eastern side of Hobcaw Barony.

vials. The first transect sampling was conducted two weeks after Hugo. Subsequent samplings were conducted approximately 4, 8, 16, 32 and 52 weeks after the storm. During the first sampling, the height of Spartina detritus in tree branches, the location and thickness of detritus deposits on the forest floor and the extent and height of detritus rub marks on tree trunks were recorded.

Measurements of the concentrations of Na, K, Ca and Mg in water samples were obtained by atomic absorption spectroscopy using a Perkin Elmer Zeeman 5100 spectrophotometer (American Public Health Association, 1985). Samples were diluted with distilled, deionized water to appropriate concentration ranges for analyses. Sample conductivities were measured with a VWR Scientific conductivity meter (Model 604). Salinities were computed from sample conductivities and temperatures using the equations given in Cox et al. (1967).

RESULTS AND DISCUSSION

Hurricane Hugo was classified as a Category 4 storm with a 100 year return period (Stauble et al., 1991). Hugo struck South Carolina near high tide, which is the reason for the massive property damage along the coast. Fortunately, the hurricane-associated rainfall was relatively low. Sullivans Island, just east of Charleston, received only 21 mm of rain, and Myrtle Beach, to the north of North Inlet, received only 6 mm. An unofficial rain gauge at North Inlet, SC, registered 65 mm. Had it rained more, coastal flooding would have been far greater.

The North Inlet Long-Term Ecological Research Site experienced high winds (steady 40 m s^{-1} winds from 30° N for more than 12 hours preceding landfall; (Gardner et al., 1991) as well as a storm surge (3 to 4 m above mean sea level) which traversed the

Holocene barrier islands, flooded the Spartina marshes and temporarily caused ocean water to flood over a large area of coastal forest. Approximately 30% of the forest ecosystem lies at elevations below the 3 m contour and thus was subject to the effects of salt water inundation (Figure 2). Fragments of Spartina caught in tree branches indicate that the maximum surge elevation was 3 to 4 m above MSL, or approximately 2 to 3 m above mean high tide.

The different sources of data available allowed examination of the hurricane-related impacts at a variety of scales (satellite, low altitude aerial and ground). Ground observations were essential for ascribing observed damage to causal mechanisms. We examine these issues in the remainder of the discussion and speculate on the applicability of the various techniques for studying additional long-term impacts.

HURRICANE HUGO DAMAGE ASSESSMENT
(Landsat TM — October 1989)

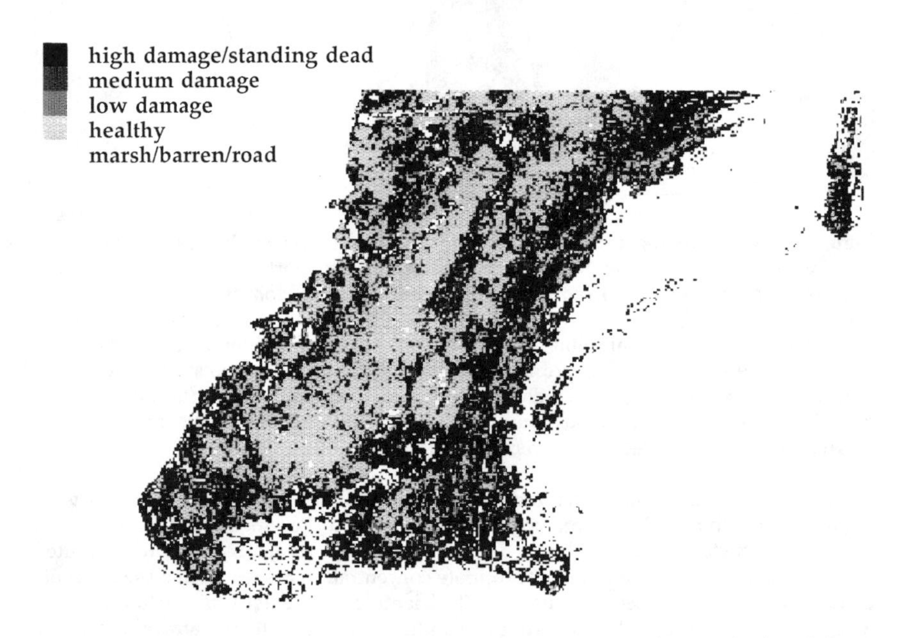

Figure 3. Classified Landsat TM image showing extent and magnitude of forest damage.

SATELLITE OBSERVATIONS

Geomorphic modification of the barrier island system and salt marsh creek network was not evident in the satellite imagery because of the relatively small size of the impacted zones and the time the image was taken, which corresponded to high tide. Visual comparison of the classified image to aerial photographs revealed common landscape patterns in the forested area. Extensive areas of high damage were located along the eastern, southern and southwestern portions of Hobcaw Barony (Figure 3). These areas were most exposed and received the full impact of high winds and storm surge. Other areas of high damage located on the western edge of the forest or in the interior of Hobcaw Barony generally corresponded to areas of high elevation (ridges), tall old-growth stands or low elevation (swales) which would also be most susceptible to wind damage or salt stress.

Background standing water and saturated soil conditions influenced spectral signatures. Storm surge and elevated levels of precipitation (more than 18 cm of rain in the two week period immediately after Hurricane Hugo) left large standing pools in the forest which contribute to higher reflectance values. Although the marsh extends inland and is patchily distributed throughout Hobcaw Barony, areas of standing dead and mixed pixels were sometimes classified as marsh. Another confounding factor is the high amount of sediment present in the marsh creeks. Areas of marsh partially covered by water and channels draining into Winyah Bay were frequently classified as land, presumeably because of the extremely high sediment load which was observed to be present and their similarity to heavily damaged areas which were also covered by water. Additional processing and comparison with aerial photography or images from other periods alleviated these problems.

AERIAL OBSERVATIONS

Geomorphological Changes

Despite its intensity, Hurricane Hugo had only a modest impact on the geomorphology of the undeveloped coastal landscape at North Inlet, SC. Pre- and post-Hugo aerial photographs show no change in the salt marsh creek network, nor can changes be seen in the size or shape of sandbars within the creeks. No breaches developed on the island, but several new, small washover fans were formed. These lobate fans extend 50 to 100 m from the dune line into the back barrier area and were deposited on older but recently formed fans in areas where the islands are thin and devoid of large shrubs and trees. Existing dunes on the ocean side of the new washover fans were leveled. The large ebb-tidal delta at the mouth of North Inlet appeared to experience erosion as a result of the large volume of surge water draining on the following ebb tide. Hugo's failure to have a more dramatic geomorphic effect was probably related to the rapid approach of the storm along a path perpendicular to the coast. This allowed minimal time for the surge to build and for wave attack to modify the shoreface. More detailed descriptions of the geomorphological changes associated with Hugo are presented elsewhere (Gardner et al., 1991).

Forest Damage

Since Hobcaw Barony is dominated by pine and much of the die-off associated with salt stress was not fully evident in the October photographs, we present only the results from the Febuary 1990 aerial survey. Approximately 45% of the forest received damage (high, 12.5%; medium, 11.4%; and low, 13%; Figure 4). In addition, 8.2% of the forest experienced heavy damage which could be attributed to the storm surge and subsequent salt stress and die-off. Over 50% of the forest appeared healthy in the February photographs. Classification statistics revealed close agreement between the aerial and satellite damage assessments. However, damage to understory species could not be discriminated.

GROUND OBSERVATIONS

Above normal salt concentrations were found in shallow groundwater samples from sites up to about the 3.0 m contour (MSL). In general, salt concentrations decreased inland from the forest-marsh boundary and with the passage of time. Trees along the forest-marsh boundary and in swales between relict beach ridges experienced mortality and chronic stress. Although more surge water infiltrated ridge soils than

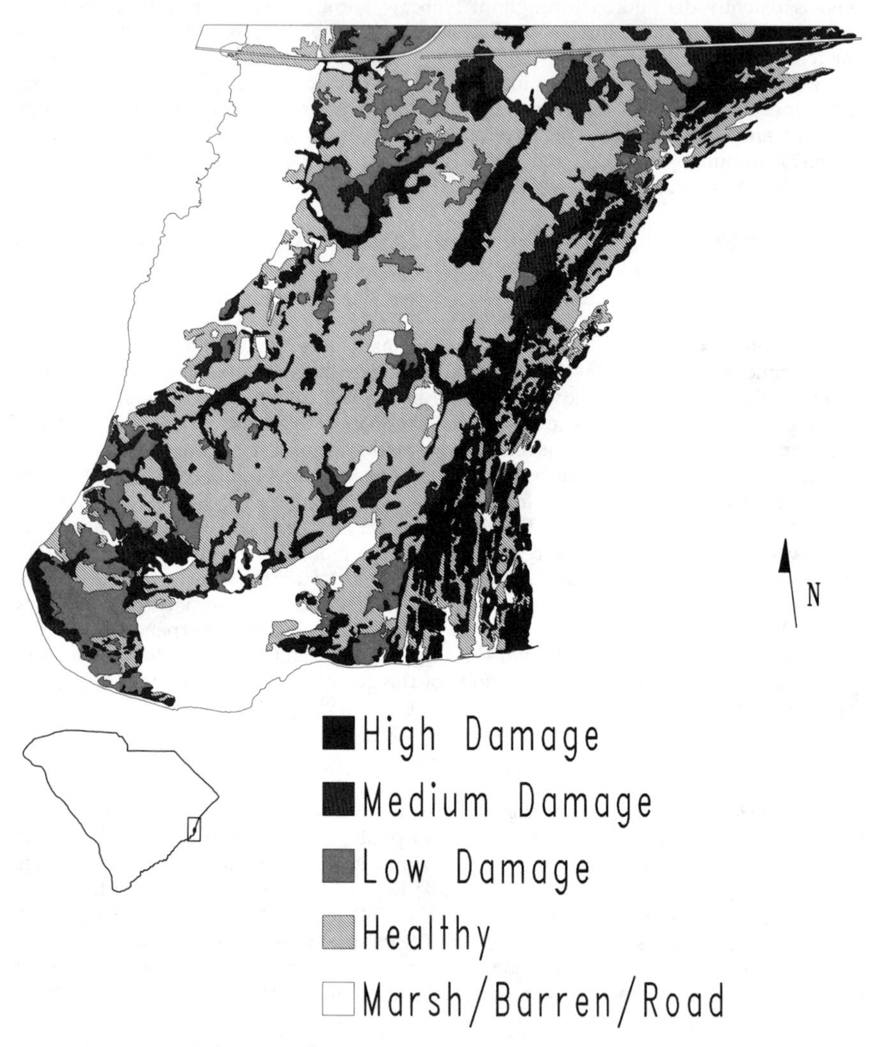

Figure 4. Map showing extent and magnitude of forest damage (based on low altitude color infrared photography).

swale soils, trees have survived better on ridges. Several mechanisms account for the acute and chronic impacts.

The effects of wind, salt spray and salinity stress associated with the storm surge varied throughout the Hobcaw Barony forest. On the eastern side of the property, leaves and needles on trees and shrubs located near the marsh edge turned brown or dropped within days of the hurricane, probably as a result of wind stress and salt spray. Similar effects were less evident on the western side of the property where trees received some protection from the high winds and any spray was probably brackish or fresh due to the position adjacent to the lower salinity waters of Winyah Bay. Areas of heavy damage (tree blowdown, limb breakage, defoliation, etc.) were scattered throughout the forest although the most extensive areas which were affected were generally confined to the eastern side of Hobcaw Barony. The direction of the storm surge and the low gently sloping topography on the eastern side of Hobcaw Barony combined to result in extensive salt-related stress and mortality up to the 3 m contour (above mean sea level). The spatial pattern of salt stress and evidence of surge effects were relatively complex. For example, along Transect 3 (Figure 2), extensive rub marks on trees from branches carried on the surge and detrital deposition on trees and shrubs were largely confined to the area within 50 m of the marsh. Extensive wrack deposits comprised mainly of Spartina, pine needles and branches (up to approximately 40 cm in thickness) were also deposited in this area (Figure 5). However, significant wrack deposits comprised of Spartina, pine needles and gum leaves were also found 200-250 m inland. At distances 300-750 m inland wrack deposits consisting primarily of pine, leaves and other woody debris were observed. Little Spartina wrack was observed within this zone and no wrack was found at the station 794 m inland. Different tree and shrub species appeared to exhibit varying sensitivity to salt stress. Leaves on understory species consisting of blueberry (Vaccinium spp.), red bay, wax myrtle (Lagerstroemia indica) and others were brown at distances up to 750 m from the marsh. Loblolly pine needles were brown up to 50 m inland. Gums appeared dead or severely stressed up to 300 m inland. Cypress trees located in a swale 300 m inland were either browned or partially defoliated. Canopy trees appeared healthy at distances more than 500 m inland. Generally, salt induced mortality was distributed throughout the eastern side of the forest and appeared equal in magnitude to the wind damage in this area (based on the Febuary 1990 aerial photography). Mortality was most severe along swales and at the marsh-forest boundary.

Salinization of forest soils.

Profiles of Na concentration, water table depth and station elevation along Transect 3 are shown in Figure 5. Two weeks after Hugo (10 October 89), all stations in the surge-affected forest had Na concentrations in excess of 100 mg l^{-1} with several in excess of 1000 mg l^{-1}. Maximum concentrations approached 3000 mg l^{-1} or approximately 30% that of seawater. Generally, Na concentrations decreased both inland and with the passage of time. However, at stations 250 and 300 m inland (located in a swale), Na concentrations increased between the October and November 1989 collections. At the time of the storm surge, water table elevations in swales were high and most of the salt water initially infiltrated ridge soils. Subsequent rainfall events led to the downslope movement of saline water and corresponding increases in Na concentrations observed in swale groundwater. By October 1990, Na at most stations had declined by an order of magnitude.

SUMMARY AND CONCLUSIONS

Satellite, aerial and ground observations indicate that the geomorphic effects of Hurricane Hugo were focussed on the barrier island complex, whereas ecological effects were most evident in the nearby coastal forest. Aerial and ground observations revealed no apparent effect on the Spartina alterniflora marsh, marsh-creek network or the extensive intertidal oyster reefs.

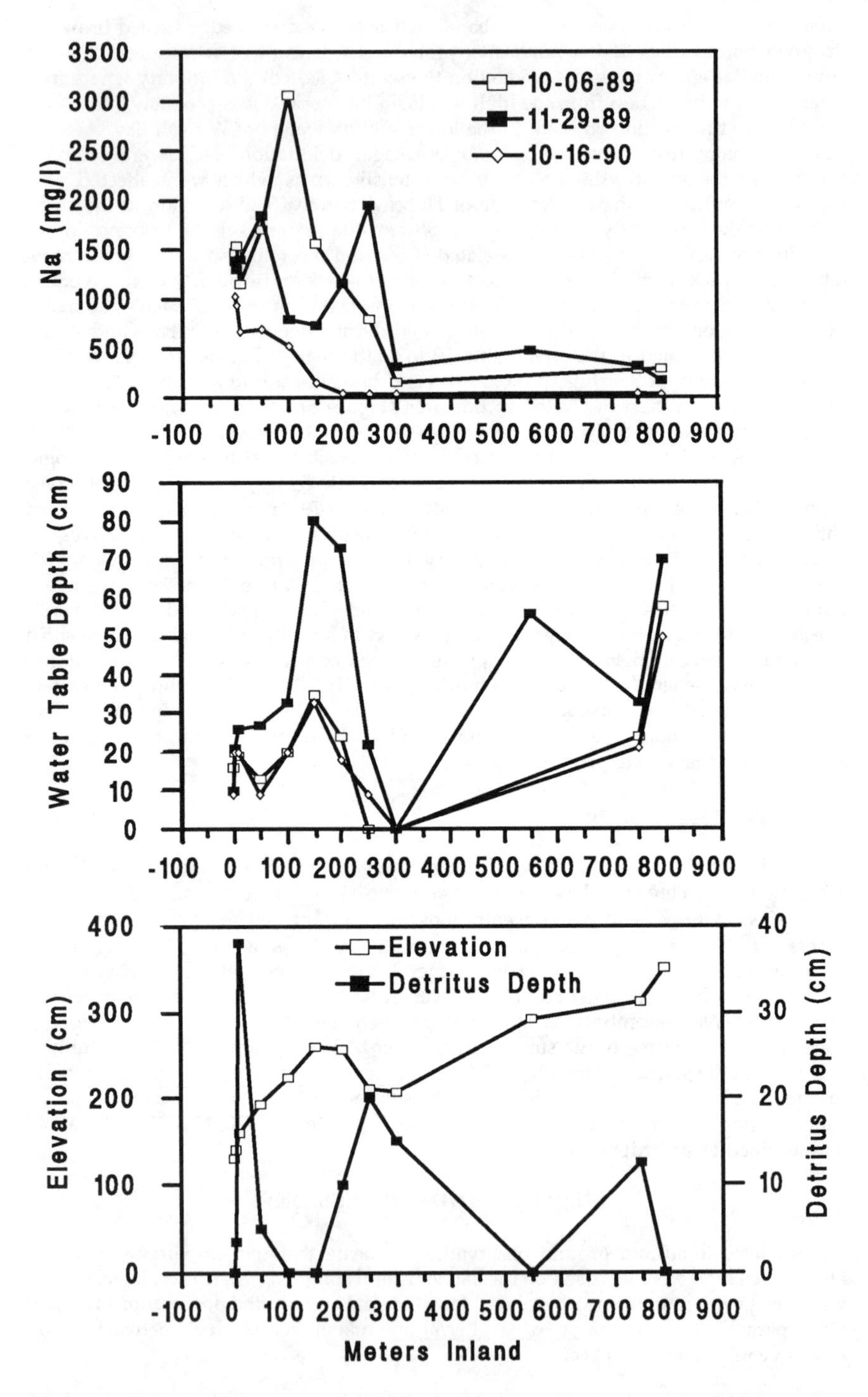

Figure 5. Variation of groundwater Na concentration, water table depth, ground elevation and thickness of detrital deposits along Transect 3.

Despite the size and intensity of the storm and the massive property damage along the coast, Hugo's impact on the geomorphology of the undeveloped landscape at North Inlet was minimal. The relatively small size of the affected area precluded identification through analysis of satellite imagery. Low altitude aerial photography revealed modification of the North Island shoreline including loss of dunes, new or expanded washover fans and changes to the ebb tidal delta (Gardner et al., 1991). Volumetric beach losses were greater near the point of landfall (Birkemeier et al., 1991). Coastal shore protection features generally received damage or were destroyed; all fishing piers along the affected coast were destroyed (Stauble et al., 1991). Between Dewees Island and the Santee Delta, coastal dunes were eroded, washover terraces widened, inlet width expanded and a new inlet was formed (Sexton and Hayes, 1991). Aerial observations were particularly well suited for identifying these types of damage.

Analysis of aerial photographs revealed that Hobcaw Barony consisted of a mosaic of forest patches which differed in severity of damage. Heavily damaged areas ranged in size from a few trees to 500,000 m^2. Despite the fact that the 30 x 30 m spatial resoultion of Landsat TM represents averaged reflectance values for that pixel, there was surprising agreement between the damage estimates obtained by aerial photography and satellite imagery. The accuracy matrix and additional direct comparisons between specific areas of the satellite scene and aerial photos revealed that uniform areas, such as stands of healthy trees or 100% blowdown, were generally more accurately classified than small areas or where complex borders produced mixed pixels or misclassification. However, the general agreement between the satellite imagery and aerial photography and the low cost of satellite imagery relative to aerial surveys of similar sized areas support the utilization of satellite imagery for regional damage assessments. Analysis of additional images from subsequent years and different seasons will provide further insight into the long-term impacts of this catastrophic event.

Ground observations revealed the type and extent of the damage to forest vegetation as well as the mechanisms leading to salt-related stress and mortality. The sensitivity of individual species (particularly understory species) and the types of damage exhibited (limb breakage, browning, defoliation, etc.) were most apparent during the transect sampling. The mechanisms leading to salt damage in the swales (initial infiltration of ridge soils followed by precipitation-induced leaching of salts into the swale soils) were ascertained only as a result of extensive transect sampling of groundwater over an extended period.

Recovery of the coastal forest will be a long term and uncertain process, particularly in the surge affected area. Additional direct and indirect mortality as a result of soil salinization and subsequent attack by pine bark beetles is anticipated. Remotely sensed imagery should prove useful for distinguishing large areas experiencing these impacts as well as subtle spectral changes which may be indicative of tree stress. Additional aerial surveys will be necessary for locating small stands of trees which die or are affected by localized outbreaks of insects. Furthermore, it should be possible to ascertain whether areas exhibiting new damage lie in close proximity to swales and thus were subject to initial soil salinization. Ground observations should prove useful for identifying the types of forest damage and the species of insects which infest specific stands. It may not be possible, however, to attribute death or stress to a specific cause since insect outbreaks may be confounded by the wind or salt damage which may have been the primary cause of mortality.

ACKNOWLEDGEMENTS

Funds for this study were provided by NSF-SGER grant BSR-9001807, NSF grant BSR-8514326 and NOAA grant NA90AA-D-SG672. Betsy Haskin, William Johnson, Dan

Taylor, Peggy Anderson, Susan Service and Steve Hutchinson provided assistance in the field. GPS Pathfinder is a trademark of Trimble Navigation, Limited. ARC/INFO is a trademark of Environmental Systems Research Institute, Inc. ERDAS is a trademark of ERDAS, Inc. This paper is Contribution Number 887 of the Belle W. Baruch Institute for Marine Biology and Coastal Research.

REFERENCES

American Public Health Association, 1985. Standard Methods for the Examination of Water and Wastewater, Washington, DC: American Public Health Association.

Birkemeier, W.A., E.W. Bichner, B.L. Scarborough, M.A. McConathy and W.C. Eiser, 1991. Nearshore profile response caused by Hurricane Hugo, Journal of Coastal Research, Special Issue # 8:113-128.

Brennan, J.W., 1991. Meteorological summary of Hurricane Hugo, Journal of Coastal Research, Special Issue # 8:1-12.

Callahan, J.T., 1984. Long-term ecological research, BioScience, 34:363-367.

Cox, R. A., F. Culkin and J. P. Riley, 1967. The electrical conductivity/chlorinity relationship in natural seawater, Deep Sea Research, 14:203-220.

Dame, R.F., T. Chrzanowski, K. Bildstein, B. Kjerfve, H. McKellar, D. Nelson, J. Spurrier, S. Stancyk, H. Stevenson, J. Vernberg and R. Zingmark, 1986. The outwelling hypothesis and North Inlet, South Carolina, Marine Ecology Progress Series, 33:217-229.

Franklin, J.F., C.S. Bledsoe and J.T Callahan, 1990. Contributions of the long-term ecological research program, BioScience, 40:509-523.

Gardner, L.R. and M. Bohn, 1980. Geomorphic and hydraulic evolution of tidal creeks on a slowly subsiding beach ridge plain, North Inlet, SC, Marine Geology, 34: 91-97.

Gardner, L.R., W.K. Michener, B. Kjerfve and D.A. Karinshak, 1991. The geomorphic effects of Hurricane Hugo on an undeveloped coastal landscape at North Inlet, South Carolina, Journal of Coastal Research, Special Issue # 8:181-186.

Hook, D.D., M.A. Buford and T.M. Williams, 1991. Impact of Hurricane Hugo on the South Carolina coastal plain forest, Journal of Coastal Research, Special Issue # 8:291-300.

Jensen, J.R., 1986. Introductory Digital Image Processing: A Remote Sensing Perspective, Prentice-Hall, Englewood Cliffs, New Jersey, 379 p.

Michener, W.K., A.B. Miller and R. Nottrott, 1990. Long-Term Ecological Research Network Core Data Set Catalog, Belle W. Baruch Institute for Marine Biology and Coastal Research, University of South Carolina, Columbia, SC, 322 p.

Miller, A.B., W.K. Michener, A.H. Barnard and F.J. Vernberg, 1989. Publications of the Belle W. Baruch Institute for Marine Biology and Coastal Research (1969-1989), University of South Carolina Press, Columbia, SC, 149 p.

National Hurricane Center, 1989. In house report, National Hurricane Center, Coral Gables, Miami, FL.

Sexton, W.J. and M.O. Hayes, 1991. The geologic impact of Hurricane Hugo and post-storm shoreline recovery along the undeveloped coastline of South Carolina, Dewees Island to the Santee Delta, Journal of Coastal Research, Special Issue # 8:275-290.

Sparks, P.R., 1991. Wind conditions in Hurricane Hugo and their effect on buildings in coastal South Carolina, Journal of Coastal Research, Special Issue # 8:13-24.

Stauble, D.K., W.C. Seabergh and L.Z. Hayes, 1991. Effects of Hurricane Hugo on the South Carolina coast, Journal of Coastal Research, Special Issue # 8:129-162.

Williams, T. M., C.A. Gresham, E.R. Blood and R.L. Hedden, 1989. Impact of Southern pine beetle infestation on nutrient retention mechanisms in loblolly pine, Proceedings of the 5th Biennial Southern Silvicultural Research Conference, USDAFS Southern Forest Experiment Station, New Orleans, LA, edited by J.H. Miller, General Technical Report SO, pp. 465-472.

GIS-RELATED MODELING OF IMPACTS OF SEA LEVEL RISE ON COASTAL AREAS

Jae K. Lee and Richard A. Park
School of Public and Environmental Affairs
Indiana University
Bloomington, Indiana 47405

Paul W. Mausel and Robert C. Howe
Department of Geography and Geology
Indiana State University
Terre Haute, Indiana 47809

ABSTRACT

Impacts of future sea-level rise on coastal areas in northeastern Florida, near Jacksonville, were estimated using an integrated system, which included remote sensing and geographic information systems and a rule-based model, SLAMM3. ERDAS and pc-ARC/INFO systems were linked with SLAMM3 to provide an efficient modeling environment to project impacts of future sea-level rise. Multispectral SPOT data were used to characterize coastal zone conditions through computer classification into land covers and land uses. Digital elevation data were interpolated from digitized contours. The land covers and elevation data were then merged with site-specific data on tidal ranges, subsidence rates, and local fetch. The SLAMM3 model evaluated the input data to predict responses of coastal wetlands and lowlands to inundation and erosion by sea level rise, and determined transfers from one habitat to another on a cell-by-cell basis. Significant changes in coastal wetlands and lowlands were predicted from using different scenarios of sea-level rise. The integration of remote sensing and geographic information systems with a simulation model was shown to be an efficient way to analyze future impacts of sea-level rise on coastal areas.

INTRODUCTION

With increasing concern about global warming, sea-level rise has become a critical issue. If a global warming trend occurs over the next century, eustatic sea-level rise will be one major environmental effect, with serious impacts on coastal environments. Flooding, erosion, and saltwater intrusion can disrupt coastal wetlands (Titus and Barth 1984), converting marshes, mangroves and other swamps to open water. The vegetated wetlands may migrate inland when the adjacent lowlands are flooded.

The value of wetlands is critical in terms of their intrinsic qualities and ecological services (OTA 1984). A precise estimation of the impacts of sea-level rise on coastal wetlands is necessary to support policy decisions associated with coastal resource management.

In present study, the possible impacts of sea-level rise on coastal wetlands and lowlands were estimated using a simulation model, SLAMM3 (Sea Level Affecting Marsh Model version 3.0). The model was linked to geographic information systems (GIS) in this study in order to efficiently manipulate and manage the databases required for simulation modeling.

MODEL DESCRIPTION OF SLAMM3

The objective of the modeling was regional-scale simulation of the dominant processes involved in vegetated wetland conversions and related shoreline reconfigurations during long-term sea-level rise. SLAMM3 differs from other wetland models (Wiegert et al. 1975, Costanza et al. 1987, 1990, Browder et al. 1985, 1989) by its ability to predict high-resolution map distributions of wetland cover under conditions of accelerated sea level rise and by its applicability to the diverse wetlands of the contiguous coastal United States.

In essence, SLAMM3 is a knowledge-based simulation model that uses a complex decision tree and scalars to represent qualitative relationships. The basic structure of the model is cell-based storing land-cover/land-use data, elevation data, and site characteristic data for each grid cell in unique form (Table 1). Simulation modeling is performed cell-by-cell; thus, the grid cell is the spatial unit of the model. At present, SLAMM3 supports three different sizes of grid cell: 125 x 125 m, 250 x 250 m, and 500 x 500 m. The appropriateness of the process resolutions of the model for these grid sizes was recently tested and verified (Lee 1991). At this time, SLAMM3 models fourteen different categories of coastal feature for regional-scale modeling (Table 2).

Table 1. Database required for SLAMM3 Modeling

- land cover and land use data
- elevation data
- site characteristic data
  - subsidence rate
  - tidal ranges
  - wind direction
  - location of dikes

Table 2. Land Categories Considered for Modeling within SLAMM3.

- developed dry land
- undeveloped dry land
- non-beach sand
- hardwood swamp
- cypress swamp
- freshwater marsh
- high saltmarsh
- low saltmarsh
- mangrove
- beach/tidal flat
- rocky intertidal
- non-ocean water
- ocean water
- others

The basic constructs for each of the processes are organized in two sections, the inundation model and the map-based spatial model.

The Inundation Model

The colonization of newly inundated dry land by wetland vegetation and loss of wetlands due to further inundation is based on a straightforward geometric relationship, with lag effects for some conversions. Five processes are considered as part of the inundation model:

- Relative sea level change, including subsidence, sedimentation, and accretion
- Conversions between classes
- Protection by coastal engineering structures
- Death and colonization
- Change to tropical conditions

The Spatial Model

In addition, second-order effects occur due to changes in the spatial relationships among the coastal elements. Accordingly, SLAMM3 incorporates a map-based model component to consider five spatially important processes:

- Erosion of wetlands due to increased fetch for waves
- Exposure to open ocean and subsequent erosion of wetlands
- Beach erosion
- Overwash
- Erosion of sandy lowlands

More details on these processes are found Park et al. (1989, 1990) and Lee (1991).

STUDY AREA

The study area is located in the extensive coastal lowlands in parts of Nassau and Duval Counties, Florida (Figure 1). This area typically consists of barrier islands, marshes, level plains, and a series of five terraces resulting from the most recent advances and retreats of the sea during the late Pleistocene (Karause and Randolph 1989). A variety of wetland and dry land vegetation characterizes this low, gentle-to-flat topographic area, where most elevations are under 50 ft.

Marshes are very well developed in this area, due to a combination of high tidal ranges (approximately 5.6 ft on the ocean side and 3.75 ft in sheltered areas) and high sedimentation rates. Because of these high tidal ranges and availability of lowlands for marsh colonization, the wetlands in this area will be more persistent in the face of rising sea level than other U.S. coastal wetlands (Park et al. 1989). An area of approximately 900 km^2 was selected for mapping and modeling for this study.

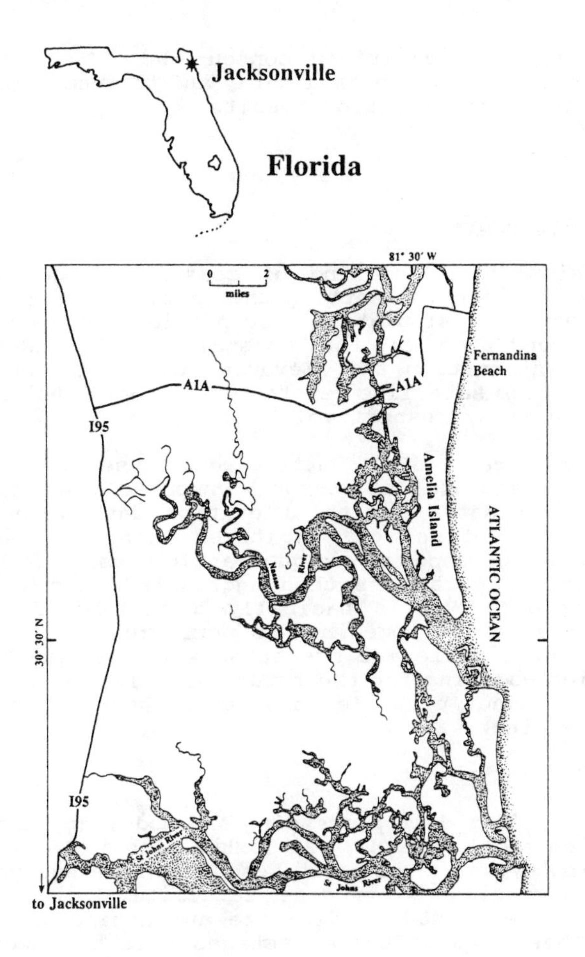

Figure 1. Map of the Study Site

DATA USED

Multispectral SPOT high resolution visible (HRV) data were used to generate land cover and land use data over the study area through computer classification. The data were acquired in 20 x 20 m ground resolution in 3 spectral bands on September 28, 1986. A set of ancillary data was used to help the classification of SPOT data; U.S.G.S topographic quadrangle maps (1:24,000 scale) and land-use maps (1;100,000 scale), National Wetland Inventory Maps (1:24,000 scale), and National High Altitude Photography (NHAP) color infra-red photography (1:58,000 scale). These data were used to generate digitized contour data, to help develop training samples of spectral data, and to interactively edit the initial classification results.

METHODS

General Procedure

The simulation modeling of coastal changes using the SLAMM3 model requires setting the initial conditions of the coastal area for simulation. To provide the model with the initial condition of the coastal area, land-cover and land-use information and elevation data for the site were generated from satellite remote sensing data and topographic quadrangle maps, respectively.

Multispectral SPOT data were used to generate land-cover/land-use data through computer classification of the spectral data. Digital elevation data were generated from interpolation of digitized contour lines from 7.5-minute USGS topographic quadrangle maps. The land-cover and elevation data were then aggregated into 125 x 125-m grid, since SLAMM3 is basically a cell-based simulation model. Simulation modeling was performed at the grid-size scale. Three different scenarios were implemented in the simulation modeling for the future sea-level rise by 2100 in order to project possible impacts of the sea-level rise in the study site.

System Generation

For efficient data processing and modeling, the computer simulation model, SLAMM3, was linked with a GIS and an image processing system. The Earth Resources Data Analysis System (ERDAS) was used to process satellite digital data and pc-ARC/INFO was used to digitize and manipulate elevation data. These two software packages were implemented on a 80386-based personal computer and linked with SLAMM3 by developing a set of interface routines. Several programs were written in FORTRAN 77 to manipulate the datasets and to interface the GIS and image processing systems with the SLAMM3 model. Figure 2 illustrates the system configuration used in this study.

The ERDAS and pc-ARC/INFO systems were used to generate, store, and manipulate digital land-cover and elevation data. Interface routines integrated these data into suitable forms for simulation modeling, using SLAMM3. The simulation results were then incorporated into ERDAS and pc-ARC/INFO to generate output maps.

DATA PROCESSING AND SIMULATION MODELING

Processing of Remotely Sensed Data

First, the multispectral SPOT imagery was geometrically rectified at approximately 0.04 root-mean-square error (RMSE) value based on the UTM coordinates. Bilinear

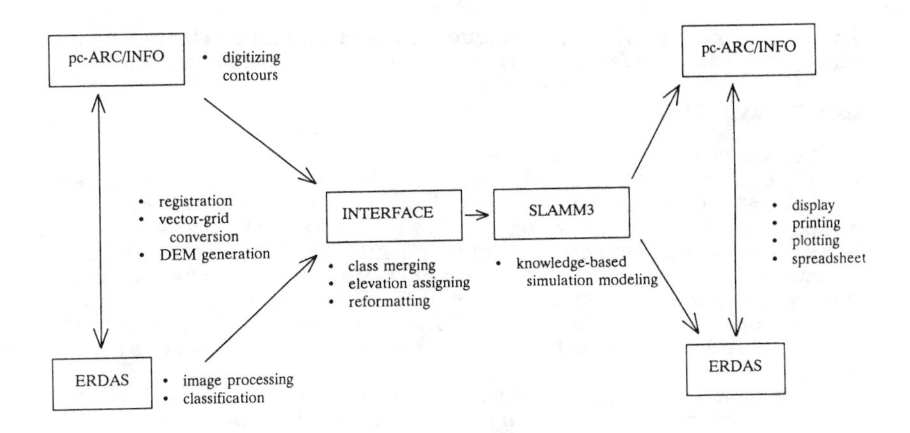

Figure 2. System Configuration used for the Study

interpolation was applied for the resampling of gray values. The bilinear interpolation is a reasonable compromise of computational time and image distortion (Gonzalez and Wintz 1987).

The geocorrected SPOT data were then transformed into normalized difference vegetation index (NDVI) and principal component form, since the transformation of spectral data into these forms often enhances spectral separation of coastal wetlands features (Budd and Milton 1982, Gross et al. 1987). Since principal component analysis provides data reduction, only the first two components were used for classification along with the NDVI data. Thus, the three-band transformed SPOT data were used for spectral classification into coastal features.

The classification of SPOT data in this study was divided into two phases. The first phase was the spectral classification of the transformed SPOT data into various land-cover classes, using the traditional supervised approach. Then, the classes were merged and redefined into classes required by SLAMM3 model. From the initial spectral classification, twenty-two different land-cover classes were developed with a maximum-likelihood classifier, and these classes were merged and redefined into ten different classes with particular attention to wetland and lowland classes. The SLAMM3 model can simulate up to fourteen different coastal land-cover categories (Table 2), but some of the classes did not exist in the study site.

Accuracy assessment of the classification map was performed by developing a ground-truth map from NHAP photographs and field checking. Overall accuracy of 87.8 percent was achieved from the classification of SPOT data. Most classes were classified with an accuracy of approximately 90 percent or above. The non-beach sand class was spectrally confusing with the developed dry land class including residential, commercial, and industrial area; and, thus, it had a classification accuracy of 76.4 percent.

Generation of Elevation Data

Because USGS Digital Elevation Model (DEM) data were not available for the study site, the elevation data were generated by manual digitizing of contour lines from topographic quadrangle maps (1:24,000 scale). The contour lines were digitized on the pc-ARC/INFO system as line features, and converted to raster form. Then, interpolation was performed on the raster data to create continuous elevational surface data. A linear interpolation algorithm was applied to the dataset for the following reasons:

- The linear interpolation is a computationally simple algorithm which is suitable for the personal computer;
- The study area is coastal wetlands and lowlands where the topography is very flat and change in slope is not a major factor in the interpolation processes.

A 3x3 average filtering was applied to the interpolated elevation data in order to reduce observed vertical discontinuity. Finally, the elevation data were registered with the classification map on a pixel-by-pixel basis using UTM coordinates.

Further Processing and Interfacing

The land-cover and elevation data were further processed before simulation modeling. First, the land-cover data were aggregated into a grid of 125 x 125 m. This is the highest resolution grid size supported by the SLAMM3 model. The aggregated land-cover data were stored in terms of percent cover within each grid cell. Then, elevation data were combined with the aggregated land-cover data in such a way that minimum and maximum elevation data were developed for each class existing within the grid cell. Finally, the integrated data were reorganized into a suitable form for linking with the SLAMM3 model. Additional site data were added to the database for simulation modeling: tidal ranges, wind direction, subsidence rate, and location of dikes.

Simulation Modeling

The data prepared for simulation were submitted to the SLAMM3 simulation model. Three different scenarios for future sea-level rise were implemented to estimate the possible impacts on the study site in 2100: 0.13 based on the historic trend of sea-level rise, 0.5 m, and 1 m by 2100. All the simulations were performed with existing developed areas protected against the sea-level rise.

RESULTS

Figure 3 illustrates predicted land-loss by sea-level rise in 2100. In this study, the loss of undeveloped dry

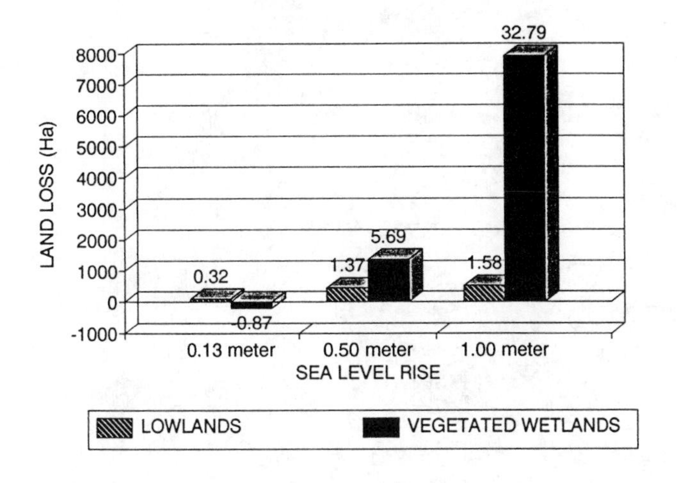

Figure 3. Land Loss from Simulation Modeling.

land and non-beach sand was included in the lowlands. The vegetated wetlands include all marsh classes and swamp. With 0.13-m (historic trend) and 0.5-m rises, small loss of land was predicted, while significant loss was predicted with a 1-m rise. With the historic trend rise, the vegetated wetlands increased slightly in size. That is attributed to the high accretion rate that was assumed for salt marshes. However, with increasing rates of sea-level rise to 0.5 m and 1 m, the loss of wetlands and lowlands became more obvious. Figures 4 and 5 illustrate the initial conditions and simulation results. Landward displacement in marsh zonation was observed from the simulation maps. Freshwater marsh was converted to saltmarsh by increasing salinity and saltmarsh was converted to tidal flats and open water with continuing inundation. The developed dry land did not change because it was assumed to be protected in this study.

SUMMARY AND CONCLUSION

Simulation modeling of coastal landscape change was performed to estimate the impacts of future sea-level rise on coastal areas. The simulation predicted relatively small changes with continued historic rise and a 0.5-m rise by 2100. Those small changes were thought to be due to such local factors of the site as high tidal ranges, high rates of sedimentation and marsh colonization,and low erosion. However, with a 1-m rise, a significant loss of marshes was predicted.

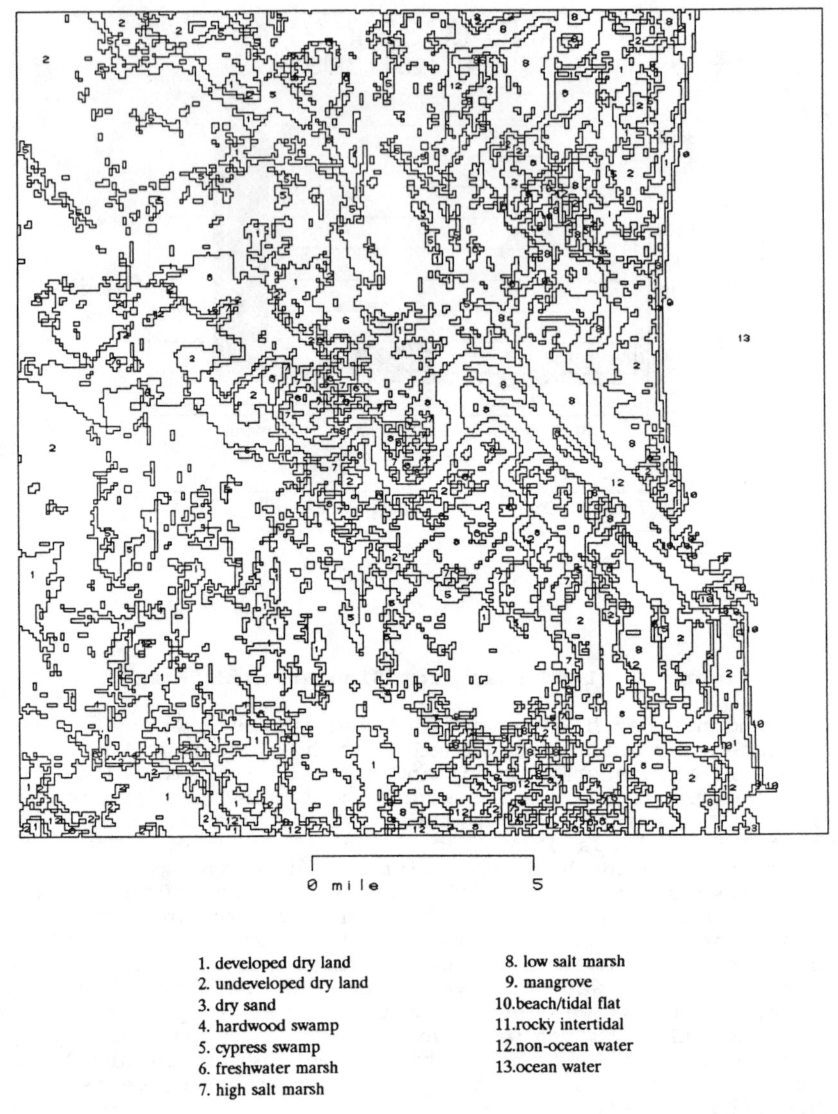

Figure 4. Map of the initial conditions of the site in 1986

These results have implications in future coastal fishery production. In a related study in the Gulf of Mexico, Park (1991) found that brown shrimp, which depend heavily on saltmarshes, could decline precipitously with a 1-m rise. Investigators at the National Marine Fisheries Service Galveston Laboratory have observed that shrimp production may increase temporarily with slightly accelerated sea-level rise, but that it may decline sharply with higher rates of rise (Zimmerman et al. 1991).

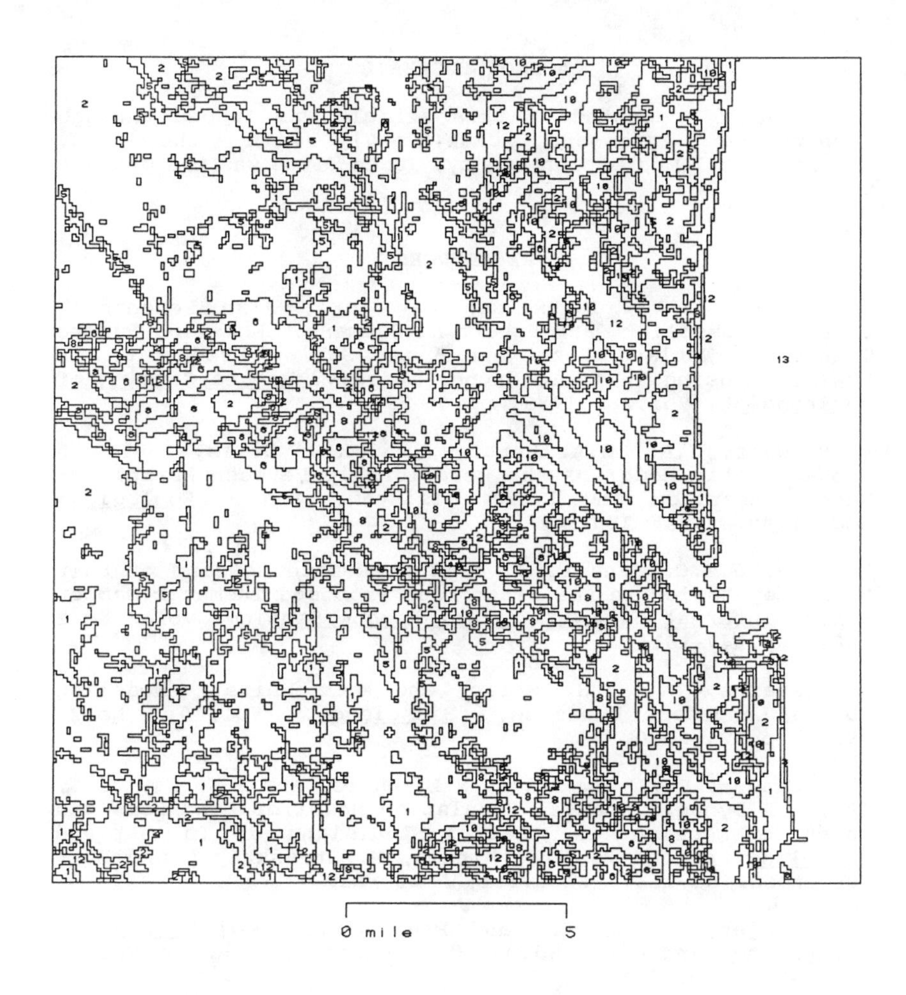

Figure 5. Simulation Map with 1-m Rise by 2100; legend is the same as Figure 4.

The integration of remote sensing and GIS with a computer simulation model in this study proved to be a very efficient modeling environment. All the procedures from initial data processing to simulation output generation were performed on the system in a time-efficient manner. The role of GIS within the system was essential for data input, storage, manipulation, and model preparation, with an efficient linkage between the model and various operational functions. Further refinement of the integrated system is in progress in order to incorporate flood and storm-surge data.

ACKNOWLEDGMENTS

The work presented in this article was funded through Cooperative Agreement CR814578-01 with the U.S. Environmental Protection Agency; James Titus was the project monitor.

REFERENCES

Browder, J.A., L. Nelson May, Jr., Alan Rosenthal, James G. Gosselink, and Robert H. Baumann. Modeling Future Trends in Wetland Loss and Brown Shrimp Production in Losisiana using Thematic Mapper Imagery. Remote Sensing of Environment. Vol.28. 1989. pp.45-59.

Browder, J.A., H.A. Bartley, and K.S. Davis, 1985. A Probabilistic Model of the Relationship Between Marshland-Water Interface and Marsh Disintegration. Ecological Modelling 29:245-260.

Budd, J.T.C., and E. J. Milton. Remote Sensing of Salt Marsh Vegetation in the First Four Proposed Thematic Mapper Bands. International Journal of Remote Sensing. Vol.3 No.2 1982. pp.147-161.

Costanza, R., F.H. Sklar, and M.L. White. Modeling Costal Landscape Dynamics. Bioscience. Vol.40. No.2. 1990. pp.91-1070.

Costanza, R., F.H. Sklar, M.L. White, and J.W. Day, Jr., 1987. A Dynamic Spatial Simulation Model of Land Loss and Marsh Succession in Coastal Louisiana. In Wetland Modelling, edited by W.J. Mitsch, M. Straškraba, and S.E. Jørgensen. 99-114. Elsevier, Amsterdam.

Gonzalez, Rafael C., and Paul Wintz. Digital Image Processing. 2nd ed. Reading, Massachusetts: Addison-Wesley Publishing Co. 1987.

Gross, M.F., M.A. Hardsky, V. Klemas, and P.L. Wolf. Quantification of Biomass of the Marsh Grass Spartina Alterniflora Loisel using Landsat Thematic Mapper Imagery. Photogrammetric Engineering and Remote Sensing. Vol.53 No.11. 1987. pp.1577-1583.

Krause, Richard E., and Robert B. Randolph. Hydrology of the Floridan Aquifer System in Southeast Georgia and Adjacent Parts of Florida and South Carolina. U.S. Geological Survey Professional Paper 1403-D. 1989. D65 pp.

Lee, Jae K. Effects of Spatial Resolution on Simulation Modeling of Coastal Changes. Unpublished Ph.D. Dissertation. Series III, No.521. Terre Haute, Indiana: Indiana State University. 1991. xi+171 pp.

National Research Council (NRC). Responding to Changes in Sea Level: Engineering Implications. Washington, DC: National Academy Press. 1987. 148 pp.

Office of Technology Assessment (OTA). Wetlands: Their Use and Regulation. Washington, DC: Office of Technology Assessment. 1984.

Park, Richard A. Testimony before the Subcommittee on Health and Environment. U.S. House of Representatives. Congressional Record. 1991.

Park, R.A., J.K. Lee, P.W. Mausel, and R.C. Howe. Predicting Impacts of Sea Level Rise with a GIS-based Simulation Model. Proceedings of the 2nd Annual State of Indiana GIS Conference. 15-16 Nov., Indianapolis, Indiana. 1990.

Park, R.A., M.S. Trehan, P.W. Mausel, and R.C. Howe. The Effects of Sea Level Rise on U.S. Coastal Wetlands. In:The Potential Effects of Global Climate Change on the United States:Appendix B - Sea Level Rise, edited by J.B. Smith and D.A. Tirpak, 1-1 to 1-55. EPA-230-05-89-052. Washington, DC: U.S. Environmental Protection Agency. 1989.

Titus, James G. Greenhouse Effect & Coastal Wetland Policy: How Americans Could Abandon an Area the Size of Massachusetts at Minimum Cost. Environmental Management. Vol.15 No.1. 1991. pp.39-58.

Titus, James G., and Michael C. Barth. An Overview of the Causes and Effects of Sea Level Rise. In:Greenhouse Effect and Sea Level Rise: A Challenge for This Generation, edited by M.C. Barth and J.G. Titus, New York, New York: Von Nostrand Reinold Co. 1984. pp.1-56.

Wiegert, R.G., R.R. Christian, J.L. Gallagher, J.R. Hall, R.D.H. Jones, and R.L. Wetzel, 1975. A Preliminary Ecosystem Model of a Coastal Georgia Spartina Marsh. In Estuarine Research, Vol. 1, edited by L.E. Cronin, 583-401. Academic Press, New York.

Zimmerman, R.J., T.J. Minello, E.F. Klima, and J.M. Nance. Effects of Accelerated Sea-Level Rise on Coastal Production. Proceedings of Coastal Zone 91. in press.

THE REALITIES OF THE DATA CONVERSION PRICE QUOTE

William B. Reid
Vice President, Corporate Development
Baymont, Inc.
14100 58th Street North
Clearwater, FL 34620
(813) 539-1661
(813) 539-1749 FAX

BIOGRAPHICAL SKETCH

Mr. Reid has more than 30 years of experience in the information management field. His Automated Mapping/Facilities Management/Geographic Information Systems (AM/FM/GIS) experience includes management of production, sales and administrative departments for a conversion service company, and in the direction of business development activity including the production cost estimates and proposal preparation.

ABSTRACT

When the users of an Automated Mapping/Facilities Management/Geographic Information System (AM/FM/GIS) enter the realm of data conversion, they may find the wide range of conversion price quotes somewhat baffling. The assumption is that if the user prepares an adequate Request For Proposal, with the technical requirements and schedule sufficiently outlined, most qualified conversion service companies will submit relatively similar price bids. It has been the unfortunate experience of many AM/FM/GIS users, however, to be the recipients of price quotes that have a high degree of disparity. This leaves the user with the often confusing and difficult task of determining why the price quotes are so different, and which one represents the best value.

This paper describes a formula that AM/FM/GIS users can develop for evaluating their particular conversion price quotes. The formula incorporates the use of various factors to bring each of the varying price quotes to the same relative level -- in effect allowing the user to compare apples to apples. Making the dissimilar price quotes comparative includes effectively checking references, determining past performance in terms of quality and schedule, and generating "risk" factors for the bidder's price quote. Having quantified the elements involved, the user determines a Probable Real Cost for each price quote which allows for much more effective and accurate price comparisons.

INTRODUCTION

Values

> It's unwise to pay too much, but it's unwise to pay too little.
> When you pay too much you lose a little money, that is all.
> When you pay too little, you sometimes lose everything, because the thing you bought was incapable of doing the thing you bought it to do.
> The common law of business balance prohibits paying a little and getting a lot. It can't be done.
> If you deal with the lowest bidder, it's well to add something for the risk you run.
> And if you do that, you will have enough to pay for something better...
>
> John Ruskin

When the users of an AM/FM/GIS enter the realm of data conversion, they may find the wide range of conversion price quotes somewhat baffling.

Often the user assumes that if the Request for Proposal sufficiently outlines the technical requirements and project schedule, all qualified conversion service companies will submit relatively similar price bids.

It has been the unfortunate experience of many AM/FM/GIS users, however, to be struck with conversion price quotes that have a high degree of disparity. Because more sophisticated project teams understand that automatic low bid award is not the best way to go about conversion group selection, the wide range of bids complicates an already demanding, high-risk task.

This leaves the user team with the often confusing and difficult task of determining why the price quotes are so different, and which one represents the best value.

Each of the various conversion proposals may include a unique methodology and a wide range of marketing promises that can inhibit effective price-to-service comparisons. This paper proposes an empirical formula which has been developed over a period of several years. User teams can use this "formula" to help evaluate their conversion price quotes.

Before using the formula that is described in the following paragraphs, a careful evaluation of the methodologies in the various proposals should be made to ensure that the services proposed will yield the same end result. If it is determined that one proposal offers more or less than the others, an adjustment to the Actual Quoted Price must be made. This evaluation should look for evidence that the bidder is either offering what he thinks the user wants or needs (as opposed to what was asked for), or that the bidder may be planning to "work the change order process" in order to modify the quoted price.

The formula incorporates the use of various adjustments to bring each of the varying price quotes to the same relative level. This, in effect, allows the team to compare apples to apples. The price quote formula uses the two most important conversion issues as a foundation -- quality and schedule.

Essentially, the formula helps to paint a more accurate picture of each bidder's quote by assessing quality and schedule adjustments based on prior performance. If a bidder's quality has been questionable on past projects, for example, then that bidder's quote would be penalized to reflect the increased probability that quality may again be a problem. Conversely, the quotation of a company which has demonstrated exceptional adherence to schedule would be adjusted to reflect the increased probability that schedule adherence would continue to be good. See Figure 1.

Conversion Services Pricing Formula

(AQP x QRA) + SRA = PRC

AQP = Actual Quoted Price
QRA = Quality Risk Adjustment
SRA = Schedule Risk Adjustment
PRC = Probable Real Cost

Figure 1

The first step in building a conversion price formula is to contact the references of the service companies which provided the proposals. Most references will be happy to provide fellow users with information pertaining to the quality and schedule of their project's conversion process. This process may take several weeks, however, so users should build this time into their selection schedules.

Several questions should be asked of at least three references per service company. Each of these questions will result in a "score" for a different part of the complete formula. Using the pricing formula, let's evaluate three hypothetical conversion service companies. For the example, we will assume that the total project cost is $5 million and has a five-year payback period. This will allow user's to generate a number for Estimated Weekly Savings (EWS). See Figure 2. In this case, ignoring the "cost of money" and other factors for the sake of illustrative simplicity, it is $19,230.

System Payback Calculation

Project cost = $5 million
Payback period = 5 years

Estimated Weekly Savings (EWS) =

$1million per year ÷ 52 weeks = **$19,230 per week**

Figure 2

The first calculation user teams should do is the Quality Risk Adjustment (QRA). This calculation includes determining how far off the first delivery of converted data was from the specified quality standard.

Delivery Accuracy

	Standard	Actual	Difference	# of Deliveries
Company A	97%	77%	20% or .20	4
Company B	97%	77%	20% or .20	2
Company C	97%	95%	2% or .02	2

Figure 3

In this case (see Figure 3) the specified quality standard is 97 percent. Service company A's accuracy rate on their first delivery was 77 percent, and they reworked the data three subsequent times before they met the 97 percent accuracy rate. Company B's accuracy also was 77 percent, but they achieved 97 percent in only one redelivery. Company C's first delivery was 95 percent accurate, and they got to the standard in one redelivery.

Subtract the actual first-delivery accuracy rate from the standard and determine its percentile. Multiply the percentile by the total number of deliveries and divide by 10 (a number that has been found appropriate -- the user may adjust the number if it is warranted). Then add one to get the QRA of each company. Enter that figure into the formula. See Figure 4.

Determing the Quality Risk Adjustment (QRA)

QRF = 1 + ((Difference from Accuracy Standard x Number of Deliveries) ÷ 10)

Company A : 1 + ((.20 x 4) ÷ 10) = **1.08**

Company B: 1 + ((.20 x 2) ÷ 10) = **1.04**

Company C: 1 + ((.02 x 2) ÷ 10) = **1.004**

Figure 4

The QRA, if high, indicates that the service company may be lacking in efficient quality control procedures and effective project management. Delivery dates should be firmly established before conversion begins, and service companies should take steps to understand the conversion specifications and achieve high levels of accuracy <u>before</u> the first delivery.

The next step is to address each company's adherence to schedule. Determine the average of the delivery delays in weeks. If there were multiple deliveries, add together the number of weeks late for each delivery and divide by the number of deliveries. See Figure 5. The late period should include the total amount of time it took to deliver <u>accepted</u> data.

Determining (average delay weeks)

Company A		Company B		Company C	
Project	Weeks late	Project	Weeks late	Project	Weeks late
1	1	1	1	1	1
2	1	2	3	2	1
3	3	3	4	3	0
weeks late:	5 5÷3=		8 8÷3=		2 2÷3=
Ave. delay =	**1.67**		**2.67**		**.67**

Figure 5

Company A, over three projects, has a sum of five weeks late and an average of 1.67 weeks late. Company B has an average of 2.67 weeks late while Company C has an average of only .67 weeks late overall.

Schedule overruns, either on first delivery or because of a rework cycle, cost the user lost productivity that was calculated into the original payback period. Only when the converted data is delivered and working on the user's system can the payback begin.

Multiply the sum of weeks by the EWS to get the Schedule Risk Adjustment (SRA). Plug that number into the calculation. See Figure 6.

The Schedule Risk Adjustment (SRA)

(Average Delay in Weeks) x (Estimated Weekly Savings) = SRF

Figure 6

Enter the Actual Quoted Price (AQP) into the formula as indicated in Figure 7. Multiply the AQP by the Quality Risk Adjustment (QRA), then add the Schedule Risk Adjustment (SRA) to derive the Probable Real Cost (PRC) -- a number that users should carefully consider.

Determining Probable Real Cost (PRC)

(AQP x QRA) + SRA = PRC

	PRC	Difference from bid
Company A		
($5 million x 1.08) + $32,114.1 =	**$5,432,114.1**	**$432,114.1**
Company B		
($5 million x 1.04) + $51,344.1 =	**$5,251,344.1**	**$251,344.1**
Company C		
($5 million x 1.002) + $12,884.1 =	**$5,032,884.1**	**$ 32,884.1**

Figure 7

Certainly each AM/FM/GIS project is different, and the project team is the best group to decide what conversion approach and what associated services are best for their project. The Probable Real Cost (PRC) is not an absolute. It simply is a good indication, based on the past performance of the bidders, of what engaging their services may really cost your project.

COORDINATION OF SURVEYING, MAPPING, AND RELATED SPATIAL DATA ACTIVITIES

Doyle G. Frederick
Chairman, Federal Geographic Data Committee
U.S. Geological Survey
102 National Center
Reston, Virginia 22092

ABSTRACT

On October 19, 1990, the Office of Management and Budget issued the revised Circular A-16, titled "Coordination of Surveying, Mapping, and Related Spatial Data Activities." The revised Circular A-16 expands the breadth of coordination of spatial data* and assigns leadership roles to Federal departments for coordinating activities related to these data. The revised Circular A-16 also establishes a new interagency coordinating committee named the Federal Geographic Data Committee. The committee has established subcommittees and working groups and is beginning to coordinate different categories of data and to work on issues of standards, technology, and liaison with the non-Federal community.

OPPORTUNITIES AND CHALLENGES

Today, agencies are required to quickly respond to increasingly complex problems involving a wide variety of geographically referenced data sets, such as environmental, natural resource, or socioeconomic. New and growing administrative and regulatory responsibilities assigned to agencies also have placed tremendous pressure on existing information delivery systems. Computerized spatial data handling technologies, such as geographic information systems (GIS), have emerged as cost-effective tools for solving complex geographic problems and are assisting decisionmakers in finding solutions to real world management challenges. These powerful data handling and analysis technologies are acutely dependent on the availability and quality of digital spatial data.

Digital spatial data are a critical linkage among user organizations, hardware and software systems, and applications. Moreover, development of the necessary digital spatial data bases is invariably the largest cost factor in the computer-assisted analysis of complex issues. The immediate need for, and the large cost of, high-quality digital spatial data are creating opportunities to build innovative partnerships among government institutions and the public and private sectors to avoid wasteful duplication of effort and to yield high-quality data for the mutual benefit of all interested parties.

---

* Spatial data are geographically referenced features that are described by geographic positions and attributes in an analog and (or) computer-readable (digital) form.

Building innovative partnerships and improved coordination processes also will increase the Nation's ability to deal with future complex societal issues and to compete more effectively in the world marketplace through the eventual development of a national spatial geographic data infrastructure, with linkages at all levels of society (Federal, State and local governments and the private sector). This national infrastructure, linked by criteria and standards, will facilitate sharing and efficient transfer of digital spatial data between producers and users, which will increase the availability and timeliness of information such as new street and road networks, demographic and cultural patterns, and land use and land cover changes.

The Office of Management and Budget (OMB) realized that greater efficiency and effectiveness are offered by the use of spatial data and related technologies, but was concerned about the potential for waste caused by duplication of effort and a lack of coordination. In 1989, the OMB asked the Federal Interagency Coordinating Committee on Digital Cartography (FICCDC) to analyze the need to expand the coordination of Federal spatial data use and to review and recommend potential revisions to Circular A-16. The FICCDC recommended that: (1) the breadth of coordination carried out by the committee be increased by the addition of other types of spatial data, such as geologic, resource (including soils, wetlands, and vegetation), cultural and demographic, and ground transportation; (2) the name of the committee be changed to the Federal Geographic Data Committee (FGDC) to reflect this broader coordination responsibility; and (3) the new committee and its responsibilities be incorporated within a revised and expanded OMB Circular A-16. The FICCDC recommendations of a revised Circular A-16 were reviewed and discussed by Federal agencies and the non-Federal spatial data community. On October 19, 1990, the revised Circular A-16 was signed by OMB Director Richard Darman.

THE FEDERAL GEOGRAPHIC DATA COMMITTEE

The FGDC "supports surveying and mapping activities, aids geographic information systems use, and assists land managers, technical support organizations, and other users in meeting their program objectives through:

- "Promoting the development, maintenance, and management of distributed data base systems that are national in scope for surveying, mapping, and related spatial data;
- "Encouraging the development and implementation of standards, exchange formats, specifications, procedures, and guidelines;
- "Promoting technology development, transfer, and exchange;
- "Promoting interaction with other existing Federal coordinating activities that have interest in the generation, collection, use, and transfer of spatial data;
- "Publishing periodic technical and management articles and reports;
- "Performing special studies and providing special reports and briefings to OMB on major initiatives to facilitate

understanding of the relationship of spatial data technologies with agency programs; and

- "Ensuring that activities related to Circular A-16 support national security, national defense, and emergency preparedness programs." (Office of Management and Budget, 1990)

Committee Goals and Objectives

The committee's goals include increasing the Nation's ability to compete more effectively in the world marketplace and to resolve complex issues and improving the efficiency and effectiveness of Federal programs. These goals recognize that the United States needs timely and accurate data to compete with the information-based economies of developed nations. Geographic data allow information to be used in new and powerful ways. Geographic data and related technologies are important tools for understanding complex national and global environmental, resources, and societal issues and for proposing and evaluating possible solutions. Customers of Federal programs have increasingly high expectations of service from Federal agencies and are pressing to have these services delivered more quickly. Geographic data offer opportunities to deliver current services at lower costs and to provide new services.

The committee has four major objectives to accomplish these goals: promote the development of a national digital spatial data infrastructure, reduce duplication and waste, promote sharing of digital spatial data, and promote wise use of spatial data technologies. The development of a national digital spatial data infrastructure provides a cooperative framework for sharing data among Federal, State, and local government agencies and the private sector. Reducing duplication and waste includes avoiding the duplication of work already complete or in progress and identifying high-priority areas where scarce resources should be applied. Sharing of digital spatial data will be promoted by providing information on data development and availability, by encouraging data exchange through the development and implementation of standards, and by understanding the application of data in different fields. The committee will provide forums for information exchange on technologies and their use and will encourage the development of procedures to assess the benefits and effectiveness of technologies to promote the wise use of spatial data technologies.

Examples of activities of the FICCDC to be continued and enhanced by the FGDC include: (1) developing a National Geographic Data System (a system of independently held and maintained Federal digital spatial data bases, linked by standards and criteria); (2) updating the FICCDC technical report "A Process for Evaluating Geographic Information Systems"; and (3) continuing the inventory and producing revisions to the "Summary of GIS Use in the Federal Government".

Committee Membership and Structure

Circular A-16 names the Departments of Agriculture, Commerce, Defense, Energy, Housing and Urban Development, Interior, State, and Transportation; the Federal Emergency Management Agency; the Environmental Protection Agency; the National Aeronautics and Space Administration; and the National Archives and Records Administration as members of the FGDC. The committee subsequently admitted the

Tennessee Valley Authority and the Library of Congress. The Department of the Interior chairs the FGDC. Other Federal departments and independent agencies with activities or interest in surveying, mapping, or related spatial data should be represented and can request membership by writing to the Secretary of the Interior.

The FGDC will establish, in consultation with other Federal agencies and appropriate organizations, the standards, procedures, interagency agreements, and other mechanisms to carry out its coordinating responsibilities. The committee may recommend to OMB additions to, revisions of, or deletions from Circular A-16 and supporting documents. Subcommittees, working groups, and advisory committees may be convened. The new subcommittees and working groups build on coordinating mechanisms established under the previous version of the circular and the FICCDC working groups.

The FGDC established a series of subcommittees, working groups, and a Coordination Group to undertake the responsibilities assigned by the circular (fig. 1). Subcommittees coordinate activities and develop standards for the categories of data to be coordinated under the circular. Each subcommittee is chaired by a lead agency identified in the circular for the category. Members includes agencies that have program needs involving that category. There may be further subdivisions of these groups to handle specific issues. As additional spatial data categories are identified, lead agencies will be designated and subcommittees will be established to develop standards and coordinate activities.

Working groups deal with issues common to all spatial data categories and have been established for three such issues: standards, technology, and liaison with State and local governments, academia, and the private sector. Members include Federal agencies interested in these issues. The Coordination Group provides a means for the subcommittees and working groups to interact and coordinate their activities. The Coordination Group is composed of the chairs of the subcommittees and working groups, the committee secretariat, and representatives of FGDC member departments and independent agencies and complementary coordination groups.

SPATIAL DATA CATEGORIES AND LEAD AGENCIES

Surveying, mapping, and related spatial data categories are characterized by spatial data that are a part of, or frequently analyzed with a reference to, the base topographic data and geodetic network of the Nation. These national spatial data categories are of interest to many agencies and represent a substantial part of the national spatial geographic information resource.

The revised Circular A-16 expands the coordination to 10 categories of spatial data: base cartographic, cadastral, cultural and demographic, geodetic, geologic, ground transportation, certain international boundaries, soils, vegetation, and wetlands data. Most of these spatial data categories are collected, maintained, and disseminated under national programs established by law. The circular also initiates coordination for other national spatial data categories for which there are broad interest and resources spent by FGDC member departments and independent agencies.

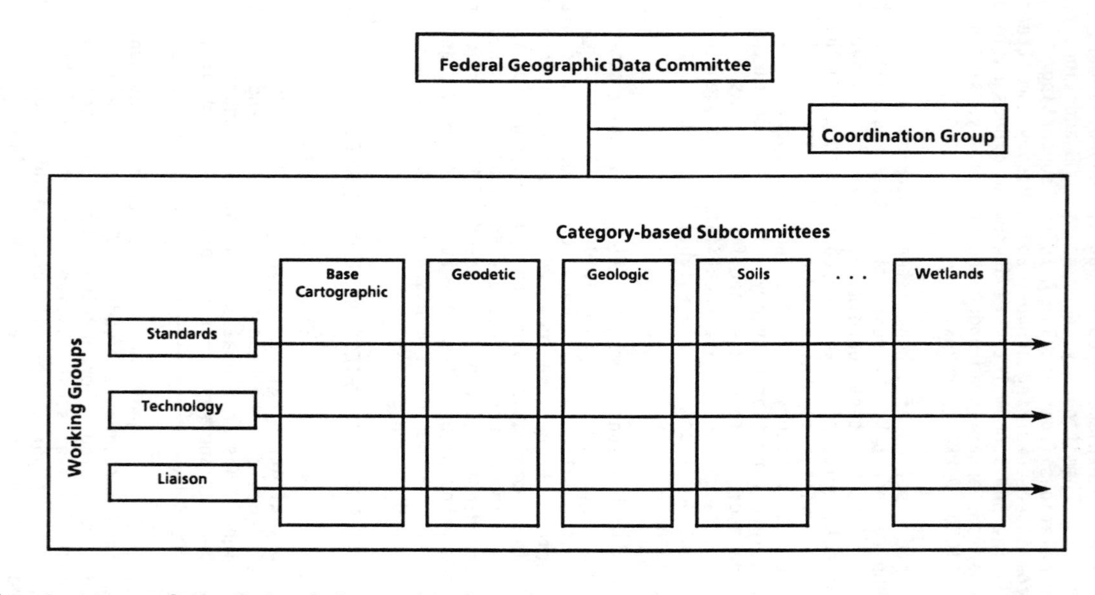

Figure 1. The structure of the Federal Geographic Data Committee. The Federal Geographic Data Committee guides and oversees the activities of the subcommittees and working groups. The subcommittees coordinate activities related to each spatial data category. The working groups deal with issues common to all spatial data categories and promote consistency among the subcommittees on these issues. The Coordination Group provides a means for the subcommittees and working groups to interact and coordinate their activities.

Circular A-16 assigns to the Departments of Agriculture, Commerce, Interior, State, and Transportation the responsibility to lead the coordination of these national spatial data categories. This responsibility includes facilitating information exchange and data transfer; establishing and implementing standards of quality, content, and transfer capability; and coordinating the collection of spatial data to minimize duplication of effort. Each department has authority described in its mission or implied as part of its program responsibilities. The departments delegated this responsibility to the organizations listed in table 1.

This leadership is carried out under the policy guidance and oversight of the new FGDC. All lead agencies except the Department of State will carry out this responsibility by chairing FGDC subcommittees. Because of the unique nature of its coordination responsibilities for the portrayal of international boundaries, the Department of State will use other means to coordinate these activities.

INTERACTION WITH STATE AND LOCAL GOVERNMENTS AND THE PRIVATE SECTOR

Another important focus of the FGDC will be to provide guidance and promote cooperation among Federal, State, and local government agencies and between the public and private sectors in the collecting, producing, and sharing of spatial data. The Liaison Working Group is developing strategies to assess opportunities for coordination with non-Federal spatial data users. The objective of this activity is a partnership among Federal, State, and local government agencies and the private sector in the development of a national spatial geographic data infrastructure.

Table 1. Spatial data categories and lead agencies under the revised OMB Circular A-16.

Category	Lead agency
Base cartographic	U.S. Geological Survey
Cadastral	Bureau of Land Management
Cultural and demographic	Bureau of the Census
Geodetic	Charting and Geodetic Survey
Geologic	U.S. Geological Survey
Ground transportation	Federal Highway Administration
International boundaries	Department of State
Soils	Soil Conservation Service
Vegetation	U.S. Forest Service
Wetlands	U.S. Fish and Wildlife Service

INTERACTION WITH OTHER COORDINATION GROUPS

The revised Circular A-16 builds on the FICCDC and earlier versions of the circular. The authors of the revised circular recognized the existence of other coordination activities related to spatial data and avoided duplicating these efforts. The FGDC's coordination with these complementary coordination groups will ensure the compatibility of activities. Discussions with representatives of other coordinating groups have resulted in suggestions of joint technical meetings, close coordination among the committees' secretariats, and involvement of representatives of the coordinating groups at meetings.

SUMMARY

The revised OMB Circular A-16 greatly expands the breadth of coordination of Federal spatial data activities and encourages the cooperation of Federal, State, and local government agencies and the private sector in the development of the Nation's spatial information resources. The successful development of this resource depends on all of these groups working together.

The FGDC has established a structure for coordinating spatial data activities. The subcommittees and working groups are coordinating agencies' activities, developing standards, facilitating data exchange, and investigating means of closer cooperation among government institutions and between the public and private sector.

REFERENCES

Office of Management and Budget, 1990, Coordination of Surveying, Mapping, and Related Spatial Data Activities (Circular A-16): Washington, D.C.

ACTIVITIES OF THE FEDERAL GEOGRAPHIC DATA COMMITTEE STANDARDS WORKING GROUP

Stephen C. Guptill
U.S. Geological Survey
519 National Center
Reston, VA 22092

ABSTRACT

The revised Office of Management and Budget Circular A-16 encourages the development of a national spatial data infrastructure. This information resource will facilitate sharing and efficient transfer of spatial data between producers and users and will increase the availability and timeliness of these data. The circular establishes an interagency coordinating committee, named the Federal Geographic Data Committee (FGDC), to promote the development, use, sharing, and dissemination of spatial data.

Standards are a major underpinning to the Nation's spatial data infrastructure. The FGDC Standards Working Group develops the standards needed for this infrastructure, including the National Geographic Data System concept, profiles for implementing the Spatial Data Transfer Standard, and frameworks for defining geographic features and attributes.

The National Geographic Data System is a set of independently held and maintained Federal data bases, national in scope, that meet certain minimum standards and are supported by their sponsoring agency. This system offers a means of coordinating Federal holdings of spatial data and will provide data that can be used with confidence by users.

The Standards Working Group develops Federal profiles for implementing the Spatial Data Transfer Standard. A profile is designed to handle a particular type of data in a standard way. The FGDC has developed a profile for topologically structured, vector-based spatial data and is considering a profile for raster data.

The working group also develops frameworks for defining geographic features and attributes. These frameworks will assist and provide consistency among the FGDC groups working on standards for particular categories of data.

CHARTER OF THE STANDARDS WORKING GROUP

The Standards Working Group actively promotes the exchange of information and ideas on common data standards for content, quality, and format, and develops methods for transferring digital cartographic and geographic data from one system to another. The responsibilities of the Federal Geographic Data Committee (FGDC) Standards Working Group are to encourage the development and implementation of standards, exchange formats, specifications, procedures, and guidelines; to promote the development, maintenance, and management of the National Geographic Data System (NGDS); and to establish a mechanism that would allow for FGDC agencies to agree on a framework for defining geographic features, attributes, and attribute values.

Specific tasks of the Standards Working Group include:

A. Promote the development of the NGDS and solicit comments and support from FGDC agencies.

B. Promote and coordinate development of Federal profiles for the Spatial Data Transfer Standard (SDTS) that would narrow the SDTS options used by Federal agencies.

C. Develop the methods and working arrangements for interacting with the FGDC subcommittees and the procedures for resolving conflicts in data content, data standards, and conformance with the SDTS.

D. Promote and coordinate the expansion of the spatial features and attributes section of SDTS.

E. Promote and coordinate the expanded use of the data quality section of SDTS.

F. Evaluate data definitions and standards used by the United Nations and other international organizations.

G. Support higher order or crosscutting activities established or recognized by the FGDC.

To accomplish these tasks, the Standards Working Group has focused its efforts on two major activities: the NGDS and the specification of a Spatial Data Transfer Standard Federal Profile for Topologically Structured Vector Data.

THE NGDS

The NGDS is a proposed method for improved coordination, organization, and management of Federal geographic data bases. The NGDS was developed by the Standards Working Group of the FGDC, formerly the Federal Interagency Coordinating Committee on Digital Cartography.

Although the U.S. Geological Survey had previously established the National Digital Cartographic Data Base (NDCDB), various other agencies are developing other digital spatial data bases, national in scope, which are not necessarily part of the NDCDB. The NGDS is being proposed to encompass other geographic thematic categories that can be managed in the broader context by geographic information systems technology.

The NGDS is a system of independently held and maintained Federal digital geographic data bases. The NGDS encompasses traditional cartographic data categories as well as geographic thematic data. The initial data categories are addressed in the 1990 revision of Office of Management and Budget (OMB) Circular A-16.

OMB Circular A-16 describes the responsibilities of Federal agencies with co-ordination of all survey and mapping activities financed, in whole or in part, by Federal funds. The following geographic data categories and corresponding lead agencies have been designated in Circular A-16:

- Base topographic mapping - U.S. Geological Survey
- Geology - U.S. Geological Survey
- Public land survey - Bureau of Land Management
- Wetlands - Fish and Wildlife Service
- Geodetic survey - National Oceanic & Atmospheric Administration
- Cultural and demographic - Bureau of the Census
- International boundaries - Department of State
- Soils - Soil Conservation Service
- Vegetation - Department of Agriculture
- Ground transportation - Department of Transportation

The primary objective of the NGDS is to promote and enhance the use of these federally developed geographic data bases. This use will improve the efficiency and effectiveness of Federal agencies in managing the Nation's natural resources, protecting the environment, developing the economy, supporting the national security, preparing for national emergencies, and facilitating the preservation of digital geographic information for historical purposes.

At least three strategies will be used to help achieve the primary objective of promoting and enhancing the use of federally developed geographic data bases.

1. Inform potential users about the availability and content of digital geographic data to avoid duplication of efforts.
2. Encourage and hasten the use of the SDTS to facilitate data transfer and sharing.
3. Develop and use additional standards for data content and quality to achieve more consistent and compatible data among Federal agencies and other users.

Regarding the first strategy, a centralized NGDS data base is envisioned to maintain and provide access to a collection of data dictionaries and directories describing the content of the various geographic data bases. Open access to this metadata base is a major component in making information obtainable on data availability and content.

The second and third strategies are addressed by the set of criteria that are used to accept an agency's data base as a participant in the NGDS. The scope of the NGDS includes digital spatial data bases of Federal agencies willing to meet the "Criteria for Inclusion" in the NGDS.

Candidate data bases must be available in conformance with the SDTS. The SDTS Federal Profile for Topologically Structured Vector Data will be used to the maximum extent possible. The SDTS was designed to accommodate the transfer of digital spatial data, and so any data from a candidate NGDS data base must be available as a SDTS data set.

In addition to the use of the SDTS for transfer and distribution, a NGDS data base must be in conformance with certain minimum content requirements, including the use of standardized feature sets. Often identifications of important geographic phenomena have not been formally standardized. Different codes, names, or labels are used by Federal agencies to identify the same phenomena in their geographic data bases. Incumbent upon the lead agency of a data category is the preparation of standardized feature set definitions and the maintenance of these feature sets using the principles described in various information resources management guidelines.

The SDTS specifies the form and components of a data quality report. The report consists of five sections covering lineage, positional accuracy, attribute accuracy, logical consistency, and completeness. An SDTS quality report on each NGDS data set is required to ensure data integrity and to allow an adequate evaluation of "fitness for use."

In addition, participating agencies need to have established mechanisms for:

- Distribution
- Maintenance
- Reporting status and progress
- Quality control
- Collection specifications
- Classification specifications
- Inclusion of other agency requirements in setting priorities for population of the data bases
- Records disposition

A formal review and approval process will be followed for (1) recommending additional lead agency designations for various data categories and (2) conformance with the SDTS and other specified criteria. The responsibilities for review and approval of data bases included in the NGDS rest with the FGDC and OMB. When acting on matters of the NGDS, the FGDC Steering Committee is considered the NGDS Review Board.

Agencies that request geographic data from the NGDS must work with the appropriate lead agency to acquire the data. If a second agency can provide the necessary data to meet a requirement, then both the lead agency and the second agency must negotiate an agreement that will provide the necessary data in conformance with the criteria for inclusion.

When data in another Federal agency's data bases are in conflict with the lead Federal agency's data base, then the lead agency must take the initiative to resolve the differences. The lead agency would be designated as the authority for the resolution of differences. Inconsistencies and differences among two or more lead agency data bases are also identified and resolved if possible. The FGDC Standards Working Group will assist in this process as necessary. Inconsistencies within the NGDS data bases may then be reduced to a minimum.

THE SDTS FEDERAL PROFILE FOR TOPOLOGICALLY STRUCTURED VECTOR DATA

The SDTS offers several potential advantages to users of spatial information: (1) it provides a set of simple cartographic objects from which digital cartographic feature representations can be built; (2) it allows the transfer of digital spatial information between incompatible systems while preserving the meaning of the information; (3) it supplies data quality information to users to permit evaluation of the fitness of data for a particular use; (4) it offers the opportunity to lower project costs by sharing data; and (5) it supports efforts to update data using multiple sources.

Basic Components of the SDTS

The SDTS consists of four components. Part I, Definitions and References, includes a conceptual model of spatial data and definitions of fundamental cartographic objects and key terms that serve as conceptual building blocks for the constructs presented in the standard. Part II, Transfer Specification, defines the logical file structure for the transfer of the data. Part III, Data Quality, specifies the form of the quality report and requires the data provider to supply detailed information about the data set being transferred in order for the user to evaluate the fitness of that data for a particular use. Part IV, Cartographic Features, presents a domain of cartographic entities with standard definitions.

Part I, Definitions and References, defines a set of primitive and simple cartographic objects in zero, one, and two dimensions with which digital cartographic feature representations can be built. Four different data models are supported in the standard. They are a raster (or grid) data model and three variations of vector models: geometry only, topology only, or geometry and topology combined. Definitions are also presented for key conceptual terms that are used throughout the standard.

Part II, Transfer Specification, is the method whereby spatial data from one data handling system can be easily transferred to another data handling system. The standard provides a modular specification of data types and formats for the full set of digital information to be transferred. Of necessity, the transfer specification is complex because it must handle both vector and raster data. ISO 8211 (1986), "Specification for a Data Descriptive File for Information Interchange," is a general purpose interchange standard that is used as the implementation mechanism. ISO 8211 is adopted for Federal use as described in Federal Information Processing Standard Publication 123.

The basic concept underlying Part III, Data Quality, is "truth in labeling." The standard requires a report of what is known about the quality of the included data. Five components are available to define data quality: lineage, positional accuracy, attribute accuracy, logical consistency, and completeness.

Part IV, Cartographic Features, provides a model and definitions for cartographic features. The model consists of entities, attributes, and attribute values. A list of topographic and hydrographic entities and attributes is provided with the standard; however, the standard also provides a mechanism for user-supplied definitions to accompany the data transfer.

Developing an SDTS Federal Profile

A profile, in general terms, is a clearly defined and limited subset of the SDTS designed for use with a specific type of data. The SDTS contains a full range of capabilities and options designed to handle a wide spectrum of geographic and cartographic data structures and content. The standard specifies a series of transfer modules, each containing a collection of module records. Each module record contains data fields that have been grouped together because of the purpose and (or) function of that information. A transfer may consist of a single physical file containing multiple modules or of multiple physical files each containing one or more modules. Decoding software to handle this full range of elements would be difficult to design. An alternative way to use the SDTS is to define a profile with a limited number of SDTS options and design decoding software to handle just those options.

The SDTS Federal Profile for Topologically Structured Vector Data has been developed for use with geographic data encoded using planar graph topology. Geographic data describe real world features, rather than a symbolized map graphic. The data may be derived from a cartographic product (map), but the purpose of the data is not to convey the map graphic, but rather information about the geographic features displayed on the map.

The SDTS Federal Profile provides for the transfer of files, records, fields, and subfields in conformance with the following objectives:

a. to encode in a standard format;
b. to provide for machine and media independence;
c. to accompany the data with their description;
d. to preserve all meaning and relationships of the data; and
e. to keep both field and records to an appropriate maximum length.

An SDTS profile consists of specifications, each making a specific choice (or a very limited number of choices) for an encoding possibility not addressed, left optional, or left with a number of choices within the SDTS.

There are many areas within which choices can be made. The SDTS contains a few profile specification subfields that provide initial choices at developing subsets based on the presence or absence of the various object types. But the same kind of choices can be made at the record, field, and subfield levels. Certain ISO 8211 specific decisions can be made. File names, ordering, primitive data types, number of constructs—all can be limited or predefined to contribute to the rigor of an SDTS profile specification.

Detailed specifications for the profile have been developed and are being reviewed by the members of the Standards Working Group. The Bureau of the Census plans to provide a new extract of selected geographic and cartographic information from its TIGER system in a format that conforms to the specifications of the SDTS Federal Profile. The plans are to submit the SDTS Federal Profile to the National Institute for Standards and Testing for Federal Information Processing System approval as an annex to the SDTS.

SUMMARY

Building innovative partnerships and improved coordination processes will increase the Nation's ability to deal with future complex societal issues and to compete more effectively in the world marketplace through the eventual development of a national geographic data infrastructure, with linkages at all levels of society (Federal, State, and local governments and the private sector). This national infrastructure, linked by criteria and standards, will enable sharing and efficient transfer of digital geographic data between producers and users. The activities of the FGDC Standards Working Group in the development of the NGDS and related spatial data standards is seen as the first step in creating this information infrastructure.

GIS EDUCATION AND TRAINING: AN AUSTRALIAN PERSPECTIVE

Barry J. Garner and Qiming Zhou
School of Geography, University of New South Wales
P.O. Box 1, Kensington, NSW 2033, AUSTRALIA

ABSTRACT

The development of technology *per se* is a necessary but not sufficient condition for effectively exploiting the potential for GIS. The extent of adoption of new technologies depends upon the general level of public awareness of the importance of the new technology. Eduction and training are essential for this. By way of background, this paper reviews GIS development in Australia and the responses to education and training needs. Particular reference is then given to the experiences gained over the past ten years in developing courses and curricula at the University of New South Wales - now one of the leading centres of GIS research and teaching in Australia.

INTRODUCTION

Geographical information systems represent one of the fastest growing areas in information technology both from the viewpoint of developments in the technology itself as well as in the rapid pace of its innovation and adoption. Although the development of the technology per se is a necessary condition, it is not itself sufficient to ensure that the full potential of GIS is exploited effectively. The pace and extent of the adoption of GIS and its extension into a much broader range of areas of application depends ultimately upon the level of awareness of the potential benefits that may be gained from their application.

The lack of awareness of the significance of GIS and spatial data handling generally has been singled out by many commentators as one of the most important reasons for its low level of "take-up" and the still relatively small size of its "core" market. Additionally it has been argued that the lack of awareness of the current state of the art in handling geographical information is not only holding back its wider use but results also in the duplication of investment, effort and, importantly, repetition of the learning process between and within organisations (DOE, 1987). Although significant advances have been made since Lord Chorley's committee of Enquiry into the handling of geographic information published its findings in the UK, many of the key problems relating to lack of awareness are still apparent today. An important reason for this is, of course, the continued expansion of the use of the technology into new, diverse markets and its adoption by new groups of users.

The full exploitation of geographic information requires that potential users are made aware of the benefits of GIS and that there exists a body of adequately trained personnel at all levels to implement, manage, and operate systems. Education and training are essential for this. That this is being increasingly recognised in most countries is reflected by the debate in now an already considerable body of literature addressing GIS educational issues and training needs. This paper contributes to this debate by reporting on some of the more significant recent initiatives in GIS education and training in Australia. By way of background, a brief overview of LIS/GIS developments in Australia is presented. Against this, the finding of a recent national survey of LIS/GIS research and education needs are summarised and selected responses are discussed. Emphasis is placed on the different disciplinary approaches to GIS education being adopted in Australia.

GIS DEVELOPMENTS IN AUSTRALIA

Australia has been especially responsive to the advances made during the past decade in the development of computer systems for handling geographic data. The application of LIS and GIS has been widely embraced and is now well established throughout the country in both the public and private sectors. This is particularly the case in the areas of LIS and the use of GIS in natural resources and environmental management but less so thus far for social, economic, and planning applications - areas for which the adoption of spatial data handling is still at an early stage compared to some other countries (Garner, 1990).

To date, the most active players have been in the public sector, particularly the state's government departments and agencies. Under the federal structure of government in Australia, the six state and two territory governments have jurisdiction over all the major areas for which land and geographical data are important: biophysical resources, social services, planning and economic development, transportation, utilities and infrastructure, and especially property ownership, land titles, registration, and land taxation. The role of the federal government in Canberra is by and large one of coordination, regulation, and setting national policies and priorities. The division of powers and responsibilities between these two upper levels of government has resulted in some fragmentation in the way the new technologies have been embraced. As a result it is difficult to provide national perspectives given that to date few nation-wide programs have been put in place. The most notable exception relates to LIS as distinct from GIS (O'Callaghan and Garner, 1991; Williamson, 1986). Good overviews of current developments in all areas of LIS/GIS application and development in Australia are found in the Proceedings of the annual Urban and Regional Planning Information Systems conferences that have been held since 1976 (AURISA, 1976ff).

Land Information Systems

From a national perspective, the most visible and well coordinated area of spatial data handling has undoubtedly been that of land information management and administration. Compared to many other countries, Australia has a sophisticated and tightly controlled system of recording land ownership based in large part on the Torrens system of land registration. Land parcel records are complete and comprehensively maintained; the title to land is supported by government guarantees and land and property taxes are a significant source of state government revenues. The information embedded in the cadastre is thus central to state land administration practices in Australia and is basic to the requirements of local governments and their utility authorities (Williamson, 1982).

Building on initiatives taken in the late 1970s, each of the six states and the two territories have put in place progressively the administrative and management structures to develop and implement statewide computer-based LIS during the 1980s. Currently each jurisdiction has operational cadastre-based systems for legal and fiscal administration although there are still some differences between their level of coverage, structure, and implementation. The development of these statewide LIS is the only area in which Australia can be considered to have a national program in the application of spatial information technology. This has been coordinated since 1986 by the Australian

Land Information Council (ALIC) which in 1987 endorsed a set of policies and procedures which now form the National Strategy for Land Information Management (ALIC, 1987).

The principal aims of the National Strategy are to encourage cost-efficient access to and use of land data; to provide an operational basis for effective decision making about the social and economic use of land; to promote development of technology; and to provide institutional mechanisms for efficient data transfer and standards. The strategy has been viewed as a major accomplishment in promoting inter-government cooperation in Australia where land-related issues have traditionally been a sensitive matter between the federal and state governments. Importantly, the strategy explicitly recognises the need for a substantial improvement nationally in the opportunities for education and training for the land - as distinct from the geographic - information industry which it is fair to say is dominated in Australia by the discipline of surveying.

Geographical Information Systems

Australia's geography, particularly its natural resource wealth, fragile environment, and major environmental problems and conflicts, is ideally suited to the application of GIS. Compared to the initiatives in LIS however, the adoption of GIS has been somewhat less significant nationally to date although it is growing rapidly. All states are currently engaged in building GIS for a range of environmental applications, especially for forest management, water resources, soil conservation (especially for farm planning and catchment management), mineral resources exploitation, nature conservation, the management of national parks, and coastal monitoring (O'Callaghan and Garner, 1991). The systems that have been put in place to date, however, are mostly at an early stage of implementation, are generally uncoordinated, and thus far are not formally integrated with the national LIS effort. Given the dynamic and evolving concept of the National Strategy however, it is ultimately envisaged that this will be enlarged to encompass environmental and socio-economic considerations.

A significant step in this direction was taken in 1989 when ALIC convened the first National Workshop on Natural Resources Data Management. Recent initiatives by the federal government have also highlighted the need for better mechanisms for the national coordination of environmental data and stressed the need for an increased emphasis on environmental and natural resources applications of GIS. As part of this effort, the federal government has now established two new agencies - the National Resources Information Centre (NRIC) within the Department of Primary Industries and Energy and the Environmental Resources Information Network (ERIN) within the portfolio embracing arts, sport, the environment, tourism, and territories. Both will rely heavily on GIS technology in their applications work.

In contrast, there have been few comparable federal initiatives directed at the management of human resources and the built environment. GIS developments in these areas are still restricted to the activities by the individual states although there are already clear indications that there is rapid growth in the adoption of GIS in the private sector for which favourable taxation arrangements for investments in information technology in recent years are already beginning to bear fruit. The finance and banking sector in Australia ranks among the world's leaders in the adoption of information technology and its application to the spatial domain is already visible. Similar trends are emerging in retailing and to a lesser extent in transportation and market research.

THE EDUCATIONAL NEEDS

The comparatively high rate of diffusion of GIS technologies in Australia has resulted - as elsewhere - in a shortage of trained and skilled personnel. Governments are finding it difficult to find staff to build, operate, and manage systems; the private sector is short of trained staff; vendors have difficulty in recruiting qualified technical experts and knowledgeable representatives; and the educational sector in Australia shares the same problems as elsewhere in attracting academics capable of providing instruction in the theory and practice of LIS/GIS at all levels. The need for increased emphasis on GIS education and training is as apparent and urgent in Australia as it is throughout the world. The problems and issues of adequately responding to the challenge, as well as the experiences, are likewise those typically shared elsewhere as might be anticipated (see DOE, 1987).

The National Survey

The need for better educated and trained personnel in LIS/GIS was highlighted in a national survey undertaken in 1989 - the only one to date - undertaken by a Working Party of the Australasian Urban and Regional Information Systems Association (AURISA, 1989). The major tasks of the Working Party were to:

- determine the research issues and the research facilities and programs relating to LIS/GIS in Australia;
- determine the educational requirements and document the educational programs and facilities relating to LIS/GIS in Australia;
- identify the feasibility of implementing a national strategy on education and research in LIS/GIS.

Although the final report identified a number of educational initiatives that might profitably be promoted and suggested policies that should be implemented, the response to these this far has not been particularly encouraging outside a few educational institutions. Consequently, there is little evidence today that there is a truly national response to the challenges GIS education pose.

Not withstanding, the Report documented the extent to which teaching and research in LIS/GIS was then being undertaken at Australian Universities and Colleges. That this is widespread is evidenced by the finding that some 35 different educational institutions were actively engaged in teaching and research in fields and areas relating to LIS/GIS and that within these some 50 separate departments and schools were involved. Principal among these were departments of geography, surveying, computer science, planning, landscape architecture, and environmental studies. Many of these were offering specialist courses at both undergraduate and postgraduate levels but few institutions could be considered to be significant centres for LIS/GIS education. Those that are include Curtin University in Western Australia, the University of Melbourne in Victoria, Queensland University, the Centre for Resources and Environmental Studies at the Australian National University in Canberra, and the University of New South Wales in Sydney.

Various Approaches

The need for adequate provision for education (and training) in GIS is widely recognised and clearly understood. How best to service this need is, however, still a matter of considerable debate as evidenced from even a cursory review of recent literature in the field. The significant differences in philosophical viewpoints would appear to be equally matched by differences in what is perceived to be the appropriate educational content. Interestingly, questions of how to teach GIS are addressed in the literature much less often than matters relating to what to teach, when it should be taught, and the sequencing of instruction.

The reasons for this are not hard to find. In large part they relate to the nature of GIS itself coupled with the fact that the interest in the new technology spans very many disciplines, industries, and application areas. Consequently it is unrealistic to expect that there is a single solution to the thorny problems of GIS education. This is borne out by the varied nature of the approaches being adopted. In any case, it may be argued that the rapid rate of adoption of GIS technology has forced us into a situation of responding as a matter of urgency, as best as we are able, to cope with what can only be described as a situation akin to 'crisis management'. It will undoubtedly take a long time before truly adequate curricula in GIS education, regardless of discipline, are firmly put in place - the commendable efforts to date of the National Centre for Geographic Information Analysis in this regard notwithstanding.

The challenges and difficulties of devising suitable educational programs in GIS or for that matter the 'new' cartography, are now reasonably well documented (see Taylor, 1985; Unwin, 1989) and by way of response several "model" curricula have been proposed (see Poiker, 1985; Banting, 1987; Drummond et al, 1989; Nyerges and Chrisman, 1989; Unwin, 1990). In summarising the difficulties of providing adequate GIS education, Unwin (1990) identifies at least five characteristic problems:

- The difficulty in defining formal educational aims and objectives resulting from differing disciplinary-based conceptions of the field.

- The problems resulting from alternative definitions of the scope of GIS and the way these relate to curriculum content, especially those relating to issues of 'breadth' as opposed to 'depth' of treatment (see Table 1).

- The relatively high cost of GIS education in resulting from its dependence on state-of-art technology despite the continued trend towards cheaper hardware and software.

- The far varied nature of student backgrounds and deficiencies in formal instruction in key disciplinary areas.

- The difficulties resulting from the recency of the field and the rapid rate of development of GIS R & D which is reflected in an acute shortage of skilled educators, the paucity of instructional materials and ease of accessing them, and importantly, the institutional barriers to the implementation of what inevitably must ultimately be interdisciplinary or multidisciplinary educational programs.

Given the technical underpinnings of GIS, an important additional problem relates to the level of computing skills that is, or should be, required. Some may argue that GIS education should be built on a solid grounding in computer science added to which is a sufficient understanding of GIS. Others have argued that the focus in instruction should be on the theory, concepts, and applications of GIS first and foremost coupled to which is a sufficient understanding of pertinent materials from computer science.

Table 1 Viewpoints on the aims of a GIS syllabus

Viewpoint	Aim	Shallow	Deep
Data driven (input and output)	Understanding of GIS capabilities at a 'black box' level	Point, line, area topology	Data structures and data base design
Function driven	Understanding how GIS work at an abstract level	GIS operations and commands	Design of spatial and other algorithms
Systems driven	Knowing how to drive a GIS	User training	Software engineering
Project Applications driven	Understanding about GIS can do in a specific area of application	Demonstrations of working systems	'Mapping' problems into GIS terms

Source: Unwin, 1990 (Tables 1 and 2, pp 459).

AUSTRALIAN RESPONSES

The response to the need to introduce higher levels of GIS education in Australia has been a positive one during the past five years. Several major programs have now been put in place at Universities and colleges in most states although a strong and identifiable focus on GIS research, education, and training has only been developed to date at a handful of these. The approach adopted at these centres reflects the disciplines that have taken the lead - notably surveying, computer science, and geography.

There is no equivalent in Australia to the National Centre for Geographic Information Analysis in the United States or to the Regional Research Centre initiative of the Economic and Social Research council in the United Kingdom. In part because of this, there is still a noticeable lack of coordination in GIS education and research nationally despite the establishment of what may be considered regional centres. The notable exception is the Australian Key Centre for Land Information Systems based at the University of Queensland. To date, however, that centre has not assumed a national role or taken a national lead in setting agendas for GIS education although it has emerged as a key centre for GIS training, particularly for personnel from the Asian region.

Computer Science and Surveying

Given the strong emphasis to date in Australia on land information systems and management, and the computer science underpinnings of GIS, it is not surprising that LIS/GIS education to some degree has focussed on the disciplines of surveying and computer science in Australia. Most computer science departments have now introduced some instruction in GIS, albeit from an essentially technical standpoint although generally this is considered a minor component within overall programs. The notable exception is at the Curtin University in Perth, Western Australia.

Formal courses emphasising a significant component of GIS were introduced in the School of Computing Science at Curtin University in 1988 (Kessell and Moore, 1989). Since then an undergraduate stream in GIS has been put in place and a graduate diploma with a multidisciplinary basis has been introduced. The success of these has recently resulted in a separate department for LIS/GIS with the School of Computer Science. The stated aims of these programs is to produce a computing scientist first in contrast to a GIS specialist with some knowledge of computing. This is reflected in the structure of the undergraduate program offered in which GIS is a stream in years 2 and 3 of the four year course building on a mathematics and computational geometry core. Identifiable GIS and related subjects include spatial data bases, geographic modelling, photogrammetry and remote sensing, computer graphics, and map design and production as well as an introductory treatment of GIS per se.

The Graduate Diploma in contrast is designed for a progressional audience seeking training or retraining in the GIS field. The focus is more directly on developing a thorough understanding of practical and theoretical elements in the structure, development, implementation, management, and application of GIS suitable for students from a diverse range of disciplinary backgrounds and particularly professionals already working in the LIS/GIS field. The School also offers a full program of 2-3 day short courses on GIS topics for specialist audiences as well as a major 6-month intensive GIS training course aimed primarily at overseas students in conjunction with the Asian Development Bank and Bakosurtanal - the National Mapping and Surveys Agency for Indonesia. Although Curtin University currently must be considered the leading computer science based centre for GIS education and training in Australia, initiatives are now being undertaken at the University of New South Wales and the University of Technology in Sydney.

Teaching in LIS/GIS is considerably more strongly established within Schools and Departments of Surveying in Australia than in other disciplines. Some instruction is included in surveying degrees at each of the six surveying departments although the undoubted focus is the Department of Surveying and Land Information at the University of Melbourne which has been active in the field since 1985 (Williamson and Hunter, 1990). There too an undergraduate program, Graduate Diploma, and short courses and continuing education activities are now well established. As can be expected the structure, content, and orientation reflects the surveying perspective based on a balance between science, measurement science, land management, environmental management and other associated applications. The undergraduate program leading to the B.Surv. degree is based on six streams in which those in mathematics and computing, surveying science and professional studies still dominate. Subjects in land information management and technology only account for 25 percent of the overall program and includes introductory treatments of LIS-GIS, spatial analysis, cartography, LIS-GIS, and remote sensing together with related land-based subjects typically included in traditional surveying degrees in Australia. The Graduate Diploma, offered jointly with the School of Environmental Planning, is primarily designed to fill the need for those from diverse academic backgrounds wishing to obtain working knowledge of the theory, technology, and various applications of GIS within the broader discipline of the management of spatial data. The emphasis in the Diploma is much more firmly placed on GIS related topics with electives typical of those traditionally offered in Landscape Architecture.

Geography

Whereas in many countries geographers have taken a leading role in the design and implementation of GIS courses and education, somewhat surprisingly perhaps this has not yet been the case to the same extent in Australia. In part this probably reflects the placement of the discipline at most universities in Arts or Social Science Faculties and the budgetary constraints this imposes on infrastructure provisions. In part it reflects the shortage of trained academic staff. Developments at an early stage are however now evident within several departments although only those at the University of Queensland and the University of New South Wales could be considered to be significant.

THE UNIVERSITY OF NEW SOUTH WALES

Development in GIS education in the School of Geography at this University started in a simple way ten years ago with the introduction of the first course in computer mapping and data display following visits by T. Waugh, the originator of the GIMMS mapping package and D. Marble who was instrumental in providing advice and promoting the opportunities provided by the then relatively recent new developments. At about the same time the University established its Centre for Remote Sensing, the first of its kind at an Australian University and the beginnings of a major emphasis on spatial information technology began to emerge embracing computer science, librarianship, surveying, landscape architecture, and geography - all of which are today involved in different ways in various aspects of spatial data handling.

Geography at the University is unique in the discipline nationally because it is placed in the Faculty of Applied Science rather than in Arts or Science as it typically is elsewhere. Consequently its academic activities are built on a strong basis in science and technology consistent with the philosophy of the Faculty itself. The destructive emphasis in undergraduate and postgraduate programs is on applied geography embracing both physical and economic branches of the discipline. This more than anything else has driven the strong emphasis now placed on computing, remote sensing, and GIS in its programs and has accounted for the significance of these in its overall academic profile.

In contrast to the developments in GIS education and training in computer science and surveying, the emphasis in the geography courses is placed primarily on GIS applications particularly for natural resources management and environmental monitoring and to a lesser but increasing extent on applications in urban and regional analysis - areas in which all students receive systematic instruction in the 4-year course leading to the BSc degree. Mathematics is a compulsory subject for all students which strengthens their ability for a more analytical approach in which spatial and environmental modelling is significant. As a result, a greater emphasis is placed on the cartographic modelling underpinnings of GIS capabilities and their use in spatial and environmental analysis than is typical of GIS course offered in computer science and surveying.

Undergraduate programs

Following a revision to the curriculum, all students now follow a sequence of courses over 3 years in computer mapping and data display, GIS, and remote sensing. Typically the structure adopted is to provide a good grounding in the theory, concepts, and technical aspects of GIS followed by courses focussing on GIS applications. These are increasingly being integrated with teaching in the main systematic branches of geography - biogeography, geomorphology, pedology, and urban and regional analysis. The course is organised to permit students to engage in advanced studies in GIS in the final year. A significant development is the emphasis on the integration of the diverse technologies - especially remote sensing and GIS in which the School of Geography is among the leaders within Australia (see Zhou and Garner, 1990).

This strong emphasis on GIS education has been made possible by the infrastructure in place. This included 10 unit tektronix 4107 graphics teaching laboratory and a 15 unit networked SUN workstation laboratory in addition to a range of supporting graphics equipment. The MAP package is used to introduce students to the basic structure and operations of GIS and ARC-INFO for more advanced instruction. GIMMS is currently still the main software for teaching the basics of thematic mapping and data display. ERDAS is used for instruction in remote sensing and image processing and in-house software named REMAP is used both for teaching and research involving the integration of remote sensing and GIS and to develop an understanding of image-based GIS (Zhou, 1989a, 1989b).

Postgraduate Programs

A coursework Masters degree and a Graduate Diploma in remote sensing have been offered at the University since 1983. However, it was not until 1987 that an equivalent program was offered in LIS/GIS. Initially this was offered jointly between the Schools of Surveying and Geography largely because of a lack of adequate qualified teachers. The principal aim of both programs was essentially to provide a solid introduction to basic concepts and areas of application of the new technologies. Students were drawn from a wide range of disciplinary backgrounds and most are in full time employment. A significant member of students completing the courses have come from countries in the Asian region.

Since 1990, the program in LIS/GIS has been restructured and is now offered through the School of Geography. The former emphasis on LIS has been dropped and instruction now focuses entirely on the theory, concepts and applications of GIS and integration of GIS with remote sensing. Subjects included may also be taken as electives by students taking postgraduate degrees in other disciplines, notably in computer science and landscape architecture. The success of teaching may be gauged through enrolments which for the past three years has exceeded 50 students in the Principles of GIS subject.

Postgraduate teaching in GIS and in remote sensing is supplemented by a regular program of short courses designed to retool and upgrade partitioners in the field. The University is now one of the leading centres in Australia in this aspect of training in spatial information technology.

CONCLUSION

The rapid rate of development of spatial information technologies, particularly LIS/GIS, in Australia has raised the general problems for education and training as in other countries. The responses have to date been many and varied, differing between disciplines, and focused nationally at one or two key institutions despite the fact that a concern with LIS/GIS education is now widespread across the country. The structure and content of formal programs at the undergraduate and postgraduate levels reflects the disciplinary perception of needs notably between computer science, surveying, and geography. For the latter discipline, the University of New South Wales has played an important role in GIS education and training during the past decade. Now that basic programs have been put in place, it is anticipated that the School of Geography will emerge as a key centre for GIS education and training in Australia servicing the Asian region. Steps towards achieving this goal have already been put in place with sponsorship from ESRI Australia and Sun Microsystems under the Federal governments Partnerships for Development Program.

REFERENCES

ALIC (1987), National Strategy in Land Information Management. Australian Land Information Council Secretariat, Canberra.

AURISA (1976ff), URPIS - Proceedings of the Urban and Regional Planning Information Systems Annual Conferences. Australasian Urban and Regional Information Systems Association Inc., Canberra.

AURISA (1989), Report of National Working Party on Education and Research in Land and Geographic Information Systems. Technical Monograph 3, Australasian Urban and Regional Information Systems Association Inc., Canberra.

Banting, D.R. (1987), GIS in an undergraduate geography program, Proceedings, International Geographic Information systems Symposium: the Research Agenda, vol. 1; 297-304. Arlington: NASA.

DOE (1987), Handling Geographic Information. (The Chorley Report). Department of the Environment, London: HMSO.

Drummond, J., Muller, J-C., and Stefanovic, P. (1989), Geographic Information Systems in teaching at ITC, Proceedings, Auto-Carto 9; 38-46.

Garner, B.J. (1990), GIS for urban and regional planning and analysis in Australia, in L. Worral (ed.), Geographic Information Systems: Developments and Applications. Bellhaven Press: London. 41-64.

Kessell, S.R. and Moore, D. (1989), Meeting diverse needs in GIS training and education, URPIS 17, AURISA: Perth. 231-244.

Nyerges, T.L., and Chrisman, N.R., (1989), A framework for model curricula development in cartography and geographic information system, Professional Geographer, 41; 283-293.

O'Callaghan, J. and Garner, B.J. (1991), Land and geographical information systems in Australia, in D. Maguire, D. Rhind, and M. Goodchild (eds).

Poiker, T.K. (1985), Geographical information systems in the geography curriculum, Operational Geographer, 8; 38-41.

Taylor, D.F.R., (1985), The calculational challenges of a New Cartography, in D.F.R. Taylor (ed.), (Education and Training in Contemporary Cartography; 3-25. New York: Wiley.

Unwin, D.J. (1989), Curriculum for Teaching Geographical Information Systems. Department of Geography, University of Leicester, UK.

Unwin, D.J. (1990), A syllabus for teaching geographical information systems, Int. J. Geographical Information Systems, 4(4); 457-465.

Williamson, I.P. and Hunter, G. (1990), Education and research in Land and geographic information systems at the University of Melbourne, URPIS 18, AURISA: Canberra. 19-26.

Williamson, I. (1982), The role of the cadastre in a statewide land information system, URPIS 10, AURISA: Canberra. 285-305.

Williamson, I. (1986), Trends in land information system administration in Australia, Proceedings AUTOCARTO London. 149-161.

Zhou, Q. (1989a), The Integration of Remote Sensing and Geographical Information Systems for Land Resources Management in the Australian Arid Zone. PhD thesis, School of Geography, University of New South Wales.

Zhou, Q. (1989b), A method for integrating remote sensing and geographic information systems, Photogrammetric Engineering and Remote Sensing, 55(5):591-6.

Zhou, Q. and Garner, B.J. (1990), GIS and remote sensing: towards the better integration of data for land resource management, URPIS 18, AURISA: Canberra. 185-194.

DESIGN AND IMPLEMENTATION OF A USER INTERFACE FOR DECISION SUPPORT IN NATURAL RESOURCE MANAGEMENT

Gregory A. Elmes
Department of Geology and Geography
West Virginia University
Morgantown WV 26506

ABSTRACT

The user interface plays a critical role in the adoption and successful application of GIS technology. GypsES, a knowledge-based spatial decision support system for the management of gypsy moth in forested environments, has to overcome many user-oriented problems if it is to make a useful contribution to the pest management decision process. The user interface developed for the GypsES project integrates components considered necessary for its successful application, including the user's calendar of needs; appropriate devices for communication; and the common 'look and feel' of the system components. Among the considerations in design, development and implementation of the graphic user interface are the X-Windows access to GRASS code to permit the use of windows, icons, menus and pointing devices; its availability on multiple platforms; the exclusive use of low level functions to maintain portability and consistency; and the modularization of components to permit separate development teams to contribute to the project.

INTRODUCTION

With the acknowledgment that questions of GIS functionality may be solved technically, the user interface is gaining more attention. In 1991, software releases and NCGIA's research initiative on the user interface illustrate growing commercial and intellectual interest in a recognized but previously underdeveloped aspect of GIS. A successful GIS user interface is directly linked to the adoption and productive use of GIS technology. In the first generation of GIS success has often come at the expense of the user, in terms of expensive training, interpretation of arcane commands from voluminous, if not fully enlightening, documentation, and hours on the telephone to software support. This paper discusses the user interface development strategy of a knowledge-based information system, from the creation of a hypertext prototype to the encoding of a fully functional software package.

GypsES is a knowledge-based, spatial decision support system (SDSS) designed to aid in the management of gypsy moth in diverse forested environments. Information and management practices obtained from regions that have experienced gypsy moth infestation are included in the knowledge based 'advisors' of the system, to be transferred to newly infested regions where little or no management experience exists. Typical potential users range from an environmental manager with no support staff and multiple responsibilities, to entire administrative teams specifically assigned to gypsy moth control. The system can be tailored to different users and user situations through a user profile initiated at the start of a session. The user profile manages a set of rules pertaining to the appropriate map sets, type of analysis, scale of resolution, and other operational details.

The first section of the paper introduces the question of system evaluation and measures of success in user interface design. The general description of the system and the objectives of the interface design are the focus of the second section of the paper. The third and last section discusses the role of knowledge-based GIS in the user interface design.

SUPPORT SYSTEM EVALUATION

Decision support systems are designed to assist decision-takers to manage complex, ill-specified or poorly-structured problems, using well-structured models and routines. Spatial decision support systems have a corresponding role for problems that are explicitly spatial in nature. According to Beaumont (1991), " ... the ultimate appraisal of any DDS must be in terms of its enhancement to the effectiveness of the decisions and to the efficiency of the decision-making process". The twofold makeup of appraisal, so defined, is not easily dissected for analysis. In consequence, evaluation of a DSS requires detailed knowledge of the outcomes of the decision-making process for comparison with outcomes from a control situation. Evaluation also requires detailed knowledge of the decision-making process in reality and how it is represented in the decision support system. Neither requirement is accomplished simply or in the sort term. IN fact totally controlled evaluation is practically impossible to accomplish. While the focus of this paper is primarily on the second element, understanding and representing the decision-making process, it is understood at the outset that the dual elements of decision effectiveness and process efficiency are closely related and interdependent.

Usability encompasses ease and effectiveness of system use and generally implies that the human-computer interaction is easy. Poor usability will lead to distancing of the user from the SDSS along a continuum from use via an analyst through partial use to rejection of the system. High levels of usability will encourage the decision-makers themselves to explore the full range of functions provided by the system. In addition, Armstrong (1991) argues that SDSS interfaces must provide an adaptive ability for the generation of alternatives, and also that group activities should be supported. While flexibility is desirable it often necessitates higher levels of learning with respect to options and procedures. The nature of the personal computer or workstation mitigates against a natural group-oriented level of decision support yet to allow the technology to dictate work practices will not only decrease usability but will violate the principle of an accurate representation of the decision-process.

Currently the GypsES system is being evaluated in two field tests selected for differences in their management information environments, as well as in location and qualities of the physical environment. One site is a low-lying county on the fringe of metropolitan Washington D.C. with a longstanding insect pest problem. Management structures have developed locally to cope with the problem. The other test site is a ranger district of a National Forest that has neither much experience of the pest problem, nor an existing management system for gypsy moth related decision-making. Each test will last for one year and will record the degree of integration of the GypsES into day-to-day pest management activities. Means of testing usability range from participant observation, interviews directed at specific functionalities, log files, and audit files generated during system use.

DESIGN STRATEGY

The goal of this section is to outline the software engineering process of the user interface development strategy of GypsES. We followed a relatively informally specified software development lifecycle approach to the design, development and implementation of GypsES as befits a situation that is in part well-understood and in part lies on a research frontier. From its inception in 1988, the design strategy has emphasized identification of the management procedures of potential users, and the means of accessing appropriate SDSS functions to meet the requirements. The design process called for documentation and review, resulting in a functional requirements document using methods outlined in Abbot (1986) and Bloch (1987). Additionally each knowledge base is described by the specialist development team and a data requirements document has been outlined.

In order to organize the information in GypsES, the elements of the system are ordered in a hierarchy of nested objects and attributes. These are: The Land - the world view of GypsES; Administrative Units - spatially bounded areas administered under a single regime with respect to gypsy moth management; Management Units - regions of uniform management objectives (e.g. timber, recreation, wildlife); and Treatment units - identifiable, finite areas uniform with respect to gypsy moth management decisions. GypsES will assist in the decision-making related to the following functions at the Administrative Unit level: define spatial extent, define fiscal resources, select treatment units, priorities and schedule treatment, and design monitoring system and sampling protocols. At the Management Unit level the system will provide risk and hazard assessments, project defoliation potentials and identify candidate treatment units. At the Treatment Unit level GypsES identifies intervention tactics, directs the implementation of intervention, and evaluates treatment decisions.

The nature of the user interface plays a critical role in the successful adoption and continued application of GIS technology. The user is unaware of the formal hierarchical structure but relies instead on direct manipulation of objects of interest, e.g. forest stands and compartments, opportunity areas, city blocks, counties, etc,. Issues that have been considered in the design of the GypsES user interface include: a) the managers' seasonal calendar of activities, b) control over analytical procedures, c) physical communication devices, d) the consistency of 'look and feel' between different components of the system; e) the interchange of data and information between system and user; f) error trapping and recovery from mistakes, g) storage and processing of meta-data, and h) control over the design of product output formats. Because of the wide range and poorly understood nature of these user interface issues, a strong incentive existed for research to be conducted on operational questions during field trials.

GYPSY MOTH YEAR

A strong organizational concept in the management of gypsy moth is the calendar year. The existence of a regular sequence of activities, occurring at more or less the same time each year enabled us to develop a management strategy in circumstances where no single strategy has become accepted practice. While not every user will proceed through exactly the same sequence, the gypsy moth year provides a common ground with which most managers will have some experience. In consequence the logical control over analytical functions in GypsES is strongly organized around this temporal framework. One paradox of the interface design process is that, in this case, we anticipate many applications where detailed management practices do not exist. The system must simultaneously present an acceptable management process while retaining flexibility for individual preferences and predilections.

HYPERTEXT PROTOTYPE

Initially, design ideas for the GypsES user interface were developed in a hypertext prototype (Yuill et.al., 1990). Hypertext was a valuable tool for simulating alternative organizational and interface designs. Interactive demonstrations were held to elicit suggestions for improvement from selected groups of potential users. The use of Hypertext was such that improvements were implemented quickly, frequently with the user able to direct the changes. The hypertext prototype also proved useful as a central focus for development teams working in widely separated locations. Frequently the prototype revealed discrepancies in concept, missing steps and unspoken underlying assumptions.

GRAPHIC USER INTERFACE

As GypsES' central role is to provide information from linked knowledge-based modules, significant emphasis has been placed on the relationship of the visual appearance of the user interface to system functionality (see Figure 1). The graphic user interface integrates several features considered to be for the valuable for the successful adoption of the technology in managerial environments having little or no previous GIS and computing experience. These include cognitive cues such as relative placement of operating devices on the screen, e.g. buttons, menus, and windows; the use of color to indicate current location and 'depth' within the system; navigational aids and reminders, attention-demanding status and error messages, and consistency of interface devices between components.

By means of a hierarchical series of flow charts the GypsES user interface guides an inexperienced user through the most efficient use of the system. Experienced users are provided with tools to maximize system accessibility.

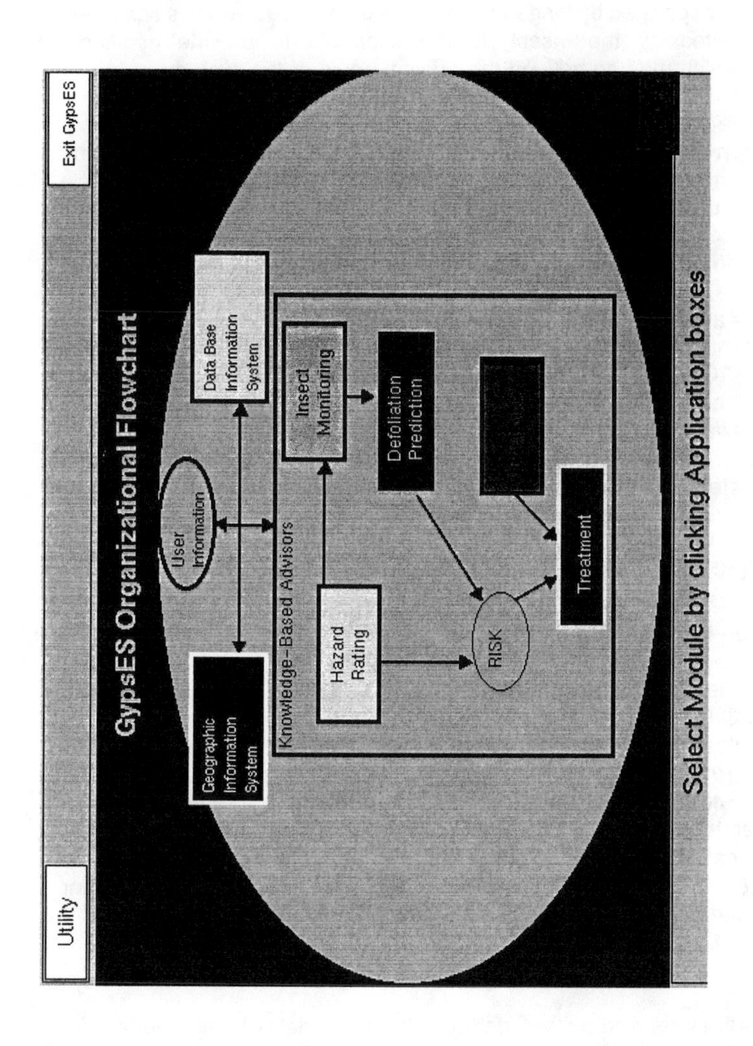

Figure 1. Organizational interface to GypsES knowledge-based advisors

The role of guidance through the system conflicts with the need for fully flexible access to GIS tools. How much GIS and computer expertise a user requires to user GypsES is an issue to be evaluated during field testing. Elements of the user interface that may increase the usability for a novice or infrequent user may quickly pall and decrease the level of usability for an expert or frequent user. The nature of the gypsy moth problem is such that we plan for intensive periods of use, separated by long periods of low demand. Given this schedule, which is dictated by the insect, it is reasonable to provide additional functionality which may go beyond the original goals of the system.

To this end the user interface also provides concurrent access to multi-tasking. Several design decisions were made to ameliorate the problems generated by rapid developments in GIS-related hardware and software industries. The adoption of public domain software has ruled the development process. Among the considerations in the user interface design have been the implementation of an X-Windows interface to GRASS files to permit the use of windows, icons, menus and pointing devices (WIMPS) in the GIS context. The use of menus and icons is extended to selected GIS operations such as object selection, identification, reclassification, and zoom and pan. GypsES runs under INTERACTIVE ™ UNIX and Macintosh AU/X using X-Windows 11.4 as the graphical environment. Presentation of the same user interface across multiple hardware platforms is facilitated by the exclusive use of low level functions between the user interface, GRASS code, and graphics functions to ensure consistency. We have avoided the use of the embellishments of higher-level functions such as Open-Look and Motif widgets to preserve portability.

KNOWLEDGE-BASED GIS AND THE INTERFACE

A distinction is necessary between the functioning of an expert GIS and an intelligent GIS. Maidment (1991) identifies an expert GIS as one that applies rule-based reasoning to spatial data with the goal of spatial problem solving. An expert GIS is at the heart of a SDSS and is analogous to the knowledge-based modules or advisors of GypsES. In contrast an intelligent GIS incorporates rule based reasoning applied to the user interface and to the control of GIS functions. An intelligent GIS is necessary to create the illusion that the user is manipulating the real world phenomena with which she is familiar. So-called 'transparent' user interfaces permit the decision-maker to think about the dimensions of the real world problem, not about the computer or the data and model structures (Gould, 1991).

The purposes of intelligent GIS functions within GypsES are to:

1) document, store, and assess data quality (meta-data) as appropriate for particular decisions and classes of user,
2) provide a means to measure accuracy of data and generated information, and convey the accuracy qualities effectively to the user,
3) provide rule bases for map layer analysis for selected operations, i.e. act as GIS power user,
4) protect the data and GIS operations from user error/misuse, basic security and integrity functions operate alongside error trapping and recovery,

5) provide a graphical user interface for easy access to decision support tools and resultant information (Lanter, 1991),
and,
6) provide tracking through scenarios, and a record of the decision framework, thus allowing modification of decision-outcomes and sensitivity analysis.

META-DATA

The user interface incorporates the means to ensure that data cannot be accidentally corrupted by the selection of improper operations, or by inappropriate changes in the scale of resolution. The means to store meta-data, including the data dictionary, data lineage, and to present error estimations to the user, are essential knowledge-based functions of the user interface. The complexity of knowledge required to handle all aspects of meta-data presents serious questions of system overhead. Implementation of expert components has been prioritized to focus initially on error handling and solution tracking. Error is inherent in the data source, the data capture operation, in GIS analysis and at display time. Rules governing each component are processed within frame-based modules.

Automatic documentation of operations is enabled to permit evaluation of the use of system functions. Unfortunately such a log provides only a partial perspective of the system's usage. Given that long term participant observation is impractical, the trial users will record their observations about the operation of the system in written logs and discuss them with the study team. We are also interested in identifying those elements of the decision process which are best left to the user. The computer need not become the single element in a SDSS.

NOVICE OR EXPERT?

Systems such as GypsES may not only be the first point of contact a user has with a GIS or SDSS, but may be the first contact with any computer system. While this issue will diminish with time, it will never entirely vanish. Unless more emphasis is placed on ensuring computer literacy as a basic component in education than at present, computer-avoidance will continue to exist at higher levels than those currently immersed in computer technology are prepared to admit. Is it possible to embed a future version of GypsES in the work place so completely that the user is unaware of the presence of a computer? Or conversely should we anticipate that the computer workstation using standard graphic interface conventions will become as familiar an object in everyday experience as the telephone and automobile?

There is considerable debate between the developers as to the desirability of the development of a single pathway through the interface. The choice of an appropriate metaphor for the interface provokes a philosophical debate somewhat akin to differences between rights and privileges. Associated with such a dilemma is the question of classes of user, public access and democratic use of public information. Are we able to separate the SDSS function from the GIS function in an organizational sense? Armstrong (1991) has raised the undesirability of solution tracking for fear that the decision process would be *'en clair'*, revealing flaws and dissent. GypsES' central role is providing information to gypsy moth managers from multiple knowledge-based advisors but the designers have no control over the political forum. In federal or state agency applications, sunshine laws may require public access to the data and the decision process. The designers of the user interface have the ability to conceal or reveal information simply by controlling the ease with which an untrained person can use the system. Because the majority of anticipated users will not have GIS training, significant emphasis has been placed on the appearance and functionality of the user interface. Such access may have unintended consequences for the decision-makers.

ACKNOWLEDGEMENTS

This research is made possible by grants from the USDA-Forest Service, Northeastern Forest Experiment Station, Gypsy Moth Research Program; and State and Private Forestry, Technology Development Program, Northeastern Area. Cooperators include Dr. Mark Twery, Ms. Patrice Janiga, Mr. Dan Twardus and Mr. John Ghent (USDA - Forest Service); Drs. Michael Saunders and Michael Foster (Pennsylvania State University); Drs. William Ravlin, Jesse Logan and Lukas Schaub (Virginia Tech).

REFERENCES

Abbot, R.J., 1986, An Integrated Approach to Software Development, John Wiley and Sons, New York. 334pp.

Armstrong M.P., 1991, User Interfaces to Support Locational Decision Making, Position Papers for the Specialist Meeting of NCGIA Research Initiative 13, "User Interfaces for Geographic Information Systems",(D.M Mark and A.U.Frank, Co-Leaders), June 23-26, 1991, Buffalo, NY.

Beaumont J.R., 1991, Spatial Decision Support Systems- A Personal Agenda, Environment and Planning A, Vol 23,3, pp. 311-317.

Bloch C., 1987, Geo-Referenced Information Network Functional Requirements, The Research Libraries Group, Inc., Jordan Quadrangle, Stanford, CA.

Gould M., 1991, The GIS User: Make No Assumptions, Proceedings of the Eleventh Annual ESRI User Conference, Vol 2, pp 519-523.

Maidment D., 1991, Panel Discussion, Eleventh Annual ESRI User Conference.

Lanter D. P., 1990, Lineage in GIS, NCGIA Technical Paper 90-6 University of California, Santa Barbara.

Yuill C. B., Millette T. L., Elmes G. A. and Twery M. J., 1990, Prototyping an Intelligent Geographic Information System using HyperText, Technical Papers, 1990 ACSM-ASPRS Annual Convention, Vol 3, pp 307-316.

GIS AND REMOTE SENSING TECHNIQUES FOR MONITORING WETLAND AND PRAIRIE COMMUNITIES AT THE UNIVERSITY OF WISCONSIN ARBORETUM

Brian L. Cosentino, Raad A. Saleh, David J. Savory, Jana S. Stewart, Robert A. Wasserman

Environmental Remote Sensing Center
Institute for Environmental Studies
University of Wisconsin-Madison
1225 West Dayton Street
Madison, Wisconsin 53706

ABSTRACT

The University of Wisconsin Arboretum, a 509ha site within Madison, Wisconsin, is currently exploring ways to better manage ecological data and to assess the potential of computerized spatial information technologies, GIS and remote sensing, for monitoring and analysis of plant communities. The wetland and prairie investigation is one of four demonstration pilot projects conducted by the 1990-1991 Environmental Monitoring Practicum at the UW-Madison for the Arboretum.

The objectives of this pilot project were to : 1) demonstrate methods for creating vegetation maps from digital aerial photography, and 2) demonstrate GIS overlay analysis for correlation of plant spatial variation and environmental factors.

Color infrared (IR) 70mm and large format (9" x 9") color transparencies were digitized with a scanning microdensitometer. The large format images were rectified to the Wisconsin State Plane Coordinate System NAD83 using control points derived by photogrammetric methods with GPS control. The rectified data served as an image base for registering the 70mm imagery.

Two study sites were used for the first objective. In Curtis Prairie, approximately 26ha in area, both automated and manual image classification techniques were tested for mapping prairie plant communities. An unsupervised classification yielded plant specificity at the association level (e.g., xeric/mesic prairie grasses, forbs and shrubs). On-screen manual interpretation provided species-level classification. Wingra Marsh, approximately 20ha in area, served as the second study site. An enhanced image base for on-screen vegetation mapping was prepared by merging color, large format imagery with a 70mm color IR image using an Intensity-Hue-Saturation color space transformation. The modified image base augmented vegetation discrimination as well as demonstrate a method for extending the utility of aerial photography.

The Curtis Prairie study site was also used for the second objective. Several raster map overlays were compiled such as soil water capacity, pH, and elevation. A coefficient of areal correspondence was calculated between the map overlays and the Curtis Prairie vegetation maps prepared for objective one. A wide range of site factor-vegetation relationships were noted (e.g., xeric/mesic plants and soil pH).

This pilot project has demonstrated that GIS and remote sensing techniques can be effective tools for ecological research and management. As a result, the UW Arboretum has expressed interest in incorporating these techniques in research and management activities.

IMPLEMENTING GIS TECHNOLOGY TO MANAGE THE NATION'S AGRICULTURAL PRODUCER RECORDS

Lane C. Price, National Leader for
GIS and Remote Sensing Applications
USDA Soil Conservation Service
P.O. Box 2890, Washington, DC 20013

ABSTRACT

The success of the U. S. Department of Agriculture (USDA) to carry out its responsibilities depends upon not only the ability to efficiently share data among the Department's agencies, but also the ability to capture, manipulate, analyze, and maintain spatial (geographically referenced) data. This need for improved spatial data handling is perhaps greater than ever before, due in part to 1985 and 1990 farm bill responsibilities, as well as new initiatives in the water quality arena. Geographic informations systems (GIS) technology offers some tremendous advantages over USDA's conventional use and management of such geographic information as field boundaries for producers and soils maps.

The USDA Soil Conservation Service (SCS) currently has an implementation strategy of utilizing GIS in many of its 3000 county-level field offices across the country. The Agricultural Stabilization and Conservation Service (ASCS) is also investigating the use of GIS in its county offices. Putting GIS "in the hands of the decision-maker" at the county level poses many challenges to SCS and ASCS, including 1) making GIS "friendly" enough for non-GIS experts, 2) building the geographic databases at scales suitable for USDA's farm and field level analysis, 3) sharing the geographic and attribute data between USDA agencies, and 4) developing a relatively low cost system required for such a large decentralized implementation.

INTRODUCTION

Several USDA agencies use farm/tract/field boundaries and associated producer records to carry out their specific agency missions. These agencies include ASCS, SCS, Farmers Home Administration (FmHA), Federal Crop Insurance Corporation (FCIC), and Agricultural Extension Service (AES). This paper focuses on the use of GIS by SCS and ASCS, as these two agencies share the bulk of the responsibility for capturing and maintaining field boundary lines and much of the related natural resources and cultural attribute data.

SCS and Producer Records

SCS is considered a technical arm of USDA, with the mission of providing national leadership in the conservation of our Nation's natural resources. SCS accomplishes this mission by working one-on-one with landusers to develop Conservation Plans. A landuser's Plan is a tabular report of conservation practices needed for each field, an implementation schedule, and a series of farm level maps. The maps have traditionally included themes such as field boundaries, soils, and other resource themes over a photographic image. It is on these individual farm maps (usually at scales of 1:7920 to 1:12000) that SCS has traditionally tracked the location of field boundaries and conservation program activities.

SCS is also mapping detailed soils information (at scales of 1:12000 to 1:24000) nationwide. This geographic database, approximately 70 percent complete, has also traditionally been recorded and stored on conventional photography. Other mapping programs within SCS include the mapping of wetlands (for farm bill implementation), watershed delineations, and national inventory samples.

ASCS and Producer Records

ASCS is responsible for administering programs which provide for commodity loans and price support payments to farmers, commodity purchases from farmers and processors, acreage reduction, cropland set-aside, conservation cost-sharing, and emergency assistance.

Like SCS, ASCS also maintains a record of the location of the field boundaries for the county's producers on aerial photography. Unlike SCS, ASCS stores the field boundaries by photograph (typically at 1:660 scale), rather than by individual. ASCS records associated data for each field, including acreage, cropping history, tract designation, and other information.

Compliance programs also require ASCS to measure the actual cropped acreage annually by field for certain crops. ASCS accomplishes this task with 35mm slide photography. These non-rectified annual images are typically projected onto the hardcopy rectified base photography for making planimetered measurements by field.

A Decentralized System

The missions of both agencies is essentially accomplished at the county-office level where all the work with individual farmers or ranchers is performed. This decentralized field-based delivery system, unquestionably the strength of USDA programs, dictates the environment under which GIS can be efficiently utilized.

The information management requirements of the two agencies is ever increasing. Both agencies deal extensively with both tabular and geographic data. Because of the lack of spatial data automation, USDA agencies which deal with producers are duplicating the capture and maintenance of geographic data such as field boundaries. Inconsistencies in hardware, software, and databases has made even the sharing of tabular data (such as participants names and addresses) difficult.

GEOGRAPHIC DATA MANAGEMENT NEEDS

An example geographic data sharing and analysis required of USDA in the 1990's is obvious from the farm bill legislations. To implement the conservation provisions of the Food Security Act of 1985 and the Food, Agriculture, Conservation, and Trade Act of 1990, USDA must inventory, manage, and analyze natural resources and cultural data for the Nation's agricultural lands to determine a producer's eligibility for USDA program benefits.

To maintain eligibility for benefits by ASCS and other agencies, the 1985 farm bill requires agricultural producers to have an approved SCS Conservation Plan for all their highly erodible fields by 1990, and have the plans implemented by 1995. (Highly erodible fields are those with greater than 50 acres of HEL or greater than one-third HEL.) The bill also denies program benefits for producers who drain and plant crops into wetlands. The bill also established criteria for producers to remove sensitive lands (based on soil erodibility, slope, and cropping history) from crop production for annual incentive payments. The 1990 farm bill extended the 1985 bill, targeting the most highly erodible and environmentally sensitive lands.

This legislation essentially requires that producer records from ASCS and resources data from SCS are synthesized on a field by field basis to determine eligibility for participation in programs administered by ASCS and other agencies, such as FmHA and FCIC. Unfortunately, almost none of these maps are recorded on a planimetrically accurate base, so simple map overlays (eg. soils and fields) are not possible. Since none of the geographic data is in digital format, all of the transactions between agencies occur by photocopying a photograph for a single farm and hand-delivery to other agencies. Not only is error propagation likely as data delineated on one photograph is recompiled to another photograph, but the process is very time-consuming and costly. For example, SCS estimated that work on the conservation provisions of the 1985 farm bill alone would require about 70 percent of its field staff's time until 1995 (1). A significant portion of this effort has been geographic data analysis and management: making HEL and wetland determinations, disseminating that information to other USDA agencies for program administration, and developing farm bill Conservation Plans.

Similar types of geographic analysis are required for other on-going programs as well, such as ASCS's Agricultural Conservation Program (ACP), Special Projects, and Compliance programs; and SCS's conservation technical assistance programs, resource inventories, watersheds and floodplain management studies, Great Plains conservation programs, Resource Conservation and Development projects, and soil surveys.

A GIS IMPLEMENTATION STRATEGY

A GIS is often defined as a system of hardware and software which allows the user to capture, store, manage, analyze, and display geographic data, such as maps or photography. In simple terms, it can be described as a relational database management system, in which geographic "layers" (rather than tables of text data) are managed and manipulated. A GIS is "relational" because each layer is referenced to the Earth. As virtually all the data recorded by SCS and ASCS is geographic in nature or linked to a geographic entity (ie. a field or tract), GIS could be viewed as a potential coordinating mechanism between agencies.

The strategy presented in this paper is primarily that underway by the Soil Conservation Service, and outlines the agency's plan to address the major challenges of implementing GIS in its field offices.

MAKING GIS USER FRIENDLY

Few SCS field offices will ever have a full-time GIS specialist. Perhaps the greatest challenge to putting GIS technology in USDA field offices is overcoming the real (and perceived) difficulty in operating a GIS.

SCS Selects GRASS

SCS selected the Geographic Resources Analysis Support System (GRASS) in September 1988 as the agency supported GIS. As of July 1991, SCS has about 125 sites running GRASS, most of which are at the State Office, NTC, and National Headquarters, levels. Approximately 40 field offices currently have GRASS, all running on 386-based microcomputers. ASCS is currently evaluating GRASS at its Aerial Photography Field Office in Salt Lake City, Utah. GRASS was selected by SCS because it was low cost (public domain), written in "C" language in the UNIX operating system, was easily portable to SCS hardware, well-designed and supported, and perhaps most important, easy to use. The decision to support GRASS was also due in large part to SCS's requirements for an "analysis-oriented" GIS.

GIS Interfaces

SCS is striving for a system where the entire field office staff will be GRASS users, regardless of GIS expertise. Integral in this strategy is the development and use of transparent interfaces linking the GRASS GIS to other software, models, and databases.

The CAMPS-GRASS Interface

The primary GIS interface to be used at the SCS field office level will be the CAMPS-GRASS Interface. CAMPS (the Computer Assisted Management and Planning System) is menu-driven software which manages such field office tasks as maintaining client operating records and planning data; writing and storing conservation plans; designing conservation practices; developing long range plans of work; and generating statistical, tabular, or narrative reports from client, soils, range, agronomic, climate, or other databases. A pilot test of the prototype software (Version 1.2) is now complete and an operational version will be developed during the winter of 1991-92.

Components of the CAMPS-GRASS Interface. The goal of the Interface is for the field office staff to be able to generate the most frequently required products very quickly, with no knowledge of advanced GIS principles or GRASS commands. The major components or functions of Version 1.2 of the Interface are:

(1) A module exists allowing the user to digitize field boundaries, and link these fields to a producer's fields. The Interface transparently creates and stores a window for the landuser, and (by analyzing the tracts dimensions) computes the most appropriate cell size for rasterizing the fields vector map. Field and soiltype-by-field acreage determinations can be automatically generated, and stored in CAMPS databases. Labels files, such as field numbers, are also transparently created for the user. Linking GRASS polygons to CAMPS data is accomplished with a point and click interface.

(2) A module exists allowing the quick generation of Conservation Plan Maps and Soils Maps (both screen and hardcopy) for a farm or tract. Windowing and masking functions are transparent to the user, and the user has complete control over display options such as vector overlays, label overlays, and color selections.

(3) A module exists to create and display various thematic maps from CAMPS databases for a farm, watershed, county, or other defined area. Examples include soils interpretive maps, landuse maps derived from producers field data, highly erodible land maps, and others.

Specifications. The CAMPS-GRASS Interface requires the AT&T 6386 computer and the normal peripherals required for GIS: digitizer, color printer/plotter, color display monitor and card, etc. Approximately 300 MB of disk storage is recommended for field databases, software, and operating system. Version 1.2 of the Interface is written in UNIX shell scripts, with a few commands in "C" language, and requires Prelude DBMS from VentureCom, Inc.

Future releases. Future releases of the Interface, after current pilot testing, will generally correspond to new CAMPS releases. Major planned enhancements include a re-write totally in "C" language utilizing the CAMPS Common User Interface (CUI) and the Vermont Views screen generation software; CAMPS 2.0 data structures; conversion to Informix SQL DBMS; easier development of the fields datalayer using on-screen digitizing with imagery; and perhaps most significantly, a more integrated use of GIS functions throughout CAMPS. This integration will include more specific applications, such as "expertly" locating practices such as cross-fencing in rangeland, computing runoff for the design of engineering practices, or identifying critical water quality problem sources -- all transparent to the user.

Other Interfaces

Other existing GIS Interfaces include the GRASS-SSURGO Interface and the GRASS-STATSGO Interface, both GRASS interfaces to SCS digital soils databases. STATSGO is the State Soil Survey Geographic database and is a generalization of the detailed soils maps useful for state and multi-county planning. SSURGO (Soil Survey Geographic database) is the detailed soils data typically digitized at 1:24,000.

Other potential GIS interfaces to databases which are at some stage of evaluation, planning, or actual development are: National Resources Inventory (NRI) data, range data, cultural features data, engineering survey data, etc. Under discussion or early planning, are GIS interfaces to various natural resource models, including Agricultural Non-Point Source Pollution (AgNPS), Water Erosion Prediction Program (WEPP), and others.

BUILDING DATABASES

For the effective management of producer records and basic resources information in the USDA field office, three geographic data layers are essential: field boundaries, soils, and orthophoto imagery.

Soils

Soils mapping for the U.S. is the responsibility of the Soil Conservation Service. Approximately five percent of the country is digitized to meet SCS's vector soils digitizing specifications. Soils digitization is typically contracted or digitized by the SCS State Offices. Soils digitization is hindered by the large number of soil surveys recorded on non-ortho imagery. All new surveys are mapped on orthophotography.

Field Boundaries

Unlike digital soils for which SCS receives numerous requests, field boundaries are somewhat specific to USDA. The only counties with digitized field boundaries at this time are those few which have participated in pilot tests for GIS implementation. Because local landowners are the primary data source and because of its dynamic nature, field boundaries will probably be digitized and maintained at the field office level. This means that software must facilitate an extremely fast and simple boundary changes and tract reconstitutions.

Orthophotography

An essential component to GIS implementation for managing producer records is the availability of an accurate base on which to map field boundaries. The critical features of this base map are 1) high level of positional accuracy to facilitate map overlays with other data, 2) a scale suitable for delineation of small fields (1:12,000 or larger), and 3) an image backdrop to provide distinction of cultural features required for delineating fields.

Orthophotography is a planimetrically accurate product with an photoimage background. Orthophotography can be generated digitally with a scanned aerial photograph, a digital elevation model (DEM), and camera parameter information. A 1993 budget initiative is being proposed by SCS, ASCS, and the U. S. Geological Survey (USGS) to develop digital orthophotography for the United States over a five year period. The proposal calls for the use of NAPP (National Aerial Photography Program) imagery, and a scan resolution of 25 or 50 microns (translating to a ground resolution of 1 or 2 meters). Being digital, hardcopy products could easily be output at a locally desirable scale and format (USGS quadrangle, Public Land Survey section, etc) for offices without GIS. The digital imagery would be distributed to GRASS sites, and serve as another layer in the GIS.

The primary functions of the orthophotography would be to 1) serve as a backdrop for orientation when dealing with producers, 2) serve as a highly-accurate hardcopy base map for mapping, digitizing, and tracking field and tract information, soils, or other data, and 3) facilitate very quick field boundary database development with on-screen digitizing. From the CAMPS-GRASS Interface pilot test sites, it is estimated that digital orthophotography and on-screen digitizing will not only result in a much more accurate product, but will also reduce digitizing time by over 60 percent. (The currently used procedure involves manually recompiling field boundaries from a conventional ASCS photograph to a 1:24,000 USGS quadrangle map.)

Storage of the digital orthophotography will be on read/write optical disk. At 2 meter ground resolution, a 7.5-minute quadrangle of imagery requires about 48 megabytes of storage in GRASS. Costs for generating orthophotography are very dynamic, becoming less and less expensive as technology improves and volume increases. The USGS will currently produce a digital 3.75-minute quadrangle of orthoimagery from NAPP for $170 if a DEM is available, or $420 if no DEM exists.

Other Data Layers

Other geographic data layers that are extremely important for SCS field offices include transportation and hydrography, landcover, landuse, watersheds, and elevation.

DATA SHARING

With the high cost of database development, few agencies can work in a vacuum. Sharing data is an essential component of any strategy to implement GIS. The "custodian" concept, where a specific agency is designated the primary responsibility to digitize and maintain a particular data layer, is used frequently in the GIS community. With this strategy, one agency should be assigned the lead role in developing and maintaining field boundary delineations. ASCS field boundaries are currently considered the official producer record for farm bill activities, and therefore ASCS would probably be the logical agency to accept this responsibility. SCS is already charged by the federal government with the responsibility for the soils layer.

There are several issues relating to sharing spatial databases between ASCS and SCS. One is hardware incompatibility. For the most part, SCS, FmHA, and FCIC have standardized on the UNIX operating system on AT&T 3B2s and 386 computers. ASCS utilizes the IBM hardware and operating system.

Incompatible GIS software is also an issue. Although most GIS software (including GRASS) can export and import data from other systems using one of the common data exchange formats (such as DLG-3), it does require extra effort to share data.

Another issue is communication between agency offices. Among the alternatives for communication include a local area network, modem to modem communications, data sharing via a compatible removable media, or combination of alternatives. For agencies such as FmHA, there is probably little need for a GIS in the office; an alternative may be an overnight remote login to utilize the GIS analysis tools and data on the SCS computer. In many cases, ASCS, SCS, FmHA, and other agencies are co-located in the same building.

EMPHASIS ON LOW COST

The Soil Conservation Service has taken an extremely low-cost approach to GIS implementation. The selection of GRASS (a public domain system) to avoid the typical $10,000 to $100,000 per site software fees of commercial systems is an example.

To keep costs low, SCS is also implementing GIS on the existing field office computer platforms -- primarily the 386 computers. Peripherals, such as printers and graphics displays, are selected carefully and integrated into the system to keep costs to minimum.

Within SCS, the pace of GIS implementation is dictated by the State Offices. With no national budget for GIS acquisition, peripherals are added and data is developed as budgets permit and as benefits are recognized.

BENEFITS OF GIS IMPLEMENTATION

Below is only a partial list of benefits from the implementation of GIS technology to SCS and ASCS:

Benefits to ASCS

1. By digitally storing field boundaries as a separate "layer" of data in a GIS, ASCS will not have to redraft all the field boundary lines for a county each time new photography is received.

2. The tracking of field boundaries, tracts, and farms becomes less confusing because updates are made to a digital database, rather than manually editing previous annotations on a photograph.

3. Obtaining photography at a scale suitable for local conditions, field sizes, etc. is no longer an issue. The maps and imagery in a GIS can be displayed at any desirable scale. When hardcopy is required, any scale or format is conceivable possible from the digital data.

4. GIS would facilitate better communication of information with the producer, with the ability to select desirable scale and overlay information when working with each client.

5. GIS would allow quicker and more accurate acreage and/or field measurement. In contrast with planimeter-based measurement from a projected 35mm slide, on-screen digitizing allows the redisplay of measurement lines for better documentation to the producer.

6. Products and services could be generated (with little or no effort) that were never before possible. Examples include interpretive maps and reports, such as program participation across the county, coincidence between ASCS data and other geographic data such as drought conditions, projected impact of new legislation or definitional changes to program participation.

Benefits to SCS

1. GIS would provide the ability to combine soils and soils interpretive maps with other geographic data quickly and accurately.

2. GIS facilitates easier tracking of SCS clients and may improve program planning and workload analysis.

3. GIS interfaces allow quicker generation of Conservation Plan Maps. Also the ability to generate a complete set of thematic maps from soils or other databases specific to the objectives of each landuser.

4. GIS offers improved tools for analyzing resources data in the conservation planning process.

5. By spatially referencing field level data, GIS allows data collected during the one-on-one planning process to be aggregated to a watershed, community, county, or larger areas. This is extremely desirable for watershed level treatment programs, resources inventories, workload analysis, and other uses, and prevents the current duplication of data collected.

6. GIS, with its spatial component, allows SCS to realistically consider initiatives focusing on surface and subsurface water quality, or other concerns requiring spatial analysis.

7. GIS has many tools to automate analytical processes now performed manually, such as computation of storm water runoff, erosion estimation, etc.

8. GIS could reduce the amount of time-consuming tabular data currently recorded. For example, SCS offices now measure the acres of each soil type by field, identify a watershed for each field, and similar information. GIS can compute this information on an as needed basis, eliminating the requirement to manually measure or even store.

10. Improved ability to respond to new legislation (such as farm bills) integrating producer information with natural resources data. For example, the determinations of percentage of HEL and acres of HEL by field for a 2500 acre watershed (with digitized field boundaries and soils) can be determined in about 30 seconds using GRASS GIS.

CONCLUSION

SCS and other county-based USDA agencies face some major challenges to implementing technologies such as GIS. As GIS matures, it appears that most of the more difficult issues are not technical in nature, but rather management priorities for the agencies' resources.

In summary, there are those in SCS who believe GIS provides the structure and the environment for organizing and analyzing resource data in the computer. As part of an initiative by Secretary of Agriculture Edward Madigan, GIS technology is being explored as a tool for improving services to USDA clientele, and making it easier for USDA's customers to apply for programs and obtain information. GIS may prove to be the means to coordinate the sharing of data between agencies, and the efficient conduct of agricultural programs of the future.

REFERENCES

1. Helms, Douglas. 1990. New Authorities and New Roles. Implementing the Conservation Provisions of the FSA of 1985, SWCS, Ankeny, Iowa, p. 21.

GIS APPLICATIONS SOFTWARE IN AN "OPEN SYSTEMS" ENVIRONMENT: HOW STANDARD IS IT?

William J. Szymanski
Computer Sciences Corporation
Geographic Information Systems Center of Excellence
15245 Shady Grove Road
Rockville, Maryland 20850

ABSTRACT

The implementation of distributed Geographic Information Systems (GIS) that adhere to open system standards is increasing, which is creating a growing trend in standardization to make GIS applications software more portable, more readily available, and more cost-effective. Any company performing or contracting for the development of customized GIS applications software can benefit from this trend.

Standards affect applications software components for user interface; spatial and alphanumeric data formatting, management, analysis and output, and network communications. Standards also influence the off-the-shelf capabilities of commercially available GIS packages. Adherence to standards in the development or procurement of GIS applications software improves applications portability, maintainability, and interoperability. Standards also affect the users' acceptance of a GIS.

This paper discusses specific issues relating to standardizing GIS applications in today's systems. It reviews the software components involved in building a standardized GIS application by examining a generic GIS workstation software architecture. Finally, it summarizes GIS standardization issues and presents a strategy for making GIS applications more standard.

INTRODUCTION

This paper addresses software standards primarily as they impact GIS applications software development and management. Applications software consists of the source code programs that reside outside the GIS package. This applications software uses individual data to make the system perform the required processing. It also provides any customized processing not implemented using commercial, off-the-shelf (COTS) components. The need for standards is increasing as heterogeneous, distributed systems become more prevalent.

OVERVIEW OF STANDARDS AND OPEN SYSTEMS

The dictionary (Bardhart 1970) defines a standard as "anything taken by general consent as a basis of comparison; an approved model." In the world of computing, standards provide an approved model for the interface and capabilities of a component. For example, the X11.3 standard provides the X-protocol for transmission and receipt of graphical display data in a distributed system.

Standardized, network-based technologies enable the distributed open systems that are becoming more prevalent in GIS applications. But what are open systems? *UNIX World* (February 1991) recently asked UNIX industry leaders this question and got many different, but somewhat similar answers. Peter Cunningham, President of UNIX International, provided one of the most concise answers. He defines open systems by the following criteria:

- Scalability - the ability to add and delete system resources on an as-needed basis to resize a system to meet current needs.

- Portability - the ability to move components, especially software programs, between systems.

- Interoperability - the ability for diverse systems and components to work together without that cooperation being apparent to the user. To the user, it seems that the entire system is sitting on the desk.

- Compatibility - the ability to swap individual system components from different vendors without having to customize those components.

Open systems are structured for change. Because responding to change is a real challenge for any organization, GIS systems must be able to respond to change to optimally support an organization's ongoing geoprocessing requirements. A static system simply cannot effectively support a dynamic organization.

GIS SOFTWARE COMPONENTS IN A STANDARD ENVIRONMENT

Standards affect applications software for user interface, spatial and alphanumeric data input, management, analysis and retrieval, operating system support, and network communications. A model for a truly standardized GIS application, therefore, must encompass user interface, attribute database interface(s) and management, spatial database interface, manipulation, processing, analysis, output operating system services, and network services.

GIS Implementation Model Becomes More Standardized

Figure 1 shows a GIS workstation architecture model with the primary software components identified. Each of these components is discussed briefly below. Figure 1 also shows the same workstation architecture with a set of appropriate standards labeled at each component interface.

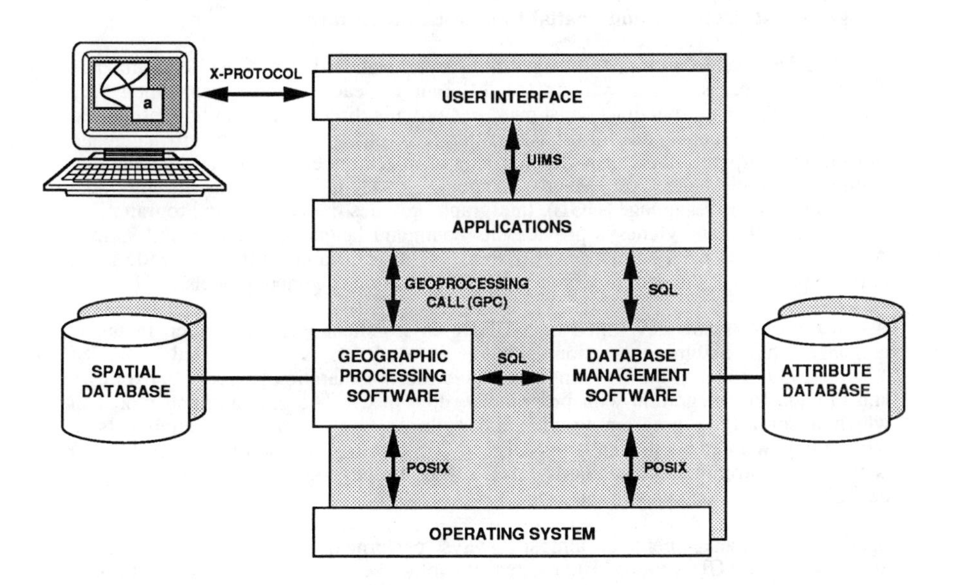

Figure 1. Standards on a Workstation Architecture

In a typical open system, the workstation software architecture revolves around layered software components whose interfaces are defined by the standards discussed above. Applications software and data represent the two components most likely to be customized to the specific needs of an organization. These components, especially data, often represent years of investment.

Many workstation components are likely to be COTS components that are used "as is" or with little modification. When these components have standardized interfaces and perform standardized functions, they can be interchanged to provide an open systems workstation architecture.

Programming customization occurs in the applications software component. This customization might be a tailoring of colors, symbology, or map themes in a simple example or a more complex, full-scale applications programming in a third generation language such as C or FORTRAN with full linkage to integrated RDBMS, GIS, and user interface components in a distributed environment.

Graphical User Interfaces Are Often X-Compliant

With the advent of intelligent graphical workstations, driven most recently by RISC technologies, many GIS vendors have ported their software to these workstations to take advantage of the increased data, processing capacity and power, and their graphical resolution and color display throughput capabilities. X-Windows is evolving as the most common industry and GIS user interface standard on these workstations. Within X-Windows, OSF/MOTIF and UI/Open Look standards are competing as standardized "look and feel" elements.

Several other very robust user interfaces such as Presentation Manager and Windows 3.0 are currently available commercially, but X-Windows is the leader in large-scale GIS applications. However, the only real consensus at this point is that the user interface of the future will remain highly graphical and window-based.

Geographical Processor and Spatial Database Management

Software designed to provide GIS capabilities is not standardized. The GIS commands that provide access to the systems are different for each vendor, even when they accomplish the same function. Most packages provide direct command line interfaces, some provide menu- and form-based interfaces, many provide macro language capabilities, and still others provide programmer interfaces exclusively. For example, Arc/Info provides the Arc Macro Language (AML), GeoVision provides the Geographic Query Language (GQL), Intergraph provides the ParaMetric Programming Language (PPL), and Genasys provides its command language interface and scripts. All have different commands, different syntaxes, and different underlying implementations of the geoprocessing and spatial data management models.

Applications software developed in a UNIX environment is often written in the C language when a third generation language is used. C is standardized by ANSI X3.159-1989. Most of the GIS languages have fourth generation language capabilities and provide for integration with programs written in C. The programming model in which a controlling program to integrate multiple COTS software components is written in C with embedded calls to SQL and GIS fourth generation languages (4GLs) is becoming prominent. Standards in GIS 4GL processing have not been in wide demand and are a long way in the future.

Spatial data management is almost always performed by proprietary software developed by the GIS vendor. But because the interface to the data is provided through the proprietary geoprocessor software via standardized import/export geodata formats such as DLG-3, the new Spatial Data Transfer Standard (SDTS), and Standard Interface Format (SIF), these differences do not present a problem. These proprietary geoprocessor differences occur because GIS vendors must implement very sophisticated algorithms for managing spatial data because of the high volume of data, complex GIS functions that must be performed on that data, and demanding user interface performance requirements levied for the rapid retrieval and display of spatial data. Spatial data management and spatial data processing algorithms are the two most proprietary areas of GIS packages.

Attribute Processors and Database Management Follow the Relational Model

The other data component of GIS packages--attribute data management--is usually much more open. Even though a few GIS packages, such as TerraView, utilize proprietary attribute data management, most GIS packages now provide a standard SQL interface or binding so that attribute data can be stored in a database managed by a COTS RDBMS adhering to ANSI standards for SQL data definition and manipulation. Some packages even incorporate COTS DBMS packages into their GIS. GeoVision, for example, is built around the Oracle RDBMS for attribute data processing. Arc/Info incorporates the Info DBMS and, optionally, supplies bindings or gateways to specific external DBMSs such as Sybase, Ingres, and Oracle.

POSIX-Compliant Operating Systems Are Prevalent

Operating system software provides the services interface to access the power of the "iron." Most of the GIS vendors have ported their packages to a version of the UNIX operating system. The standard for operating systems is currently the POSIX 1003.1, which addresses system services only. Other standards in the 1003 series from IEEE are still being finalized and should eventually be widely adopted.

UNIX is closest to matching the POSIX standards exactly because a great deal of POSIX was modeled after UNIX. GIS packages, historically, have been hosted on a wide variety of operating systems, including DOS. For years, DOS has been a *de facto* standard for many users. With UNIX operating widely on personal computers, intelligent workstations, and minicomputers, scalability (i.e., the ability to run an application on processors of different capacities) is provided by many GIS packages. The GIS vendor community has done a good job adapting to operating system standards. UNIX is becoming the *de facto* standard for the GIS operating system of the 1990s.

Network Interfaces Enable Distributed Systems

Network interfaces and connections are becoming more standardized. ISO/OSI protocols and GOSIP, Telenet, Ethernet, Token Ring, and TCP/IP all have a role to play in GIS environments. GOSIP standards are currently prevalent in Federal Government procurements. Gateways, bridges, routers, dual protocol stacks, and hybridized "brouters" all help link distributed applications and data, thereby enabling distributed geoprocessing systems to utilize hardware components from multiple vendors.

STANDARDIZATION ISSUES

GIS Applications Software Development Gets Easier...

Higher order or 4GLs, such as those bundled around SQL, and the many macro languages found in GIS products (such as Arc/Info AML, Intergraph PPL, GeoVision GQL, SPANS Macro language, Genasys Genius, etc.) make programmers' lives easier by permitting applications development at a higher level of abstraction from the fundamental components of hardware and software processing. Application Programmer Interfaces (APIs) now exist for most GISs, almost all major RDBMSs, communications components, and user interface management components. APIs and their associated 4GLs simplify programming a single component or linking a component to another component. 4GLs are often limited in performance, logic constructs, and procedural features; however, the advantages of 4GLs far outweigh these disadvantages and the use of a 4GL interface to program a GIS is the best solution in many cases.

...and More Complex

Applications development is made more difficult by the power of the integrated components. Imagine a user entering a command at an X-terminal to query a distributed database to find all class 2 roads in the Commonwealth of Virginia and display them on a base map. Ensuring all the processes that make this type of transaction function together effectively takes time, skill, and testing.

Such an example becomes conceptually complex for even experienced system architects and software analysts/programmers to fully understand. Even after breaking the transaction down into smaller components, it involves a number of different software concepts for database management, GIS processing, network communications, and user interface control. Although once the application program design is allocated to individual components, the APIs and 4GLs discussed above make the development of these individual components much easier. Standards help limit complexity by providing consistency across a given interface (e.g., SQL provides the same core language to all compliant RDBMS packages even though vendor-specific extensions to the language are not standardized).

As Figure 2 shows, the class 2 roads query example involves a multidimensional software model of at least four dimensions. This model involves at least three layers of X-Windows components following X-protocols for server and client communications (where the X-server is at the display and the X-client is at a remote processor); a window manager; a high-level user interface management system (UIMS) on top of the X-library; seven layers of OSI protocols for communications (somewhere in the system there must reside a communications server process); a database client linked to the X-server and perhaps communicating with several database servers elsewhere on the network; a GIS processing and display engine controlled as an X-client process; and a process (probably as another X-client) to integrate the results and send the display information back to the user. Understanding individual components is greatly enhanced by two-dimensional models, structured program design, and disciplined implementation and testing. As the example shows, however, linking multiple components results in a very complex processing environment that cannot be easily understood, implemented, or debugged by novice programmers developing GIS applications, even in a standardized environment.

Marketplace Influences on GIS Standards

Ultimately, the marketplace chooses standards. As standards are set and tested via the marketplace, selection occurs. Some standards are clearly more widely accepted and prominent than other standards. Among the widely accepted and emerging standards affecting geoprocessing systems are SQL, UNIX as a POSIX-compliant operating system, TCP/IP, Ethernet, X-Windows, and C.

The standards accepted by the marketplace are often not the standards that vendors and standardizing bodies would choose. For example, the marketplace's acceptance of relational database standards hurt companies such as Cullinet, whose IDMS/R database fundamentally followed the Codasyl data model and only adopted a relational "front-end." Farther outside the GIS community, the Ada language provides an example of an unadopted standard. Even though Ada is widespread now, it was years before federal procurement guidelines to use Ada had widespread effect.

***De Facto* Standards**

With the current increased attention to standards, several types of standards have emerged. DOS, for example, became the *de facto* standard for personal computer operating systems during the 1980s. SIF is an example of a graphical data exchange format that has been adopted by most GIS vendors because of the sizable amounts of spatial data available through the large number of Intergraph installations. The Digital Exchange Format (DXF) is common because of the large number of AutoCAD installations. In fact, most data exchange formats used in GIS are *de facto* standards (TIGER, DLG-3, DEM, etc.). Standards such as these and SQL are tangential to the GIS community, which responds to these standards after they are set. Currently, AML could be considered the *de facto* geoprocessing standard because it has a large part of the GIS marketshare.

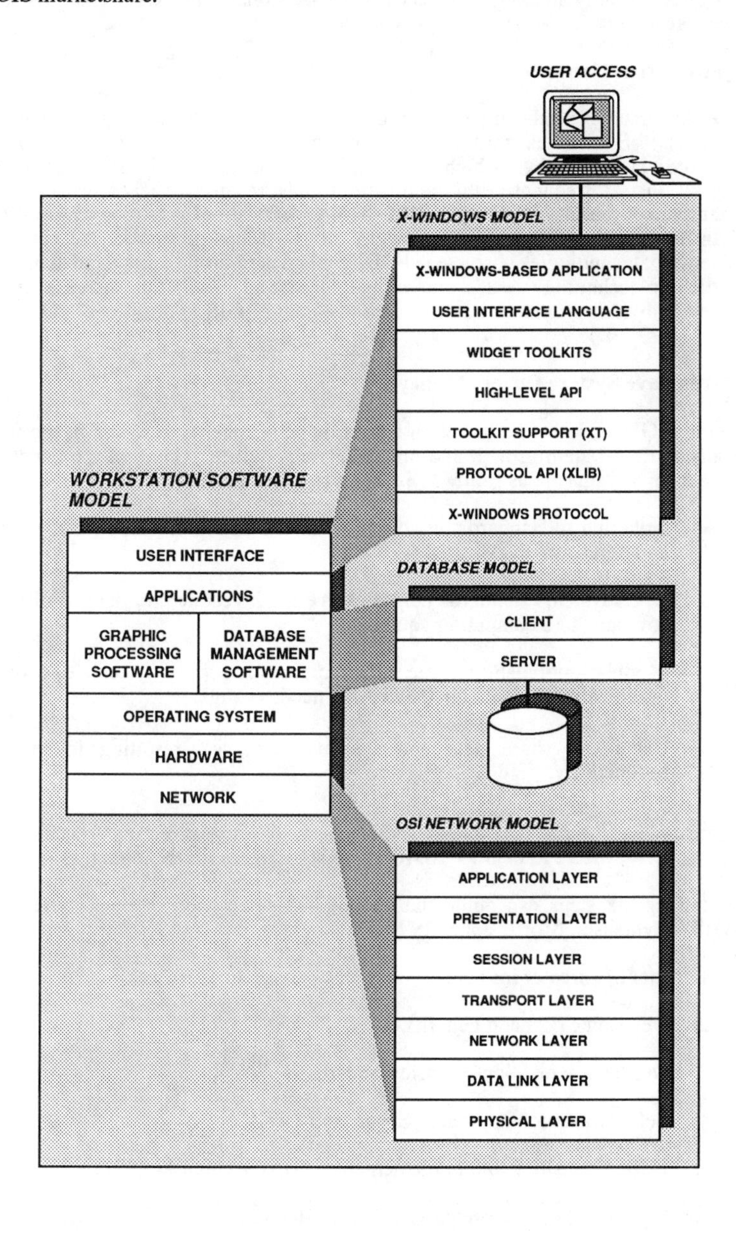

Figure 2. Multidimensional Software Model

Standard Standards

"Standard standards" are proactively determined by standards organizations such as the Open Software Foundation (OSF), Unix International (UI), IEEE, and International Standards Organization (ISO). There is no guarantee that these standards will be accepted. Currently, the UNIX industry press is having fun writing about which standard user interface will emerge as the winner, MOTIF or OPEN LOOK. More vendors are currently making products MOTIF-compliant, but OPEN LOOK is the leader because of its large base of SUN workstations.

Standards Often Compete

The user interface example shows that there are competing standards that impact GIS installations. Sometimes this competition ends because of a shakeout in the marketplace, such as that which occurred between relational and nonrelational databases. Just as frequently, the competition ends because of leaps in technology. For example, in the last few years, RISC-based workstation technology has changed GIS. In the next few years, FDDI networks at 100 Mbps may make the current 10 Mbps network technologies obsolete if FDDI continues to drop in price and availability and switching technology issues are completely solved. Will this increase in cost-effective bandwidth provide a boon to capacity-hungry GIS applications? Only time will tell.

Standards Save System Buyers Money

Achieving a GIS architecture based on standards pays benefits now and in the future, especially during system growth and maintenance activities. Some of the benefits of adopting open systems are as follows (Szymanski, Morris-Jones 1991):

- Adhering to standards broadens automated data processing (ADP) options and saves money.
- Open systems standards help preserve the investment in data, applications software, and computing equipment.
- Modular upgradability of system components permits systems to evolve as organizational, functional, and ADP needs change.
- Users can share data and systems resources, resulting in increased productivity.

STRATEGY FOR STANDARDIZATION

The following six steps can guide the overall implementation of open systems and open GIS (Szymanski, Morris-Jones 1991):

1. Identify area of influence.
2. Determine cost and benefits.
3. Define organization-specific standards.
4. Select transition approach (Szymanski 1991).
5. Define transition and implementation plan (Szymanski 1991).
6. Implement open systems and enforce standards.

Applications Standardization

On an applications level, a different strategy is appropriate during software design and implementation. This strategy has the following cornerstones:

Examine Organizational Issues. Identify and analyze standards already existing in your organization. Examine how your organization obtains application software. Who builds it--users, consultants, systems integrators, a centralized data processing shop? Is it purchased as COTS packages, such as a coordinate geometry package, and then customized? A centralized data processing support organization has a much better chance of enforcing organization-wide use of SQL for alphanumeric data processing than does a civil engineering department or tax records section that uses a GIS for a specialized purpose.

Adhering to the prevalent standards already existing within an organization, if they are standards recognized by enough vendors, is almost always beneficial. Users learn faster when systems look the same, data is shared much more readily when formats are compatible between applications components, and spatial data structures and processing are kept compatible.

Evaluate the Impact of Adhering to Standards Across the Total Expected Life Cycle of the Software. If standards do not seem relevant now, consider the industry trend toward downsizing, the rapid changes in hardware capabilities, and the length of time many of yesterday's systems needed to endure. Applications software and proprietary data must be protected from obsolescence at all costs. The short-term approach, when all components of a new system architecture are bright, shiny, and modern, may preclude applications software and data management options desperately needed in 5 years when data volumes are up 500 percent and more users than ever expected demand access to GIS capabilities. One of the best ways to protect an organization's investment in software and data is by adhering to standards.

Determine Organization-Specific Standards. This is a key consideration, especially since the spatial data analysis program calls are product specific. An application program written in AML to call Arc/Info will function well for a long time; but if for some reason the organization's GIS is switched or someone in the organization buys a package other than Arc/Info, the GIS calls will need to be rewritten in any customized applications software in the language associated with the new product if that applications code is to be reused.

Use Structured, Modular Applications Programming Techniques. Although top-down design and implementation techniques have been adopted by large programming shops for years, many GIS applications are still built by people whose primary job is not programming. This effort is individually commendable, but the resulting "hacked" code is often difficult to upgrade, maintain, and port to the new platforms even if it contains standard elements. If standards are not implemented in a disciplined, modular manner, upgradability is severely limited almost the same as if standards were not used.

CONCLUSIONS

Areas of GISs that apply broadly to all systems are becoming fairly well standardized. ANSI standard SQL exists for alphanumeric database management and X-Windows Release 11.3 is prominent in GIS user interfaces in a distributed environment, although the further details of user interface paradymes and programming interfaces are much less standardized with MOTIF and OPEN LOOK contending with the X environment.

Spatial data formats are well standardized because spatial data transfers can be performed by adhering to any number of widely accepted standards such as SIF, DLG-3, DXF, and TIGER. The SDTS is in the process of being formalized as a FIPS standard, but its acceptance ultimately hinges on the marketplace. Conversely, access to spatial data once it is in a GIS package and the program or command calls to perform geoprocessing (analysis, manipulation, and output) on that data remains highly proprietary and vendor specific; this impacts geoprocessor upgrades. If a different GIS package is required when a system is upgraded, geoprocessing language calls must be rewritten. If the geoprocessing calls need to be preserved, then the geoprocessor upgrade is limited to the current GIS vendor.

The development of a standardized geoprocessing interface remains a challenge to the GIS community. The need for standardized geoprocessing is not currently strong, but as GIS matures and becomes more integrated as a component in a total information system architecture, that need will increase with the advent of total spatial processing architectures and increasingly widespread implementation of distributed information systems with geoprocessing components.

REFERENCES

Bardhart, C.L., 1970, *The American Collegiate Dictionary,* Random House, New York, p. 1177.

Szymanski, W.J. and Morris-Jones, D., 1991, "The Role of Open Systems in GIS: Costs, Risks, and Payoffs," AASHTO GIS-T Conference Proceedings, (in publication).

Szymanski, W.J., 1991, "Accomplishing an Uninterrupted Transition and Phase-in of a New GIS: Issues, Concerns, and Strategy," URISA 1991 Conference Proceedings (in publication).

No author, February 1991, "Open Says Me," *UNIX World,* quoted from Peter Cunningham, President and Chief Executive Officer of UNIX International, p. 48-52.

"STATUS AND DESCRIPTION OF THE DIGITAL GEOGRAPHIC INFORMATION EXCHANGE STANDARD (DIGEST)"

Dr. Charles A. Roswell
Mr. Gary A. Hacker
Defense Mapping Agency
PRS/Mail Stop A-13
8613 Lee Highway
Fairfax, Virginia 22031-2137

BIOGRAPHICAL SKETCH

Dr. Roswell is Chief of the Specifications and Standards Division, Defense Mapping Agency Plans and Requirements Directorate, which is responsible for activities concerning the development and maintenance of Department of Defense and international military specifications and standards for analog and digital geographic information. Mr. Hacker has been a member of the Digital Geographic Information Working Group Technical Committee/Panel of Experts for four years.

ABSTRACT

The Digital Geographic Information Exchange Standard (DIGEST) is a suite of standards being prepared by the Digital Geographic Information Working Group (DGIWG). The DGIWG is composed of representatives from the defense mapping establishments of 11 NATO nations and Supreme Headquarters Allied Powers Europe (SHAPE). DGIWG has been meeting regularly since 1983 to develop the DIGEST which provides guidelines for exchanging raster, vector, and matrix data among the member nations. DIGEST edition 1.0 was issued in June 1991. This paper gives the status of and describes the various parts of DIGEST, including feature-oriented vector, relational vector, raster, and matrix implementations. It also describes DIGEST conformance to ISO standards, in particular ISO 8211 and ISO 8824, DIGEST's Feature and Attribute Coding Catalog (FACC), and the various NATO-desired dataset specifications under development by the DGIWG.

BACKGROUND

The Digital Geographic Information Exchange Standard (DIGEST) has been developed by the Digital Geographic Information Working Group (DGIWG). The DGIWG Steering Committee, established in 1983, consists of representatives from the defense and civilian mapping establishments of Belgium, Canada, Denmark, France, Germany, Italy, the Netherlands, Norway, Spain, the United Kingdom, the United States, as well as the Mapping, Charting, and Geodesy (MC&G) officer of the Supreme Headquarters Allied Powers Europe (SHAPE). The technical development of DIGEST was undertaken by DGIWG's Technical Committee and Panel of Experts (POE). The POE consists primarily of representatives from France, Germany, Italy, the United Kingdom, the United States, with representatives from other member nations periodically. Since 1986, the POE has met two to three times per year for two to three weeks per session.

DESCRIPTION OF DIGEST

DIGEST specifies exchange standards for raster, vector, and matrix (gridded) data in three encapsulations--ISO 8211 to support exchange of raster, matrix, and feature-oriented vector data; ISO 8824 to support exchange of all data sets via telecommunications methods; and the Vector Relational Format (VRF) to support exchange of vector data in a geo-relational format. (VRF is the DIGEST implementation of the Vector Product Format (VPF) developed by DMA to support the Digital Chart of the World and future DMA vector products.

DIGEST includes the Feature and Attribute Coding Catalog (FACC), a comprehensive list of features, attributes and attribute values which provides a relatively complete selection of geographic information pertinent to military applications. Although FACC is comprehensive, it is understood that not all features and attributes can possibly be coded; therefore, DIGEST allows user-defined data dictionaries as long as these data dictionaries conform to the specified coding rules.

RASTER DATA IMPLEMENTATIONS

DIGEST currently supports exchange of raster data in three forms--Standardized Raster Graphics (SRG), ARC Standardized Raster Product (ASRP) and UTM/UPS Standardized Raster Product (USRP).discussed separately below.

SRG is simply a high resolution raster facsimile of a paper map and chart. The entire map image (including margins and legend information) is captured in raster format, while selected control points are retained in a registration point file. The main purpose for SRG is to support raster update of maps and rapid reproduction in a field environment. Germany and Italy have expressed interests in producing and exchanging SRG data.

ASRP and USRP are also raster facsimiles of paper maps and charts. However, ASRP and USRP data are transformed into the Equal Arc Second Raster Chart/Map (ARC) and the Universal Transverse Mercator/Universal Polar Stereographic (UTM/USP) projection systems, respectively. The marginal information from the paper maps is eliminated from both ASRP and USRP, and the data are merged to present the user with an apparently seamless image, thereby solving the problem of always fighting the war at the intersection of 4 maps. ASRP is closely related to DMA's ARC Digitized Raster Graphics (ADRG); in fact, the only difference is that ADRG pixels are assigned 24 bits of color, whereas ASRP (and USRP) use a technique called Extended Color Coding which uses transition colors to avoid the aliasing problems associated with traditional color coding techniques.

The UK, Germany, the US, Canada, Spain, and Italy (Air Force) are interested in exchanging ASRP data, while France and Italy (Army) prefer USRP. First Edition SRG, ASRP and USRP product specifications were issued in June 1991.

MATRIX DATA IMPLEMENTATION

Although a matrix data format is specified in DIGEST, most member nations currently require only elevation data in matrix form, and these requirements are currently satisfied by the Digital Terrain Elevation Data (DTED) format specified by an existing NATO Standardization Agreement. Future matrix products will conform to the DIGEST implementation. Unlike raster data, there have been no unique product specifications defined for matrix data.

VECTOR DATA IMPLEMENTATIONS

DIGEST allows the exchange of several types of vector data, including spaghetti, chain-node, and full topological structures. In addition, DIGEST allows any of these types to be placed in either a feature-oriented format or a geo-relational format, as specified in annexes A or C of DIGEST. Member nations have indicated initially that they will use the feature-oriented implementation primarily for exchange of data among producing agencies, whereas they will use the geo-relational implementation for providing data directly to operational users.

NATO DATASET SPECIFICATION DEVELOPMENT

The DGIWG is currently developing dataset specifications in response to requirements identified by the 1990 NATO Geographic Conference for digital geographic information to support military planning and operations. The requirements include specific datasets for Terrain Analysis, Transportation and Logistics planning, Simulation, Aeronautical Information, Background Display, Toponymy, and Neutral Data. NATO has designated DIGEST as the exchange standard to be used for all future exchange of data among NATO member producers and users of digital geographic information. Plans are to promulgate DIGEST as a NATO Standardization Agreement (STANAG) as soon as possible.

ONGOING AND FUTURE ACTIVITIES

With several digital geographic information standards coming to the fore, DMA and USGS representatives are establishing a group to explore the possibility of creating a DIGEST relational profile within the Spatial Data Transfer Standard (SDTS).
This group's charter is to identify and resolve any differences that exist between the two standards and prepare formal change recommendation documentation for each standard to be put before the respective approval boards.

DIGEST representatives are also meeting with representatives of the International Hydrographic Organization to similarly identify and resolve differences between DIGEST and DX90.

CONCLUSIONS

- DIGEST is a comprehensive suite of standards designed to provide common information exchange among a variety of producers and users of digital geographic information.

- DIGEST provides the capability to exchange raster, vector, and matrix data.

- DIGEST includes a comprehensive feature and attribute coding catalog (FACC) used to uniquely identify most geographic phenomenon

- Several countries are currently producing raster data in accordance with DIGEST raster specifications.

- Development is proceeding on specific vector datasets to support NATO military requirements.

- Although differences exist between DIGEST and other proposed exchange standards, efforts are being made to identify and resolve those differences as soon as possible.

- DMA is committed to exchanging data among NATO member nations and military forces via DIGEST.

A PROPOSAL FOR IMPROVED COORDINATION OF FEDERAL GIS ACTIVITIES

John M. Palatiello
Principal
John M. Palatiello & Associates
12020 Sunrise Valley Drive
Suite 100
Reston, VA 22091
(703) 391-2739

BIOGRAPHICAL SKETCH

John M. Palatiello is principal of a consulting firm formed in 1987 to provide public relations, government relations and marketing services to surveying and mapping firms and organizations. He received a B.A. in Political Science from The American University, Washington, DC, served eight years as a staff assistant to members of the U.S. Congress and was the first ACSM-ASPRS Government Affairs Director. He is also Executive Director of the Management Association for Private Photogrammetric Surveyors (MAPPS), the only trade organization of private geographic information firms.

ABSTRACT

The Federal Government has significant need for geographically referenced information accurately portrayed in graphical and digital formats on maps, charts, surveys and other media. Numerous Federal agencies are involved in the development and dissemination of geographic information. Current Federal activities lack coordination. Improved coordination and consistency in the use of this data will result in the elimination of duplicate effort and economy and efficiency in government. State and local government has a need for larger scale, more detailed mapping data than is provided by most Federal agencies. Moreover, a notable private sector capability in geographic information services can and should be better utilized by Federal agencies. A government-wide policy is needed to assure better organization, enhance compatibility, reduce duplication, eliminate competition with the private sector and increase U.S. exports of geographic information services.

INTRODUCTION

Historically, the Federal government has been a dominant surveyor, cartographer and geographic information specialist. The time has come to redefine the role of government to reduce competition with the private sector and to better focus activities on areas where government should be legitimately involved.

This paper reviews previous studies on Federal surveying, mapping and geographic information activity and proposes a plan to enhance GIS; define respective roles and improve information exchange and sharing among and between Federal, State and local government agencies and the private sector; create a funding system; and increase international exports.

IMPROVED MANAGEMENT OF FEDERAL GIS RESOURCES

Geographic Information Systems[1] in the Federal Government lack coordination and compatibility standards, duplicate effort among agencies and tender competition with and duplication of the private sector.

Since 1973, studies by the Office of Management and Budget, National Academy of Sciences (NAS) and others advocated consolidation of civilian surveying and mapping activities in a single Federal agency or designation of a lead agency.[2] Many such findings were made before computer mapping and GIS became common technologies. The NAS recently said "technology has changed to allow complex systems to develop: this is an appropriate time to acknowledge that institutions and organizations involved in this information infrastructure will also have to change...it is time for the administration to require greater inter-departmental cooperation and to designate a lead agency for GIS."[3]

The hardware and software used in GIS applications varies within and across agencies.[4] This points to the need for better coordination.

---

[1] The General Accounting Office has defined a geographic information system as one which "consists of computer hardware and software capable of manipulating, analyzing and presenting spatially-referenced information. This is information associated with a specific place on the earth, such as the geographic location and the characteristics of a lake, road, or stand of trees. (General Accounting Office, "Geographic Information Systems: Status at Selected Agencies", B-240065, August 1, 1990).

[2] SEE, for example, Office of Management and Budget, "Report of the Federal Mapping Task Force on Mapping, Charting, Geodesy and Surveying", July 1973; and National Academy of Sciences, "Federal Surveying and Mapping: An Organizational Review", 1981.

[3] National Academy of Sciences, "Spatial Data Needs: The Future of the National Mapping Program", 1990.

[4] Op.Cit., GAO

Congress ordered a comprehensive Land Information Systems (LIS) study of "the need for and cost and benefits associated with improvements in the existing methods of land surveying and mapping and of collecting, storing, retrieving, disseminating, and using information about Federal and other lands." [5]

In that study, the inconsistency in GIS or LIS among various agencies was again highlighted.

> even though critical information may exist for a particular parcel of land, it is often difficult to retrieve efficiently and often impossible to relate to relevant information in other land information files. The land data held by one agency frequently are "unavailable" to another agency, not because of jurisdiction, but because of the method used for record keeping or because of the data format.[6]

The study team did not reach a consensus on the earlier recommendation for a single Federal surveying and mapping agency. However, the final report cited the need for better coordination and the establishment of standards and guidelines.

> (the study team) considered several alternatives that would provide coordination for LIS activities at all levels of government ... the study team believes that active coordination of surveying, mapping, and land information functions is the most practical approach to ensure an effective nationwide program.

* * *

Land information systems must provide for maximum data exchangeability. Certain standards and guidelines must be implemented to provide needed compatibility, since the systems will not operate under a single control.

The previously noted recent NAS study on GIS activities of the USGS National Mapping Division of the made a number of significant findings.

> If ours is to be an information-based economy that is competitive on a global basis, there is a critical need for a coordinated and efficient national information infrastructure to facilitate the sharing and communication of

---

[5] Section 8 of Public Law 100-409, August 20, 1988.

[6] U.S. Department of the Interior, "A Study of Land Information", November 1990.

information resources. This must include a geographic or spatial data component dealing essentially with where things are to support all manner of ... activities. The most important function of the USGS's National Mapping Division (NMD) in the future will be to act as the federal coordinator of the national geographic data infrastructure.

* * *

continue and, if possible, expand its efforts in establishing and promulgating digital spatial data quality standards, including standards for larger-scale data sets and maps.

* * *

creation of the National Digital Cartographic Data Base by (1) increasing emphasis on work sharing and cost sharing programs, (2) developing, prototyping, testing and implementing a digital data donor program throughout the public and private sectors, and (3) allocating adequate NMD resources to information management and user/donor coordination, and, if necessary increasing these relative to traditional data production programs.

* * *

In summary, private sector mapping has usually been done where more detailed information is needed ... Commercial mapping firms have been traditionally well organized politically to protect their "right" to do any mapping at scales larger than 1:24,000 ... The expanding use of GIS technology is creating demands for more detailed uniform information over ever-larger areas...The need now is national in scope, and it seems reasonable to consider the possibility of meeting these emerging needs at the federal level. Direct federal responsibility is probably not the solution, but cooperative, directed funding of some sort may be in order to develop shared data bases of widespread utility with national standards.[7]

In part to implement these findings, $1 million was included in the proposed FY 1992 Federal budget for a USGS Map and Digital Data Cooperative Program. That funding was rejected by Congress. The Management

---

[7] National Academy of Sciences, "Spatial Data Needs: The Future of the National Mapping Program", 1990.

Association for Private Photogrammetric Surveyors (MAPPS) had concern for the proposal.

> While we believe this program may have merit, and we have been told by USGS officials the program will result in contracting opportunities for private firms, we believe there must first be better enforcement of and compliance with existing Federal procedures designed to protect the private sector, including small business, from unfair government competition.[8]

MAPPS' concern was based on the fact that the Federal government annually spends $1 billion on surveying and mapping activities, but contracts only about 5 percent to the private sector.[9] Numerous government studies advocate more of these services be performed by the private sector and OMB targeted mapping for increased contracting out in its FY 1990 budget.[10]

OMB PROPOSAL

Based on that initiative, the OMB developed a proposal for privatization of certain GIS related activities.[11]

This proposal would create a public/private partnership to implement the the NAS concept. It is also similar to strategies implemented in Ontario, Canada to "bring together the unique capabilities of both public and private sector partners to market this leading-edge technology to the world marketplace" [12] and Queensland, Australia, "to generate export earnings ... (in an) industry ... in which enterprise and production are concentrated in the private sector, with the Department

---

[8] MAPPS, Statement to the Subcommittee on Interior and Related Agencies, House Committee on Appropriations, March 18, 1991.

[9] Palatiello, J.M., "Uncle Sam Makes Government Market for Surveying and Mapping More Attractive", P.O.B., 1989.

10 Office of Management and Budget, Budget of the Government of the United States, Fiscal Year 1990, p.2-39/40, 1989.

[11] Office of Management and Budget, Privatization Initiatives in the 1991 Budget, p. 18, March 1990.

[12] Ministry of Consumer and Commercial Relations, Province of Ontario LAnd Registration and Information System (POLARIS), News Release, February 15, 1991.

providing support, planning and coordination as required".[13]

These foreign initiatives are quickly putting the United States at a competitive disadvantage in the global market and may have an adverse impact on America's ability to export its GIS capabilities. The competition between U.S. government and the private sector has not been limited to domestic mapping. In fiscal year 1989, the USGS/NMD estimated providing $22.2 million in mapping services for foreign countries, financed by the agency for International Development (AID).[14] None was contracted to U.S. firms.

The text of a public/private partnership proposal, drafted by the OMB staff, follows:

"GEOGRAPHICAL INFORMATION SYSTEMS DEVELOPMENT
Preliminary Draft

Technological advances in the geographical information industry have changed the way cartographic and related information is used in virtually all aspects of national and community needs. It is a vital component of city planning, weather forecasting, analyzing demographic trends, and agricultural strategies, to name a few areas. The ways in which this information is collected, produced, disseminated and used is changing, and the industry is becoming interdependent and complex. It could, in fact, be defined as part of our nation's infrastructure.

Simple topographical maps of land, streets, river and mountains have given way to sophisticated satellite imagery, and to digitalized schematics of underground sewer, power and electric lines. Paper maps have given way to computerized graphics, capable of being accessed in a growing arena.

We have the opportunity to build a cohesive geographical information system, to structure the future of a vital commercial enterprise to best suit the needs of the nation. We need to better define the proper role of government, which can act as a catalyst in securing America's place as a world leader in the geographic information industry.

<u>The Issues</u>: As geographical information technology

---

[13] Sunmap, Corporate Plan, Department of Geographic Information, Brisbane, Queensland, Australia, 1989.

[14] U.S. Small Business Administration, "Report on Government Competition for Services Provided to Foreign Governments," May, 1989.

grows, we need to define national needs and create a strategic plan for the industry's future, which addresses both public and private sector contributions and uses.

Both the public and private sectors have built up an industry, and often are in competition for foreign and domestic contracts.

Major technological advances in the field of geographical information systems will render current methods of production and dissemination obsolete.

The role of the Federal government in this industry will need to be redefined. It may be more properly the role of government to promote the industry and provide research and development, rather than to act as map maker and geographer.

Background: OMB began looking at this issue last year when we decided to include map making and cartographic functions on agency A-76 inventories. The government has large map making capabilities, spending over one billion dollars annually and employing over 8,000 FTE. By contracting with the growing private sector map making industry, we have the opportunity to save money and reduce government competition with the private sector, as prescribed in OMB Circular A-76.

In addition, government agencies with map making capacity compete for work from foreign, state and some local governments, as well as from other Federal sources. This, too, reduces the opportunities for private map makers. Legislation was introduced but not passed last year which sought to restrict government competition in the foreign arena, as an amendment to the SBA reauthorization bill; it passed this year (6/28) as a Stallings-sponsored amendment to the Foreign Assistance Reform bill.

A number of Federal agencies have the capacity to produce digital cartographic data, obtainable from the private sector. OMB has recognized the need to coordinate these programs to avoid duplicity (in February OMB Director Darman continued the function of the Federal Interagency Coordinating Committee on Digital Cartography), but has never addressed the cost of in-house capacity versus contracting with the private sector. Digital cartography is certainly a part of the geographic information industry.

And as far back as 1973, OMB conducted a study of Federal mapping requirements in a "Task Force on Mapping, Charting, Geodesy and Surveying." The study found, in part, that the government was not relying sufficiently on the private sector.

These are all part of our geographical information system, and could be combined under a strategy to cooperate with the private sector to coordinate the future of the industry.

<u>What Needs to be Done</u>: We have developed a preliminary five-pronged approach to meeting national needs.

1. **Elevate the Issue.** In the past we have addressed components of geographical information systems individually. The private sector solicits Congressional support; OMB places mapping and cartographic functions on A-76 inventories; an interagency task force tries to avoid duplicity in a particular arena. Incorporating these separate but inter-related functions into a single, over-all agenda gives us the opportunity to treat geographical information as the national issue it is. A national approach would allow proper focus of Federal involvement in geographical information systems development.

2. **Coordinate Federal Agencies Involved.** One way to elevate the issue is through an interagency group dedicated to geographical information systems development. The group would be made up of technical a well as policy experts, and charged with defining the scope of the need. They would be responsible for analyzing current methodologies and trends, and charting a strategic plan for the industry's future. In addition, they would address public and private sector roles in the industry, and arrive at a clear definition of the responsibility of the government, taking into account the growth of private sector capabilities, the cost to the government to keep geographic functions in-house, and competing interests between the private industry and the public sector.

This could be accomplished via an existing intergovernmental entity, such as the Council on Competitiveness. While the council has tentatively decided to concentrate on US access to the 1992 unified European market, they might be willing to address this single-topic issue as a means to ensure America's competitiveness in the geographical information arena.

Alternatively, an interagency task force or Presidential Commission could be convened to address the future of the industry. Because virtually every facet of what we do can be enhanced by geographic information, the development of a secure, competitive industry is important to every State and community, as well as the Federal government.

Finally, OMB could commission a study to meet the objectives outlined above.

3. **Ensure Private Sector Support.** Because geographical information systems involve both public and private interests, we must ensure that the private sector plays a key role in its development. This would be an example of a public-private partnership to create a public-

private solution. The Management Association for Private Photogrammetric Surveyors (MAPPS) is already an active participant in seeking legislation to reduce government competition with the private sector in cartographic areas. Preliminary discussions with the association indicate they may be willing to address the broader issue of geographical information, or form a group to do so. In addition, the American Consulting Engineers Council, the Contract Services Association, the Electronic Industries Association, the Professional Services Council and the Western Federation of Professional Surveyors have voiced support for the initiative.

4. **Encourage Congressional Support.** One of the main functions of the private sector would be to encourage congressional support for the initiative. They could target supporters of America's competitiveness in addition to members who represent districts with private firms and customers. This is another bipartisan issue which could win across the aisle approval, if couched in terms of competitiveness.

In addition, we should work on legislation designed to reduced government competition with private industry, building on last year's (unsuccessful) amendment to the SBA authorization bill and this year's attempt.

5. **Redefine the Role of Government.** To date, the government has been a map maker and cartographer, an active participant in the industry. Federal agencies have aggressively marketed their services to foreign governments. They routinely receive requests from foreign governments for professional and technical services in this area. For example, the US Geological Survey has had recent projects in Chad, Zaire, Egypt and Saudi Arabia. This is despite our policy not to provide a service or product if such product or service can be obtained from private enterprise through ordinary business channels (A-76).

We need to redefine the role of government, not only in terms of reducing this government competition with a healthy, viable private industry, but also in terms of what the proper role of the government should be. In Queensland, Australia, for example, the government serves as a promoter of Queensland private firms, both domestically and abroad. They also have outlined a broad strategy designed to bring Queensland into the world market for geographical information systems.

The intergovernmental working group, as described in 2 above, should be tasked to address this issue, and determine if the government should turn from a production mode to a promotion mode, or possibly provide research and development functions for the industry."

CONCLUSION: A PUBLIC-PRIVATE PARTNERSHIP

This concept can be advanced by; (1) OMB revision of Circular A-16 or a separate directive; (2) Congressional enactment of legislation establishing Federal GIS policy and procedures; or (3) An agency seeking designation as the lead GIS agency and soliciting cooperation and support of other agencies. If the GIS community in Federal, State and local government and the private sector agree on this concept, convincing OMB to implement strategy 1 would be feasible and strategy 2 possible. Turf battles make strategy 3 improbable.

A public-private partnership has been advocated by MAPPS for several years to end the acrimony between the public and private sectors and break the intransigence of Federal agencies on the contracting-out issue. Such a partnership would yield universal benefits. Through a focus on funding, the GIS market would grow. By creating a strong GIS role for the private sector, firms can utilize their ability to participate in the political process to assure such funding by Congress and State legislatures. A well defined government role provides agencies direction and resources to provide an important service to the citizens. The citizens gain from current, accurate geographic information that supports efficient environmental protection and resource management, equitable taxation, economy in government services and other benefits of GIS applications. The Nation as a whole also profits from income and tax revenues generated by increased GIS exports.

Distinct roles and responsibilities in GIS is needed. They must be well understood and observed. Such roles are suggested as follows:

FEDERAL GOVERNMENT
manage and disseminate information
establish standards
coordinate user requirements
conduct research and development
commercialize technology resulting from R&D
provide project funding
promote U.S. private sector in exports
award and administer contracts

STATE/LOCAL GOVERNMENT
manage and disseminate local information
develop requirements
award and administer contracts
provide project funding
organize local project participants

PRIVATE SECTOR
collect data
develop data bases
manage projects
sales and marketing

A SURVEY AND ANALYSIS OF STATE GIS LEGISLATION

H. Bishop Dansby, J.D.
President
American Cadastre, Inc.
501 Sovran Bank Building
Harrisonburg, VA 22801
(703) 434-6275

ABSTRACT

A number of states have passed legislation that is either directly related to GIS or to the issues that have have been raised by the arrival of this technology. A framework of legal and institutional issues that could or should be addressed by the legislation is developed. Issues include organizational structure, funding, sale of data to the public (for commercial and non-commercial uses), data accuracy and liability, and confidentiality and privacy. A survey of the GIS legislation in the various states is conducted. The legislation is then analyzed and evaluated in the context of the recommended scope and framework of issues. An appropriate scope of GIS related legislation is discussed and recommended. The paper should be useful for state policy makers to a) clarify what the scope and content of legislation should be to address issues raised by GIS implementation in state and local government, and b) compare their state's legislative efforts with those of other states.

INTRODUCTION

The United States is a big country with many jurisdictions, and GIS is a complex, if not nebulous, subject. This paper attempts in a few pages to survey the groundswell of GIS activity on the legislative front. Necessarily, much of the coverage will be cursory and less than completely rigorous. Nevertheless, the exercise should be informative and useful as a basis for any of a number of more focused studies.

OVERVIEW OF CURRENT GIS LEGISLATION

At least twenty states have some form of GIS legislation as of the date of the writing of this paper. The phrase GIS legislation is used here to include legislation that specifically refers to GIS or of the equivalent of GIS, such as automated mapping or land information systems. There are, of course, many pieces of legislation that impose requirements that are appropriately responded to by the use of GIS technology, for example, growth management or planning legislation. That type of legislation is generally not included here, unless it does make specific reference to GIS. In at least one instance discussion has included legislation that has established an agency that has the responsibility for GIS in a state, even though the legislation itself does not make specific reference to GIS or the equivalent.

At the outset, one could legitimately ask why any legislation should specifically refer to GIS. If one thinks of GIS as a tool, then there is no particular reason for legislation, which generally is goal oriented, to refer to the tools necessary to achieve the goal. Further, if one thinks of GIS as a subset of information systems, again, there would be no requirement to specifically mention GIS in legislation pertaining to information systems.

Nevertheless, and somewhat surprisingly, many legislatures have chosen to make specific reference to this technology, and such references have all been made within the last couple of years. Earlier legislation that made reference to the functional equivalent of GIS tended to use other terms.

The origins of GIS legislation vary from one state to another, and tend to reflect the interest and priorities of those states. The types or locations of GIS legislation include at least the following:

- Open records laws
- Laws pertaining to education
- Specific GIS organization legislation
- General appropriation of funds
- Data processing laws
- Health and safety laws
- Laws dealing with the census, redistricting, etc.
- Laws to regulate natural resources
- Establishment of task forces to study GIS or other issues
- Planning and land use laws

As one might expect, states such as Iowa and Idaho have GIS legislation in connection with water resources. A number of states include GIS as a mandatory tool for planning and land use requirements, including the states of Maine, Vermont, New Hampshire and Washington. The State of Rhode Island and Alaska have included GIS as part of legislation related to higher education, such as the establishment of GIS labs in state universities.

Only a very few states have legislation that was conceived solely to devise a comprehensive GIS strategy. These would certainly include Virginia and Wisconsin, and could arguably include Utah, Mississippi, Kentucky (and perhaps others, depending on one's definition of comprehensive). It should be mentioned that many states have active GIS endeavors within state and local government and even could be said to have GIS plans, but may not have GIS legislation.

LEGISLATIVE ELEMENTS SUPPORTIVE OF GIS

Development of GIS in a state is a complex subject. The needs and objectives of states and their constituent jurisdictions are bound to vary from one state to another. Legislation can be important to define state plane coordinate systems, parcel identifier numbers, and many other standards. However, there are several major areas that are probably common to all state programs for GIS development. Indeed, it is likely that the proper approach to GIS will avoid specific pieces of legislation to address technical issues, and, instead, will seek to put in place the appropriate organizational and legal structure. This structure, in turn, will be able to institute and maintain the various technical, organizational and legal elements necessary for successful GIS development.

Areas that should be addressed in legislation designed to establish an organized approach to GIS in state government and its local governments include at least the following:

- Organizational Structure
- Funding Mechanism
- Distribution Policy
- Privacy Considerations
- Governmental Liability

Organizational Structure. The state that has studied GIS development from the perspective of a state wide program longer and in more depth than any other has probably been the State of Wisconsin. North Carolina has some of the oldest legislation on the books, most of which preceded the popular use of terms such as GIS or LIS. The Commonwealth of Virginia passed one of the first statutes that was developed in the modern era of LIS/GIS and that sought, after deliberate study, to set up an organized approach to LIS/GIS development in the Commonwealth.

Actually, among the twenty-five states that now have some sort of GIS legislation, only a few have legislation that sets out to establish a comprehensive approach to GIS and GIS related activities. One of the main considerations is whether the program will include state government only, local government only, or both state and local government. Some states emphasize local and regional governments, while others treat local governments as an afterthought. Consider Florida legislation that created the agency called the Florida Growth Management Data Network Coordinating Council. Although GIS is not mentioned specifically in this legislation, it is logically included within the Council's scope, and, in fact, the Council is responsible for state wide coordination of GIS in Florida.

> 282.403 Coordinating council; creation; membership; duties
>
> (1) The Executive Office of the Governor shall establish the Florida Growth Management Data Network Coordinating Council. The coordinating council shall review and recommend to the Administration Commission solutions and policy alternatives to establish and govern the Florida Growth Management Data Communications Network. In formulating solutions and policy alternatives, the council shall consider the needs, interest, and expertise of other public agencies, such as the state universities, water management districts, regional planning councils, and *municipal and county governments*, and of the private sector, and allow for their participation in the network.
>
> Fla. Stat. § 282.403 (1990). (Emphasis added.)

As seen, although municipal and county governments are mentioned, the emphasis seems to be on state agencies.

The Florida legislation is interesting from an organizational point of view for another reason. It does not single out GIS from information systems, in general, and yet, as stated, the Council established is, in fact, in charge of GIS coordination in that state. This unequivocally frees GIS from "mapping" and puts it in the general world of automated information (or, for that matter, information).

The coordinating function is generally a major part of any comprehensive approach to GIS in state government. State agencies and local governments tend to be quite independent, so that the power to coordinate can be limited unless leverage is built into the legislation. The Virginia legislation gives the coordinator very broad scope, both in the subject areas and levels of government, but he is given absolutely no leverage or enforcement power.

Mississippi uses the "stick:"

> No geographic information system or multipurpose cadastre shall be contracted for, purchased, leased, or created by any county or municipality

> unless the county or municipality shall first submit its plan for a geographic information system and multipurpose cadastre to the Mississippi Central Data Processing Authority for its approval, and all bids or proposals for such a geographic information system or multipurpose cadastre shall be submitted to and evaluated by the Mississippi Central Data Processing Authority before any bid or proposal is accepted . . .
>
> Miss. Code Ann. § 25-58-1 (1990).

As does Florida:

> If any specified agency fails to comply with this act without good cause, the Executive Office of the Governor may withhold releases of appropriations of those portions of the agency's operating budget that pertain to the collection and analysis of data related to growth management.
>
> Fla. Stat. § 282.403 (1990). (Emphasis added.)

While North Carolina uses the "carrot:"

> County projects shall be eligible for assistance subject to availability of funds, compliance with administrative regulations, and conformity with one or more of the project outlines as follows:
> (1) Base Maps . . .
>
> N.C. Gen. Stat. § 102-17.

Funding Mechanism. Thus far states have used at least the following methods to fund GIS development:

- General revenue funds
- Sale of GIS data and products
- Sale of services
- Bonding authority
- Recording fee surcharge

Table 1 shows some of the funding approaches used by the states. Of course, most states use more than one funding mechanism and all states use general funding as part of their source of funds. All of the statutes providing for the charging for data or products and recording surcharges have been passed within the last year and are, therefore, new and untested.

In the case of GIS, two of the funding approaches — sale of data/products and recording surcharges — are of particular interest. The sale of data and products is particularly attractive because (a) development of the data bases are expensive and cost recovery is important and (b) the potential for GIS is so great that there is an almost irresistible temptation for the government to go beyond its own bare necessities and provide data and products useful to the commercial sector. [Stated another way, it is difficult to move into modern GIS without developing the capacity that can exceed the minimum requirements of government. Once the basic data layers and processing equipment are in place, the analytical capabilities of GIS begs for additional applications.]

Real property instrument recording surcharges are of interest for several reasons. First, they would appear to be very equitable, in the sense that they draw funds from the primary "user" of land information. Ultimately, it is the real estate transaction that drives the land information system and it is the parcel to which land information must relate. On the other hand, it can be argued that commercial users of a modern GIS database should pay as well, in that such users get value from the whole system set up to collect and maintain the data and from obtaining the data in bulk.

Funding Approach	State	Legislation
Sale of services, data or Products	Connecticut	1991 CT H.B. 5807, Permits towns to charge a fee for use of the geographic information system established and maintained by the town.
	Vermont	1991 VT S.B. 216 Creates an organizational structure for the Vermont Office of Geographic Information Services and for the advisory board and provides for fees to be charged for providing geographic information system products and services.
	Alaska	Sec. 3. AS 09.25.110 (c) If the production of records for one requester in a calendar month exceeds five person-hours, the public agency shall require the requester to pay the personnel costs required during the month to complete the search and copying tasks. Sec. 09.25.115. ELECTRONIC SERVICES AND PRODUCTS. (b) The fee for electronic services and products must be based on recovery of the actual incremental costs of providing the electronic services and products, and a reasonable portion of the costs associated with building and maintaining the information system of the public agency.
	Iowa	Section 22.2-3, Code 189 . . . a government body which maintains a geographic computer data base is not required to permit access to or use of the data base by any person except upon terms and conditions acceptable to the governing body. The governing body shall establish reasonable rates and procedures for the retrieval of specified records, which are not confidential records, stored in the data base upon the request of any person.

	Kentucky	KRS § 61.970 (Michie 1990) § 61.970. Use of database or geographic information system. (1) A person who requests a copy of all or any part of a database or a geographic information system, in any form for a commercial purpose shall provide a certified statement stating the commercial purpose for which it shall be used. (2) Such person shall enter into a contract with the owner of the database or the geographic information system. The contract shall permit use of the database or the geographic information system for the stated commercial purpose for a specified fee. The fee shall be based on the: (a) Cost to the public agency of time, equipment, and personnel in the production of the database or the geographic information system; (b) Cost to the public agency of the creation, purchase, or other acquisition of the database or the geographic information system; and (c) Value of the commercial purpose for which the database or geographic information system is to be used.
	New York	1991 NY S.B. 6087 Authorizes the Erie County Water Authority to enter into contracts for computerized mapping or geographic information systems and to sell data and services therefrom.
Recording Fees	Wisconsin	§59.88 (5) Land Record Modernization Funding. (a) . . [A] register of deeds shall submit to the land information board $4 from the fee for recording the first page of each instrument . . . (b) A county may retain $2 of the $4 submitted under par. (a) . . . if . . : 1. The county has established a land information office . . . . . 3. The county uses the fees . . . to develop, implement and maintain the county-wide plan for land records modernization.
Bonds	Mississippi	Miss. Code Ann. § 25-58-3 (1990): The board of supervisors of any county and the governing authorities of any municipality are hereby authorized and empowered, to borrow money, pursuant to the provisions of this section to create the geographic information system and prepare the multipurpose cadastre ...

Table 1 - GIS Funding Provisions

Distribution Policy

State	Legislation
North Carolina	1991 NC S.B. 583 SYNOPSIS: Provides that if Catawba County has geographical information systems databases and data files developed and operated by that county and county makes electronic access and hard copy access available at reasonable cost to public by terminals and other output devices located in public facilities, person who receives electronic copy (whether on magnetic tape, disk, etc.) under public records law and not as provided by electronic or hard copy access may not resell that information/use it for commercial purpose.
Alaska	Sec. 4. AS 09.25 (e) Each public agency shall notify the state library distribution and data access center established under AS 14.56.090 of the electronic services and products offered by the public agency to the public under this section. The notification must include a summary of the available format options and the fees charged.
Wisconsin	1989 Wis. Act 31 (3) BOARD DUTIES. The board shall direct and supervise the land information program and serve as the state clearinghouse for access to land information.
Kentucky	§ 61.970. Use of database or geographic information system. (1) A person who requests a copy of all or any part of a database or a geographic information system, in any form for a commercial purpose shall provide a certified statement stating the commercial purpose for which it shall be used. (2) Such person shall enter into a contract with the owner of the database or the geographic information system. The contract shall permit use of the database or the geographic information system for the stated commercial purpose for a specified fee. § 61.975. Public records available for inspection --Fees for copying or contracts for electronic use. (1) Public records stored on a database or a geographic information system which are subject to KRS 61.870 to 61.884, but are not requested for a commercial purpose, shall be made available for inspection to the public upon request at the offices of the public agency.

Table 2
- GIS Data Distribution Provisions

Privacy. For a discussion of privacy in GIS legislation, see Dansby, H. Bishop, "Informational Privacy and GIS", 1991 URISA Proceedings.

Liability Provisions. The only state found that had specific provisions for limiting liability for GIS or similar data was Alaska. Even in this case, the limitations were for contractual liability and constituted only a directive as opposed to creating a right that the state would not otherwise have.

> (d) Public agencies shall include in a contract for electronic services and products provisions that . . .
>
> (2) limit the liability of the public agency providing the services and products.
>
> AS 09.25.115 (d)(2)

A more important liability consideration is the less predictable tort or negligence liability that could result from inaccurate data. It does not appear that any state has addressed this issue.

CONCLUSIONS

GIS legislation around the United States is remarkable both in its frequency and its variety. It is clear that GIS legislation has been spawned by different political environments and needs of the various states. There is a surprising lack of legislation that sets out to establish a comprehensive strategy for GIS development in state and local governments. Finally, a number of states have had executive orders or similar resolutions endorsing GIS and issuing some vague mandate for GIS study or development.

At this point, there does not appear to be a strong correlation between GIS legislation and GIS activities within the states, although one would speculate this would probably change as the technology and implementation of the technology within the states evolve.

If it is accepted that certain elements are necessary and useful to be included in GIS legislation, such as organizational structure, cost recovery, liability provisions, funding mechanism, privacy provisions, and access provisions, then all of the GIS legislation that exists is found lacking, and the vast majority is greatly lacking.

Although it is venturesome to make recommendations with respect to such a diverse and complex subject, it is recommended that a) states return to the fundamentals of GIS and develop legislation that will support those fundamentals, generally referred to as the organizational, institutional, and technical issues, and b) that they do a better job of aligning the organizational structure responsible for GIS development with political realities, to accomplish objectives such as coordination with enforcement.

DEVELOPMENT OF AN INTERDISCIPLINARY EARTH SYSTEM SCIENCE ACADEMIC CURRICULUM AND AN EARTH OBSERVING SYSTEM RESEARCH PROJECT

Dianne J. Love, Kevin L. Schultz, Frederick K. Wilson, and Kambhampati S. Murty

Earth Observing System (EOS) Program
Department of Biology, Jackson State University,
P. O. Box 18540, Jackson, MS 39217

ABSTRACT

The National Aeronautics and Space Administration (NASA) awarded a contract to the Department of Biology at Jackson State University (JSU) to develop activities which support the goals and objectives of NASA's Earth Observing System (EOS) Program - their principal contribution to the U.S. Global Change Research Program. The primary goal of JSU's EOS Program is to promote increased participation of minorities in NASA's Earth System Science (ESS) initiative and thereby increase the pool of minorities choosing careers in ESS or related fields. The JSU/NASA/EOS program consists of four main components: (1) establishment of a pre-college outreach program, (2) development of a high school teachers outreach program, (3) establishment of an undergraduate academic curriculum, and (4) initiation of an EOS research project which focuses on the coastal ecosystem dynamics along the Mississippi Gulf Coast. This new program will provide an opportunity to: (1) produce highly trained and qualified minority scientists in ESS, (2) provide professional role models to minority students, (3) significantly contribute to an enhanced understanding of the total Earth system. Pertinent specifics of each component will be addressed.

INTRODUCTION

The Department of Biology at Jackson State University (JSU) was awarded contract by NASA to establish a program which supports NASA's Earth Observing System (EOS) Program. JSU was awarded this contract partially because of NASA's continuing efforts to encourage the development of alliances between Historically Black Colleges and Universities (HBCUs), other universities, state and local governments, and the private sector, and to increase participation of HBCUs and other minority institutions in federally sponsored programs. NASA's EOS Program is the primary component of their "Mission to Planet Earth" contribution to the U.S. Global Change Research Program (GCRP). The GCRP is an integrated international scientific effort designed to document global change, with EOS serving as the cornerstone of this long-term program. Both the EOS and the GCRP missions were instituted to develop an improved scientific and predictive understanding of the means by which the biogeochemical, physical, and social processes interact with one another to create, control, and maintain the Earth's environment at varying spatial scales through time (i.e., by describing how its component parts and their interactions have evolved, how they function, and how they may be expected to continue to evolve through time, especially on a global scale). In conjunction with other elements of the Mission to Planet Earth Program, EOS will utilize a suite of remote sensors and other appropriate instruments to provide systematic, continuous, and comprehensive global observations of Earth. These instruments and observations will be used to establish baseline data about the Earth, to quantify changes in the Earth system, and to monitor this complex system (NASA/GSFC, 1991).

INSTITUTIONAL BACKGROUND

Jackson State University is one of the few historically minority institutions in the nation with programs in marine science, meteorology, and GIS/RS. JSU also has a recently approved Ph.D. program in environmental science. Jackson State is committed to the development of these programs as part of its overall long-term plan, which is to produce highly trained minority men and women in every sphere of education. These programs have made significant contributions in preparing minorities for the environmental, marine, and atmospheric sciences, the GIS/RS work force, and for matriculation to other graduate programs. The present JSU/NASA/EOS program is designed to enable these programs to continue their mission. Ultimately, this program will motivate and prepare minority high school students and JSU undergraduates and graduates to pursue advanced degrees and/or professional careers in ESS.

JSU/NASA/EOS PROGRAM GOALS AND OBJECTIVES

The primary goal of the JSU/NASA/EOS Program is to increase the pool of trained minority students choosing careers in Earth System Science and related fields. The two primary approaches used to achieve this goal are: (1) development of an ESS academic curriculum, and (2) initiation of a long-term EOS-oriented interdisciplinary research project. The program consists of four parts: a pre-college outreach program, a high school teachers outreach program, an undergraduate academic curriculum and training program, and an interdisciplinary EOS research project. The objectives for each of these components follow.

Pre-College Program Objectives

1. To identify and recruit approximately twenty Mississippi minority high school students who have shown evidence of high ability in their academic work and who have a strong interest in the enrichment program in the stated disciplines.

2. To provide these students with training in and exposure to ESS, mathematics, computer science, writing skills, geographic information systems, and remote sensing.

3. To expose and provide the students with new and challenging approaches to the study of ESS by focusing on the interdisciplinary nature of this field from the biological, chemical, and physical perspectives.

4. To expose the students to professionals in ESS and to help them explore career options and research opportunities in this field.

5. To enhance the admission and retention of these students into the University.

6. To provide academic scholarships for promising high school seniors and JSU undergraduates to pursue a bachelor's degree in the ESS curriculum.

High School Teachers Program Objectives

1. To apprise selected high school teachers of career options and opportunities in ESS.

2. To promote recommendation of the ESS academic curriculum to students.

3. To provide these teachers with background materials on the pre-college program and to facilitate program publicity.

4. To enhance the educational training of high school teachers via summer placement at national laboratories.

Undergraduate Program Objectives

1. To provide early research experience with dedicated mentors and role models to high ability undergraduate students.

2. To offer academic enrichment activities, counseling, and mentoring to enhance student desire, interest, and confidence to achieve in ESS.

3. To provide opportunities to minority students to take courses and participate in external research during summer months at institutions and laboratory installations with records of training minority students and with strong research programs, such as the John C. Stennis Space Center, Goddard Space Flight Center, and Jet Propulsion Laboratory.

4. To produce graduates in chemistry, marine science, environmental science, atmospheric science, computer science, and GIS/RS who are qualified to pursue postgraduate studies and compete successfully for advanced degrees in their respective disciplines.

Research Project Objectives

1. To enhance the research capabilities of both faculty and students participating in the ESS program at Jackson State University.

2. To increase the number of minority students entering graduate programs in fields or careers related to ESS.

3. To increase the number of trained professionals in fields related to ESS.

4. To strengthen the existing alliances between the JSU/NASA/EOS Program and the public and private sectors.

PRE-COLLEGE OUTREACH PROGRAM

This program consists of a four-week, residential summer enrichment program for about twenty minority high school students at Jackson State University. The summer program includes an interdisciplinary introductory course in Earth System Science (ESS) to expose, encourage, and solidify student interest in ESS. This introductory course is a menagerie of facts, techniques, and methodologies gathered from biological, chemical, oceanographic, atmospheric, and GIS/RS backgrounds. A balanced perspective is important in any subject, especially ESS, because it touches on the realms of many other disciplines. Therefore, in this course, we have endeavored to merge these disciplines and examine ESS in its entirety. There are three additional courses which augment the introductory course and which are designed to enhance each student's basic skills. These subject areas involve mathematics, English, and computer science. Also, there is an enrichment component which consists of seminars by invited guest speakers from public and private sectors, field trips to educational and research facilities and installations, and a training session entitled Introduction to Instrumentation for Research.

During the summer of 1991, the JSU/NASA/EOS Program sponsored a four-week instructional workshop for thirteen minority high school students. The schedule included all of the aforementioned enrichment activities. The academic progress of each student was evaluated through tests and exercises given by each instructor. Students received a grade for each course and also a composite grade for the overall program. To assist in program evaluation, a questionnaire consisting of two major sections (i.e., Likert scale assessments and open-ended questions) was developed. Each participant was asked to respond to a variety of evaluation statements about different aspects of the program to indicate the degree of their agreement or disagreement with that particular statement. There were five open-ended questions which solicited feedback from the participants about specific aspects of the program. Prior to their dismissal from the program, participants completed the questionnnaire. Preliminary results indicate that the students gave the overall program a satisfactory mark. Participants suggested that the program be complemented by providing introductory and advanced sessions for each course. These sessions will be dependent on the grade level of the students.

HIGH SCHOOL TEACHERS OUTREACH PROGRAM

This component consists of a two-week informational workshop for approximately twenty minority high school teachers. The program includes seminars by personnel from ESS related industries, laboratories, and other disciplines. Also, materials containing career options and opportunities in ESS, including curriculum enhancement materials which may be used as course enhancement tools to the basic science curriculum taught in the secondary school system, are distributed to the participants. In addition to seminars, the workshop consists of a tour of the research facilities located in the John A. Peoples, Jr., Science Building and the Just Science Hall at Jackson State University and demonstrations of GIS/RS hardware and software. The workshop is also comprised of a field excursion (concurrent with the Pre-College Program's field trip) to the Gulf Coast Research Laboratory, the John C. Stennis Space Center, the Marine Education Center, and other educational and research facilities and installations.

Also during the summer of 1991, ten minority teachers participated in JSU's EOS High School Teachers Summer Outreach Program. In addition to the aforementioned activities, the schedule of this two-week informational workshop included lectures and hands-on laboratory exercises on current issues and new ideas in environmental science, marine science, chemistry, atmospheric science, computer science, and GIS/RS. A questionnaire was also developed similar to the one used for the Pre-College Program, which participants completed prior to their dismissal from the program. Preliminary results suggest that the teachers were highly pleased with the entire program, especially the field trips, and are extremely interested in participating in similar programs in the future.

UNDERGRADUATE PROGRAM

The ESS academic curriculum is an interdisciplinary honors undergraduate training program and concentration within the Department of Biology, involving marine science, atmospheric science, environmental science, chemistry, computer science, and GIS/RS. Because of the program's flexibility, students are able to pursue their respective discipline curriculum for the bachelor's degree, to obtain credits in research, and to pursue a concentration in ESS. A certificate program is currently being developed to formalize the ESS curriculum. Each student will be awarded a certificate upon fulfilling the ESS curriculum requirements. Students must complete at least sixteen credit hours from the ESS curriculum course list; eight of these credit hours are required, with the remaining eight hours to be chosen from highly recommended and restricted elective courses. All of these courses are designed to strengthen each student's research background. Further, courses must be chosen which do not form the core curriculum for their major.

Currently, the lack of positive reinforcement is one of the factors causing a decline in the number of students pursuing undergraduate and graduate degrees with a focus on research. During the academic year, students will be involved in research under the supervision of JSU faculty sponsors. This research will begin with the first semester of the fellowship. Partial credit may be given during each semester, however, the faculty trainer is required to meet with the trainee for three hours per week to insure maximum training benefit. Each trainee will be required to enroll in an Introduction to Research course for one credit hour during the second semester of the training experience. The course will serve to introduce the trainees to all phases of scientific research, i.e., identification of problem, formulation of hypothesis, design and execution of experiment, data collection and analysis, and interpretation and presentation of results (oral and written).

All trainees and trainers will participate in a monthly seminar which consists of the following objectives: (1) to allow for interchange of ideas and research findings, (2) to discuss theories and methodologies, (3) to provide group identity to program participants, (4) to provide periodic interaction between the faculty and the participants, and (5) to provide participants opportunities to informally discuss their research. At the end of each year, a technical paper will be written by each trainee and presented orally in a seminar. The Introduction to Research course and associated research project are two factors which determine direction and the ultimate curriculum which a particular trainee follows. All trainees will be required to attend seminars and review primary references of speakers prior to attending. The intent of this provision is to make the student sensitive to research reporting and to the presentation and discussion of technical material in a group setting. The training program will include lectures or special seminars by nationally recognized experts during the academic year. During the summer, trainees may enroll in an Independent Study course for two credit hours regardless of where they do their summer research internship, pending the approval of the internship sponsor.

An important part of each individual training program will be trainee's summer training or study experience. Trainees will be identified and accepted near the end of his/her sophomore year. They will participate in the program during summers following their freshman year. Most trainees will receive a summer training experience at one of the outstanding institutions (e.g., industrial and national research laboratories and academic research institutions) whose researchers have agreed to participate in the program. Some trainees will remain at JSU for their summer research experience to fully exploit facilities and expertise readily available.

All trainees will attend at least one national professional meeting or conference in fields related to their research training program goals. They will be encouraged to present papers at these meetings whenever possible. However, each trainee will be required to present at least one paper at a national meeting during his/her research training tenure.

Complete records of trainee activities, reports, research papers, evaluations, etc., will be maintained in the office of the EOS Program Director. The training program will be evaluated periodically by participating faculty, trainees, a Steering Committee, and an Advisory Committee to maintain high standards of performance and to provide for continued upgrading of the program.

JSU/NASA/EOS RESEARCH PROJECT DESIGN

Both natural and anthropogenic processes have greatly influenced, and possibly accelerated, the rate of environmental change along the dynamic Mississippi Gulf Coast. The major causes of environmental changes result from human activities on the Earth's surface and natural events or processes (both past and present) (Wheeler, 1990). The Gulf of Mexico coastal marshes and estuaries have often been perceived as infinite resources to be exploited at will. However, during the last few decades, the marshes located along the Gulf have begun to show signs of habitat degradation. For example, Louisiana coastal wetland loss has progressed at an average rate of about 100 km^2/yr (Gagliano, *et al.*, 1981) due to reduction in sediment supply and to human intervention in natural processes.

The JSU/NASA/EOS research project will study the dynamic terrestrial, marine, and atmospheric processes and their interactions with human activities along the Mississippi Gulf Coast. This study will be conducted first on a local scale (e.g., selected Mississippi river systems which flow into the Gulf of Mexico via the Mississippi Gulf Coast) and then on a regional scale. The primary aim of the research project is to address how land cover changes and atmospheric phenomena affect coastal ecosystem dynamics, especially biological resources. This study will use data derived from a suite of remote sensors of varying spatial scales and temporal periods to detect, quantify, and analyze the nature and extent of coastal ecosystem changes (Light, 1990). The entire project is categorized according to three major components (i.e., terrestrial, marine, and atmospheric). These components also comprise the three research teams. Each research team will examine the relationships between the reciprocal effects and consequences of human activities and a combination of the following environmental processes and phenomena: landscape ecological patterns and processes, biological productivity, nutrient dynamics, toxicology, physical oceanography, and meteorological phenomena. Each major component is briefly summarized in the following paragraphs.

Terrestrial

The terrestrial research team is interested in examining landscape patterns and processes, biological productivity, nutrient dynamics, and toxicology. This research team, for example, will concentrate on: landscape ecology/land cover changes; sediment/pollutant transport; vegetation discrimination, shifts, and stress; biomass estimation/biological productivity; and invertebrate population dynamics. The primary goal of this research component is to develop a comprehensive model which will predict and monitor the effects of environmental changes at varying scales on terrestrial ecosystem dynamics. This study will identify and delineate changes such as coastal land loss and land cover patterns as a result of environmental processes. Biological productivity of forests adjacent to the coastal and intertidal zones will be measured to estimate biomass budgets. This team will also investigate the nutrient dynamics related to carbon, nitrogen, and phosphorus budgets to foster a better understanding of the relationships between these budgets and landscape patterns and processes, such as coastal land loss and sediment/pollutant transport. The toxicology subproject will focus on identifying sources for forest canopy vigor changes and examining the short- and long-term effects of atmospheric pollutants on forest vegetation (Smith, 1989). The model developed from this project will be used to establish appropriate management practices governing resource use and protection. Results from this component will enhance our understanding of landscape ecological processes influencing the coastal environment.

Marine

The marine research team is interested in examining biological productivity, nutrient dynamics, toxicology, and physical oceanography. Specifically, this team will concentrate on: the abundance and distribution of selected biophysical resources, pollution effects on biota, sea level changes, biogeochemical cycling/nutrient dynamics, biomass estimation/biological productivity, and temperature regimes. The major goals of this research project are to determine the reciprocal effects of human activities and environmental changes on the marine ecosystem (Bernstein and Stevens, 1986) and to develop a holistic model which predicts the effects of these different processes on marine ecosystem dynamics. The biological productivity of the marine food web will be measured, focusing on phytoplankton, penaeid shrimp, and fish. The physical oceanographic parameters and processes that will be investigated include sea surface temperature, salinity, dissolved oxygen, chlorophyll, coastal upwellings, and water body variability. Nutrient dynamics related to the carbon, nitrogen, and phosphorus budgets in sediments and at different trophic levels of biota will be examined, focusing on the relationship between the sources and the sinks of carbon, nitrogen, and phosphorus in this coastal ecosystem (Seshavatharam and Murty, 1989). A biogeochemical dynamics database will be developed to improve the determination of bacterial biomass production and to assess the physiological status of the Gulf Coast region ecosystems. The effects of hydrologic pollutants, especially heavy metals such as mercury, zinc, and cadmium, on marine biological productivity will be examined (Delaune, *et al.*, 1990; Koeck, *et al.*, 1990). This team will also determine the phytotoxicity of "OUST" (sulfomethuron methyl) on the growth and reproductive rate of Lemna spp. (duckweed) found in freshwater swamps and ponds. Both spatial and temporal distribution trends and patterns will be identified for selected marine resources. The relationships between these parameters and biological productivity will be the focus of this research team's investigations. These results will eventually contribute to a better understanding of marine processes, establishment of optimal resource management policies, and development of new fisheries.

Atmospheric

The atmospheric research team is also interested in examining landscape patterns and processes, meteorological phenomena, and physical oceanography. This research team will concentrate on: air temperature regimes; precipitation; solar radiation; cloud cover; water vapor content; atmospheric impact on land cover; atmospheric conversion of ambient aerosols; and other atmospheric, chemical, and climatic changes. The goals of the atmospheric research team are to model the effects of various meteorological phenomena (e.g., hurricanes, severe storms, and coastal fronts) on the coastal landscape and to determine the atmospheric conversion of ambient aerosols. This research team will focus on determining the feasibility of using satellite observations to improve forecasting of heavy precipitation events, determining the impact of these events on the coastal landscape processes, and investigating the effects of chemical reactions caused by various atmospheric conditions in the nearshore and offshore environments. This team will also examine the relationships between sulfur dioxide conversions, seasonal changes, and specific meteorological conditions and events. The results of this study will be used to develop a model which improves our knowledge and understanding of atmospheric processes.

Integration of GIS/RS

Existing geographic information system (GIS) and remote sensing (RS) techniques and methodologies will be tailored and applied for rapid inventory, integration, and analysis of the aforementioned diverse and disparate data sets. These methodologies will be coupled with modeling approaches to derive local, regional, and ultimately, global scale models for prognosis, prediction, and monitoring of coastal ecosystem dynamics (Browder, *et al.*, 1988). Such an integrated approach will better serve user-specific purposes both in breadth and depth.

EXPECTED OUTCOME

The expected results of this new program are: (1) an increase in minorities pursuing postsecondary education in ESS, (2) an increase in minority undergraduates pursuing postgraduate degrees and careers in ESS, (3) an increase in the pool of highly trained and qualified minority scientists with postgraduate research experience in ESS, and (4) an enhancement of the research capabilities of participating interdisciplinary ESS faculty. Ultimately, the previously outlined interdisciplinary research approach for studying Earth system processes as a unit flux will achieve the goals and objectives of NASA's EOS program by: (1) yielding an enhanced scientific understanding of the Earth's terrestrial-marine-atmospheric processes and interactions at varying spatial scales through time, (2) serving as a basis for local, regional, and ultimately global scale studies, and (3) providing valuable insight into the development of an integrated predictive global change model.

REFERENCES

Bernstein, R. L., and R. S. Stevens. 1986. Ocean Remote Sensing, Space Science and Applications, Progress and Potential. IEEE Press: New York, NY.

Browder, J. A., L. N. May, A. Rosenthal, R. H. Baumann, and J. G. Gosselink. 1988. Utilizing Remote Sensing of Thematic Mapper Data to Improve Our Understanding of Estuarine Processes and Their Influence on the Productivity of Estuarine-Dependent Fisheries. Report to NASA, Goddard Space Flight Center. Center for Wetland Resources, Louisiana State University, Baton Rouge, LA.

Delaune, R. D., R. P. Gambrell., J. H. Pardue, and W. H. Patrick. 1990. Fate of Petroleum Hydrocarbons and Toxic Organics in Louisiana (USA) Coastal Environments. Estuaries. 13(1): 72-80.

Gagliano, S. M., K. J. Meyer-Arendt, and K. M. Wicker. 1981. Land Loss in the Mississippi River Deltaic Plain. Transactions of the Gulf Coast Association of Geological Societies. 31: 295-300.

Koeck, M., et al. 1990. Accumulation of Heavy Metals in Animals. Part 2. Heavy Metal Contamination of Fish in Styrica Waters (Austria). J. Hyg. Epidimiol Microbiol. 33(4): 529-535.

Light, D. L. 1990. Characteristics of Remote Sensors for Mapping and Earth Science Applications. Photogrammetric Engineering and Remote Sensing. 56(12): 1613-1623.

National Aeronautics and Space Administration (NASA), Goddard Space Flight Center (GSFC). 1991. EOS Reference Handbook. NASA/GSFC: Greenbelt, MD. 147 pp.

Seshavatharam, V., and K. S. Murty. 1989. Nitrogen and Phosphorus Contents of Water, Sediment, and Some Aquatic Macrophytes of Lake Kondakarla, India. Tropical Ecology. 30(1): 41-47.

Smith, W. H. 1989. Air Pollution and Forests: Interaction Between Air Contaminants and Forest Ecosystems. Second Ed. Springer-Verlag: New York, NY. 618 pp.

Wheeler, D. L. 1990. Scientists Studying "The Greenhouse Effect" Challenge Fears of Global Warming. J. For. 88(7): 34-36.

CERTIFICATION IN GIS: MORE TROUBLE THAN IT'S WORTH?

Nancy J. Obermeyer, Ph.D.
Department of Geography and Geology
Indiana State University
Terre Haute, IN 47809

ABSTRACT

A recurring idea in discussions within the GIS community is the need to develop a means of assuring the qualifications of GIS professionals. The most commonly suggested strategy to achieve this objective is the development of an approved program of training for individuals wishing to enter the field: the final rite of passage would be an examination. Similar certification procedures are well established in medicine, law and planning. The development of a certification process for GIS professionals will not be a simple task. Many problems must be resolved before certification can become a reality. We must first answer several basic questions such as: What skills and level of understanding are needed to become qualified in GIS? Do project managers and GIS programmers need the same skills and understanding? Who will design the examination? How will it be graded? This paper examines these and other relevant issues, taking cues from the experiences of other certification processes in an effort to provide insight into the task that lays ahead if, indeed, we choose this option.

INTRODUCTION

The idea of developing a certification process for GIS professionals has been bandied about in GIS circles recently. This perceived need grows, in large part, from concern to assure a consistently high level of quality within the field -- without a doubt, a noble and worthwhile objective. Certification is well established in other fields, including medicine, law, planning and even cosmetology. As those of us in the GIS community consider whether or not to adopt certification for our own professionals, we may learn some lessons from the experiences of other professions.

This paper begins with an exposition of the roots and rationale of certification which are found in the basic concepts of expertise and professionalism. Following this background material is a description of various aspects of certification drawing examples from the fields of law, medicine, planning and cosmetology. The next section suggests several questions for consideration within the GIS community regarding certification for GIS professionals. Finally, the conclusion urges discussion of this issue within the GIS community.

BACKGROUND

To understand the concept of certification, we must begin by exploring its roots, which are found in the ideas of expertise and professionalism. Expertise and professionalism are closely related, but they are by no means synonymous. Much of what we know about these concepts is derived from Weber's theory of bureaucracy (1968).

Expertise

Webster's Dictionary defines expertise as "specialized knowledge or skill: mastery." We can and do think of expertise in all (or nearly all) aspects of life, from dog training to neurosurgery to geographic information systems, and the specialized knowledge or skill required to demonstrate expertise is unique to each field.

In addition to being field-specific, expertise is inherently time-specific. For example, leeches were once a common and acceptable course of treatment within the medical community. Today, lasers and ultrasound are important medical tools over which modern physicians must develop mastery if they are to be considered experts. As innovations pervade a field, the specialized knowledge of that field shifts to include them, as well as to eliminate obsolete techniques and ideas.

Not surprisingly, expertise forms the backbone of professionalism. It is the unique composition of specialized knowledge that makes professionalism possible within a specific field. Fields that cannot achieve consensus on exactly what mix of specialized knowledge is required to demonstrate expertise in the field have been known to question their own validity as a profession. Normally, this occurs within professions with diverse missions and members. Public administration is one such field.

A small body of specialized knowledge is not necessarily an impediment to evolution of a profession, particularly if that body of knowledge is important to others. For example, beauticians and cosmetologists have been very successful as a profession. This group has managed to put in place certification processes in many states, thereby gaining public acknowledgement and support of the expertise of members of the profession.

In addition to its role as the backbone of professionalism, expertise is often cited as a key consideration in the development of certification processes. The development of some sort of certification process is frequently perceived and promoted as a means to assure that people who claim to be an expert within a particular field actually possess the necessary expertise to make this claim. Certification thus is designed, in part, to set a standard of competency for the field (whatever it is). Indeed, this rationale is frequently mentioned as a driving force for development of a certification process within the GIS community.

While many noble goals are associated with the development of expertise (for example, setting a standard of competency), Cayer and Weschler (1988: 45) also note that the expertise of professions and the concomitant control over information may lead to a concentration of power within the profession. Similarly, Habermas (1970) suggests that experts may use their specialized knowledge to build a technocracy, thus gaining hegemony within their field. Weber (1968) likewise raises concerns about the elevation of technical experts to the status of a mandarin caste. There is a thin line between the concentration of expertise necessary to assure competency within a field and the use of expertise to create a technocracy. While the assurance of competency is of great value to society at large, the creation of a technocracy usually benefits the technocrats at the expense of regular citizens.

Professionalism

The development and maintenance of a specialized body of knowledge (expertise), as has been noted, is a critical prerequisite for the evolution of professionalism within a field. Pugh (1989) identifies expertise as one of the most important attributes that characterize a profession, and further stipulates that within this body of knowledge there must exist "... propositions generally accepted by scholars and practitioners" in the field (p. 2). The existence of such a body of knowledge makes possible the evolution of additional attributes that ultimately result in the coalescence of a profession (Pugh 1989).

As the profession coalesces, it develops what Pugh calls a "social ideal" (p. 2) that defines the mission of the profession. Eventually, as the profession evolves, members of the professional community join together formally to create a professional organization. Frequently, professional organizations establish a journal, newsletter, or some other medium (or media) for communication among the professional membership.

These publications become integral as a means to continue the development and maintenance of the profession's expertise, as members share new ideas and refine (and sometimes eliminate) old ones. In addition, these publications facilitate the development of a common language -- sometimes better described as jargon. This common language serves a valuable function within the profession: it serves as an efficient short-hand means to communicate professionally. In addition, more ominously, it helps to identify who is a member and who is not, and may be used to make entry into the profession more difficult.

In order to understand the profession's common language and enter the field, typically some sort of specialized training is required, often followed by successful completion of some sort of licensing examination leading to certification. This process leads directly to the establishment of a professional hegemony, as self-defined experts join together to determine what steps qualify non-experts to enter their ranks. Berlant (1975) goes even further in his description of the medical profession, describing the American Medical Association as a virtual monopoly.

Over time, each self-defined profession develops its own particular culture and lore, disseminated by a combination of professional journals and gatherings (conferences), specialized education and training, certification or licensing procedures and so on. An important element in this professional culture is what Pugh (1989:3) calls the professional "... hall of fame, a gallery of luminaries." Individuals become part of this hall of fame by performing works in support of the profession, including theoretical and scholarly contributions, teaching and mentoring activities and general advocacy on behalf of the profession.

Pugh (1989:3) also notes one final component of a profession: the development of a "code of ethics." Codes of ethics may be viewed as taking standards of practice a step further: not only does the profession assume responsibility for a standard of technical competency among practitioners, but it endeavors to assure that its members will use their expertise ethically at all times. Professions may adopt any of several mechanisms to encourage ethical practice, including peer pressure and sanctions such as fines, suspensions, or even expulsion from the profession.

Within the context of professionalism, the development of a certification process is a logical step in the process of self-definition. As we review the attributes of a profession, it is apparent that the GIS community is well on its way to becoming a profession. We have formed a professional association, albeit an unusual amalgam of pre-existing organizations, and we have several publications by which we communicate, including the *International Journal of GIS* and *GIS World.* We meet regularly. Members from among our group, the National Center for Geographic Information and Analysis, have developed a core curriculum for GIS. There are several esteemed individuals within our profession who would qualify for a "GIS Hall of Fame." Calls for data standards, concerns about accuracy and full disclosure regarding data may be interpreted as first steps toward a GIS code of ethics. It is not surprising, then, that some of our members should suggest the institution of a certification process at this time.

CERTIFICATION IN OTHER FIELDS

The idea of certification is not new. Indeed, many professions, including medicine, law, planning and cosmetology, have long-established certification procedures. These certification processes are characterized by varying degrees of effectiveness. This section describes some of the successes and failures of each of these processes as a means to provide insight into certification in GIS.

Although the concept of professionalism and certification as an internal, self-regulatory (as well as self-promotional) process is implied by the preceding descriptions, it is important to make this aspect explicit. It is within this context that we understand the effectiveness of the certification process. On the one hand, effectiveness is defined by the competent, appropriate and ethical service of members of the profession to the larger society. On the other hand, effectiveness is defined by the success of members of the profession in the marketplace, and the ability of the profession to insure that practitioners it has certified are hired in preference to practitioners who do not have the profession's official seal of approval.

Development of a Monopoly: The American Medical Association

One of the most successful of all professional organizations is the American Medical Association (AMA). Formed in 1847, its objective was to upgrade the educational standards in medicine (Berlant 1975:226). At the urging of Nathan Smith Davis, whose ideas formed the basis for early AMA policy, the AMA adopted a system that separated teaching and licensing. Medical schools continued to function as before, while licensing was instituted at the state level, with medical societies having responsibility for appointing individuals to the state licensing boards. In addition, the AMA established a policy that required *both* a diploma and a license in order to enter the practice of medicine. Previously either a diploma from a medical school *or* a license had been sufficient to allow doctors to set up a practice (Berlant 1975).

"Protection of the public against quacks," Berlant (1975:227) notes, was the primary justification for the development of what eventually became a monopoly in medical services by the AMA. Following the German tradition of state licensing superimposed on university examinations, Davis effectively promoted legislation that ultimately pressured medical schools into a specific line of development, what Berlant describes as "orthodox medicine" (p. 227). The imposition of this legislation effectively made orthodox medicine (i.e., that promoted by the AMA) the only medicine that would receive the legislature's official seal of approval. What had previously been professional dominance by the AMA became hegemony; with the support of state legislatures, this hegemony became a virtual monopoly (Berlant 1975).

In the years after its establishment, the American Medical Association became synonymous with quality health care. Practitioners and products alike proudly displayed the AMA's seal of approval. In recent years, however, concern about the monopolistic domination by the AMA has arisen. Complaints about the arrogance of doctors have become commonplace. Until relatively recently, one dared not question the medical judgment of the physician: he or she alone possessed the medical expertise to save a life. The AMA had come to resemble the technocracy described by Habermas (1970). Berlant (1975) suggests that the legislative seal of approval played an important role in the evolution of the AMA into an exceedingly powerful monopoly. We might also conclude that the complexity of medicine itself, along with the universal importance of medical care, were critical factors in the evolution of the AMA into a monopoly.

Professional Autonomy: The American Institute of Certified Planners

If the American Medical Association stands out as a professional organization whose certification process assured a virtual monopoly for its members, then the American Institute of Certified Planners (AICP) is an example of what happens when certification is a totally voluntary exercise.

In his discussion of the AICP certification examination, Rasmussen (1986:7) argues that "Certification by the professional society means planners themselves control their own occupational standards. In contrast, state licensing or registration gives the state legislatures statutory control over occupational standards by regulating the right to practice planning or to use the title of planner." Rasmussen seems to imply that the current arrangement promotes maximum autonomy for the AICP. However, comparing the AICP to the AMA suggests that this autonomy comes with a price tag.

As described above, the AMA exercises considerable authority over the appointment of members to state medical licensure boards. This authority, Berlant argues, has resulted in development of a virtual monopoly by the AMA over the entire medical practice. One aspect of this monopoly control is that only certified physicians may practice medicine. By contrast, the planning profession is open to any individual who cares to call himself or herself a planner.

While some jobs in planning (most commonly academic positions) require that candidates have AICP certification, most do not. Surprisingly, many do not even require completion of a degree program in planning. For example, in the June 15, 1991 issue of the American Planning Association's *JobMart*, of thirty-seven positions advertised, not one *required* AICP certification, and only one gave *preference* to AICP planners. Furthermore, barely half (eighteen) required a degree in planning, and ten listed planning as one of several acceptable degrees. The nine remaining positions did not specify educational requirements, or did not specify planning among their lists of preferred programs of study.

From the perspective of the planning professional, the value of AICP certification seems limited at best. Furthermore, one could logically challenge Rasmussen's implication that the AICP exercises autonomous control over the quality of planning practitioners. In fact, if one need not achieve AICP certification in order to work as a planner, and indeed, that *most* planning positions do not require AICP certification, then it is clear that the AICP exercises almost no de facto quality control in the planning marketplace. Rasmussen himself admits that only about twenty percent of the members of the American Planning Association are also members of the AICP (and, by implication, have passed its certification examination).

The AICP certification procedure is intended as a means to assure quality among planning practitioners. It is probable that planners with AICP certification possess high quality skills and experience. However, in practice, the AICP exercises little control over the practice of planning in the United States. Without some mechanism (either voluntary or coercive) to limit entry into planning practice exclusively to AICP members, the AICP will be unable to assure quality across the board within the planning profession.

The differences between the American Medical Association, which has a virtual monopoly on medical practice in the U.S., and the American Institute of Certified Planners are readily apparent. In the first instance, their approaches are very different, as the AMA assiduously courted legislators in its early stages to assure the dominance of its practice of medicine over all other types. The AICP, by contrast, has resisted what it perceives as the threat of legislative control of its profession. The outcomes are dramatically different: the AMA has achieved a stranglehold on medical care in the U.S.; the AICP is a lofty goal to which many planners may aspire, but which few attain (with little or no apparent career damage, it might be added).

QUESTIONS WE SHOULD ASK ABOUT CERTIFICATION

Implementing a certification process for GIS will not be a simple task. It is not clear, based in part on the experiences of the AICP, if it is necessarily even a worthwhile task. In order to decide if the potential advantages of certification outweigh the potential problems, we should begin by asking several questions, and discussing our responses within the GIS community. I have listed some of the key questions below. Since most discussions of certification emphasize examinations, I will initially raise questions about establishing a test.

Who Will Develop the Examination?

The core of the GIS community can be found for the most part within the membership of the organizations that have joined together to sponsor this GIS/LIS conference: the AAG, ACSM, AM/FM International, ASPRS and URISA. However, all members of all these organizations do not identify with the GIS community. To put it another way, only a fraction of the membership of each of these organizations self-identify with the GIS community. Moreover, each of these organizations as a whole has a specific organizational mission, different from that of the others, and not specifically dedicated to GIS.

Within this framework, have we the necessary cohesion to identify a core of individuals who can develop an appropriate examination for GIS professionals? Perhaps it would be possible to empower a committee comprised of representatives of each of these five GIS/LIS sponsoring organizations to begin work on a certification examination. However, there is a lack of a formal structure to legitimate such an arrangement. For example, within the AAG (the sponsoring organization of which I am a member), regular elections are held and all administrative actions are carried out by elected representatives, thereby assuring accountability to the membership.

Currently the GIS community does not have such a formalized structure. (Perhaps on paper the GIS community exists as little more than a highly developed and refined mailing list.) In spite of this, we are surprisingly well-informed, cohesive, and even congenial. Moreover, the GIS community is extremely active. Perhaps these characteristics will enable us to work together to achieve consensus regarding the establishment of an examination development group and appropriate mechanisms for soliciting input from members of the community. Or perhaps it is time to organize formally.

The constitution of the committee developing a certification is not a trivial matter. Such a committee must have the respect of the GIS community at large and in some way reflect our diverse interests. Such a group will be perceived as legitimate, and its products will likewise be perceived as legitimate. On the other hand, an examination devised by a committee that fails to meet these qualifications will be subject to challenge as illegitimate.

How Will the Examination Approach the Diverse Applications of GIS?

That the annual conference of the GIS community has five separate sponsors provides ample evidence of the diversity of the applications of GIS. The conference program describes an incredible array of tasks to which GIS is put, including emergency management, forestry and wildlife management, public utilities and so on. How would a GIS certification examination approach these diverse applications?

This question takes on additional relevance when we acknowledge that those who achieve certification will practice within the context of a specific organizational mission. How will we assure that the GIS professional using the technology in a forestry (or emergency management, or a public utility and so on) context will be qualified for his or her assignment? Can any testing procedure take account of such diversity? The alternative is to test on the basis of a common denominator in GIS, which brings us to our next question.

What Are the Core Skills and Knowledge Tested by the Certification Examination?

While a certification examination may be not be amenable for use in testing a variety of GIS applications, such an exam must cover the core GIS skills and knowledge needed by a professional in the real world. The NCGIA Core Curriculum has gained widespread acceptance within the GIS community as a guideline for teaching GIS. It may be useful as a guide to identifying specific skills and knowledge to be covered on the certification exam. In addition, the group gathered to develop the exam will have a great deal of influence in identifying essential skills and knowledge in GIS.

Should a certification exam cover historical development of GIS? What about elements of spatial analysis? Database management? Specific commands within a GIS program? Issues related to managing a GIS project? Ethical and legal issues?

GIS is a broad area; identifying its core is more than a notion. In addition to the diversity of applications described above, we know that GIS professionals practice at different levels, from digitizing and basic data entry to project management. The core skills and knowledge to be tested by a certification examination must at least acknowledge, if not reflect, this variation in practice among GIS professionals.

How Will the Certification Examination Assure High Standards among GIS Professionals?

It is clear that the driving force behind suggestions to develop a certification procedure for GIS is the noble desire within the GIS community to assure high standards among GIS professionals. However, as we have seen in the cases of the American Medical Association and the American Institute of Certified Planners, simply instituting a testing procedure provides no such assurances.

The AMA strategy of garnering legislative support to force doctors to pass a certification exam in order to earn their licenses to practice has proven very successful in the sense that the AMA exercises virtually monopoly control over the qualifications of practicing physicians. (Don't forget the role of the AMA in appointing state medical licensing boards.) Such monopoly control, which is indeed an extreme case of unbridled technocracy, is clearly not an appropriate goal for the GIS community. However, it seems necessary to develop some system to assure that a GIS certification body exercises adequate control of GIS professionals.

When we look at the situation in the planning profession, it is apparent that if there is no way to assure that the large majority of jobs within the profession go to certified professionals, there is little or no advantage to the individual practitioner to earn certification. Under these circumstances, the certifying body has little, if any, control over the quality of the profession as a whole.

How might the GIS community assure the necessary control to assure quality? The development of state licensing for GIS professionals is a possibility. Other professional bodies besides the medical community have taken advantage of this alternative, including the legal profession as well as beauticians and cosmetologists. These latter two examples suggest that the profession need not necessarily be associated with advanced education and high income in order to develop state-required licensing procedures.

It is not clear that the GIS community -- not to mention state legislatures -- would embrace this approach, and even if they did, it would take time to put in place a comprehensive state licensing system. Given the cohesiveness of the GIS community, it may be feasible and preferable to develop a voluntary system among potential GIS employers, whereby they would agree to hire only certified GIS professionals. Negotiating agreements with key GIS employers, such as the federal government, state governments and large corporations may set the stage for general acceptance of such a voluntary system. In addition, the institution of a job listing service within the GIS community that would announce positions for certified GIS professionals could be a supplemental tool in gaining the support of members of the GIS community.

The basic task will be to convince the majority of GIS employers that a GIS certified professional is better qualified and will perform better on the job than one who is not certified. If there is no difference, the employer will have no incentive to hire the certified person, especially (as is sometimes the case) the certified person comes at a premium price.

On the flip side, if there are more costs (study time, test fee, anxiety, and so on) than benefits (a better job or higher salary) to the professional in becoming certified, few will endure the trouble.

I therefore see this question as critical to any discussions about whether or not we should, as a community of professionals, establish a certification procedure, since it goes to the heart of whether or not we will be able to achieve the stated goal of improving the overall quality of GIS professionals. If a GIS certification process does not gain universal (or at least near-universal) support, it will be ineffective as a tool to assure the quality of GIS practitioners.

Other Questions

There are several additional questions that bear raising within the context of this discussion.

What Kind of Test?

If the GIS community should decide to adopt some sort of a certification examination, it will face the challenge of developing an appropriate test format. Typically certification examinations feature multiple choice questions for the ease they present in grading and the consistent results they provide. However, problems with multiple choice exams abound. Some people simply have trouble with multiple choice exams, and numerous researchers have identified cultural bias in standardized tests generally that must not be overlooked. Essay exams may provide more and better insight into the qualifications of professionals, but it takes time to grade them. And what about a test at the workstation?

How Will the Costs of the Exam be Allocated?

Administering a certification exam will cost money. Expenses include preparing and printing the exam, proctoring the exam, grading the exam, notifying exam-takers of their grades, and maintaining records. Often exam fees are used to help pay for these activities. Membership fees may also be used to fund part of the cost. Record-keeping, for example, could be covered by an employee of the professional organization who is already responsible for some other similar task. Alternatively, members of the professional community may donate their time, thus keeping costs low.

Will a New Administrative Bureaucracy be Needed?

Given the growth in the field of GIS, if the idea of a certification process catches on, it may precipitate the need for an administrative structure to manage the process. As noted in the preceding section, a certification process will not run itself. Presumably exams will be given at regular (possibly annual) intervals. Even if members are willing to contribute their time to this endeavor, the need for accountability will require that someone have ultimate responsibility for assuring that the process runs smoothly. If an administrative bureaucracy is needed, that bureaucracy will need office space.

This is a sample of the questions that we as a community must discuss as a prelude to making a decision on whether or not a GIS certification process is desirable. It is clear that the stated goal of assuring that GIS professionals will be fully qualified for their assignments is a worthwhile goal. How we as a community of professionals choose to achieve this goal is another question entirely.

ALTERNATIVES TO A CERTIFICATION EXAM

If our goal is to assure that GIS professionals will meet a minimum standard of qualification, then there are other means to achieve this goal which may be used alone or in combination with one or more other alternatives. Among these alternatives is to continue as we now exist, within an informal, yet relatively cohesive network of GIS professionals. This framework, similar to formal professional organizations, relies on peer pressure as a means to assure appropriate and ethical professional behavior. We keep each other honest.

It is also possible to formalize our current arrangement. Establishing a formal professional organization will make it increasingly possible to impose sanctions on members whose professional behavior is inconsistent with community norms and conventions. The disadvantage of this alternative is that most of us are already dues-paying members in at least one or two other professional organizations and may have difficulty coming up with dues to join a professional organization devoted to GIS. As a result, a *formal* GIS community might be somewhat smaller than our current informal group.

Another alternative to institution of a certification exam would be to develop an accreditation system for GIS education and training programs. Possibly using the NCGIA Core Curriculum as a starting point, the GIS community could develop a set of performance criteria that GIS programs would be required to meet in order to earn certification. There are a couple of advantages to instituting accreditation at the instructional level (as opposed to the individual level). First, there are fewer programs in GIS than individuals interested in the field, making the programs an easier target. In addition, many instructors in GIS programs are already active members of the GIS community. Reaching them will likely be easier than reaching individuals who come from a wider variety of backgrounds.

One final alternative to a certification exam is the development of a code of professional ethics for GIS practitioners. This alternative has the advantage of being completely voluntary. Moreover, it is the kind of action that recalls Mom and apple pie, and therefore should be relatively popular within the community. Regardless of what we decide with respect to certification, development of a code of ethics merits consideration.

CONCLUSION: THE NEED FOR DISCUSSION

I have deliberately avoided saying whether or not we need to institute a certification examination within the GIS community. Instead, I have tried to raise relevant questions that we may discuss among ourselves. These questions do not have right or wrong answers, but that does not mean that they are unresolvable, or that they do not lend themselves to consensus.

I have also tried to provide some examples of successes and failures in other professional certification processes. Ironically, both major examples contained elements of both success and failure. These experiences should help heighten our own awareness of the difficulties inherent in establishing a certification process.

Discussions on the subject of certification exams within the GIS community will undoubtedly raise additional questions not included in this paper. My goal was not to be all-inclusive. If I have opened the discussion, I will have met my objective.

REFERENCES

Parish, J. (editor). 1991, JobMart, American Planning Association, Chicago.

Berlant, J.L. 1975, Profession and Monopoly: A Study of Medicine in the United States and Great Britain, University of California Press, Berkeley.

Cayer, N.J. and Weschler, L.F. 1988, Public Administration: Social Change and Adaptive Management, St. Martin's Press, New York.

Habermas, J. 1970, Toward a Rational Society: Student Protest, Science, and Politics (translated by J.J. Shapiro), Beacon Press, Boston.

Pugh, D.L. 1989, Professionalism in Public Administration: Problems, Perspectives, and the Role of ASPA: Public Administration Review, Vol. 49, pp. 1-8.

Rasmussen, P.W. 1986, What Should We Do With the AICP Exam?: Journal of the American Planning Association, Vol. 52, pp. 7-8.

Weber, M. 1968, Economy and Society (editors, G. Roth and C. Wittich), University of California Press, Berkeley.

Webster's II New Riverside University Dictionary. 1976, Houghton Mifflin Company, Boston.

INTERFACING GIS EDUCATION WITH GRANT AND CONTRACT RESEARCH

Richard D. Wright
Department of Geography
San Diego State University
San Diego, CA 92182

ABSTRACT

The hardware and software costs of a comprehensive GIS curriculum are greater than most universities can afford. To help defray program costs, GIS educators are relying in increasing numbers on outside grants and contracts for funds to meet the needs of their programs. The Department of Geography at San Diego State University, through its Center for Earth Systems Analysis Research (CESAR), has developed a successful model for using outside funding to support its GIS education mission.

This paper describes the GIS curriculum and facilities at San Diego State University and the way in which the Department of Geography has interfaced grant and contract research with GIS education. It critically examines the pros and cons of integrating GIS education with outside grants and contracts and offers recommendations for those institutions that may be considering a similar course of action.

INSTITUTIONAL SETTING

San Diego State University is a comprehensive university with a student body of about 33,000. GIS is the exclusive responsibility of the Department of Geography which has 20.5 faculty positions to serve 50 graduate majors and 165 undergraduate majors. Approximately 5,000 students per year are enrolled in geography courses. Primary specialities in the Department include GIS, cartography, remote sensing (RS), spatial statistics, urban and behavioral geography, biogeography, environmental geography, and Latin America. The Bachelor of Arts, Master of Arts, and Doctor of Philosophy (offered jointly with the University of California, Santa Barbara) as well as a Certificate in GIS are awarded in Geography.

CURRICULUM CONTEXT

Interfacing GIS education with funded research must be accomplished within a curricular framework and thus will be influenced by the length of the academic term (semester or quarter), the number of student contact hours normally expected to meet course requirements, the mode of instruction (lecture, independent study, laboratory, discussion, or field), the course prerequisites, and course level (lower division, upper division, or graduate). At SDSU the GIS curriculum consists mainly of three-unit, lecture/laboratory courses offered at the upper division and graduate levels on a semester basis. Enrollments in GIS classes amount to about 60 persons per year.

Prerequisites include one or more courses from cartography and computer programming.

The following courses are included in the GIS offerings:

GIS Courses

Geographic Information Systems (junior/senior)
GIS Applications (senior/graduate)
Internship/Special Study (senior/graduate)
Advanced Geographic Information Systems (graduate)
Geographic Information Systems Laboratory (graduate)
Seminar in Techniques of Spatial Analysis (graduate)
Research and Thesis (graduate)

Prerequisite Courses

Map Investigation (junior)
Aerial Photograph Interpretation (junior)
Maps and Graphic Methods (junior)
Remote Sensing of Environment (junior/senior)
Spatial Data Analysis (junior)
Computer Programming (freshman/sophomore)

Ancillary Courses

Automated Cartography (senior/graduate)
Quantitative Methods in Geographic Research (senior/graduate)
Intermediate Remote Sensing of Environment (senior/graduate)
Field Geography (senior/graduate)
Advanced Automated Cartography (graduate)
Advanced Automated Cartography Laboratory (graduate)
Advanced Quantitative Methods in Geography (graduate)
Advanced Remote Sensing (graduate)
Advanced Remote Sensing Laboratory (graduate)
Advanced Field Research (graduate)

PHYSICAL FACILITIES

Room Layout

As the room layout illustrated in Figure 1 shows, the Department's GIS instructional/research facilities consists of three main rooms containing computer equipment. The two large rooms on the west, separated from each other by a CPU room and offices for the technical Manager and other staff, are devoted primarily to research activities in GIS digital image processing (DIP). These rooms also accommodate specialized graduate and advanced undergraduate courses and independent research. The large room on the east, the spatial analysis laboratory (SAL), is used mainly as an instructional laboratory for upper division and graduate courses in GIS, RS, spatial statistics and automated cartography. A darkroom and an office for the Department Cartographer round out these new laboratory facilities. These facilities are described in greater detail elsewhere (Wright and Griffin 1991).

The rooms are well designed for the intended uses. A 30" x 48" computer table for each piece of equipment, two sets of four 30" x 72" tables for workspace and small group discussions, map storage cabinets, supplies storage cabinets, bookcases for manuals, light tables, numerous bulletin boards, white boards, adjustable chairs and a projection screen comprise the furnishings.

Hardware

The Department has put together an impressive facility in terms of hardware capability (Table 1). On the west side of the building, the computer hardware consists of clusters of Tektronix graphics terminals, 486 PCs and SUN and IBM workstations. The adjoining SAL contains 386 PCs and Macintosh computers (Mac IIs and MacII Cxi's), eventually to be expanded to 20 students stations.

All hardware and software in the Center are internally and externally linked via the Ethernet. These linkages include connections to the in-house VAX 11/780 as well as to the University's VAX 6000-320.

Software

The Department's software holdings for GIS, DIP, automated cartography, and spatial statistics are extensive (Table 2). Heading the list of software are ARC/INFO and ERDAS. The former is available on PCs, on workstations, and on the two VAXs, whereas the latter is installed on PCs and the SUN workstations. A significant feature of the laboratory is the integration of ARC/INFO and ERDAS on some stations via the "live link." ARC/INFO gives the Department vector GIS capability while raster capability is provided by ERDAS, SPANS, OSU-MAP, IDRISI, and the MAP II Map Processor. ARC/INFO, SPANS, and ERDAS are employed for both research and instruction, whereas OSU-MAP, IDRISI, and MAPII are primarily for instruction.

Operation of the Center

During the academic year CESAR is open everyday. Summer hours are limited to weekdays. From 8:00 am to 5:00 pm Monday through Friday the Center is supervised by the Technical Manager and his assistant. In the evenings and on weekends the Center is overseen by graduate assistants who are also working on CESAR research projects. CESAR is dedicated for use by students enrolled in courses in the Geography Department (typically GIS/RS and related classes) and Geography faculty and graduate students working on research projects that include funding to cover some of the costs of using CESAR. Use by faculty outside the Department is limited to those who are involved in joint research with members of the Geography Department.

RELEVANT FACTORS

Integrating GIS education with funded research requires careful consideration of many issues. These are as follows:

First, is the computer hardware capability of the laboratory adequate for a dual instructional/funded project approach? For instruction it may be preferable to install large numbers (20 or more) of one type of computer with a minimum configuration. But research may be better served by a facility with small numbers (4-8) of several different types of equipment, each with a large capacity. The advantage of the latter approach is that it maximizes the flexibility and adaptability of the facility for responding to technological change and different research/instructional needs. A negative aspect, however, is that fewer students can be served. Having a wide variety of computer hardware, with different operating systems, also present more difficult management problems.

Second, is adequate software available to meet basic instructional and project needs and are the software requirements comparable? Some types of software, e.g. OSU-MAP and IDRISI, might be ideal for teaching GIS concepts, but may not be appropriate for a project requiring vector GIS capability.

Third, is accessibility to the GIS facilities adequate in relation to the number of students and the amount of time needed by students to carry out their course responsibilities? Clearly, the expectations for students to complete their project work must be in line with equipment availability. At SDSU, for example, students can gain access to CESAR virtually 24 hours per day. All graduate research assistants involved in CESAR activities are issued keys on a semester basis. Undergraduates can enter the facility after hours by checking out a key for 24 to 48 hours from the Geography Department Office

Fourth, to what degree are project requirements consistent with GIS course objectives? If the project tasks can be structured to meet educational goals, then a good match is possible. Course objectives should be clearly preeminent over those of the project; if not, the education content could be unintentionally sacrificed under the pressures of project exigencies.

Fifth, what is the degree of correspondence between course and project timelines? The two schedules must be capable of being overlapped substantially in ways that do not negatively impact education. If this cannot be done, the effort to combine project and education should be abandoned.

Sixth, are the students' abilities, knowledge, motivation, and availability sufficient to create a quality product within the framework of the course schedule and the project task requirements. These variables are closely related to the class level. For example, students in an introductory GIS class could reasonably handle responsibilities such as simple data automation, but should not be expected to be proficient in advanced data manipulation processes, e.g. rubber sheeting. Students should not be involved in project research until they have completed a basic course in GIS.

Seventh, is it feasible for students to visit the project site? Whenever possible, fieldwork to collect data and conduct ground checking should be an integral part of a GIS project so that students can more easily visualize the relationship between real world objects and their representation in digital form. Thus the value of a project in meeting education goals will be determined in part by its accessibility for visitations by students. Projects near the university will have greater potential educational value than those that are located so far away that it is impractical, because of time and cost considerations, to transport students to the site.

Eighth, is there sufficient instructor time available to manage a combined GIS class and a funded project? The time required to direct integrated activities may, in fact, be far greater than if each is managed separately. As an example, in a combined situation more attention will need to be given to the characteristics of the project teams. The success or failure of a GIS class and the associated funded project will depend on variables such as the number and mix of students on each team, the equitable division of tasks among team members, the provision of equal opportunities for all team members to learn and the organization of each team to maximize efficiency. The educational and project benefits must clearly offset the substantial management costs associated with a combined effort.

Ninth, does the course grading system provide a proper balance between work on the project (applied GIS) and performance on examinations covering readings and lectures (GIS concepts and theory)? Without careful course planning and management, the time required for project work can greatly exceed that spent by the student in completing other components of the class.

ADVANTAGES

Linking GIS education with funded projects can result in win-win situations for students, faculty, and clients.

Instructional-Related Benefits

Looking first at the instructor and students, the following benefits are noteworthy:

Learning by doing can be an effective educational approach. Participating on a "real world" project that involves an interesting application of GIS can help students gain a clearer, more in-depth comprehension of GIS concepts.

The typical GIS project is most efficiently conducted through the team approach involving several specialists. Participating as a project member can provide students with an excellent opportunity to develop their skills both as leaders and as team players.

Working on a project that has clearly defined objectives, tasks, and tangible final products is an excellent method of generating interest in GIS and motivating students to learn.

Funded research projects provide opportunities for qualified students to obtain financial support while working on projects that enhance their education. The convenience of working part-time in the department of their major is an important plus in their education.

Students participating in real world projects gain valuable experience, insights into GIS as a profession, and contacts with prospective employers. These are obvious advantages to the student seeking a GIS position after graduation.

Students enrolled in GIS-related courses may benefit from the availability of materials obtained through a GIS project. For example, aerial photographs purchased in support of a GIS project may also be used for laboratory exercises in a course on aerial photograph interpretation.

Overhead fees generated from funded research can provide a source of monies for maintaining existing equipment and for updating technological capability. In this way, students and faculty are afforded the opportunity of using state-of-the-art hardware and software for their research and educational activities.

Finally, faculty and students can benefit through the faculty member's research on GIS in "real world" situations. This experience can help to enrich the instructor's classroom presentations, thereby making his lectures more meaningful and interesting.

Project-Related Benefits

Involvement with an educational institution that uses student labor to fulfill project requirements can offer several benefits to a public or private entity. The most significant of these advantages are:

Access to a skilled labor pool for carrying out a wide range of labor intensive tasks at moderate costs. This is important in today's job market with its scarcity of GIS specialists that continues to be a bottleneck in the growth of the GIS industry.

Access to faculty experts on GIS. This is especially important for those projects that have a substantial research component.

Access to costly technology. Although prices for hardware and software have dropped dramatically in recent years, the high capital and operating costs of a comprehensive GIS setup continue to make the technology unavailable to many organizations. It is still good business for many organizations to turn to universities for assistance on GIS projects, or at least until the volume of work is sufficient to warrant the purchase of in-house GIS capability.

DISADVANTAGES

There are several potential problems associated with GIS education and funded projects that may make it difficult, or impossible, to satisfy either educational or project objectives, or both.

Instructional-Related Difficulties

Students working on a project can become too focused on a small part of the GIS process. To avoid this type of problem, an instructor needs to be certain that each student participates in all project phases, i.e., in data collection and compilation, data automation, data editing and management, analysis and modeling, and presentation of the results.

The timing and length of the project may be such that it is impossible to directly involve students in the process from beginning to end. If this is the case, it is important that students be briefed on the project at the beginning of the class so they he/she can understand its objectives, what has been done, and what remains to be done.

Many projects require substantial data gathering and

compilation that cannot be accomplished in a few weeks. Clearly, if such a project is to be brought to a point by the end of the GIS class where some conclusions are possible, it is essential that significant project work be done prior to the beginning of class. In this way, students can move the project along when the term begins.

Pressures to complete a project on time may negatively impact the creative process by reducing the time available for learning through making mistakes and for experimenting with different techniques and approaches of problem solving. Under severe time constraints the tendency is to employ conservative, well-travelled procedures instead of attempting to break new ground.

Project-Related Problems

Conducting a GIS project while simultaneously trying to meet educational objectives can be difficult. Careful and continuous attention must be given to project management. Contingency planning is especially important in dealing with the uncertainties associated with involving students on projects and in keeping both the class and the project on track and on schedule. Here are a few typical problems.

Students enrolled in the GIS class may meet the course prerequisites but may still not be familiar with the details of the hardware and software used for a particular project. Thus, it is often necessary to devote time to training. In some instances, the time spent by the graduate research student in training project team members to do a particular task may be greater than if the trainer had handled the task himself.

Turnover and distractions may have negative project impact. A significant characteristic of student labor is its temporary nature. Change is a given. Turnover requires constant education training of new students and their integration onto the project team. Also, students have many other responsibilities that may distract them from carrying out their project tasks. This must be factored in when scheduling project tasks.

Involving four or five students invariably leads to problems of project control. Strong management is necessary to deal with personnel problems. For example, not all students will have a strong commitment to the project even after efforts have been made to involve them more fully. It is also important to have a clearly defined and well understood assignment of responsibilities, both for assigning grades and maintaining project control.

Another component of project supervision is the need to develop and institute a program of quality control. Given the difficulties associated with using students on projects, it is especially important that error management (identification and correction of errors) receive high priority. Errors not corrected on "real world" projects extend beyond the confines of the classroom.

CONCLUSIONS

GIS education and funded projects can be integrated under certain circumstances to the benefit of the goals of student learning and client needs. However, detailed planning, careful organization, and strong management of the GIS course and the project are essential if the desired results are to be achieved.

REFERENCES

Wright, R.D., and Griffin, E.C., 1991 (forthcoming), The Center for Earth Systems Analysis Research (CESAR): An Integrated Geographic Information Systems and Digital Image Processing Laboratory, in GIS Teaching Facilities: Six Case Studies on the Acquisition and Management of Laboratories, NCGIA Technical Paper. The development and characteristics of the SDSU GIS laboratory facilities are described in detail in this report compiled and edited by Steve Palladino, NCGIA, UC Santa Barbara.

FLOOR PLAN

CENTER FOR EARTH SYSTEMS ANALYSIS RESEARCH (CESAR)
AND
SPATIAL ANALYSIS LABORATORY (SAL)

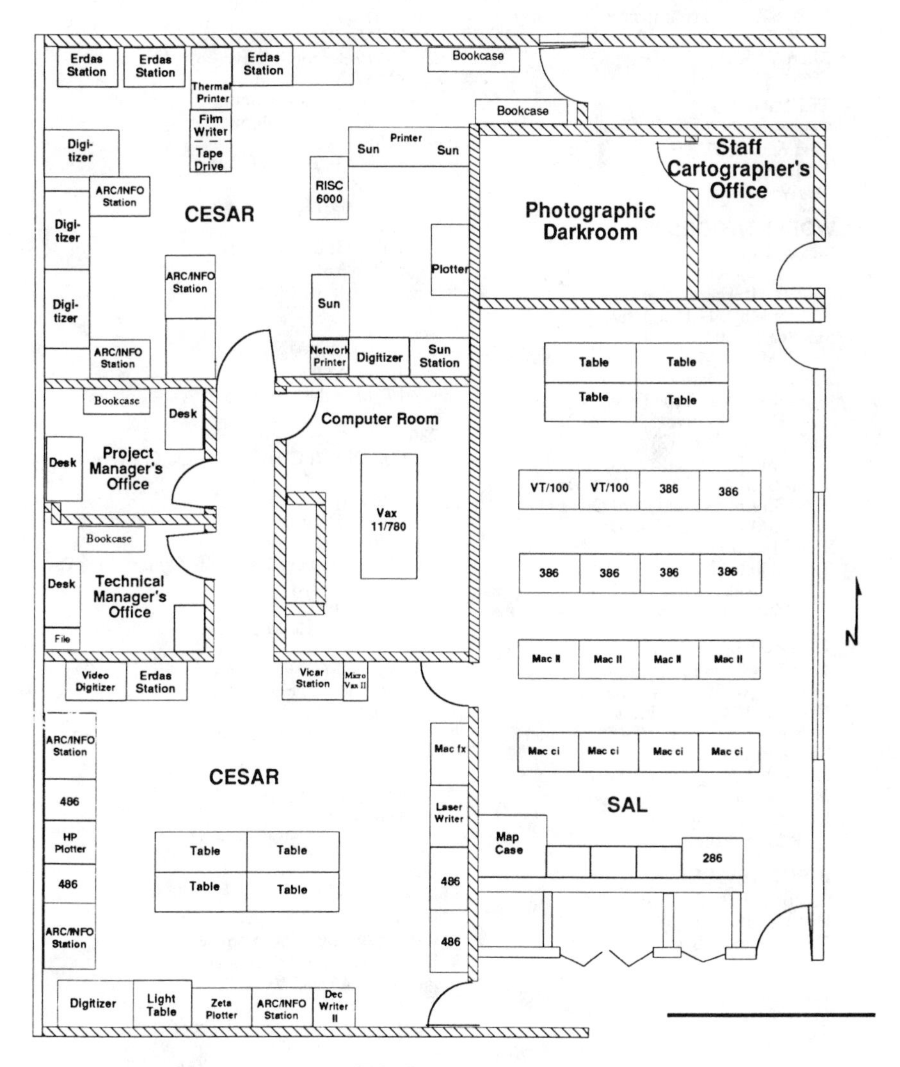

FIGURE 1

HARDWARE

CENTER FOR EARTH SYSTEMS ANALYSIS RESEARCH (CESAR)
AND
SPATIAL ANALYSIS LABORATORY (SAL)

MINICOMPUTERS

VAX 6000-320 (Campus Computing)
VAX 780
MicroVAX II

TERMINALS

Tektronix 4207 (5)
Tektronix 4107
VT100's (7)
DecWriterII

WORKSTATIONS

Sun 4/110
- 16Mb RAM
- 300Mb Hard Drive (2)
- 150Mb 1/4" Tape Drive

Sun SparkStation2
- 32Mb RAM
- 1Gb Sun Hard Drive
- 150Mb 1/4" Tape Drive
- CD ROM Drive

Sun SparcStation 1 (2)
- 16Mb RAM
- 600Mb AnDATAco Hard Drive
- 1Gb AnDATAco Hard Drive (1)
- 105Mb Conner Hard Drive

IBM RISC 6000
- 16Mb RAM
- 1Gb AnDATAco Hard Drive
- 150Mb 1/4" Tape Drive

OUTPUT DEVICES

Tektronix 4696 Inkjet Printer
Tektronix 4396DX Thermal Printer
Calcomp 1044 8 Pen Plotter
Hewlett Packard 7550A 8 Pen Plotter
Zeta 8 Pen Plotter
LaserWriterII (Postscript) (2)
Personal LaserWriter (Postscript)
Dot Matrix Printer (4)
QCR Film Writer

INPUT DEVICES

Calcomp 9100 Digitizer (4)
Hitachi Digitizer
Cohu Photodigitizer

MICROCOMPUTERS

486's (3)
- 200Mb IDE Hard Drive
- 20Mb Bernoulli Drive
- 8Mb RAM
- 14" Sony 1304 Monitor
- 19" Mitsubishi Monitor

486 (2M
- 200Mb IDE Hard Drive
- 8Mb RAM
- 14" Sony 1304 Monitor

486 (3)
- 200Mb IDE Hard Drive
- 8Mb RAM
- 16" NEC 4D Monitor

386's (6)
- 100Mb IDE Hard Drive
- 4Mb RAM
- 14" Mitsubishi Diamond Scan Monitor

286
- Copy2PC (PC To MAC Data Conversion)

Macintosh IIfx
- 160Mb Hard Drive
- 8Mb RAM
- 650 Mb Iomega LaserSafe Optical Disk Drive
- Color Scanner
- Digitizing Tablet
- 19" SuperMac Monitor

Macintosh II (4)
- 40Mb Hard Drive
- 2Mb RAM
- 19" SuperMac Monitor

Macintosh IIci (4)
- 80Mb Hard Drive
- 5Mb RAM
- 19" SuperMac Monitor

Other

Albedometer
Biovision Infrared Video Camera
Everest Infrared Thermometer (1)
Exotech Handheld Radiometer (2)
Panasonic 4 Head VCR
Tape Drive (9 Track)
Vista Truevision Frame Grabber

TABLE
1

SOFTWARE

CENTER FOR EARTH SYSTEMS ANALYSIS RESEARCH (CESAR)
AND
SPATIAL ANALYSIS LABORATORY (SAL)

VAX

ARC/INFO 5.1
Disspla
Erdas 7.2
Erdas/Arc Live Link
LandTrak
Vicar

BMDP
C compiler
Fortran compiler
Minitab
Oracle
SPSS

MACINTOSH

Autocad
Canvas
Claris CAD
Cricket Graph
GIS Tutor
Excel
Exstatix
MacDraw Pro
MacProject
Map II Map Processor
Mapmaker
MacDraw II
MacGIS
MacWrite
Pixel Paint
Statview
Super 3D
Word

WORKSTATIONS

APART
ARC/INFO 5.1
Erdas 7.5
Erdas/Arc Live Link
IPW
SPANS

C compiler
Fortran compiler

PC

ARC/INFO 3.4D
Autocad
Erdas 7.5
Erdas/Arc Live Link
GISPlus
GEO/SQL
Idrisi
Map Collection
MapInfo
OSU Map
SPANS
Transcad

BMDP
Fortran Compiler
Grapher
Lotus 123
Manifest
Oracle
SPSS PC
Surfer
QEMM
Windows
Word Perfect

TABLE 2

CONFLICT PROVENTION IN LAND USE PLANNING USING A GIS-BASED SUPPORT SYSTEM

Tim Lesser, Wei-Ning Xiang, Owen Furuseth, John McGee, Jian Lu
Department of Geography and Earth Sciences
University of North Carolina at Charlotte
University Boulevard, McEniry Building
Charlotte, North Carolina 28223

ABSTRACT

In land use planning, conflicts between several different department entities are ubiquitous. With the competition of data and incompatible ideas or interests, conventional planning becomes a series of conflictive management techniques. This approach forces planners to be more reactive in their planning decision making process. An attractive strategy, other than conflict management, is conflict provention. This proactive approach emphasizes the removal of potential causes of conflict; and attempts to create and promote cooperation among departments.

INTRODUCTION

There are a variety of ways to deal with conflicts. Main approaches are either defensive or offensive in nature, as well as "the grey area" (somewhere in between these approaches) of where most land use practices occur. The defensive approach is usually regarded as ordinary aspects of planning as conflict management. No action is taken until a conflict arises and certain techniques to correct the conflict become reactionary. Such techniques include court settlement, mediation and intervention. The offensive approach to land use planning is regarded as conflict provention. There are three main aspects that are emphasized:

1) To predict potential conflicts
2) To search for strategies to avoid conflicts
3) To promote cooperations among parties involved

From a planning perspective, the concept conflict provention should obviously be the more attractive option, although the majority of planning practices again seem to lie within the realm of greyness, in between reactionary and proactionary decision making processes.

In conflict provention, an outside entity acts as a facilitator to play an essential role in the resolution process. Parties that may be involved in a potential conflict are often unable to predict or have problem insights within their own planning arena. Therefore, an outside knowledgeable and analytical skilled party is required to exercise the process of conflict provention. The outside party, who possesses these skills, is able to assist in the resolution of potential conflict - to gain and to develop analytical foresights and strategies to avoid potential conflictive situations.

This paper focuses on how a Geographic Information System land use methodology is implemented as a vehicle to promote conflict provention through cooperation of the parties involved. The GIS (ARC/INFO) based support system integrates a series of decision making models to assist in the resolution of potential planning conflicts. Three different data sets (physical, ecological, human) are incorporated into land suitability and land allocation models. In turn, a series of land suitability maps are generated from these decision models. The maps are utilized to aid competing departments to be analytical about possible conflict resolutions.

The GIS methodology is being applied to the City of Concord, North Carolina. This case study, in cooperation with the UNC-Charlotte Department of Geography and Earth Sciences, demonstrates through different land use suitability scenarios the optimum potential sites for the city's long term multi-criteria annexation goals. Overall, the main goal is promote a cooperative proactive environment not just from the planning aspect, but also from a community standpoint; to fulfill their goals.

THE CASE STUDY

The City of Concord, North Carolina is located in the County of Cabarrus ten miles northeast of Charlotte. The city influence area and geographical boundaries for the GIS system encompasses the current municipal limits as well as areas which may be annexed by the city. The areal coverage for the GIS is approximately 60 square miles.

The purpose of the GIS is to assist the Concord Department of Planning and Community Development in carrying out its short and long-term programmatic objectives. The system is developed to enhance the collection and storage of relevant planning data, including physical, economic, and social information, and facilitate the compilation and analysis of different data sets. It can reduce the time necessary to produce complex maps and tabulate data. Overall and most importantly, the GIS-based system is to provide benefits to other city departments. Each department has the flexibility, from their perspective view, to incorporate any layer(s) from the database for their specific criteria needs. Moreover, the database can be easily updated as departmental views, needs, and wants change as time progresses.(figure 1).

DEPARTMENTAL VIEWS WITHIN THE PLANNING ARENA

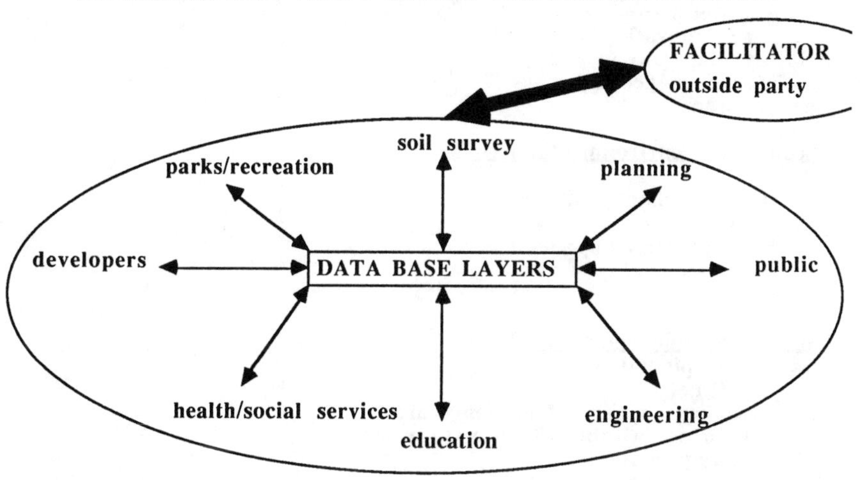

Figure 1.

Components of the project include the following:

1) All relevant data (both cartographic and tabular) are integrated with a common geographically structured data base. The scale of the data is 1: 24000 and provides block level specificity, as well as parcel landuse data specificity. Subareas and community planning districts are identified and incorporated into the system.

2) The GIS produces map displays to portrait planning attributes, including transportation systems, population growth, employment, and industrial distributions.

3) The GIS is used to carry special project "case studies" for the UNC-Charlotte Geography Department relating to growth and community planning issues.

4) Throughout the project, the UNC-C research team involved the Concord planning staff in the compilation of the data and the development of the GIS. The project provided the staff training in the management of the GIS.

5) Provide Annexation Potential areas for the City of Concord for the next ten years.

DATABASE DEVELOPMENT

The development of the database involved the collection from several sources. The 1990 Census Bureau (TIGER) provided the transportation network, hydro features, rail network, census block and tract data, as well as tabular population statistics for the influence area. The 1988 USGS topographic quadrangles provided the soil, topographic, watershed, wetland, floodplain, transmission line, telephone line, gas pipeline, and slope layers. Other data layers were compiled from individual departments and/or city and county agencies.

DATA LAYERS

Physical Data

soils
waterfeatures
watersheds
groundwater
topography
surface geology
slope

Ecological (environmental) Data

floodplain (100 year)
wetlands
endangered species
water quality

Socio-Economic (human) Data

transportation
rail net
zoning / sphere of influence areas
land use within influence study area
sewage main lines
water main lines
electric transmission lines
major underground telephone lines
waste treatment plant locations
landfill locations
school locations
hospital locations
parks and recreation areas
health clinics
land values
census blocks
census tracts
gas pipelines

DATABASE DESIGN

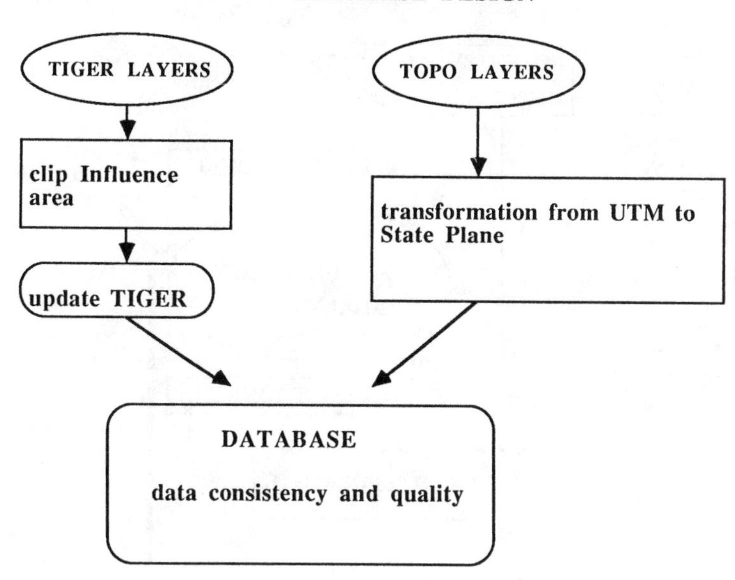

Figure 2.

Since the data layers had conflicting scales, TIGER - State Plane Coordinate System and USGS topographic quadrangles 1:24000 - Universal Transverse Mercator, projection transformation was pertinent to maintain consistency among the data for analysis purposes. When data information came from different sources other than TIGER and/or TOPO, they were compiled into the system by way of digitization or tabular file compilation. Tabular data for each layer was inputed by way of the INFO module in GIS ARC/INFO.(figure 2).

A total of six topographic quadrangles were compiled to create a mosiac of the city's influence area. The TIGER file data was CLIPPED to incorporate just the influence area which encompassed approximately 60 square miles. The topographic data (UTM) was PROJECTED into the TIGER (State Plane) scale projection. The TIGER data was then updated to 1991 standards as data availability would allow.

A 3-dimensional TIN surface was created from the topographic contours extracted from the quadrangles. A 1/4 -inch grid was placed upon the surface of the contour elevation data. A z-value was then created for each grid intersection to generally describe the elevation of the influence area. Each node represented an elevation level; a total of approximately 2750 elevation points were extracted. The software ARC/INFO created the generalized contours, as well as slope values from the same data structure.

ANALYSIS MODEL

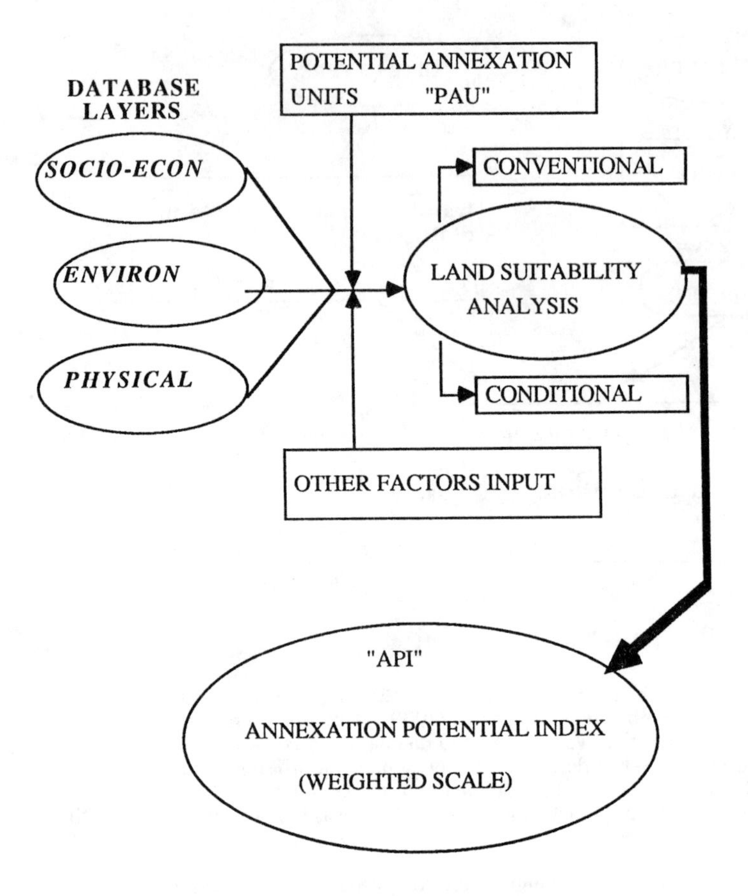

Figure 3.

ANALYSIS

The land suitability analysis consisted of the three database category layers incorporated with a unit of analysis for annexation potential, called the "PAU"or the Potential Annexation Units. The PAU developed for the potential city annexation consisted of criteria the City of Concord created to find optimium annexation land potential. Among the criteria for the PAU included; lot size, 60% developed land, legal footage (1/8 of the perimeter boundary that is considered contiguous with existing city limits), population base, and land use mix. Other factors were also considered, such as public, political, and developer input, human-made attraction stimulators - transportation networks; water and sewer utilities, and land values.

The suitability model, in addition to conventional analysis - premised on existing and pre-existing land conditions, created different annexation potential scenarios by incorporating conditional suitability analysis methods. Conditional suitability creates scenarios of potential areas if specific annexation criteria standards were to be met. Questions like "what if ?" or "how much ?" of a certain attribute would influence decisions for an area to be annexed. Lastly, an Annexation Potential Index (API) was developed from the model. The API is a low to high rating scale according to the annexation potential for each potential area being considered for annexation. As data criteria and departmental view change over time, area boundaries can be altered to be flexible to fit into the model system. The weighted areas can be re-calculated to adhere to newly incorporated criteria standards.(figure 3).

SUMMARY

In land use planning, the outside party can greatly enhance the cooperation among departments. The capabilities of GIS facilitated the collection of data for all departments to use and to share. Depending on the view perspective of the data, each department within the planning arena are able to communicate more freely due to the encompassment of information this project provided. With GIS now at their exposure, the City of Concord, NC can strengthen their capabilities to assist each other to be more analytical about their relationships and goals for the community.

REFERENCES

Aronoff, S. 1989. Geographic Information Systems: A Management Perspective. Ottawa, Ontario Canada: WDL Publication.

Burrough, P.A.1986. Principals of Geographical Information Systems for Land Resources Assessment. Oxford: Clarendon Press.

Burton, J. 1990. Conflict: Resolution and Provention. New York: St. Martin's Press

Star, J. and Estes, J. 1989. Geographic Information Systems: An Introduction Englewood Cliffs, NJ: Prentice Hall .

Xiang, Wei-ning. 1991. Integrating GIS with MCDM Models to Support Conflict Provention in Land Use Planning. Proceedings of the Eleventh Annual ESRI User Conference. pp. 495-504.

THE STRUCTURED WALK METHOD FOR URBAN LAND USE PATTERN ANALYSIS

Feng-Tyan Lin
Associate Professor
Graduate Institute of Building and Planning
National Taiwan University
One, Roosevelt Rd. Sec. 4
Taipei, Taiwan, R.O.C.
(TEL) 011886-2-3638711
(FAX) 011886-2-3638127

ABSTRACT

Urban land use patterns are composed of natural and man-made objects, where man-made objects have the property of semi-fractal, which only preserves some characteristics of fractal objects in a certain range of scales. In order to handle these complex objects, the structured walk method of fractal theory is extended so that urban land use pattern can be described and analyzed from the viewpoint of urban morphology. A PC based computer program which follows the structured walk method is implemented in the National Taiwan University. Three cases are illustrated in this article. These experiments show that this method is very powerful to reveal the complexity of spatial structures among various scales.

INTRODUCTION

In an urban environment, there is full of natural and man-made elements. Consequently, urban land use patterns are composed of objects with regular and irregular shapes. Traditionally, urban researchers analyze land use patterns by two major approaches. They draw conceptual diagrams, which are very rough, to reveal the spatial structure of land use patterns (*cf.* Lynch 1967, Fabos 1979). On the other hand, they also calculate the area of each land use to indicate the absolute size and the relative ratio in a quantitative way (*cf.* Chapin 1979).

In these traditional land use analysis approaches, the characteristics of land shapes are somewhat neglected. Conceptual diagrams reduce complex land use areas into simple shapes. These diagrams emphasize on the logical relation in spatial structures, but rarely talk about the aspect of morphology. Moreover, the quantitative analysis totally leave the consideration of shapes alone. To remedy this shortcoming, we have to find a new approach to take care of the shape factor in various land use patterns.

In 1975, Mandelbrot proposed the concept of fractal geometry to describe the shape of natural objects (Mandelbrot 1982). Later, the idea of fractal dimension was also introduced (Barnsley 1988). To measure fractal dimensions for objects in the real world, several methods, including the structured walk method, were proposed (Batty 1987, 1988). The application of fractal geometry is very wide. Particularly, the concept of fractal dimension has been used to measure the degree of roughness, brokenness or irregularity of objects (Gleick 1988).

As mentioned above, urban land use patterns are highly complex objects, which consist of natural and man-made elements. Consequently, the concept of fractal geometry has to be slightly extended and further exploited so that the characteristics of man-made objects, which basically obey the rules of Euclidean geometry, can be simultaneously handled. Batty and Longley discussed about the fractal dimensions of land use pattern in their works. However, they dealt with a type of land use at a time, and the boundaries of land use areas were basically fractal structured. In this article, we are going to deal with more complex objects consisting of natural and man-made components. By employing the structured walk method, the dimensions and morphologies of urban land use patterns will be discussed.

In the next section, the concept of fractal geometry will be extended to that of semi-fractal structure, which possesses two special characteristics, finite recurrence and unevenness. Then, the structured walk method will be introduced in the next section. Two issues, the calculation of dimension and common edges of adjacent polygons, are discussed in the following two sections. Then, three examples will be illustrated to show how the structured walk method is applied to the study of urban land use pattern. Finally, there is a brief conclusion.

SEMI-FRACTAL STRUCTURE

While a fractal structure is composed of natural objects, a semi-fractal structure is referred to a collection of natural and man-made objects. The central idea of fractal structure is that natural objects possess the characteristic of infinite self-similarity in any scale (Gleick 1988). However, in an urban environment, there are not only natural objects but also many artifacts, such as streets, blocks and pipelines. These artifacts hold two specific properties, *finite recurrence* and *unevenness*, which are quite different from that of fractal structures.

In a semi-fractal structure, the number of recurrences of self-similarity is finite. For example, residential areas in a regional map may be drawn as polygons. While residential areas are explored into details, polygons become more and more complex. Even further, polygons might be split into several smaller polygons. However, this phenomenon will terminate when basic artifacts, such as streets and walls, are explored. In other words, edges of polygons might roughly stand for boundaries of residential areas in the initial stage. While the scale of maps is increasing, edges of polygons might be substituted by poly-lines so that boundaries are able to be described more precisely. Finally, each line segment might eventually stand for a street, a wall, or some other artifacts. At that time, the phenomenon of recurrent line replacement terminates. Lines which stand for edges of basic artifacts will remain without further replacement while maps continue to scale up.

The assertion of finite recurrence might be challenged by other fractal theory research workers. Although it is a truth that the rough surface of a wall holds the property of fractal objects, we are still able to claim that the recurrence of semi-fractal structures is finite in terms of the following two reasons. First, research works who concern themselves about urban morphology will not be interested in the details of wall surfaces. It is quite sufficient for urban planners and designers to treat streets and the appearance of buildings as the most basic elements. As a matter of fact, urban researchers do really think that the center lines of roads are ideal Euclidean straight lines or some perfect curves. The other reason of the assertion is that the gap between scales of exploring the geometry of urban objects and their details is sufficient large. In other words, there is a clear cut for us to terminate the recurrent process. Therefore, it is safe to use the term of finite recurrence when the semi-fractal structure of urban objects is being considered.

The second specific property of semi-fractal structure is that the process of fractalization is unevenly and asynchronously terminated in different parts of a semi-fractal structure. Unlike to fractal objects whose parts will recursively and infinitely expand their self-similar appearances, parts of semi-fractal objects may indefinitely terminate the process of exploration in an asynchronous way. For example, a regional maps will correctly show the appearance of highways, but roughly illustrate the boundary of cities. When a city map is shown, highways keep the same appearance as those in the regional map, but streets and blocks are explored. Furthermore, in a community scale, streets and blocks terminate their fractal process, but appearances of buildings are more clear and clear. Finally, when a certain scale is reached, the fractalization of appearances of buildings and other artifacts will terminate, but shapes of natural objects, such as terrain, lakes, and rivers, still continue the process of fractalization. Summarily, semi-fractal structures have the property of unevenness.

The properties of finite-similarity and unevenness make fractal and semi-fractal structures different. For urban planners, designers and morphologists, land use patterns, which are full of natural scenes and man-made objects, are of semi-fractal structured objects. Consequently, urban researchers are particularly interested in the issues of semi-fractal dimensions and land use morphologies in various scales.

THE STRUCTURED WALK METHOD

There are many methods which have been proposed to measure the dimensions of fractal objects (Batty 1987, 1988). Among them, the structured walk method is also able to be applied to the case of semi-fractal ones. Furthermore, the structured walk method can also be used to reveal morphological structures of urban land use patterns in various scales.

Conceptually, the structured walk method employes a pair of compasses to "walk" along the edges of objects. Suppose that the distance of each "step" (*i.e.* the distance between the pair of compasses) is d, and it takes N steps to walk along all the edges, the circumference of the object is counted as $p = d \times N$. Practically, the ending point of the last step will not exactly match the starting point of the first step. As a convention, the residue is dropped. To calculate the fractal dimension, the following equation is employed,

$$\log p = \log \lambda + g(D) \log d \tag{1}$$

where P and d are circumference and step distance, λ is a constant, and $g(D)$ is the fractal dimension, respectively.

Batty and Longley wrote a computer program based on the structured walk method to calculate the fractal dimension of two places, Cardiff and Swindon, in a statistical way. Batty *et al.* first digitized maps, then employed the computer program to simulate the process of the structured walk. Because boundaries are digitized, all the curves can be approximated by these sample points and some appropriate mathematical functions. Therefore, the degree of resolution (*i.e.*, the density or distance of sample points along the boundary) mostly determines the degree of approximation. Ideally, the degree of resolution can be increased infinitely to improve the degree of approximation. In the cases of Cardiff and Swindon, boundaries are composed of fractal curves. Therefore, the increment of the degree of resolution will further explore the details of the target objects. As a result, Batty *et al.* were able to predetermine a certain degree of resolution with a satisfiable degree of approximation. Although they also mentioned that the degree of resolution would effect the result of analysis, it didn't become a serious problem in their study.

When the structured walk method is applied to the case of semi-fractal objects, the basic idea is still valid; however, two issues have to be further considered: *the calculation of dimension* and *common edges of adjacent polygons*. In the next two sections, these two issues will be further discussed.

SEMI-FRACTAL DIMENSION

As mentioned above, the dimension of fractal objects are bound within a range, and equation 1 is used to calculate the value. However, it becomes more difficult when semi-fractal objects are under consideration due to the properties of finite recurrence and unevenness. To make this issue clear, it is worthy to further examine the case of straight lines, which are major components in semi-fractal structures.

Theoretically, the fractal dimension of a straight line can be calculated by the following formula:

$$\log\left(\frac{r_i}{r_j}\right) \Big/ \log\left(\frac{n_j}{n_i}\right) \tag{2}$$

where r_i and r_j are the unit distances, n_i and n_j the number of steps of the i-th and j-th measurements (walks), respectively. Let L denote the length of the measured line, and s_i the residual length of the i-th walk. The number of steps n_i can be rewritten as

$$n_i = (L - s_i)/r_i \tag{3}$$

And formula 2 can be rewritten as

$$\log\left(\frac{r_i}{r_j}\right) \Bigg/ \left[\log\left(\frac{r_i}{r_j}\right)+\log\left(\frac{L\text{-}s_j}{L\text{-}s_i}\right)\right] \tag{4}$$

When residues s_i and s_j are sufficiently small or they are close enough, ($s_i = s_j = 0$ and $s_i = s_j \neq 0$ are two special cases), $\log(\,(L\text{-}s_j)\,/\,(L\text{-}s_i)\,)$ will highly approach zero and can be dropped. In this case, the value of formula 4 becomes constant 1. This conclusion coincides with the well accepted concept that lines are one-dimensional objects. It is noted that the calculation of fractal dimension is applied to any two measurements, and it seems that the results should be close to the theoretical dimension 1. However, the situation becomes complicate when a string of line segments are considered in a semi-fractal object.

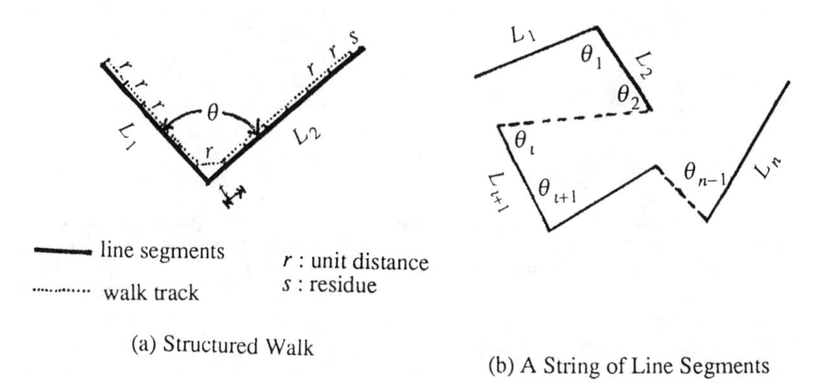

(a) Structured Walk

(b) A String of Line Segments

Figure 1 Poly-lines Measurement

In figure 1(a), a poly-line with two line segments L_1 and L_2 are considered, where l_1 and l_2 are lengths of two lines, r the unit distance of each step, t the distance when the measurement walk across from L_1 to L_2, s the residue. Therefore, the number of step is

$$n = \frac{l_1}{|r|} + \frac{l_2 - t}{|r|} + 1 \tag{5}$$

where t is a function of θ, r and l_1, Therefore, in a general case of figure 1(b), the semi-fractal dimension of a poly-line segment L (= $L_1\ L_2\ \ldots\ L_i\ \ldots L_n$) is a function of unit measurement distance r, lengths of line segments l_i, and angles between line segments L_i and L_{i+1}.

To simplify the discussion and be able to have a clear idea, we perform the structured walk method against some poly-lines. One of them is illustrated in figure 2, Its partial result is shown in table 1.

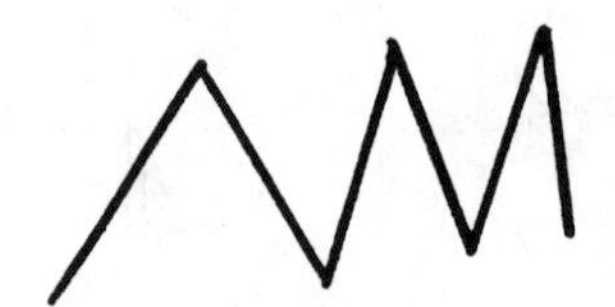

Figure 2 A String of Line Segments

Table 1 Semi-fractal Dimension of the Poly-line in Figure 2 (partial result).

unit distance (r)	number of steps (n)	absolute dimension	relative dimension
1	616	--	--
:	:	:	:
13	54	0.949049	1.316303
14	48	0.967029	1.589343
15	48	0.942392	0
16	42	0.968617	2.069018
17	42	0.947890	0
18	36	0.982478	2.696902
:	:	:	:

The semi-fractal dimension can be calculated in two different ways, absolute and relative ones. The absolute dimension is calculated with respect to a certain unit distance, called *base unit*. In table 1, we take unit distance 1 as the base unit, whose number of steps measured is 616. In this case, the absolute semi-fractal dimension of each measurement of unit distance r and number of steps n is $log(1/r) / log(n/616)$. For example, the dimension of unit distance 16 (42 steps) is 0.968617.

On the other hand, the relative dimension is calculated from two adjacent measurements. Let r_i and r_{i+1} denote the i-th and the $(i+1)$-th measurement with ascending order of unit distance. For example, in table 1, we increase the unit distance by one unit for each measurement (walk), i.e., $r_{i+1} = r_i+1$; therefore, the relative dimension is calculated as

$$log\ (r_i\ /\ (r_i+1))\ /\ log\ (n_{i+1}\ /\ n_i) \qquad (6)$$

It is noted that the absolute dimension varies within a certain range which is very close to theoretical value 1, while the relative dimension varies dramatically due to the characteristics of straight lines.

Recalled that the fractal dimensions are bound in a certain range. The dimension of semi-fractal objects are also preferred to possess the same characteristic. Therefore, the absolute semi-fractal dimension is adopted in the following discussion.

COMMON EDGES

Urban land use is highly complex. Each land use can be described as a polygon (*i.e.*, a string of closed line segments). Therefore, a land use pattern is a collection of polygon with (a string of) common edge(s). In other words, a land use pattern can be mathematically represented by a planar graph (Mitchell 1977, Preparata 1985), where each facet corresponds to a land use polygon. Any two adjacent polygons (with common edges) always belong to different land use types.

There are two different usages of the structured walk method in the case of graph structured objects. Batty *et al.* were interested in the dimension of each land use type. In this case, the measured object is a collection of individual polygons without common edges. The structured walk method is able to be directly applied to each polygon independently. However, Batty *et al.* mentioned that common edges would be counted twice for polygons (of different land use types) on each side. It means that dimensions of land use types are not really independent.

On the other hand, the graph structured object can be measured as a whole, where edges are measured only once. To do this, graph structured objects are conceptually transformed into tree structured ones, called *visitation tree*, by duplicating some semaphore nodes (explained later). Initially, all edges are unmarked. Any node could be chosen as the root. Then, the structured walk method is performed in either breadth-first or the depth-first way (Sedgewick 1983). However, every visited edge will be marked and inserted into the visitation tree by keeping the from-to relation as the parent-child relation. A node in a graph whose adjacent edges are all marked is called a *semaphore node*. Semaphore nodes are duplicated and separated. For example, in figure 3(a), suppose that edges *bc* and *bd* are all marked, and the current walk is from *a* to *b*, thus edge *ab* is also marked. In this case, node *b* is a semaphore node. As a result, node *b* is duplicated and separated. *b'* denotes the duplicated node(figure 3(b)). Correspondingly, the duplicated node *b'* is a terminal node in the visitation tree (figure 3(c)). By running this algorithm, the history of the structured walk is recorded in the visitation tree, where each edge appears exactly once. It is noted that the visitation tree is not unique. The measured dimensions may have a little deviation for different trees.

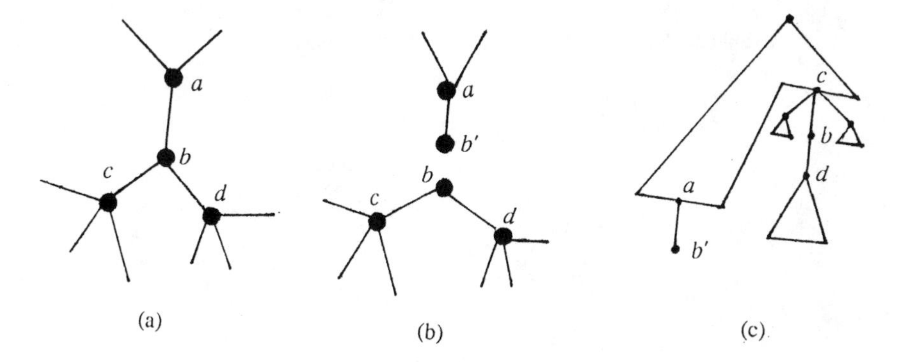

Figure 3 Visitation Tree

CASE STUDY

The structured walk method based on the previous discussion has been implemented as a PC based computer program in the National Taiwan University. The program allows the researchers to randomly pick up a node as the root of the visitation tree. Also, the user is allowed to specify unit distance for each walk. The unit distance and the number of steps of each walk are recorded. Then, the absolute semi-fractal dimension is calculated.

Several cases have been studied by running the program. In this section, three examples will be illustrated. The first example is about the dimension of the Taiwan Area with man-made administration boundary (see figure 4). The dimension is around 1.038, which is less complex than that of the coastline of Great Britain, 1.25 to 1.267.

The second example is about the land use pattern of a block in the Taipei City (see figure 5). The absolute dimension are shown in table 2, where two clusters of dimensions, 1.065 to 1.088 and 1.151 to 1.155, reveal that the land use pattern of this area can be classified into two levels. In micro-level, unit distance is between 1 to 8, the land use structure is near linear. On the other hand, in macro-level, unit distance is between 16 to 32, the block is viewed as a whole, the complexity of land uses is reflected by the high dimension value.

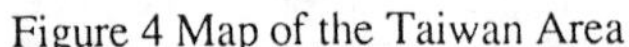

Figure 4 Map of the Taiwan Area

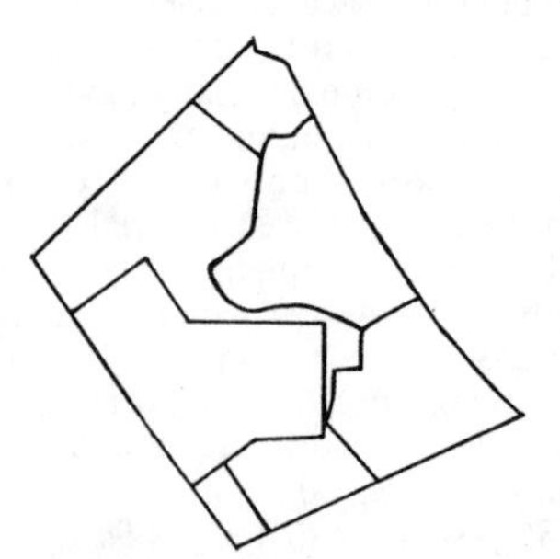

Figure 5 A City Block

Table 2 Semi-fractal Dimensions of the City Block

ratio of unit distance	absolute dimension
1:2	1.0882865
1:4	1.0654654
1:8	1.0837416
1:16	1.1512154
1:32	1.1549573

The third example illustrates that the structured walk method provides a possibility to reveal various morphological structures under different scales. Figure 8 is a land use map in the Kaohsiung Area. By drawing the walk tracks, land use patterns are shown in a new fashion. Figures 6, 7 and 8 show land use patterns associated with large, median and small unit distances, respectively. In other words, the track associated with large unit distance will illustrate the overall structure among major land use components; on the other hand, the track resulted by small unit distance will illustrate the land use pattern in details.

CONCLUSION

The concept of fractal dimension has been extended to analyze land use patterns. Particularly, artifacts in land use pattern, such as streets, blocks and buildings, have two special properties, finite recurrence and unevenness, which are quite different from that of fractal structures. Among various dimension measurement approaches, the structured walk method still works in this circumstance. However, two issues are further considered: the calculation of dimension and common edges. A computer program has been implemented. Furthermore, three examples are illustrated to show the dimension of man-made administration boundary, the difference between micro- and macro-level structures, and morphological land use patterns under various scales.

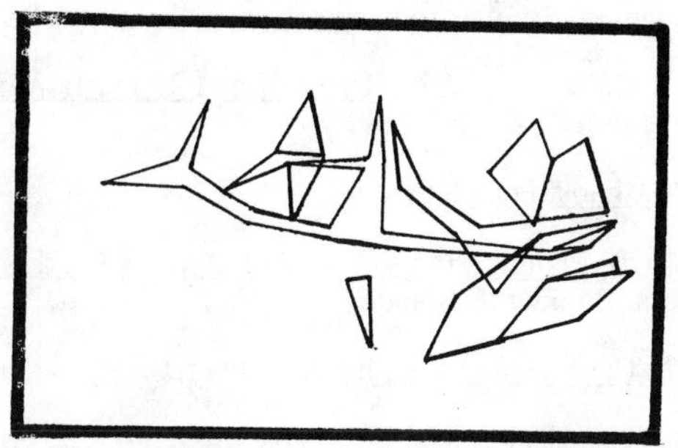

Figure 6 Land Use Pattern in Macro Level

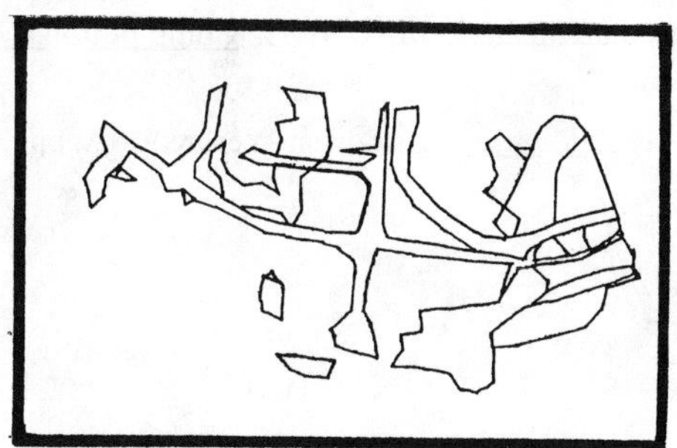

Figure 7 Land Use Pattern in Middle Level

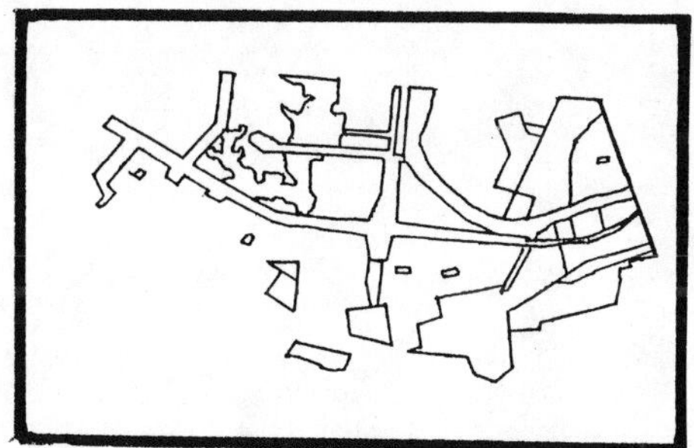

Figure 8 Land Use Pattern in Micro Level

REFERENCES

Barnsley, M. 1988, Fractals Everywhere, Academic Press.

Batty, M. and Longley, P.A., 1987 Fractal-based Description of Urban Form: Environment and Planning B: planning and design, Vol. 14, pp. 123-134.

Batty, M. and Longley, P.A., 1988, The Morphology of Urban Land Use: Environment and Planning B: planning and design, Vol. 15, pp. 461-488.

Chapin, F. and Kaiser, E. 1979, Urban Land Use Planning, third edition, University of Illinois Press.

Gleick, J. 1988, Chaos, Penguin Books, N.Y. USA.

Fabos, J. Gy. 1979, Planning the Total Landscape: A Guide to Intelligent Land Use, Westview Press, Boulder, Colorado.

Lnych, K. 1967, The Image of the City, MIT Press.

Mandelbrot, B. 1982, The Fractal Geometry of Nature, W. H. Freeman and Co., San Francisco.

Mitchell, W. J. 1977, Computer-aided Architectural Design.

Preparata, F. P. and Shamos, M. I. 1985, Computational Geometry -- An Introduction, pp. 19.

Sedgewick, R. 1983, Algorithms, Addison-Wesley, USA, pp. 387, 395, 397.